Statistical
Methods
in Research
and Production

Statistical Methods in Research and Production

Edited by
Owen L. Davies and
Peter L. Goldsmith

Fourth Revised Edition

Longman
London and New York

First published by Oliver & Boyd 1947
Second edition (revised) 1949
Second edition (revised) (reprinted) 1954
Third edition (revised and enlarged) 1957
Third edition (reprinted with corrections) 1958, 1961, 1967
Fourth edition (revised and enlarged) 1972
Fourth edition (reprinted with corrections) by Longman Group Limited 1976, 1977
Reprinted 1980
Reprinted in paper covers (with corrections) 1984
Reprinted 1986

The information in this book represents the best current advice on
the subject available to the publishers and authors. It is distributed
on the understanding that neither the publishers nor the authors
shall be responsible for the absolute correctness or sufficiency of
any information contained in it.

Longman Group Limited
Longman House, Burnt Mill, Harlow
Essex CM20 2JE, England

Associated companies throughout the world

ISBN 0 582 45087 X

Printed in Great Britain at The Bath Press, Avon

Authors

The ICI staff responsible for the three previous editions of this book were:

G. E. P. Box	L. R. Connor
W. R. Cousins	O. L. Davies
F. R. Himsworth	H. Kenney
M. Milbourn	W. Spendley
W. L. Stevens	C. C. Tanner

For this fourth revised edition the authors and revisers are:

Chapter 1	A. Baines	Mond Division
Chapter 4	W. Spendley	Agricultural Division
Chapters 7 & 8	R. H. Woodward	ICI Fibres
§8.93	G. M. Paddle	Mond Division
§9.3	A. F. Bissell	ICI Fibres
Chapter 12	{ C. Burrows D. P. Royle }	Organics Division
Remainder	P. L. Goldsmith	ICI Fibres

In addition, many other staff contributed to the preparation of this book by making individual suggestions, by commenting on chapter drafts, by typing and by proof-reading.

Preface

In scientific and industrial work the correct interpretation of data is of profound importance, since unless right conclusions are drawn the acquisition of numerical results has little or no value. Statistical methods are an indispensable tool for solving many of the problems which are encountered in the chemical industry, not only in relation to research and production but also for personnel matters such as sickness and accident rates.

This has been recognised by Imperial Chemical Industries Ltd. for many years, and the use of such methods has been continuously encouraged. As part of this policy of encouragement a team of authors was invited to write a Company book which would not only assist in spreading the use of statistical methods within the organisation, but would also help new staff to acquire a working knowledge of the subject. The authors were allowed to draw freely on Company data, so that the examples would represent genuine problems which had been met in actual practice.

It was later suggested that, since problems similar to those discussed in this text were common to many branches of industry, it would be worthwhile to publish the book externally and thus make it available to a wider public. After careful consideration this suggestion was accepted and the first edition of this book was published in 1947. Revised and enlarged editions were produced in 1949 and 1957. Over the years the book has acquired a noteworthy reputation amongst industrial personnel for its problem-orientated non-theoretical approach to statistics. It has also been used in universities and colleges to complement books on mathematical statistics and as a primary text for students of physics, chemistry and engineering.

A sequel dealing with the planning of experiments was prepared by a similarly constituted ICI team. Entitled *The Design and Analysis of Industrial Experiments*, this was first published in 1954 and a second edition appeared in 1956. More recently the Company has produced a series of five monographs on more specialised topics under the general title *Mathematical and Statistical Techniques for Industry*.

Advances in theoretical statistical methods and their application to industrial problems since the publication of the third edition, together with the additional experience gained within the industry, have now necessitated a complete revision of the book, and every opportunity has been taken in this fourth edition to bring the subject up to date. The widespread availability of electronic computers means that many more sets of data are analysed nowadays and that the associated statistical techniques have become more sophisticated. There is also a greater awareness of the part statistics can play in major management decisions; not only can the expected outcome of a proposal be determined, but a whole range of possible outcomes can be examined. The fourth edition, whilst retaining the same chapter divisions as the third, therefore includes many additional sections and there is a completely new chapter on Simulation, a method of solving complex problems which has boomed in the computer era.

Part of the description of frequency distributions in Chapter 2 has been rearranged, with an earlier introduction to the use of probability plotting papers. The sections on the coefficient of variation and on skewness in Chapter 3 have been rewritten and new sections deal with number and weight averages, the combining of coefficients of variation, the problem of outliers and the standard error of a function of a single variable.

The nature of confidence limits has been clarified in Chapter 4 and some distribution-free significance tests are introduced. The treatment of statistical inference is broadened by the addition of a lengthy section on statistical decisions. This introduces the concepts of prior and posterior probability distributions, of likelihood and utility functions, and includes a pertinent example of their application to a decision about a pilot-plant programme.

Chapters 5 and 6 are substantially the same as in the third edition, but the presentation has been improved at numerous points. The use of regression methods has increased greatly during the past decade and it is therefore appropriate that Chapters 7 and 8 should have received very thorough revision. A clearer distinction is drawn between linear functional relationships and regression analysis; the section on fitting several lines has been expanded and new sections have been added to Chapter 7 on weighted linear regression and constraining lines to pass through an origin. In multiple regression analysis, most statisticians now invert the information matrix on an electronic computer, so hand-calculation methods have been discarded from Chapter 8. Methods of building a suitable regression model, of testing its validity and the adequacy of the data are each considered at length.

New features of Chapter 9 on the analysis of counted data include a description of the negative binomial distribution and the application of the chi-squared statistic to testing the goodness of fit between a histogram and an assumed frequency distribution. Also the discussion of significance tests for association in contingency tables has been brought up to date.

Cumulative sum techniques have been the outstanding development in control chart methods for process control during the 'sixties. They are described in Chapters 10 and 11, complete with practical examples, and a new type of graph, the Double Cusum Chart, is here published for the first time.

Finally, in Chapter 12, methods of discrete simulation are outlined which have proved invaluable for solving problems as diverse as the sequencing of operations in large-scale plants and the enumeration of simple statistical distributions.

*   *   *   *   *

Sections, tables in the text, figures, formulae and references are numbered on the decimal system, but no attempt is made to keep these various sets of numbers in alignment, e.g. Table 2.4 falls in § **2.31** and Formula (*2.4*) in § **2.43**. Sections are numbered in bold type, formula numbers in italics and other numbers in roman type.

References to literature are enclosed in square brackets: thus [2.2] means "refer to Aitchison and Brown's book on *The Lognormal Distribution*". An exception is made in referring to the sequel text, which for brevity is quoted as *Design and Analysis*.

In addition to tables in the text, sixteen general tables of functions have been included at the end of the volume and labelled A, B, ..., N. Acknowledgements are due to their authors and publishers for permission to use copyright matter: full credits appear on pp. 432-3.

When a statistical term is introduced for the first time it is highlighted by the use of a capital letter. Further, certain terms, such as Normal and Population, retain capital letters throughout the book to emphasise that they are being used with their statistical rather than their everyday meanings.

Contents

		Page
	Preface	vii
1	**Introduction**	1
	1.1 Object of the book	
	1.2 Nature of statistical methods	
	1.3 Use of computers in statistics	
	1.4 This book, and others	
2	**Frequency Distributions**	9
	2.1 Histograms and probability density	
	2.2 Measures of central tendency and spread	
	2.3 Form of distribution	
	2.4 The Normal distribution	
	2.5 Population and sample	
	2.6 Samples and sampling fluctuations	
3	**Averages and Measures of Dispersion**	30
	3.1 Introduction: Condensation of data	
	3.2 Measures of location	
	3.3 Measures of dispersion	
	3.4 Measures of skewness and kurtosis	
	3.5 Outliers	
	3.6 Standard error of mean and standard deviation	
	3.7 Standard error of a function of several variables	
	App. 3A Unbiased estimate of Population variance	
4	**Statistical Inference**	58
	4.1 Introduction to confidence limits	
	4.2 Confidence limits for mean values	
	4.3 Confidence limits for standard deviations	
	4.4 The effect of non-Normality	
	4.5 Tests of significance	
	4.6 Significance of means	
	4.7 Distribution-free tests	
	4.8 The F-test	
	4.9 Statistical decisions	

App. 4A Confidence limits for a standard deviation: data not assumed Normal
App. 4B The likelihood function
App. 4C Utility functions

5 Statistical Tests: Choosing the Number of Observations   94
5.1 Introduction: Prior distributions and costs
5.2 Prior distribution and costs known
5.3 Prior distribution unknown
5.4 Prior distribution and costs unknown
5.5 Comparison of two means
5.6 Standard deviation not precisely known
5.7 Number of observations in comparing standard deviations
5.8 Sequential testing

6 Analysis of Variance   121
6.1 Introduction: Additive property of variances
6.2 Types of classification and nature of variation
6.3 Hierarchic classification with two sources of variation
6.4 Example: Variation of dyestuff yields
6.5 Hierarchic classification with three sources of variation
6.6 Cross-classifications
6.7 Comparison of means
6.8 Further aspects of the Analysis of Variance
App. 6A Combination of results from several samples
App. 6B Analysis of hierarchic data
App. 6C Confidence limits for the components of variance
App. 6D Comparison of the efficiencies of certain hierarchic designs
App. 6E Comparison of several variance estimates
App. 6F Analysis of two-way cross-classification without repeats
App. 6G Estimation of a missing value in a two-way cross-classification without repeats

7 Linear Relationships between Two Variables   178
7.1 Introduction: Functional relationship and regression
7.2 Linear regression
7.3 Conditions for valid estimation of the regression equations
7.4 Regression through the origin
7.5 Weighted linear regression
7.6 Linear functional relationships
7.7 Comparison of several regression lines
7.8 Correlation
App. 7A Fieller's Theorem

8 Multiple and Curvilinear Regression   237
8.1 Introduction
8.2 Multiple linear regression
8.3 Generalization of the method of least squares
8.4 Selection of the best subset of variables
8.5 Further aspects of model building
8.6 Curvilinear regression
8.7 Multiple quadratic regression
8.8 Interpretation and presentation of a multiple regression analysis
8.9 Miscellaneous topics: Plant records, functional relationships and nonlinear estimation

9 Frequency Data and Contingency Tables 304

9.1 Binomial distribution
9.2 Poisson distribution
9.3 Negative binomial distribution
9.4 Chi-squared distribution
9.5 Analysis of single classification data
9.6 Contingency tables
9.7 General considerations on frequency data

10 Control Charts 336

10.1 Introduction: Application in quality control
10.2 General purpose of control charts
10.3 Analysis of data by Shewhart control charts
10.4 Control charts for process control
10.5 Rate of detection of changes in average level
10.6 Cumulative sum charts
10.7 Decision making with a V-mask
10.8 Choice of charting method
10.9 Further applications of control charts: Defect charts and double
     cusum charts
     App. 10A Use of Shewhart control charts

11 Sampling and Specifications 373

11.1 Introduction: Reasons for sampling
11.2 Random sampling and systematic sampling
11.3 Sampling of attributes
11.4 Process control of attributes
11.5 Sampling of variables
11.6 Process control of variables
     App. 11A Process control by decision limit schemes

12 Simulation 409

12.1 Introduction: A blending problem
12.2 Types of simulation
12.3 Building a simulation model
12.4 Simulation experiments
12.5 The use of computers for simulation
12.6 Applications

Statistical and Mathematical Symbols 423

Tables and Charts of Statistical Functions 431

Index 473

Chapter 1

Introduction

Object of the book

1.1 The object of this book is to facilitate a better understanding and a wider use of statistical methods by staff engaged in Research and Production, particularly in the chemical industry. To this end the authors have described the appropriate statistical methods and liberally illustrated them by examples drawn from their own experience of industrial research and production investigations. Comprehensive explanations of the underlying statistical principles have been included, but formal derivations by mathematical statistics have largely been omitted, since they are not essential to understanding the methods and would have added substantially to the complexity of the book. On the other hand, descriptions of the arithmetical calculations required to use the methods are given in full, so that research and production staff may apply them to their own problems, using the book as a work guide. The book should also be helpful to new statisticians who have attended courses in mathematical statistics and are then faced with applying statistical methods to industrial problems.

Nature of statistical methods

1.2 The science of Statistics may be defined as the study of chance variations, and statistical methods are applicable wherever such variations affect the phenomena being studied—a situation frequently encountered throughout science and industry. Most measurements, for instance, are subject to an experimental error, and the behaviour of most industrial plants is subject to variations arising from a multiplicity of causes whose effects it would be impossible to trace out separately. In fact, the conditions obtaining in many experiments are such that the individual results are subject to chance variations comparable in size to the effects being studied, and it is necessary to examine a large set of data in order to draw any worthwhile conclusions.

From their very nature, statistical methods, like other mathematical devices, cannot reveal anything that is not already implicit in the data. In simple cases the experimenter's judgment should usually lead him to the same conclusions as would be arrived at by the application of statistical methods, and the latter would serve merely to test these conclusions free from any preconceived notions on the part of the observer. As the complexity of the data increases, however, it becomes more and more difficult to arrive at a reliable conclusion by unaided judgment, and some form of statistical treatment becomes essential if correct inferences are to be drawn from the data.

The use of statistical methods can improve the quality of an experimenter's work by enabling him

(i) to present his results in the simplest and clearest way;

(ii) to extract the maximum amount of information from a given set of experiments;

(iii) to draw the right conclusions, despite the variability present in the data;

and (iv) to determine the most economic way of designing a set of experiments in order to obtain the required information.

Every experimenter should have at the very least an elementary knowledge of statistical methods, so that he may be able to apply the simpler methods for himself, to follow the reasoning used in more complex studies carried out by others, and to know when he should seek help. For these purposes no more than an elementary knowledge of mathematics is required.

1.21 When an experimenter has only a few results to report, he is unlikely to confuse his readers if he just lists them. But as the number of results grows, a simple list becomes less and less meaningful and much of the additional information associated with the greater number of results will not be appreciated. Various statistical methods are discussed for summarising such data, starting in Chapter 2 with a tabulation known as a Frequency Distribution. However, many experimenters and their managers prefer to see pictorial rather than tabular summaries of data, so diagrammatic methods of presentation such as Histograms are also described.

Often the experimenter will have several sets of results to be presented and compared. Frequency distributions and histograms are too crude for such purposes and explicit quantitative measures are needed. All experimenters are familiar with the Arithmetic Average or Mean which is the most commonly used measure of the average position, or location, of a set of results. This and alternative measures of location are discussed in Chapter 3. If the experiments are concerned, for example, with the reproducibility of readings from an instrument, then clearly we are interested not only in the average value of the results but also in their variability or dispersion. Various measures of dispersion are described, including the easily calculated Range and the more complicated, but more informative, Standard Deviation and Variance.

As more and more results are obtained, the accuracy of estimation of the measures of location and dispersion improves. In Chapter 4 statistical methods are described for evaluating the precision of these estimates in terms of quantities known as Confidence Limits. These limits form the boundaries of a Confidence Interval within which the true value of the measure almost certainly lies and thus indicate what conclusion may justifiably be drawn from the data. Suppose, for example, a series of experiments is carried out to test whether a modification to a standard process results in a real improvement in a product property. This can be judged from the width of the confidence interval associated with the mean observed change in the property level, or more directly (though equivalently) by carrying out a Test of Significance. Such tests are based on the number and variability of the observations and a stated risk of a wrong conclusion.

The specialised use of the word "significance" should be noted; in statistical parlance, an observed difference is classed as significant if it is greater than can reasonably be attributed to chance variations, whether it is of practical consequence or not. For this reason, a test of significance is not wholly appropriate as a basis for action; thus experiments on a process modification are usually designed to establish whether "it is worth while" and not simply whether "it has an effect". The more comprehensive analysis necessary to settle such a question, taking account of the costs and returns and of any prior beliefs, is illustrated in Chapter 4 by an example concerning a pilot-plant programme.

The calculation of statistics such as the mean and standard deviation and the setting of confidence limits extracts the maximum amount of information from a given set of data. But there is no automatic guarantee that this is the right amount of information for firm conclusions and positive decisions. If too few experiments have been carried out, then the information will be insufficient; if too many have been done, the evidence will be conclusive but the outcome may be unimportant, and the expense will be greater than necessary. An important part of statistical methodology is the Design of Experiments to provide the right information in the most economic way. This methodology is introduced in Chapter 5 and described in some detail for the design of some simpler kinds of experiment. An extended account of this subject is available in a companion text entitled *The Design and Analysis of Industrial Experiments* [1.1]; the contraction *Design and Analysis* is used to denote this second volume when it is referred to in the present book.

1.22 Statistical methods are invaluable for problems with several sources of variation or where the effects of several different factors are to be evaluated. For example, suppose a supplier delivers consignments of a solid material packed in drums and a chemist is given the job of determining the quality of the consignments, perhaps to the extent of constructing a standard procedure for checking that the quality of each consignment passes a specification. He will need to investigate and estimate the size of each possible source of variation, such as between the drums comprising a consignment, between the samples from a given drum and between repeat tests on particular samples. To obtain this information requires a special type of experimental design, the

results from which are analysed by the statistical method known as Analysis
of Variance (Chapter 6). The information thus obtained can be combined
with data on the costs of sampling and testing to determine the most economic
way of testing future consignments.

Analysis of variance techniques can also be applied to problems involving
the effects of several factors. Suppose a plant investigator is trying to improve
the performance of his plant by studying the effects of operating variables
such as raw material ratios, reaction times, temperatures and pressures. The
one-factor-at-a-time experimental programmes which non-statistically orientated investigators may well be tempted to adopt because of their simplicity
are not only wasteful of experiments for the information obtained, but also
cannot provide any guarantee that the best answer will ever be achieved.
This is because this type of experimentation does not permit the identification
and evaluation of Interactions which may well exist among the factors. For
example, it is well known that in many chemical processes there is a dominat-
ing interaction between reaction temperature and reaction time, such that as
the level of one of these variables is increased optimum performance is
achieved at reduced values of the other.

1.23 The standard analysis of variance methods can only be used on data
from a statistically designed experiment and unfortunately the precise requirements of these designs cannot always be achieved in practice. This situation
is forced on an experimenter when his project includes variables which can
be measured but not controlled. A development officer may be trying to
determine the effect on his plant of the ambient temperature or the amount
of a particular impurity in a raw material; a physicist may be trying to
calibrate a new instrument against a control by using a selection of samples,
and a chemist may be trying to determine the effect on a reaction of a flow
rate which is easier to measure than to control at predetermined levels. In
such situations fully predetermined and controlled experiments are impossible, and the resulting data are analysed by another statistical method known
as Regression Analysis.

With this method an equation is calculated which describes how the average
value of a response variable depends on the values of the other independent
variables. The regression analysis for a linear equation between a response
and a single independent variable is discussed in Chapter 7, and special
attention is given to the case when a definite functional relationship can be
assumed to exist between them. The more complex models involving curvilinear equations and Multiple Regression on several independent variables
are described in Chapter 8.

1.24 Sometimes the quantities involved in a problem are not values of a
continuous variable but the numbers of times particular events are observed,

such as the numbers of accidents classified by type of job or the numbers of defective articles in a series of batches. These types of data follow laws different from those applicable to continuous variables. The commonest of these laws, the Binomial, Poisson and Negative Binomial Distributions, are discussed in Chapter 9, and a number of methods are given for analysing sets of counted data.

There are many industrial situations where observations naturally arise in an ordered sequence; typical examples are series of daily plant performance figures, quality tests on successive batches of product and routine standardization tests associated with instrument maintenance. The incidence of abnormal changes in systems generating such data is best detected by statistical methods which consider the whole series of data in its ordered form. Chapter 10 describes the analysis of these kinds of data by means of Control Charts; one type, the Cumulative Sum Chart, has only been developed during the last decade and is particularly suitable for detecting small changes in the average level of a variable.

Many raw materials are purchased and products sold subject to a quality specification, which incorporates a procedure for sampling and testing each consignment of material. The determination of the most economic method of sampling and testing which will provide adequate safeguards to both seller and buyer is a statistical problem. The concepts and statistical methods involved are discussed in Chapter 11, both for counted data (e.g. number of defective articles in a batch) and for continuous variables (e.g. the amount of an impurity in a raw material).

All the problems introduced so far are solved by explicit statistical procedures. However, there are many situations in which a physico/chemical system can be satisfactorily described only by a model containing random elements which is far too complex to be solved by explicit methods. For example, the study of the throughput of a multi-stage, multi-unit batch plant taking account of variable features such as process cycle times, frequency and duration of breakdowns and variations in product demand, usually involves the development of a complex model. In such cases, a direct numerical imitation of the plant can be made by processing a sequence of data according to the logical rules which govern the plant operation. This technique is known as Simulation and forms the subject of Chapter 12. Sometimes long sequences of historical plant data are available which can be processed as they stand, but more often it is necessary to sample randomly from statistical distributions of cycle times etc., the method then being known as Monte Carlo Simulation.

Use of computers in statistics

1.3 The use of statistical methods for designing experiments and analysing data ensures better and more reliable information from a given investigation, but at the same time it often involves a considerable amount of rather tedious calculation. Many of the methods discussed in this book refer to problems which can be solved in a few hours at most by using pencil and paper, a slide-rule or, better still, a desk calculating machine. However, the amount of arithmetic involved grows rapidly with the volume of data and the number of variables to be studied. In particular, the calculations associated with the methods of multiple regression analysis and simulation can be so great as to be a real limitation on the scope and detail of the investigation.

Since the arithmetic is highly repetitive, statistical calculations form ideal applications for an electronic digital computer. They were among the first types of calculation for which computer programs were developed, listing all the individual numerical steps which the computer has to undertake. The use of computers in statistical analysis is now well established and is indispensable for the more complex investigations. Virtually all computer manufacturers provide a suite of statistical programs as well as the computing equipment. Further, most industrial companies and educational establishments with any history of using statistical methods and with access to a computer have supplemented the computer manufacturer's programs with other statistical programs developed for their own special requirements.

For any but the most elementary statistical problems involving small amounts of data, the economics of using an electronic computer are clear-cut; the computer cost will be substantially less than that of the corresponding manual calculation. For the larger problems, the overall solution time will be shorter and the chance of errors is much reduced. When a chemist, physicist or engineer is faced with undertaking his own statistical analysis, there is the added advantage that he will not need a detailed knowledge of the calculation procedures required to solve his problem, as these will be built into the computer program. He only needs to submit his data in the format laid down by the author in the program specification.

Apart from reducing the cost and speeding up the application of statistical methods, the use of computers has also had fundamental effects on the methodology itself. The size and complexity of problem which can reasonably be dealt with by computer are an order of magnitude greater than by manual calculation. The thoroughness of the investigation can be greatly increased, because the power of a computer makes it more feasible to repeat analyses in order to study the choice of models and variables and the adequacy of the given data for estimating parameters in the various models. Moreover, the computer can readily be programmed to output more complete information from any particular analysis. In contrast, manual calculations are usually too tedious and time-consuming to permit such a thorough study.

Whilst the computer has improved the efficiency and widened the horizons of both the professional and the amateur statistician, it has also introduced new dangers. The choice of experimental design, the validity of the model and the adequacy of the data pose many searching questions which demand the statistician's careful attention. Answers to these questions are not yet developed in so precise and logical a form that they can be wholly built into computer programs themselves. If an amateur statistician blindly uses computer programs without a thorough understanding of the principles involved, he may easily come to conclusions which superficially look reasonable but which are in fact nonsense. Even the professional statistician must be wary; the size and complexity of problem he can deal with using the power of a large computer put a severe strain on his capacity to check and interpret the analysis. Built-in validity checks still have a long way to go and the statistician must continue to be responsible for and retain control of his investigations, accepting that the computer is a fast and inexpensive means of doing arithmetic, but is no statistician.

This book, and others

1.4 Although the chapters in this book correspond to different classes of problem with specific statistical methods for their solution, they should not be read as completely separate entities. Statistical methods are not automatic calculation procedures for drawing conclusions from experimental and other observations; they form a fundamental and powerful scientific discipline which must be used with proper understanding. The blind application of statistical methods without this understanding can all too easily result in costly decisions being taken from wrong conclusions which have been made to look impressive by statistical terminology. Thus the basic principles of statistical methodology described in Chapters 2 to 5 should be thoroughly mastered before embarking on the rest of the book. It is also essential to read Chapter 7 before tackling the more complex situations in Chapter 8, whilst Chapter 11 requires an understanding of Chapters 9 and 10. However, Chapter 12 may be studied as soon as the basic principles of statistics are thoroughly grasped.

This book is primarily concerned with explaining the principles of statistical methodology and with the statistical analysis of some of the simpler types of problem which occur in industry. It does not include the formal derivations of the methods by mathematical statistics and, for those readers interested in the theoretical development of the methods, a number of suitable textbooks are listed below. Statistics also plays a very important part in the correct planning of experiments so as to obtain the maximum information from a given number of experiments; careful planning can often effect major economies in the amount of experimental work which has to be done. The use of statistics in the planning of experiments has not been included in this book, as it forms the subject of the companion volume *Design and Analysis*.

In recent years, particularly because of the influence of electronic computation, the scope of statistics has been broadening rapidly into such fields as econometrics, operational research and systems engineering. Such areas are barely touched upon in this book, but references for further reading listed below indicate where more information can be found.

Suggestions for further reading

[1.1] DAVIES, O. L. (Editor). *The Design and Analysis of Industrial Experiments* (second edition). Longman (London and New York, 1978) for Imperial Chemical Industries PLC.

[1.2] KENDALL, M. G., STUART, A., and ORD, J. K. *The Advanced Theory of Statistics*, Vols I–III (fourth edition). Charles Griffin (High Wycombe, 1977, 1979 and 1983).

[1.3] BARNETT, V. *Comparative Statistical Inference* (second edition). John Wiley (Chichester and New York, 1982).

[1.4] RAIFFA, H., and SCHLAIFER, R. *Applied Statistical Decision Theory*. Harvard University Press (1961).

[1.5] BOX, G. E. P., HUNTER, W. G., and HUNTER, J. S. *Statistics for Experimenters: An Introduction to Design, Data Analysis and Model Building*. John Wiley (New York, 1978).

[1.6] DRAPER, N. R., and SMITH, H. *Applied Regression Analysis* (second edition). John Wiley (New York, 1981).

[1.7] WETHERILL, G. B. *Sampling Inspection and Quality Control* (second edition). Chapman and Hall (London and New York, 1977).

[1.8] GOLDBERGER, A. S. *Econometric Theory*. John Wiley (New York, 1964).

[1.9] CHURCHMAN, C. W., ACKOFF, R. L., and ARNOFF, E. L. *Introduction to Operations Research*. John Wiley (New York, 1957).

[1.10] CHECKLAND, P. *Systems Thinking, Systems Practice*. John Wiley (Chichester and New York, 1981).

Chapter 2

Frequency Distributions

Statistical methods are concerned with the study of
variation. The nature of the variation in a set of
data is shown by a frequency distribution. To
describe numerically the principal features of a
distribution, a measure of location such as the arithmetic
mean, and a measure of spread such as the standard
deviation, are used. A particularly important
theoretical distribution is the Normal distribution.
Ways of transforming data to achieve approximate
Normality are described.

Frequency table

2.1 The first step in examining a set of data is to assemble them in a form
such that the main features can be readily appreciated, particularly the degree
of spread and the manner in which the results are distributed over the range,
for example whether they are distributed uniformly or tend to be concentrated
round a central value. The results written down in order of occurrence may
vary erratically, and with a large number it is difficult to derive any general
conclusions. The simplest method of arranging the data is to divide the whole
range of variation into a number of groups, count the number of observations
falling in each group, and write these in the form of a table. As an illustration,
Table 2.1 shows the results of 178 measurements of the carbon content of a
mixed powder fed to a plant over a period of one month. The total range
was from about 4·1% to 5·2%, and this range was divided into the groups
4·10–4·19%, 4·20–4·29%, etc. Proceeding through the results in order of
occurrence a tally was built up of the number of values falling into each group.

A table of this kind, which shows the frequency of occurrence of the
different values, is known as a Frequency Table. The distribution of the
results is known as a Frequency Distribution.

It is clear from the table that:

  (i) The results cluster round a central value of about 4·7%.

 (ii) They are spread roughly symmetrically round the central value.

(iii) Small divergences from the central value are found more frequently
      than large divergences.

Ambiguity may arise in classifying some observations if the end values of the groups are not clearly defined. For example, suppose the groups had been labelled 4·1–4·2%, 4·2–4·3%, etc; then it would not be clear where a value of 4·20% should be assigned. This doubt can be removed by defining the end values of the groups according to the precision of the measurements and by avoiding overlap. Thus, when results are quoted to two places of decimals the groups 4·10–4·19%, 4·20–4·29%, etc. are unambiguous. It is also important to specify the correct central value of each group. The group 4·20–4·29% includes all carbon contents between 4·195% and 4·295%, assuming measurements can only be sensibly recorded to the nearest 0·01%. The true central value for this group is 4·245%, which is the same as the average of the group limits.

TABLE 2.1. PERCENTAGE CARBON IN A MIXED POWDER

Range of Values of % Carbon	Tally	Number of Results	Proportion of Results
4·10–4·19	/	1	0·006
4·20–4·29	//	2	0·011
4·30–4·39	7HL //	7	0·039
4·40–4·49	7HL 7HL 7HL 7HL	20	0·112
4·50–4·59	7HL 7HL 7HL 7HL ////	24	0·135
4·60–4·69	7HL 7HL 7HL 7HL 7HL 7HL /	31	0·174
4·70–4·79	7HL 7HL 7HL 7HL 7HL 7HL 7HL ///	38	0·214
4·80–4·89	7HL 7HL 7HL 7HL ////	24	0·135
4·90–4·99	7HL 7HL 7HL 7HL /	21	0·118
5·00–5·09	7HL //	7	0·039
5·10–5·19	///	3	0·017
Total		178	1·000

Histogram

2.11 The characteristics of the above frequency distribution are shown even more clearly if the results are plotted as in Figure 2.1. This block diagram is known as a Histogram. The horizontal axis is divided into segments corresponding to the ranges of the groups. On each segment a rectangle is constructed whose area is proportional to the frequency in the group. In the present example, since the range is the same for all groups, the heights of the rectangles are also proportional to the frequencies. It is usually better to make the groups of equal width; but in some cases unequal groups are more suitable, and the *area* must then be made proportional to the frequency.

For continuous variables such as carbon content, some compromises are necessary in choosing the range for each group. If the groups are too wide the histogram will fail to illustrate the finer detail of the frequency distribution. On the other hand, if the group range is too narrow, there will be rather few results falling into each group and the histogram may exhibit an irregular pattern. It is also convenient to choose a round number or simple fraction, e.g. $0 \cdot 1 \%$ for the carbon content data. A general rule which satisfies these requirements is to select a round value for the group range such that the number of occupied groups is roughly equal to the square-root of the number of results.

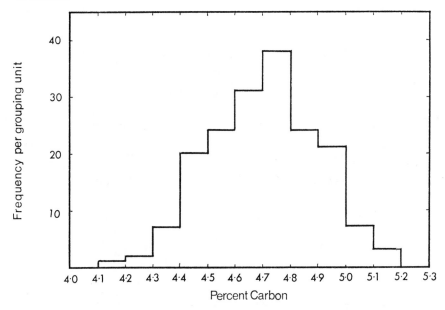

FIG. 2.1. HISTOGRAM

Frequency polygon

2.12 If the variable is not continuous but can assume only certain values, for example if it can only be a positive integer, as with numbers of failures, accidents, etc., the histogram is clearly inappropriate, since it represents the data in each group as equally distributed over the range. The appropriate method of plotting such data is simply to erect lines proportional to the frequencies. It is usual, however, to plot the frequencies as points and join the points by straight lines, thus forming, with the x-axis, a polygon, called the Frequency Polygon. This diagram is sometimes used for continuous variables also; if several diagrams are to be superimposed for comparison, histograms give a confused picture and polygons are preferable. Figure 2.2 shows the data of Table 2.1 plotted as a frequency polygon. Note that the abscissae are the central values of the groups.

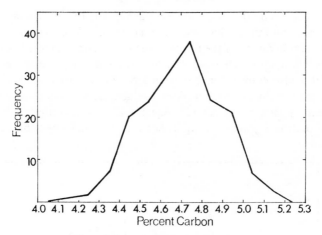

FIG. 2.2. FREQUENCY POLYGON

TABLE 2.2. CUMULATIVE FREQUENCIES BASED ON TABLE 2.1

Range of Values of % Carbon	Group Frequency	Cumulative Frequency	Cumulative Proportional Frequency
4·10–4·19	1	1	0·006
4·20–4·29	2	3	0·017
4·30–4·39	7	10	0·056
4·40–4·49	20	30	0·169
4·50–4·59	24	54	0·303
4·60–4·69	31	85	0·478
4·70–4·79	38	123	0·691
4·80–4·89	24	147	0·826
4·90–4·99	21	168	0·944
5·00–5·09	7	175	0·983
5·10–5·19	3	178	1·000
Total	178	—	—·

Cumulative distribution or ogive

2.13 In the histogram or frequency polygon we plot the group frequencies. Sometimes we may be interested not so much in the individual frequencies as in the proportion of results falling above or below given values; we then plot the cumulative frequencies. Table 2.2 shows the cumulative frequencies for the data of Table 2.1; they are obtained by summing the group frequencies up to the group in question. We see immediately that, for example, 94% of the results show a carbon content below 5·00%.

If these figures are plotted we get a Cumulative Frequency Curve, or Ogive, as in Figure 2.3. The S-shape is characteristic of most ogives, and arises whenever the frequency has a maximum value well inside the range of the variable.

When only a few results are available, say between 6 and 16, there is little merit in grouping the values and constructing a histogram. However the *individual* results can be usefully plotted on a cumulative proportional frequency diagram to give some idea of the proportion of results falling above

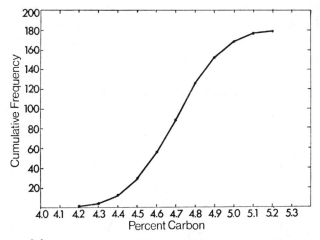

FIG. 2.3.   CUMULATIVE FREQUENCY DISTRIBUTION OR OGIVE

or below given values. For a set of N observations arranged in ascending order, the successive ordinates of the points are [2.1]:

$$\frac{1}{2N}, \frac{3}{2N}, \frac{5}{2N}, \ldots, \frac{2N-1}{2N}$$

If two results happen to be equal, the two adjacent points on the ogive are joined by a vertical line.

Computer processing

2.14 When large volumes of data are produced, as for example in chemical plant records, preparation of histograms or ogives can be very laborious. However, the effort involved in carrying out these calculations, and indeed for most of the statistical calculations described in this book, can be very much reduced by using an electronic digital computer. The data usually have to be punched on to cards or paper tape, which are the input media for most computers. Programs for standard statistical calculations have been written for nearly all machines. The user must consult the program specifications appropriate to his own installation and follow the data preparation instructions exactly.

In some histogram programs the user defines his own group limits; in others the programmer has provided a routine to calculate the limits automatically once all the data have been read into the computer. The computer output may simply be in the form of a printed table, similar to Tables 2.1 and 2.2. Alternatively, the frequency in each group may be indicated by a proportionate horizontal line opposite each central value. On computer installations equipped with graphical output facilities, a diagram like Figure 2.1 may be constructed directly. After forming the histogram, many programs go on to calculate the statistics described in Chapter 3.

Frequency curve

2.15 It is reasonable to suppose, and it is found in practice, that if the number of observations were very much larger and if the groups were made

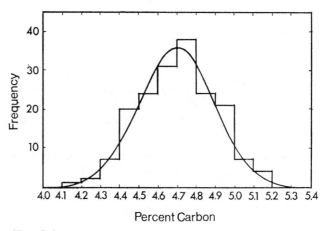

FIG. 2.4. SMOOTH CURVE SUPERIMPOSED ON HISTOGRAM

much narrower, the irregular step-form of Figure 2.1 would change to a smoother shape with smaller steps, and eventually to a smooth curve. This smooth curve may be regarded as defining the true underlying distribution. With a small number of results only an approximation to the curve is possible, since unless the groups are reasonably wide the group frequency will on the average be very small and often zero. The smoothed curve is known as a Frequency Curve. Figure 2.4 is a curve which could reasonably be assumed to represent the data of Table 2.1. If verticals are drawn at the extremes of each group, the areas enclosed between the verticals, the curve and the baseline are approximately equal to the frequencies of Table 2.1. The larger the number of observations, the more nearly will the areas under the curve approximate to the actual frequencies. For small numbers of observations the divergences can be relatively large.

A histogram or frequency curve expressed in terms of proportional frequencies instead of actual frequencies is usually a more useful representation of the distribution. The curve may then be referred to as the Probability Distribution, because the probability of the occurrence of a given observation is defined as the proportional frequency with which it occurs in a large number of observations. The area enclosed by verticals at x_1 and x_2 gives the probability that an observation will have a value between x_1 and x_2, and this probability, multiplied by the total number of observations made, is the expected number of observations for the range x_1 to x_2 in a sample of that size. Figure 2.5 is a curve (based on a theoretical distribution function) which

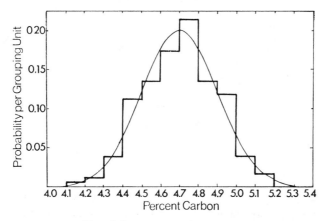

FIG. 2.5. PROBABILITY CURVE

could reasonably be assumed to represent the probability distribution of Table 2.1. Verticals have been drawn at the group limits and the histogram of Figure 2.1 superimposed in terms of proportional frequencies, i.e. actual frequencies divided by 178. Table 2.3 shows the frequencies derived from the probability curve, together with those observed; there is good agreement. The probability is highest for values of x near 4·7, and is very low for values greater than 5·0 or less than 4·4. The greater the number of observations, if the group widths are suitably chosen, the nearer will the relative frequencies approach the corresponding probabilities.

The differences between the observed and calculated frequencies are small, and the observations agree reasonably well with the assumption that their basic probability distribution has the mathematical form used in calculating the frequencies given in Table 2.3. This, however, does not prove that the actual distribution has in fact this form; other mathematical forms could be suggested that would give agreement at least as good as that shown in the table. Previous experience may, however, suggest that one particular type of distribution is more likely than another.

TABLE 2.3. CALCULATED AND OBSERVED FREQUENCIES
(BASED ON TABLE 2.1)

Range of Values	Observed Frequency	Calculated Frequency	Difference (Obs. − Calc.)
4·10–4·19	1	1	0
4·20–4·29	2	5	−3
4·30–4·39	7	9	−2
4·40–4·49	20	17	3
4·50–4·59	24	25	−1
4·60–4·69	31	32	−1
4·70–4·79	38	32	6
4·80–4·89	24	25	−1
4·90–4·99	21	17	4
5·00–5·09	7	9	−2
5·10–5·19	3	5	−2
5·20–5·29	0	1	−1
Total	178	178	0

Probability density

2.16 The height of a given rectangle in a histogram is such that the area of the rectangle is proportional to the frequency. The ordinate of a frequency curve has a similar interpretation as follows. Let x_1 and x_2 $(x_2 > x_1)$ be two values of the continuous variable x. If P_1 is the relative frequency, or probability, for all values of x up to x_1, and P_2 is the relative frequency for all values up to x_2, then $P_2 - P_1$ is the relative frequency of all values between x_1 and x_2. If x_1 and x_2 represent the extremities of a group, the area of the corresponding rectangle in the histogram is proportional to $P_2 - P_1$, and its height is therefore proportional to $(P_2 - P_1)/(x_2 - x_1)$. In the frequency curve, which represents the relative frequency distribution expected for an indefinitely large number of observations, if $(x_2 - x_1)$ is very small it can be represented by dx, and $(P_2 - P_1)$ by dP. In the limit, as $(x_2 - x_1)$ approaches zero, dP/dx represents the height, i.e. the ordinate of the curve at the point x; it is known as the Probability Density and is denoted by $p(x)$. The statistical term Variate is usually used in preference to the mathematical term Variable in this context, so as to indicate that the quantity under consideration may take any of the specified values with a given frequency. A mathematical variable, on the other hand, has no frequency distribution associated with it.

2.17 Usually the only direct information available is the set of observed results, and the problem is then to deduce the probability distribution of the variate or at least a good approximation to it. Only if a very large number of results are available can this be done; but, if a reasonable assumption about the general shape of the curve can be made, either on the basis of past experience or from general knowledge of similar situations, it may be possible to deduce the probability distribution with fair confidence. This is, in fact, a main feature of statistical methods. An assumption is made about the general form of the probability distribution, and from the observations estimates are made of the constants required to define this distribution quantitatively. It is then possible to deduce the probabilities of future events: for example, the probability that an observation will exceed a certain value; the probability that the average of a number of observations will be within given limits; the probability that the averages of two sets of observations will differ by a given amount or more; and so on.

Population

2.18 The hypothetical set of all possible observations of the type which is being investigated is known as the Population (or Universe) of values. Any finite set of these observations can be regarded as a sample drawn from this Population. As the sample size increases, its properties resemble more and more closely those of the Population, provided that the sample values may be considered as randomly selected from the Population (see § 2.6).

The Population is best represented by its probability or frequency distribution [§ 2.15], and when the variate is continuous the distribution can be represented by a smooth curve. When the variate can only take certain specific values, for example, counts of dust particles or numbers of defective items, the distribution cannot then be represented exactly by a continuous curve. It can, however, be represented graphically by means of vertical lines; a line is drawn for each of the possible values of the variate, its length being proportional to the probability of occurrence of the given value. This is known as a Discrete Distribution; the more important types of discrete distributions are considered in Chapter 9.

Measures of central tendency and spread

2.2 The curve and the histogram of Figure 2.4 have three obvious characteristics:

   (i) There is a certain value (about 4·7% carbon) which represents the "centre" of the distribution, and serves to locate it.
  (ii) The values are spread round this central value, extending over a range of about ±0·5% absolute.
 (iii) The spread is not uniform, most values being close to the central value. This reflects the shape of the curve.

We might, therefore, characterise the distribution by means of three criteria:

(*a*) The central value (4·7% approximately).

(*b*) A quantity expressing the degree of spread.

(*c*) The "shape" of the curve, i.e. the general form of the distribution.

These may or may not be sufficient to define the distribution curve completely, but as will be shown below, they very often are sufficient. The equation of the curve may be written:

$$y = p(x; \theta_1, \theta_2)$$

where θ_1 and θ_2 are constants or *parameters*, measuring the central value and the degree of spread respectively, and the function p defines the shape.

Arithmetic mean

2.21 The commonest measure of location, or central value, is the arithmetic mean (commonly abbreviated to "mean" or "average"). It is simply the sum of all the observations divided by their number:

$$\text{Arithmetic Mean} = \bar{x} = \Sigma x / N \qquad (2.1)$$

It is the most useful of such measures, but others which are better in special cases are described in §§ **3.23–3.26**. The mean of the data in Table 2.1 is 4·70%.

A distinction must be drawn between the mean of a set of observed values and the mean of the underlying probability distribution or Population. The observed mean should be regarded as an *estimate* of the true value which becomes better as the number of observations on which it is based is increased. This statement is intuitively felt to be true, and is in fact true except in unusual and unimportant circumstances.

The symbol used for the Population mean is μ; $\bar{x}$ becomes a better estimate of μ as N increases, and it approaches μ as N approaches infinity.* This is expressed symbolically by:

$$\bar{x} \to \mu \quad \text{as} \quad N \to \infty$$

Standard deviation

2.22 The most useful measure of the spread is the Standard Deviation, which for a sample of N observations $x_1, x_2, ..., x_N$ is given by the expression:

$$\text{Standard deviation} = s = \sqrt{\left\{ \sum_{i=1}^{N} (x_i - \bar{x})^2 / (N-1) \right\}} \qquad (2.2)$$

---

* In the limit $\mu = \sum_x p_x x$, where p_x is the proportion of observations having the given value of x. When the distribution is continuous, p_x becomes $p(x)dx$, where $p(x)$ is the ordinate of the frequency or probability distribution at the value x. The summation sign has then to be replaced by the integral sign, and we have $\mu = \int_{-\infty}^{+\infty} p(x)\, x\, dx$.

where $\bar{x}$ is the mean of the sample. In special cases other measures are preferable or more convenient; these are considered later.

The standard deviation is seen to be the root mean square deviation from the mean, except that the divisor is $(N-1)$ and not N. Without going into detail at this stage, it may be said that most of the observations are likely to be within the range $(\bar{x}-2s)$ to $(\bar{x}+2s)$, and practically all within the range $\bar{x}\pm3s$. For the data of Table 2.1, $s = 0.194\%$. The range $\bar{x}\pm2s$ is equal to $4.70\%\pm0.39\%$, i.e. 4.31% to 5.09%; only six of the observations lie outside these limits.

We must distinguish between the true, or Population, value of the standard deviation and that calculated from a set of observations, which is an estimate of the Population value. The symbol used for the Population standard deviation is σ*; s becomes a better estimate of σ as N increases, and

$$s \to \sigma \quad \text{as} \quad N \to \infty$$

When the data consist of replicate determinations, for example in chemical analysis, the spread of the observations is also known as the Precision.

Form of distribution

2.3 The "shape" of the distribution is obviously a more elusive concept than mean or standard deviation, but the data may give some idea of the probable shape. For example, from Table 2.1 or Figure 2.1 we can deduce that the distribution has a central peak and falls off roughly symmetrically on both sides. We must, however, on the basis of past experience or auxiliary information make some assumption about the actual form of the distribution, provided it is in reasonable accord with the observations. A number of types of distribution have been fully studied, because they, or at least close approximations to them, frequently arise in practice, and it is usual to assume that an actual distribution takes one or other of these standard forms. In this chapter the most important continuous distribution is discussed, the so-called Normal distribution, and mention is made of the rectangular and lognormal distributions; others will be described as they arise in applications.

2.31 It is clear that two distributions of different form, i.e. giving frequency curves of quite different shape, could have the same values of the mean and standard deviation. Table 2.4 shows two such distributions, each of which has a mean of zero and a standard deviation of about 2·9.

All the groups are of equal width and are depicted by their central values.

---

* In the limit $\sigma = \sqrt{\left\{\sum_x p_x(x-\mu)^2\right\}}$; for a continuous distribution:

$$\sigma = \sqrt{\left\{\int_{-\infty}^{+\infty} p(x)(x-\mu)^2 \, dx\right\}}$$

p_x and the expressions for μ are defined in the previous footnote.

In distribution A the frequency is a maximum at the two middle groups, tailing off on both sides to nearly zero at ± 9. In distribution B the frequency is constant over the range -5 to $+5$, and is zero outside these limits. Clearly, if the distribution under consideration might be either A or B, the mean and standard deviation would give an incomplete description. Usually, however, something will be known about the type of distribution. In many cases, for example, the errors of chemical analysis or variation in plant output, the

TABLE 2.4. DISTRIBUTIONS WITH THE SAME
MEAN AND STANDARD DEVIATION

Central Values of the Groups	Number of Observations	
	A	B
-8.5	3	0
-7.5	5	0
-6.5	11	0
-5.5	23	0
-4.5	40	100
-3.5	67	100
-2.5	93	100
-1.5	121	100
-0.5	137	100
0.5	137	100
1.5	121	100
2.5	93	100
3.5	67	100
4.5	40	100
5.5	23	0
6.5	11	0
7.5	5	0
8.5	3	0
Total	1,000	1,000

distribution is more likely to resemble A than B, i.e. small divergences are more frequent than large divergences. It is then reasonable to assume that the "shape" of the distribution is like that of distribution A, and with this assumption a knowledge of only the mean and standard deviation gives a good picture of the distribution.

The Normal distribution

2.4 The most commonly used of the theoretical distributions is the Normal or Gaussian distribution, which gives a bell-shaped probability curve; examples of this distribution have already been given in Tables 2.3 and 2.4.

Many of the distributions found in observational and experimental work are roughly of the Normal type, but the uses of the Normal distribution are by no means limited to data which are exactly, or very nearly, Normal. It is of the utmost importance in theoretical and applied statistics, and not only because data naturally occur in this form.

2.41 If data are Normally distributed, the probability that x will assume a value between $(x_1 - \frac{1}{2}dx)$ and $(x_1 + \frac{1}{2}dx)$ is given by:

$$dP = \frac{1}{\sigma\sqrt{(2\pi)}} \exp\{-(x_1 - \mu)^2/2\sigma^2\}dx \qquad (2.3)$$

where μ is the Population mean, σ is the Population standard deviation and exp ( ) denotes the exponential function, i.e. $\exp(x) = e^x$.

The integral of (2.3) over all values of x is equal to unity, i.e. dP is the relative frequency or probability corresponding to the range $(x_1 - \frac{1}{2}dx)$ to $(x_1 + \frac{1}{2}dx)$. The value of $p(x) = dP/dx$ is a maximum when $x = \mu$, and falls off symmetrically on both sides, as in Figure 2.5, which is a Normal curve with mean 4.70% and standard deviation 0.194%. Measurements encountered in practice are usually distributed in approximately this form. This does not necessarily mean that the distribution is in fact Normal; distributions with properties different from those of the Normal often appear similar to it when plotted as frequency curves, and even more so when the cumulative frequencies are plotted.

2.42 Equation (2.3) appears complicated, and it is by no means clear at first sight why such an equation should be assumed to represent the distribution of observed quantities. It can, however, be deduced from quite reasonable assumptions, and it has certain properties which make it of very wide applicability. The most important of these is as follows. Suppose we have a number of observations of some quantity having any probability distribution whatever (with certain exceptions which are of no practical importance). If we take the averages of pairs of observations and plot the probability distribution of these averages it will be found to resemble the Normal form more closely than does the original distribution. Repeating this process with averages of three, four, etc., the distribution approaches more and more closely to the Normal form. This is not simply a superficial resemblance; it can be proved that with almost any original distribution the distribution of sample averages tends rapidly to the Normal form as the sample size increases.

Figures 2.6(b), (c) and (d) show the distributions of the averages of samples of two, three, and four observations, from the original distribution of Figure 2.6(a). The broken line in Figure 2.6(a) represents the Rectangular Distribution, so named from the shape of the diagram, all values between the given limits being equally probable. From an original distribution which is very far from Normal, averages of even three or four observations are distributed

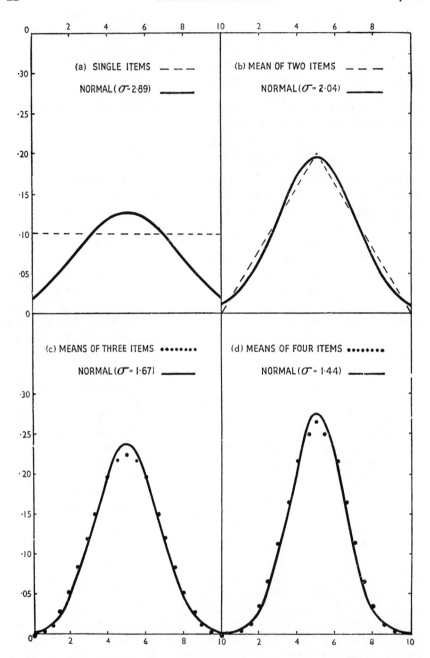

FIG. 2.6. DISTRIBUTION OF MEANS OF SAMPLES FROM A RECTANGULAR
DISTRIBUTION COMPARED WITH THE NORMAL DISTRIBUTION

quite close to the Normal form, as is evident from the Normal curves super-imposed. In Figures 2.6(c) and (d) the curves are too close to draw separately; the dots shown are for the mean of three or four observations from the rectangular distribution, the curve being that of the corresponding Normal distribution. Even the Triangular Distribution for means of two is fairly close to the Normal curve, except near the extremes or "tails" of the curve.

Similar results are found with other distributions which are far from Normal. Since distributions encountered in practice are usually much closer to Normality it is safe to assume that the means of small samples, even of three, are distributed in nearly the Normal form.

Most of the common statistical tests are based on the assumption that the data are Normally distributed, an assumption which cannot often be verified, but since in most cases the quantities required are averages, the tests are of general applicability. There are some exceptions, which will be pointed out as they arise.

Standard measure

2.43 Equation (2.3) involves two parameters, μ and σ, which must be determined before the frequencies can be calculated. It can be simplified by writing it in terms of the deviation of x from μ, using σ as the unit of measurement. This is done by making the substitution:

$$u = (x-\mu)/\sigma \tag{2.4}$$

when (2.3) becomes:

$$dP = \frac{1}{\sqrt{(2\pi)}} \exp(-u^2/2)du \tag{2.5}$$

This simple form is known as the Standardized Form of the Normal equation, and the variable expressed in the form (2.4) is said to be in Standard Measure. The equation now contains no adjustable constants, and if the value of u is known, $dP/du = p(u)$ is determined uniquely. This probability density is shown in Figure 2.7 overleaf.

Equation (2.5) gives the probability of occurrence of observations for which the standardized variate lies between the values $(u-\frac{1}{2}du)$ and $(u+\frac{1}{2}du)$. It can be integrated to give the proportion of observations expected to fall between any two values of u and hence between the corresponding two values of x. The integration cannot be performed explicitly, but the definite integral has been tabulated for a wide range of values of u.

The Normal curve is symmetrical about $u = 0$. Thus the area under the curve to the right of a given value u_1 of u, i.e. the probability of occurrence of a value greater than u_1, is equal to the area to the left of $-u_1$. Denote each of these probabilities by α. Then it is sufficient to tabulate the values of α corresponding to positive values of u_1; this is done in Table A at the end of the volume. For example, there is a probability of 0·05 that u will exceed 1·64, and hence that x will exceed $\mu+1\cdot64\sigma$.

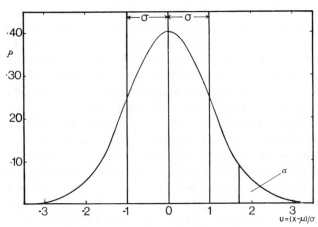

FIG. 2.7. NORMAL PROBABILITY CURVE

Extreme values

2.44 It follows that the probability that u will lie outside the range $-u_1$ to $+u_1$, i.e. that $|u| > u_1$, is 2α. This is the probability that values of x will deviate from the mean of a Normal distribution by more than u_1 multiples of the standard deviation. The complement, $(1 - 2\alpha)$, is the probability that $|u|$ does not exceed u_1, i.e. that observations will not deviate far from the mean. Some typical values of this probability are listed in Table 2.5.

TABLE 2.5.   PROBABILITY THAT $|u|$
WILL NOT EXCEED A GIVEN VALUE

u_1	P
0·5	0·383
1	0·683
1·5	0·866
2	0·955
2·5	0·988
3	0·9973
4	0·9999

It is clear that a Normally distributed variate will seldom give a value of $|u|$ greater than 3, and a value greater than 4 will be very rare, occurring only once in about 10,000 observations. It is usual to consider that if an isolated value of $|u|$ greater than 2 is obtained there is some doubt whether it represents an observation from a Normal Population with the given mean and standard deviation. The odds against finding such a value in a single trial are about 19 to 1. If, however, a series of observations is made, it must not be assumed that if the largest value of u is slightly greater than 2 it does not belong to the same Population as the others. On the average, a set of 20 observations will give one value of u greater than 2, and even in a smaller set, say 10, it is quite likely that one or two observations will give u greater than 2. It is true to say that the odds against any particular value of u exceeding 2 are approximately 19 to 1, but the odds against the largest in a set of, say, 10 are very much smaller. The probability that at least one value of u out of a set of 10 will exceed 2 is about 0·40, and for a set of 20, about 0·64.

Similarly, if an isolated observation, or one out of a small set, gives u greater than 3, this may be taken as a strong indication that it does not belong to the given Normal Population, while a value of 4 or more could be taken as definite evidence.

If the Population is only approximately Normal, care must be taken not to use the probabilities of Table 2.5 as precise measures.

Normal probability paper

2.45 The integral of (2.3) up to x_1 is the probability, $(1-\alpha)$, that x will assume a value below x_1, or equivalently that $u < u_1$. A plot of this probability for various values of x is known as the cumulative Normal distribution. If a linear scale is used for the probabilities, an S-shaped curve is obtained. However, a special plotting paper is manufactured with a probability scale transformed so that the cumulative Normal distribution curve becomes a straight line. This is called Normal Probability Paper, although the misleading term Arithmetic Probability Paper is also used.

To test whether data are Normally distributed the cumulative proportional frequencies (see for example Table 2.2) can be plotted on Normal Probability Paper. If the points fall close to a straight line it can be concluded that the data follow a Normal distribution, at least approximately. The data of Table 2.1 are plotted in this manner in Figure 2.8 overleaf.

When a linear pattern of points is obtained a straight line can be drawn between them. This line can be used to obtain approximate values for the parameters of the corresponding Normal distribution. The Population mean is estimated as the value of x corresponding to the 0·5 probability point. The slope of the line reflects the spread of the distribution; the Population standard deviation can be estimated as one-half of the difference between the values of x corresponding to the 0·84 and 0·16 probability points. Applying this approximate method to Figure 2.8 results in an estimated true mean carbon content of 4·7% and a standard deviation of 0·2%.

This method should not be used when the points show any systematic deviations from a straight line on Normal Probability Paper.

A further special sheet is manufactured known as Logarithmic Probability Paper. This has a logarithmic scale for the x variate and yields a straight

line if log x is Normally distributed.   It may be used to plot cumulative
frequencies when this form of distribution is expected, as, for example, in the
analysis of certain particle sizes.   In such a case x is said to be *lognormally*
distributed; a detailed description of the properties of this distribution is
given in [2.2].

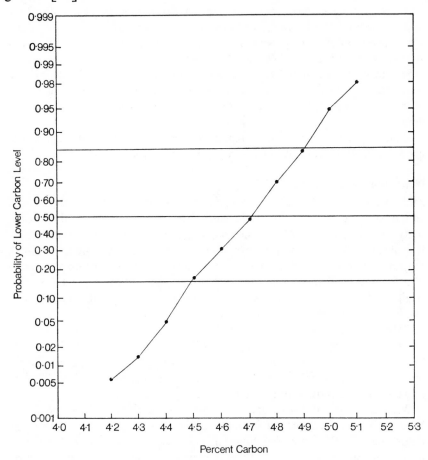

FIG. 2.8.   CUMULATIVE PROPORTION PLOTTED ON NORMAL PROBABILITY PAPER

Population and sample

2.5 A set of data obtained under uniform conditions may be regarded as a
sample from some Population.   If the exact form of the Population is known
it is possible to calculate the probability that a random sample drawn from it
will possess certain properties, for example that the mean of the sample will
be within given limits, or that its standard deviation will not exceed a given
value.   Suppose that the Population is Normal, with mean μ and standard

deviation σ; it will be shown later that the mean of a random sample of n is also Normally distributed with mean μ and standard deviation $\sigma/\sqrt{n}$. If $\mu = 100$ and $\sigma = 5$, then for samples of four the standard deviation of the mean is $5/\sqrt{4} = 2\cdot5$, and practically all samples will have mean values in the range $100 \pm 3 \times 2\cdot5$, i.e. between $92\cdot5$ and $107\cdot5$.

2.51 Problems of this kind may arise in dealing with an established process where the mean and standard deviation of the Population are accurately known. But a commoner situation, especially in experimental work, is that when only a small set of data is available it is necessary to argue in the reverse direction, i.e. to deduce the properties of the Population from those of the sample. It is easy to calculate certain properties of the sample such as the mean and standard deviation, and these can be used as estimates, though perhaps uncertain estimates, of the corresponding properties of the Population. However, these statistics do not prove that the correct form of distribution is being used and it is necessary to assume, usually from knowledge of the same or a similar type of situation, the probable form of the Population. Often the assumption is that the Population is Normal, and while this may not be strictly true it may often be justified in that it leads to valid conclusions for the reasons given in § **2.42.** A numerical procedure for testing the assumption of Normality when $N \leqslant 50$ is given in [2.3]. In certain cases some other more likely form of distribution may of course be assumed.

2.52 A distinction must be made between problems which depend on a precise knowledge of the form of the Population and those which require only an approximate knowledge. Suppose, for example, we want to know how many observations out of a total of 10,000 will exceed the mean value by more than three times the standard deviation. For a Normal Population the expected number is 13, but for Populations of different form, even if they superficially resemble the Normal, the answer may be very different. The distribution of means of samples of three from a rectangular Population appears much like the Normal (see Figure 2.6(c)); but the range is finite, and it can be shown that it is impossible for the mean of three results to differ from the Population mean by more than three times the standard deviation of the mean. For example, if the Population covers the range 0 to 1, then the mean is $0\cdot5$ and the standard deviation can be shown to be $1/\sqrt{12}$.* For samples of three the mean is again $0\cdot5$ and the standard deviation is $1/\sqrt{(3 \times 12)} = \frac{1}{6}$ [§ **3.6**]. The range covered by the mean ± 3 times its standard deviation is

---

* From the footnote on page 19:

$$\sigma^2 = \int_0^1 (x-\tfrac{1}{2})^2 \, dx = \int_0^1 (x^2+\tfrac{1}{4}-x) \, dx$$
$$= \left[x^3/3+x/4-x^2/2\right]_0^1 = (\tfrac{1}{3}+\tfrac{1}{4}-\tfrac{1}{2})-0 = \tfrac{1}{12}$$

thus $\frac{1}{2} \pm 3(\frac{1}{6}) = 0$ to 1. No result, and hence no sample mean, can lie outside this range.

On the other hand, many statistical tests, such as those described in Chapters 4 and 5, are not seriously affected, even if the Population is far from Normal; procedures such as the Analysis of Variance (see Chapter 6) and the associated statistical tests, although derived on the assumption that the results come from Normal distributions, remain valid. In such cases, therefore, it is safe to assume Normality. If, on the other hand, it is known that the distribution is of some other type, this information should preferably be used.

Samples and sampling fluctuations

2.6 Before the characteristics of a set of data are used to estimate the corresponding properties of the Population, it is necessary to know how the sample has been obtained. A Random Sample is defined as a set of observations drawn from a Population in such a way that every possible observation has an equal chance of being drawn at every trial; in other words, the probability that any observation will have a given value, or will lie within a given range of values, is proportional to the relative frequency with which that value occurs in the Population. In random values from a Normal Population, for example, values of x near the mean will occur more frequently than those far from the mean.

The properties of individual small samples may deviate more or less widely from those of the Population. These variations in the properties of small samples, due only to chance causes, are known as Random Sampling Fluctuations or Variations, the word "random" being frequently omitted. Statistics is largely a study of these fluctuations. The statistical method consists essentially of calculating, from a knowledge of the properties of the Population, the probable effects of random sampling on the properties of the sample and comparing the results of this calculation with the observations made. This concept of sampling fluctuations is therefore of great importance. It should be recognised that all measurements (with some trivial exceptions) are subject to some uncertainty, and this being the case, repeated observations or samples will not in general be identical but will vary about the expected value.

2.61 If it can be assumed that sampling is random, the observations may be used to estimate the properties of the Population. But if the sampling is not random a bias may be introduced, and the properties of the Population so estimated may show systematic deviations from the true properties. No amount of repetition will eliminate this bias. For example, if the zero value of an instrument is not properly calibrated, every measurement will be subject to this error and it will be impossible to estimate the true mean (unless a "zero correction" can be applied from some independent determination).

The existence of an unknown bias can be serious and lead to wrong conclusions.

The term Precision has already been introduced to describe the spread of experimental values themselves. The word Accuracy is used to describe closeness of agreement between an experimental value and the true value. These values may differ both because of a bias or systematic error and because of unpredictable random errors.

In a sequence of determinations an experimenter must try to ensure that the conditions remain unaltered for the replicate measurements. He should be on his guard against time trends due, for example, to change in temperature, raw material, etc., which may lead to increased variation in the results.

Transformation to Normality

2.62 Occasionally it will be found that a distribution departs so far from Normality that it would not be safe to apply the common statistical tests, and it would be a great advantage if, by a simple transformation of the variable, an approximately Normal distribution could be obtained. For example, if the weights of a collection of spheres were given and found not to be Normal, a natural transformation would be to use the cube root of the weight, in the hope that the diameters were Normally distributed. A distribution in which the standard deviation is more than 30% of the mean is often skew, but this skewness may often be removed or reduced by using the logarithm of the observation. The distribution of $\log x$ will not necessarily be Normal, but will in such cases be much closer to the Normal form, and may be close enough to justify the assumption of Normality. A few other typical transformations which may be employed to Normalise a skew distribution are the square, square root and reciprocal of the variable.

If there is some theoretical reason for assuming any particular relationship between the standard deviation and the mean, a suitable transformation is easily obtained (see *Design and Analysis*, § **2.62**). It may be worthwhile spending some time on the problem, since if an approximately Normal distribution can be obtained the statistical analysis and conclusions can be made more precise.

If such a transformation is not possible, an alternative class of statistical tests, known as Distribution-Free methods, may be applicable. Two of these will be discussed in § **4.7**.

References

[2.1] WILK, M. B., and GNANADESIKAN, R. "Probability Plotting Methods for the Analysis of Data", *Biometrika*, **55**, 1–17 (1968).

[2.2] AITCHISON, J., and BROWN, J. A. C. *The Lognormal Distribution*. Cambridge University Press (1957).

[2.3] SHAPIRO, S. S., and WILK, M. B. "An Analysis of Variance Test for Normality (Complete Samples)", *Biometrika*, **52**, 591–611 (1965).

Chapter 3

Averages and Measures of Dispersion

When a number of repeat measurements are made
these may be regarded as a sample of the results from
the Population of results which might have been
obtained. From such a sample of observations we
can calculate the sample mean and sample standard
deviation, which are estimates of the Population or
true values. The calculation and properties of these
estimates are described, and measures of their
reliability known as "standard errors" are introduced.

Introduction

3.1 When a set of data has been assembled, consisting, say, of a number of
observations of some quantity, it is necessary to condense the data into such
a form that at least the main features of the assembly are made clear. If two
or more similar sets of data are to be compared, such condensation is even
more necessary. This may be done in a qualitative way by grouping the data
and forming a frequency table or diagram as shown in Chapter 2, but for
most purposes it is necessary to have quantitative measures which adequately
represent the data and so far as possible represent the Population from which
the observations were drawn. In this chapter the more important of these
measures are described and their computation and applications illustrated.
A measure of this kind is referred to as a Statistic.

3.11 The use made of these statistics will be made clear in the chapters that
follow. As well as providing a concise summary of the information contained
in a collection of data, they are essential for the purpose of deciding whether
two sets of data show real differences, as distinct from differences which could
be ascribed to sampling fluctuations; whether two variables are correlated
with each other; and a wide variety of similar questions. They establish the
precision of any function derived from the data, i.e. they fix the limits within
which the true value will lie, the observed values differing from the true
value because of sampling fluctuations [§ **2.6**].

3.12 The statistics most commonly used to represent the properties of a
distribution fall into the following categories:

(a) Measures of Location, or Central Value, giving the location of some central or typical value. An example is the arithmetic mean.

(b) Measures of Dispersion, showing the degree of spread of the data round the central value. An example is the standard deviation.

(c) Measures of Skewness. Skewness means lack of symmetry, and measures of skewness show the extent to which the distribution departs from symmetry.

(d) Measures of Kurtosis. Kurtosis may be defined as "peakedness", and a measure of kurtosis serves to differentiate between a flat distribution curve, and a sharply peaked curve.

These four types suffice for most industrial or research applications, but other measures may be used to bring out particular features of the distribution. For example, in Quality Control it may be important to specify the 0·1 probability point, i.e. the level below which one measurement in ten falls.

Measures of location

3.2 There are three main types of measure:

(a) The means—arithmetic, geometric, harmonic.

(b) The median.

(c) The mode.

All have their uses, but the arithmetic mean is by far the most important. In special cases the others have advantages.

Arithmetic mean

3.21 The arithmetic mean has been introduced in § **2.21**; it is defined as:

$$\bar{x} = \Sigma x / N \qquad (3.1)$$

This gives the arithmetic mean of a sample of N observations. If the sample is random, $\bar{x}$ is also the best available estimate of the Population mean μ; as N increases, $\bar{x}$ becomes a more and more precise estimate of μ. The arithmetic mean is widely used as a measure of the location of a set of data, but while it is probably the most useful of all statistics, other properties of the data are of nearly equal importance. When the word "mean" appears without qualification, it almost always refers to the arithmetic mean.

Hand computation of arithmetic mean

3.22 The computation of the arithmetic mean is straightforward, but if N is large it is profitable in hand calculation to use the following devices to reduce labour and also the chance of errors:

(a) If the observations consist of several figures, they can often be converted to smaller numbers of one or two significant figures (positive or negative) by subtracting from each a constant quantity and restoring this at the end of the computation. These small numbers are easier to handle. If the constant subtracted is x_0, then:

$$\bar{x} = \Sigma x/N = x_0 + \Sigma(x - x_0)/N \qquad (3.11)$$

For example, to calculate the mean intrinsic viscosity of a polymer from the five observations:

$$0\cdot664, \ 0\cdot655, \ 0\cdot657, \ 0\cdot653, \ 0\cdot661 \quad \text{(dl/g. units)}$$

it is convenient to subtract $0\cdot650$ from each and to work with the numbers:

$$0\cdot014, \ 0\cdot005, \ 0\cdot007, \ 0\cdot003, \ 0\cdot011 \qquad (3.111)$$

Then $\Sigma(x - x_0) = 0\cdot040$ and $\bar{x} = 0\cdot650 + 0\cdot040/5 = 0\cdot658$.

(b) If the numbers all end in one or more zeros or are decimal fractions (pure or mixed), they may be divided or multiplied by a power of 10. If they are vulgar fractions and have a convenient common denominator, they are best expressed in terms of this denominator as unit. This frequently happens with British units of length, weight, etc., and computation is simplified by using, say, $\frac{1}{4}$ lb. or $\frac{1}{16}$ in. as unit. If the data are multiplied by h, so that the new variate is $x' = hx$, then:

$$\bar{x} = \frac{\Sigma x}{N} = \frac{1}{h} \frac{\Sigma x'}{N} \qquad (3.12)$$

For example, the five numbers listed in (3.111) could have been multiplied by 10^3 to give the working integers:

$$14, \ 5, \ 7, \ 3, \ 11$$

Their total is 40. Hence:

$$\bar{x} = 0\cdot650 + (40/5)/10^3 = 0\cdot658$$

(c) When N is very large, a long summation can be avoided by grouping the data [§ 2.1]. It is convenient to use between ten and twenty groups: with less than ten the loss of accuracy may become appreciable, and with more than twenty the increased accuracy does not compensate for the extra labour. If the N observations are divided into k groups containing $f_1, f_2, \ldots, f_k$ observations respectively, and with central values $x_1, x_2, \ldots, x_k$, then:

$$\bar{x} = \sum_{i=1}^{k} f_i x_i/N \qquad (3.13)$$

Devices (a), (b) and (c) may, of course, be used in combination. Consider again the carbon content data discussed in Chapter 2 and reproduced in Table 3.1. The 178 measurements are already arranged in eleven groups, each $0\cdot1\%$ wide. Devices (a) and (b) are applied automatically by defining the working variable, x', as shown in Col. 3. It will be seen that the constant subtracted is $x_0 = 4.645$ and the multiplier $h = 10$, i.e. $x' = 10(x - 4.645)$. Col. 5 is summed to give $\Sigma fx'$, from which $\bar{x}'$ is calculated. The remainder of the working is shown below the table.

TABLE 3.1.   PERCENTAGE CARBON IN A MIXED POWDER

(1)	(2)	(3)	(4)	(5)
Group Limits %	Group Centre %	x'	Group Frequency f	fx'
4.10–4.19	4.145	−5	1	−5
4.20–4.29	4.245	−4	2	−8
4.30–4.39	4.345	−3	7	−21
4.40–4.49	4.445	−2	20	−40
4.50–4.59	4.545	−1	24	−24
4.60–4.69	4.645	0	31	−98
4.70–4.79	4.745	1	38	38
4.80–4.89	4.845	2	24	48
4.90–4.99	4.945	3	21	63
5.00–5.09	5.045	4	7	28
5.10–5.19	5.145	5	3	15
		Total	178	+192 −98
				+94

$$\bar{x}' = 94/178 = 0.528$$

Hence $\bar{x} = 4.645 + 0.528/10 = 4.698\%$

If a number of samples are available, for each of which the arithmetic mean has already been calculated, the grand mean can be found as follows. Let the samples contain n_1, n_2, ..., n_k observations, with sample means $\bar{x}_1, \bar{x}_2, ..., \bar{x}_k$. Then the grand mean is:

$$\bar{x} = (n_1\bar{x}_1 + n_2\bar{x}_2 + ... + n_k\bar{x}_k)/N$$
$$= \sum_{i=1}^{k} n_i\bar{x}_i/N \qquad (3.14)$$

where $N = n_1 + n_2 + ... + n_k$.

If the samples are all of equal size, this reduces to:

$$\bar{x} = \frac{1}{k} \sum_{i=1}^{k} \bar{x}_i \qquad (3.15)$$

i.e. the grand mean is the mean of the sample means.

Geometric mean

3.23 The Geometric Mean is defined as:

$$G = (x_1 x_2 ... x_N)^{1/N} \qquad (3.2)$$

i.e. the Nth root of the product of the N observations. For calculation it is better to take logarithms:

$$\log G = \frac{1}{N} \sum_{i=1}^{N} \log x_i \qquad (3.21)$$

The use of the geometric mean thus amounts to a transformation of the variable to $\log x$, the calculation of the arithmetic mean of the new variable and a final inverse transformation. G is only computed when all the observations are positive. It can be shown that its value never exceeds $\bar{x}$.

The principal application of the geometric mean is in averaging a sequence of ratios. As an illustration, consider the annual changes in the sales of a chemical listed in Table 3.2.

TABLE 3.2.   SALES OF A CHEMICAL

Year	1965	1966	1967	1968
Sales ('000 lb.)	20·7	18·9	24·6	29·8
% of previous year	—	91·3	130·2	121·1

The arithmetic mean of the three percentages is 114·2, which seems (wrongly) to imply an average rate of increase of 14·2% per year. If sales had risen at a compound rate of 14·2% from 20·7 in 1965, they would have reached 30·8 ($\times 10^3$) lb. in 1968. Hence the true average rate of increase was less than 14·2%; it is determined by the geometric mean of the three percentages:

$$(91·3 \times 130·2 \times 121·1)^{1/3} = 112·9$$

A uniform increase of 12·9% each year leads to the actual sales of 29·8 ($\times 10^3$) lb. in 1968.

Harmonic mean

3.24 The Harmonic Mean is defined by the relation:

$$\frac{1}{H} = \frac{1}{N} \Sigma \frac{1}{x} \qquad (3.22)$$

For the same set of positive observations, H never exceeds G. Using the harmonic mean is effectively a transformation of the variable to $1/x$. It is a useful measure when the observations are expressed inversely to what is required in the average.

Suppose, for example, a Chemical Department purchases 2 dozen 1-litre flasks at 45p, costing a total of £10·80. Later it has the opportunity of buying some surplus flasks at 30p and it purchases 3 dozen of these, making the *same* total cost, £10·80. What is the average price paid in this case? It is *not* the arithmetic mean of the two prices, $37\frac{1}{2}$p, but the harmonic mean:

$$\frac{1}{H} = \frac{\text{Total number of flasks}}{\text{Total sum paid}} = \frac{24+36}{10\cdot80+10\cdot80} = \frac{1}{2}\left\{\frac{24}{10\cdot80} + \frac{36}{10\cdot80}\right\}$$

$$= \frac{1}{2}\left\{\frac{1}{\text{First price}} + \frac{1}{\text{Second price}}\right\} = \frac{1}{2}\left\{\frac{1}{0\cdot45} + \frac{1}{0\cdot30}\right\} = \frac{5}{1\cdot80}$$

i.e. $H = £1\cdot80/5 = 36$p.

Median

3.25 If the data are arranged in order of magnitude, the median is the central member of the series, i.e. there are equal numbers of observations greater than and less than the median. If N is an odd number, this definition is complete. If N is even it is ambiguous, and in this case it is usual to take the arithmetic mean of the two central values as the median.

The median of the Population can be estimated from the cumulative distribution curve; it is the value of x at which the curve intersects the 50% frequency line. This method should be used for data presented in grouped form. For a symmetrical distribution the median is identical with the arithmetic mean, but for a skew distribution it is not.

An application of the median, though not always under that name, is the use of the half-life of a radioactive element as a measure of its activity. The half-life is the time at which half the atoms present have disintegrated; it is therefore the median value of the time at which individual atoms disintegrate. Since the time for complete disintegration is infinite, the median has the practical advantage that the distribution is expressed thereby in terms of only half the atoms—those which disintegrate first. The half-life is also used as a measure of the toxicity of insecticides. If a batch of insects is exposed to the test material they do not all die at the same time, and a proportion may survive completely. The median life is the time taken for half of them to die; it is a measure of the toxicity of the material—low values mean high toxicity.

The median has two practical merits as a measure of location for small samples:

(i) It is very simply determined, especially if $N = 3$ or 5.
(ii) It is not influenced by freak extreme observations.

For these reasons the median is sometimes used to summarize the outcome of routine chemical tests which are replicated on, say, three specimens.

However, for a Normally distributed variate the median gives a less precise estimate of the centre of the distribution than the arithmetic mean, so the latter will often be preferred. On the other hand, for symmetric non-Normal distributions which are sharply peaked, the median may be the more reliable measure of location: an example of this use of the median in mechanical testing is given in [3.1].

Mode

3.26 The Mode is the value of the variate which occurs most frequently, i.e. for which the frequency is a maximum. More than one mode is possible. In most applications concerned with experimental investigations, however, distributions with more than one mode are rare. The presence of two or more modes usually means that the data are not homogeneous, i.e. that two or more distributions have been combined. Data from two or more machines set at different average values, for example, will, if the averages differ sufficiently, give a multimodal distribution.

The use of the mode is most often associated with discrete distributions, where it is simple to count which value of the variate occurs most often. It is less easy to estimate the mode for a continuous variate. However, if a large sample is available an approximate value can be obtained by plotting a frequency diagram, drawing a smooth curve through it and noting the point of maximum frequency.

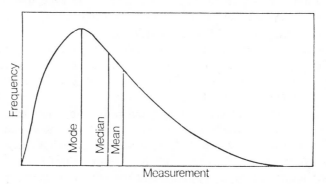

FIG. 3.1. LOCATION STATISTICS FOR DISTRIBUTION WITH
POSITIVE SKEWNESS

For a symmetric unimodal Population, the mode coincides with the arithmetic mean and the median. For asymmetric distributions these three statistics are separated, as exemplified in Figure 3.1. They are connected by the approximate empirical relation:

$$(\text{Mean} - \text{Mode}) = 3(\text{Mean} - \text{Median}) \qquad (3.3)$$

for unimodal Populations of moderate skewness. Equation (3.3) may be used to estimate the mode of a continuous variate when only a small sample is available.

Number and weight averages

3.27 In the calculation of the arithmetic mean each observation is regarded as being of equal importance. Sometimes, however, it is desirable to give greater emphasis to observations at the upper end of the scale. This leads to a class of higher-order averages which are particularly useful to characterize distributions of properties such as fibre length, particle size and the length of molecular chains.

Let x_i denote the weight of a molecule of size i ($i = 1, 2 \ldots$) and suppose that this size occurs with a frequency f_i in a polymer. To calculate the Number Average molecular weight, M_n, each molecular size is counted according to the frequency with which it is present; thus

$$M_n = \Sigma f_i x_i / \Sigma f_i \qquad (3.31)$$

is identical with the arithmetic mean, $\bar{x}$. Now the net weight of molecules of size i is $f_i x_i$, so that they form a fraction:

$$g_i = f_i x_i / \Sigma f_i x_i$$

of the total weight of the polymer. Clearly, as weight fractions, the larger molecules are given proportionally greater importance. The Weight Average molecular size, M_w, is defined as the arithmetic mean with respect to the weight frequencies, g_1, g_2, etc.:

$$M_w = \Sigma g_i x_i / \Sigma g_i$$
$$= \Sigma f_i x_i^2 / \Sigma f_i x_i \qquad (3.32)$$

Except for the trivial case when there is only one size, the weight average always exceeds the number average.

To illustrate the calculation, consider just five sizes in unit steps, $x_1 = 1$, $x_2 = 2, \ldots, x_5 = 5$, and suppose that the corresponding frequencies follow the skew pattern, $f_1 = 5, f_2 = 4, \ldots, f_5 = 1$. Then:

$$M_n = (5+8+9+8+5)/15 = 35/15 = 7/3$$
$$\text{and } M_w = (5+16+27+32+25)/35 = 105/35 = 3$$

Averages of higher order can be defined in a similar manner. Their importance is due to the fact that certain physical test methods yield one of these statistics as the characteristic value of the property. For example, in measuring the average molecular weight of a polymer, osmotic pressure methods lead to M_n whereas solution viscosity methods produce an estimate of M_w.

Measures of dispersion

3.3 The Measures of Dispersion discussed below are the Range, Variance, Standard Deviation, Coefficient of Variation, and the Mean Deviation. The variance and standard deviation, which are in substance equivalent, are of great importance, both theoretically and practically. The coefficient of variation is closely allied to the standard deviation. The range is important in special circumstances. The mean deviation is seldom used.

Variance

3.31 The Variance of a Population is the mean squared deviation of the individual values from the Population mean and is denoted by σ^2.* The symbols V and s^2 are used for the variance deduced from sample data. Owing to its convenience, the symbol V is often used to denote also the Population variance, particularly when it relates to the variance of a function of the observations as in § **3.62**.

In a sample of N drawn from a Population with mean μ, the variance of the Population is estimated by:

$$V = \frac{\Sigma(x - \mu)^2}{N} \qquad (3.4)$$

In general μ is not known, and an estimate, $\bar{x}$, based on the sample must be used. It can be shown that the sum of the squared deviations from the arithmetic mean is less than that from any other value, i.e.

$$\Sigma(x - a)^2 \text{ is a minimum when } a = \bar{x} \qquad (3.41)$$

Thus $\Sigma(x - \bar{x})^2$ is less than $\Sigma(x - \mu)^2$, so that if $\bar{x}$ is substituted for μ in (3.4) the variance is underestimated. It can be shown that this bias is removed by using $(N - 1)$ in place of N as the divisor, i.e.

$$V = \frac{\Sigma(x - \bar{x})^2}{N - 1} \qquad (3.42)$$

This gives the best estimate of the Population variance from the data available. (For proof see Appendix 3A.)

The variance measures the extent to which the data differ among themselves. If the values of x are all identical, there are no differences and the estimate of the variance is zero; if they differ only slightly from each other, the variance is small; if they differ widely, the variance is large.

Degrees of freedom

3.311 The divisor $(N - 1)$ in (3.42) represents the number of Degrees of Freedom of the estimate of variance, i.e. the number of independent comparisons that can be made among N observations. If there are N observations, $x_1, x_2, \ldots, x_N$, there are $(N - 1)$ independent comparisons such as $(x_1 - x_2)$, as can readily be verified for a small value of N. The comparisons

---

* The statistician's definition of variance is quite distinct from the accountant's use of the word to denote simply a difference between an achieved and a budgeted figure.

used in Equation (*3.42*) are $(x_1 - \bar{x})$, etc. Only $(N-1)$ of these are independent, since $\bar{x}$ is calculated from the observations, and if $\bar{x}$ and $(N-1)$ of the values of x are given, the other is determined.

If only a few results are available, the estimate of the variance, as of the mean, is of low precision. For the mean, the precision depends on the number of results used in calculating it, but for the variance the precision depends on the number of independent differences in the data from which the variance is calculated directly or indirectly.

Suppose that measurements are made of a certain quantity, the true mean and variance being unknown. It is assumed that there is no bias, i.e. no systematic error in the measurement, differences between repeats being due to random sampling variations. If one measurement is made and gives a result of, say, 91, then we have an estimate of the true, or mean, figure, but have clearly no information as to the accuracy of the measurement. We have:

$$\bar{x} = 91$$
$$V = \Sigma(x - \bar{x})^2/(N-1) = 0/0$$

This is indeterminate, as it should be. (If we had divided by N, the variance would have been 0, which is clearly false.)

If a second measurement is made, say 101, we get not only a new and better estimate of the mean, but also one comparison or difference on which to base an estimate of variance, i.e. $(x_1 - x_2) = 10$. Note that the separate differences of each of the observations from the mean are not independent, since the mean was calculated from the data; if $\bar{x} = 96$, $x_1 = 91$, then x_2 must be 101. The estimate is therefore said to have one degree of freedom.

If a third measurement is made and gives a result 90, we have:

$$\bar{x} = \tfrac{1}{3}(91 + 101 + 90) = 94$$
$$V = (3^2 + 7^2 + 4^2)/2 = 74/2 = 37$$
$$s = \sqrt{V} = \sqrt{37} = 6\cdot1$$

With three observations there are only two independent comparisons, since if two are given, say $(x_2 - x_1)$ and $(x_2 - x_3)$, the third is known, i.e. $(x_1 - x_3) = (x_2 - x_3) - (x_2 - x_1)$. As before, if the mean and two results are known, the third result is determined. There are therefore only two degrees of freedom.

If the true mean is known beforehand the case is different, since every observation then provides a comparison—its difference from the true mean—so that the number of degrees of freedom is equal to the number of observations. This will, however, rarely occur.

Similar reasoning can be applied to the case, discussed in Chapter 7, of the spread of results round a straight line (the regression line) instead of round a point (the mean). At least two results are necessary to give the position of the line, and with only two, no estimate of the spread of the results round the line can be made. Two constants are required to define the line; two degrees of freedom are thus "used up", and the number of degrees of freedom of the estimate of the variance about the line is $(N-2)$. This can be generalised to the fitting of data to any type of equation: the degrees of freedom are always the number of observations less the number of constants calculated from the data and used to determine the equation.

The idea of degrees of freedom is of fundamental importance in statistical analysis, and its meaning should be firmly grasped. The Greek letter ϕ will be used in this book to denote the number of degrees of freedom. In estimating the variance from a large number of observations no great error is made if N is used instead of $N-1$, but in the techniques used in later chapters estimates based on small numbers of degrees of freedom are widely used, and it is important to use the correct divisor.

Standard deviation

3.32 The variance is of great importance in many applications of statistics, but since it is not a linear function of the variate its numerical implication is not quickly grasped. Its square root, however, has the same dimension as the variate, and is therefore a more easily appreciated measure of dispersion. The positive square root of the variance is known as the Standard Deviation; it is, apart from the arithmetic mean, the most familiar of all statistics. After experience with its use a good idea of the "spread" of the distribution can be obtained from the value of the standard deviation. The symbol for the Population standard deviation is σ, and that for the estimate from a sample is s.

Coefficient of variation

3.33 The Coefficient of Variation is the standard deviation expressed as a percentage of the arithmetic mean:

$$C.V. = (\sigma/\mu) \times 100 \qquad (3.5)$$

Since σ and μ are both expressed in the same units as the variate, $C.V.$ is independent of the units of measurement; the position of the origin must, however, be known. It follows that the coefficient of variation is a suitable statistic to relate the spread of observations resulting from two alternative ways of measuring the same response, provided they have the same zero value. For example, if concentration fluctuations in a continuous chemical process are monitored on a chart recorder, the coefficient of variation of the trace heights (in mm., say) will be equal to that of the concentrations. Hence the observer does not need to know the factor relating concentration to distance, but it is essential that the two systems have a common origin, i.e. that the chart baseline corresponds to zero concentration.

The main use of the coefficient of variation is to compare the variability of groups of observations with widely differing mean levels. It is also invaluable when dealing with properties whose standard deviation rises in proportion to the mean. This can be illustrated by weight variations along the length of a synthetic fibre. For a 2 decitex filament the value of σ might be 0·02 decitex; for a filament of twice this weight the fluctuations produced in the manufacturing process are 0·04 decitex, and so on in proportion. All these data are summarised by the single statement that $C.V. = 1\%$.

Computation of variance and standard deviation

3.34 Equation (3.42) as it stands is not the most convenient form for computing the variance; the mean $\bar{x}$ will in general contain more significant figures than the variates, and $(x - \bar{x})$ may be a cumbersome number. The computation can be simplified by the use of the identity*:

$$\Sigma(x - \bar{x})^2 = \Sigma x^2 - S^2/N \qquad (3.43)$$

where $S = \Sigma x$. Expression (3.43) is termed the Corrected Sum of Squares.

Then:
$$V = \frac{\Sigma x^2 - S^2/N}{N - 1} \qquad (3.44)$$

This formula lightens the arithmetic involved in hand calculation. Several desk calculators are designed so that Σx and Σx^2 can be accumulated simultaneously; this saves time and avoids errors caused by failure to enter the correct values of x twice.

The devices (a), (b) and (c) given in §3.22 for hand calculation of the arithmetic mean can be used with even greater effect in computing the variance. When device (a)—subtracting a constant from each observation—is used, no correction need be applied to the variance, which is clearly the same whatever arbitrary origin is taken, since it measures the variation about the arithmetic mean. If device (b)—multiplying each variate by h—is used, then if the result of the calculation is V' it is corrected by dividing by h^2, i.e. $V = V'/h^2$. The standard deviation is similarly corrected by dividing by h. If device (c)—grouping—is adopted, V is given by:

$$V = \frac{\Sigma fx^2 - S^2/N}{N - 1} - \frac{h'^2}{12} \qquad (3.45)$$

where $S = \Sigma fx$ and h' is the width of a grouping interval. In other words, the group frequencies are multiplied by x^2, the products are summed, and then corrected for the mean by subtracting S^2/N. The term $h'^2/12$ is a small adjustment known as Sheppard's Correction, which removes a bias error introduced by the grouping operation.

---

* This identity is easily proved:
$$\Sigma(x - \bar{x})^2 = \Sigma(x^2 + \bar{x}^2 - 2x\bar{x}) = \Sigma x^2 + N\bar{x}^2 - 2\bar{x}\Sigma x$$
$$= \Sigma x^2 - N\bar{x}^2, \text{ since } \Sigma x = N\bar{x}$$
$$= \Sigma x^2 - S^2/N$$

When N is large and the values of x are given to several significant figures, the subtraction in (3.43) will cause a loss of accuracy unless a large number of digits are carried. The following alternative method of calculating the variance is preferable when an electronic computer is employed [3.2]. The first number in the set is read in and stored as a working mean, a. For each subsequent number, $(x-a)$ and $(x-a)^2$ are calculated; the values of $\Sigma(x-a)$, $\Sigma(x-a)^2$ and the number of entries to date are accumulated. When the last number has been read into the computer, calculate:

$$\bar{x} = a + \Sigma(x-a)/N$$

and

$$V = \frac{\Sigma(x-a)^2 - \{\Sigma(x-a)\}^2/N}{N-1} \qquad (3.46)$$

Examples of hand computation

3.341 For the five values of intrinsic viscosity:

$$0{\cdot}664,\ 0{\cdot}655,\ 0{\cdot}657,\ 0{\cdot}653,\ 0{\cdot}661$$

it is convenient, as in § **3.22**, to subtract $0{\cdot}650$ from each, multiply by 10^3 and work with the integers:

$$14,\ 5,\ 7,\ 3,\ 11$$

The calculation then proceeds as follows:

Working units	*Converted to original units*
Sum $(S) = +40$	
$\bar{x}' = 40/5 = 8$	$\bar{x} = 0{\cdot}650 + 8/1000 = 0{\cdot}658$ dl/g.
$\Sigma x'^2 = 400$	
$S^2/N = 320$	
$\Sigma(x'-\bar{x}')^2 = \overline{\quad 80}$	
$V(x') = 80/4 = 20$	$V(x) = 20/1000^2 \quad = 0{\cdot}00002$
$S.D.(x') = \sqrt{20} = 4{\cdot}5$	$S.D.(x) = (\sqrt{20})/1000 = 0{\cdot}0045$ dl/g.
	$C.V.(x) = \dfrac{0{\cdot}0045 \times 100}{0{\cdot}658} = 0{\cdot}68\%$

To illustrate the calculation of these dispersion statistics for grouped data, consider again the carbon content data of Table 3.1. Table 3.3 repeats the necessary figures from Table 3.1 with an additional column for fx'^2.

TABLE 3.3. PERCENTAGE CARBON IN A MIXED POWDER
(WORKING DATA)

x'	f	fx'	fx'^2
−5	1	−5	25
−4	2	−8	32
−3	7	−21	63
−2	20	−40	80
−1	24	−24	24
0	31	−98	0
1	38	38	38
2	24	48	96
3	21	63	189
4	7	28	112
5	3	15	75
Total	178	+192 −98	734
		+94	

The subsequent stages of the computation are as follows:

Working units	Converted to original units

$S = +94$

$\bar{x}' = 94/178 = 0.528$ $\bar{x} = 4.645 + 0.528/10 = 4.698$

$\Sigma fx'^2 = 734$

$S^2/N = \underline{\quad 49.64\quad}$

$\Sigma(x' - \bar{x}')^2 = 684.36$

$V(x') = 684.36/177 - 1/12$

$\quad = 3.86 - 0.08 = 3.78$ $V(x) = 3.78/10^2 = 0.0378$

$S.D.(x') = \sqrt{3.78} = 1.94$ $S.D.(x) = 1.94/10 = 0.194$

$$C.V.(x) = \frac{0.194 \times 100}{4.698} = 4.14\%$$

Combination of variances

3.342 Suppose a number of samples are available and it is required to estimate the variance and standard deviation from all samples, assuming that the sample variances differ only because of sampling fluctuations, but allowing for the possibility that the sample means may be different—i.e. the samples are assumed to be drawn from Populations with the same variance but different means. For example, when concrete is being made in batches, the mean strengths of the batches vary because of inevitable variations in the composition of the concrete, and in addition the strengths of test cubes made from one batch vary because of sampling and testing errors.

The procedure is to calculate for each sample the sum of squared deviations from the sample mean, add these for all samples and divide by $(N-k)$, where N is the total number of observations and k the number of samples:

$$V = \frac{\Sigma(x_1 - \bar{x}_1)^2 + \Sigma(x_2 - \bar{x}_2)^2 + \ldots + \Sigma(x_k - \bar{x}_k)^2}{N-k} \qquad (3.47)$$

The suffices refer to the different samples. The divisor $(N-k)$ represents the degrees of freedom of the estimate, one degree being used up for each sample mean. In this formula, the difference between N and $(N-k)$ will be much more important than that between N and $(N-1)$ in Equation (3.42), unless N is large compared with k.

If the sample variances are given, (3.47) may be rewritten as:

$$V = \frac{(n_1 - 1)V_1 + (n_2 - 1)V_2 + \ldots + (n_k - 1)V_k}{N-k} \qquad (3.48)$$

where n_1, n_2, etc. are the numbers of observations in the first, second, etc., samples. This expression is a weighted mean, each variance being weighted by its degrees of freedom.

If all the samples are of equal size ($n_1 = n_2$, etc.), then:

$$V = (V_1 + V_2 + \ldots + V_k)/k \qquad (3.49)$$

i.e. the overall variance is simply the mean of the sample variances. However, the standard deviation estimated from the sample data is $\sqrt{V}$, and not the mean of the sample standard deviations.

Finally, if each sample contains only two measurements, Equation (3.49) can be simplified to:

$$V = \Sigma w^2 / 2k$$

where w is the range, or difference, of the two results.*

Example: Efficiency of ammonia oxidation

3.343 To illustrate the estimation of a combined variance, let us consider a set of 85 measurements of the efficiency of oxidation of ammonia on one unit of a nitric acid plant, i.e. the percentage of the ammonia converted to

---

* $V = (x_1 - \bar{x})^2 + (x_2 - \bar{x})^2$

$$= \left(x_1 - \frac{x_1 + x_2}{2}\right)^2 + \left(x_2 - \frac{x_1 + x_2}{2}\right)^2$$

$$= \left(\frac{x_1 - x_2}{2}\right)^2 + \left(\frac{x_2 - x_1}{2}\right)^2 = w^2/2$$

nitric oxide. The data are in samples of five, each sample representing tests on five burners made on one day. We wish to estimate the variance within samples, i.e. within the sets of five tests done at one time, but allowing for any change in the mean from day to day. The efficiencies are quoted to one place of decimals and range from 87·0% to 98·2%. To simplify the calculations, Table 3.4 (overleaf) lists the values of the scaled variable:

$$x' = (\text{Efficiency} - 90) \times 10$$

For each sample the sum, S, and $\Sigma x'^2$, which is termed the Crude Sum of Squares, are formed. The quantity $S^2/5$ is subtracted from $\Sigma x'^2$ to give the corrected sum of squares, $\Sigma(x' - \bar{x}')^2$. These calculations are set out in Table 3.5. Note that the figures are arrayed in groups of two, to assist in summation: thus 1 51 29 would normally be written 15,129.

Since the samples are of equal size, the combined within-sample variance can be found from either of the last two columns. The degrees of freedom for each sample are $n-1 = 4$, and the degrees of freedom for the variance within all seventeen samples are $N-k = 85-17 = 68$. The calculation then proceeds as follows:

Working units	*Converted to original units*

$$V(x') = \frac{28{,}514}{68} = \frac{7{,}129}{17} = 419{\cdot}3 \qquad V(x) = 419{\cdot}3/10^2 = 4{\cdot}193$$

$$S.D.(x') = \sqrt{419{\cdot}3} = 20{\cdot}5 \qquad S.D.(x) = 20{\cdot}5/10 = 2{\cdot}05$$

$$\bar{x}' = 2{,}405/85 = 28{\cdot}3 \qquad \bar{x} = 90 + 28{\cdot}3/10 = 92{\cdot}83$$

$$C.V.(x) = \frac{2{\cdot}05 \times 100}{92{\cdot}83} = 2{\cdot}21\%$$

Combination of coefficients of variation

3.344 In the last example, it was assumed that there was a single underlying variance within samples, but that there could be real changes of mean from day to day. For properties where the standard deviation rises in proportion to the mean level, it would be more reasonable to assume that there is a single underlying coefficient of variation within samples, coupled with real changes of mean from sample to sample. In such a case we want a combined estimate of the coefficient of variation.

This can be calculated by taking the logarithms (to base 10) of the observations and computing $\sqrt{V}$ from (3.47) for the transformed data. Then [§ **3.62**]:

$$C.V.(x) = 100 \log_e 10 \times \sqrt{V(\log x)} = 230{\cdot}26 \times \sqrt{V(\log x)} \qquad (3.51)$$

Range

3.35 The Range is the simplest of all measures of dispersion. It is simply the difference between the highest and lowest values of the variate in a

TABLE 3.4. EFFICIENCY OF AMMONIA OXIDATION (WORKING DATA)

Sample No.	$x' = (\text{Efficiency} - 90) \times 10$					Range, w
1	29	18	62	1	13	61
2	9	21	20	32	23	23
3	20	24	35	20	32	15
4	30	−5	30	82	48	87
5	13	36	34	13	41	28
6	31	34	74	3	47	71
7	6	29	60	26	76	70
8	27	79	74	27	33	52
9	−10	16	18	7	42	52
10	33	16	5	2	−3	36
11	12	27	20	46	13	34
12	34	−4	7	28	−12	46
13	47	−15	15	36	28	62
14	57	45	57	62	20	42
15	50	30	27	40	33	23
16	27	22	39	63	48	41
17	34	10	14	−30	42	72

TABLE 3.5. WORKING SHEET FOR DATA OF TABLE 3.4

Sample No.	S	S^2	$S^2/5$	$\Sigma x'^2$	$\Sigma(x' - \bar{x}')^2$	Sample Variance
1	1 23	1 51 29	30 26	51 79	21 53	5 38
2	1 05	1 10 25	22 05	24 75	2 70	68
3	1 31	1 71 61	34 32	36 25	1 93	48
4	1 85	3 42 25	68 45	1 08 53	40 08	10 02
5	1 37	1 87 69	37 54	44 71	7 17	1 79
6	1 89	3 57 21	71 44	98 11	26 67	6 67
7	1 97	3 88 09	77 62	1 09 29	31 67	7 92
8	2 40	5 76 00	1 15 20	1 42 64	27 44	6 86
9	73	53 29	10 66	24 93	14 27	3 57
10	53	28 09	5 62	13 83	8 21	2 05
11	1 18	1 39 24	27 85	35 58	7 73	1 93
12	53	28 09	5 62	21 49	15 87	3 97
13	1 11	1 23 21	24 64	47 39	22 75	5 69
14	2 41	5 80 81	1 16 16	1 27 67	11 51	2 88
15	1 80	3 24 00	64 80	68 18	3 38	85
16	1 99	3 96 01	79 20	90 07	10 87	2 72
17	70	49 00	9 80	41 16	31 36	7 84
Total	24 05	—	8 01 23	10 86 37	2 85 14	71 29

sample. It is therefore very easy to calculate, and this makes its use attractive if it supplies sufficiently reliable information. The range of a large sample obviously conveys little information, since it says very little about the intermediate observations, but for small samples it has been found that the range conveys a large proportion of the information on the variation in a sample and can be used with confidence as a measure of dispersion.

The formula (3.47) for calculating the variance from a number of samples gives the best estimate of the variance. A simpler, but less precise, estimate of the standard deviation is found from the mean range instead of the mean square for samples of equal size. The formula is [3.3]:

$$s = \bar{w}/d_n \qquad (3.6)$$

where $\bar{w}$ is the mean range and d_n is a constant, depending on the size of the sample. The factors d_n are given in Table G.1. It is interesting to note that, for $n = 3$ to 12, d_n is very nearly equal to $\sqrt{n}$ (within less than 5%).

To illustrate this method, the 17 sample ranges for the ammonia oxidation data are listed in the last column of Table 3.4. The mean range, $\bar{w} = 47 \cdot 9$ in working units. From Table G.1, the conversion factor for samples of 5 is 2·326.

$$\therefore \ s(x') = 47 \cdot 9/2 \cdot 326 = 20 \cdot 6$$

This is in excellent agreement with the standard deviation found from the sum of squares, and is obtained with much less trouble.

In general, the larger the sample the less efficient is the estimate obtained from the mean range. The loss in efficiency becomes serious for large samples, and the Range Method should not be used for samples of more than 12. With samples of 6, about six sample ranges are sufficient to give a good estimate of the standard deviation. The best-known application of the range is in control chart work, where it is widely used instead of the standard deviation [Chapter 10]. Another important application is in estimating the standard deviation of a chemical analysis from a series of duplicate tests. These tests must be independent and also true duplicates; if only part of the analysis, say the final stage, is repeated, the difference between the tests corresponds to only a part, and perhaps a small part, of the real analytical error.

3.351 The range method can also be used to calculate an approximate estimate of the standard deviation for a single set of N observations. The values are divided *randomly* into small groups of n observations each, discarding a few results if N is not an exact multiple of n. (Random allocation is most important because ordered groups may yield a biased statistic.) Then the range of the observations in each group is calculated, and the mean range is divided by the appropriate d_n factor to estimate s. This method is most efficient when n is between 6 and 10 [3.4]. The calculations are quickly made and the result can be used as a rough check of the arithmetic of the full method [§ **3.34**].

The random allocation of N observations into small groups is most easily made with the aid of the random permutations given in Table J at the end of the book. Each column lists the integers 1, 2, ..., N in a random order. The first n entries in a column are the numbers of the observations to be assigned to the first group; the next n entries define the second group, and so on. When the exact value of N is not listed in Table J, the next higher column should be used, discarding the superfluous entries $N+1$, $N+2$, etc.

For example, to allocate 26 observations into three groups of 8, use the $N = 30$ column and discard the entries 27, 28, 29, 30. The resulting arrangement is:

First Group Consists of Observations Numbered: 10, 15, 18, 8, 21, 26, 23, 11
Second Group Consists of Observations Numbered: 16, 9, 22, 20, 7, 25, 14, 6
Third Group Consists of Observations Numbered: 3, 17, 2, 12, 4, 5, 19, 13

with observation nos. 24 and 1 being omitted altogether, because 26 is not an exact multiple of 8.

Mean deviation

3.36 The Mean Deviation is defined as the arithmetic mean of the deviations from the mean, all taken with the positive sign:

$$M.D. = \Sigma|x-\bar{x}|/N \qquad (3.61)$$

where $|x-\bar{x}|$ is the absolute value of $x-\bar{x}$. The mean deviation finds applications where the distribution has long "tails", i.e. where outlying results are rather frequent. It is less affected by outlying results than is the standard deviation. The squares of a few large deviations tend to outweigh those of the more numerous small deviations; in the mean deviation, which does not involve squaring, the effect of a few large values is not so great. Otherwise the mean deviation is of little practical use, and unless there are good reasons for using it the standard deviation is preferable. Apparent ease of calculation should not be considered a good reason for using the mean deviation.

Measures of skewness

3.4 A distribution will not in general be completely symmetrical; the frequency may fall away more rapidly on one side of the mode than on the other. When this is so the distribution is said to be Skew. The distribution shown in Figure 3.1 is described as Positively Skew, because the long tail is on the side of the high values of x. Similarly, if the long tail is on the side of the low values of x, the distribution is said to be Negatively Skew. Positive skewness is more common than negative; for example, the distribution of the number of items waiting in a queue and the distribution of molecular chain lengths in a polymer usually exhibit this shape.

One measure of skewness is defined by:

(Mean − Mode)/(Standard Deviation)

However, because of the difficulty of estimating the mode of a continuous distribution, the preferred measure of skewness, denoted by $\sqrt{\beta_1}$, uses the third power of the deviations from the mean. For a positively skew distribution, the cubes of the positive deviations outweigh those of negative deviations, so that $\Sigma(x-\mu)^3 > 0$. In a symmetric distribution $\Sigma(x-\mu)^3 = 0$, while for negative skewness, $\Sigma(x-\mu)^3 < 0$. The average of these third powers is called the Third Moment, denoted by μ_3:

$$\mu_3 = \Sigma(x-\mu)^3/N \qquad (3.7)$$

To obtain an index of skewness which is independent of the unit of measurement, we divide by the cube of the standard deviation:

$$\sqrt{\beta_1} = \mu_3/\sigma^3 \qquad (3.71)$$

The value of $\sqrt{\beta_1}$ calculated for a sample of observations from a symmetric distribution will not be exactly zero. Table 34B of [3.5] gives the probability of various departures from zero for a Normal distribution.

In experimental work, if a distribution is found to be skew, it is often worth while to search for a transformation of the variate which will yield an approximately symmetric distribution [§ **2.62**].

Kurtosis

3.41 The degree of "peakedness" in a distribution can be assessed from the Fourth Moment, μ_4:

$$\mu_4 = \Sigma(x-\mu)^4/N \qquad (3.72)$$

The standardized measure of kurtosis, denoted by β_2, is obtained by dividing μ_4 by the fourth power of the standard deviation:

$$\beta_2 = \mu_4/\sigma^4 \qquad (3.73)$$

β_2 is essentially positive. For a Normal distribution it is equal to 3. If the distribution is more sharply peaked than the Normal, i.e. if it has long "tails", β_2 is greater than 3; if it is flatter than the Normal, β_2 is less than 3. Unless a large set of data is available there is little point in calculating β_2, since it is greatly affected by one or two outlying results. For values of $N \geqslant 200$, Table 34C of [3.5] can be used to assess whether a sample β_2 departs significantly from the Normal distribution value.

Outliers

3.5 Sometimes a set of data contains one, or possibly more, observations which fall outside the pattern exhibited by the majority of the values. Consider, for example, six measurements of the intrinsic viscosity of a polymer:

0·664, 0·655, 0·686, 0·657, 0·653, 0·661   (dl/g.)

The 0·686 value is very much higher than the other five measurements and is described as an Outlier.

We must now decide how to deal with such an occurrence. Should we assume that the outlier is just as valid as the other observations and calculate statistics based on all the results, or should we discard the "odd" value as being unrepresentative? The latter course is the easy way out of the difficulty, but an investigator who quickly discards data which do not agree with his preconceived ideas is liable to end up with data tailored to those preconceptions, thus vitiating the study. Initially the best action is to try to discover any special circumstances which affected the outlying observation. In the example, we should check that the arithmetic leading to the 0·686 value did not contain a mistake, that the polymer sample was in no way unusual and that the viscometer was properly cleaned before the measurement was made. It may not, of course, be possible to check whether a gauge or dial was mis-read. When some assignable cause is discovered for the outlier being different from the other observations, it may safely be discarded.

If no obvious cause comes to light, a statistical test should be applied to confirm that a suspected outlier is really as extreme as it appears. A number of these tests have been devised. Those based on the range of the observations are quickest to calculate, but a more efficient criterion is to form the ratio [3.6]:

$$\frac{|\text{ Extreme Value} - \text{Overall Mean }|}{\text{Overall Standard Deviation, } s} \qquad (3.74)$$

and to reject the outlier if this ratio exceeds a certain critical level. Table I lists critical levels for various (total) sample sizes when it is assumed that the other $N-1$ observations are drawn from a Normal distribution. These test levels should not be used when the base distribution is far from Normal, e.g. if it is clearly skew.

For the intrinsic viscosity data the ratio is:

$$(0·686 - 0·663)/0·0121 = 1·93$$

and it exceeds the tabulated critical level corresponding to $\alpha = 0·05$ for a sample of size six. This implies that the probability of observing a value as extreme as 0·686 is less than 0·05. On this basis, we may therefore decide to reject the outlying observation and compute the mean and standard deviation from the other five measurements [§ **3.341**].

In chemical analysis, the practice of making three replicate determinations and then discarding the most extreme observation should be avoided. If occasional wild results are expected in these circumstances, it is better to use the median as a measure of location rather than the arithmetic mean of the two "inliers".

Standard error of the arithmetic mean

3.6 The extent to which single observations vary about the Population mean is measured by the standard deviation. For example, in a Normal distribution the probability that any particular observation differs from μ by more than twice the standard deviation is about 0·05. The mean of a number of observations is in general a more reliable estimate of the Population mean than a single observation, and this implies that the standard deviation of the mean is smaller than that of individual observations. If a number of samples, each containing n observations, are taken, the means will be distributed about μ with a certain standard deviation, which is less than that of the original data. The standard deviation of the mean (or of any other statistic) is usually known as its Standard Error. The larger the number in the sample, the more precise is the estimate of the mean, i.e. the smaller is the Standard Error of the Mean. Two important, and very general, properties of the standard error of the mean are as follows.

It can be shown, independently of any assumption about the form of the parent Population, that the standard error of the mean of n observations is given by:

$$\sigma_m = \sigma/\sqrt{n} \qquad (3.8)$$

where σ is the standard deviation of the parent Population; the proof is given in § **3.71** below. This shows how the precision of the mean is related to the sample size. The standard error is then inversely proportional to the square root of n. If for example we increase n by four times, we halve the standard error of the mean.

The second important result is that, whatever the nature of the parent Population, the distribution of the sample mean tends rapidly to the Normal form as the sample size increases. In many cases of practical importance the parent Population is unimodal and not highly skew; in these cases, for even small samples, say three or more, it may be assumed for all practical purposes that the mean is distributed Normally with a standard error of $\sigma/\sqrt{n}$. [See also § **2.42**.]

Standard error of the variance and standard deviation

3.61 It can be shown that sample estimates V of the Population variance σ^2 have, if the parent distribution is Normal, a standard error $\sigma^2\sqrt{(2/\phi)}$, where ϕ is the number of degrees of freedom. This result is of less importance than that for the standard error of the mean, since (a) it assumes that the parent Population is Normal, and (b) even when the parent Population is Normal the distribution of the variance is far from Normal except for large samples—say greater than 100. The standard deviations of samples from a Normal Population are distributed with standard error $\sigma/\sqrt{(2\phi)}$, but again the distribution of s is not Normal even when the parent Population is Normal, except for large samples—say greater than 30—and the result is not of great utility. These standard errors give a rough guide to the precision of the estimates V and s, but cannot be used as exact measures in the way that the standard error of the mean can be used. Methods for comparing the estimates V or s with standard values or with other estimates are given in Chapters 4 and 5.

Standard error of a simple function

3.62 Suppose $X = f(x)$, where f is any simple function used to transform the variate x. How does the standard error of the new variate, X, relate to the standard deviation of x?

Let δx denote the deviation of an observation from the Population mean. Then:

$$\delta X = \left(\frac{\partial X}{\partial x}\right)\delta x + \text{terms involving higher differentials}$$

the first differential being evaluated at the mean value of x. Ignoring the higher-order terms and squaring:

$$(\delta X)^2 = \left(\frac{\partial X}{\partial x}\right)^2 (\delta x)^2 \text{ approximately}$$

Our problem is solved by noting that the mean value of $(\delta x)^2$ over all observations is the variance of x, and similarly for X. Hence:

$$V(X) = \left(\frac{\partial X}{\partial x}\right)^2 V(x) \text{ and } S.E.(X) = \left|\frac{\partial X}{\partial x}\right|\sigma \qquad (3.9)$$

These approximations are reasonably accurate for any nonlinear function providing x does not have too large a coefficient of variation. A working rule is that the standard deviation should not exceed 20% of the mean.

We have already met one example of these formulae in § **3.34** when calculating the variance of a scaled variate:

$$\text{if } X = hx, \frac{\partial X}{\partial x} = h \text{ and } V(X) = h^2 V(x) \text{ exactly}$$

Another interesting example is the logarithmic transformation, $X = \log_e x$. Then:

$$V(X) = \frac{V(x)}{\bar{x}^2} = \left(\frac{C.V.(x)}{100}\right)^2 \qquad (3.91)$$

Hence the standard error of $\log_e x$ is approximately equal to the coefficient of variation of x divided by 100. This result is utilized in Equation (3.51) for combining coefficients of variation from several samples.

Covariance

3.7 Suppose two quantities, x_1 and x_2, are measured, and used to calculate a third quantity, X, which is found from x_1 and x_2 by the formula:

$$X = ax_1 \pm bx_2$$

Knowing the standard errors of x_1 and x_2 we wish to calculate that of X. Assume that we have available a large number, N, of observations of x_1 and x_2 and hence of X. Let μ, μ_1 and μ_2 be the true mean values of X, x_1 and x_2; then

$$\mu = a\mu_1 \pm b\mu_2$$

As before, denote by δx the deviation of an observation from its mean. Then:

$$\delta X = a\delta x_1 \pm b\delta x_2$$

and

$$(\delta X)^2 = a^2(\delta x_1)^2 + b^2(\delta x_2)^2 \pm 2ab\delta x_1 \delta x_2$$

The mean value of $(\delta x_1)^2$ over all observations is the variance of x_1, $V(x_1)$, and similarly for $V(x_2)$ and $V(X)$. The mean value of $\delta x_1 \delta x_2$ is defined as:

$$C(x_1, x_2) = \Sigma \delta x_1 \delta x_2 / N = \Sigma(x_1 - \mu_1)(x_2 - \mu_2)/N \qquad (3.92)$$

and termed the Covariance of x_1 and x_2. Thus:

$$V(X) = a^2 V(x_1) + b^2 V(x_2) \pm 2ab C(x_1, x_2) \qquad (3.93)$$

Covariance is analogous to variance, but instead of the square of one variate it contains the product of two variates. Covariance is a measure of the degree of correlation between x_1 and x_2, and is discussed in more detail in Chapter 7. The estimate of the covariance from a sample, the true means being unknown, is given by $(\Sigma x_1 x_2 - N\bar{x}_1 \bar{x}_2)/(N-1)$. As with the variance, the divisor is $(N-1)$, since the means must be estimated from the data.

If x_1 and x_2 in Formula (3.93) are independent observations, e.g. independent analyses, then the errors in the measurements of x_1 and x_2 will be independent. A high value of x_1 is equally likely to be associated with a high or a low value of x_2; similarly for low values of x_1. It follows that the long-term mean value of $\Sigma \delta x_1 \delta x_2$ is zero in this case and $C(x_1, x_2) = 0$. The variance of X is then:

$$V(X) = a^2 V(x_1) + b^2 V(x_2)$$

The following example illustrates the application of these formulae. The rate of consumption of material in passing through a plant is estimated from measurements of the rate at which it enters, x_1, less the rate at which it leaves, x_2. The rate of consumption is $X = x_1 - x_2$. Both x_1 and x_2 will vary with time, and they are also subject to errors of measurement. These errors are independent, so that the error variance of X is given by $V(x_1) + V(x_2)$ and its standard error by $\sqrt{\{V(x_1) + V(x_2)\}}$. Note that this statement refers only to the errors in x_1, x_2 and X. There will also be some real variation in x_1 and x_2, which will be reflected in corresponding variations in X. In order to relate the *total* variance of X to the *total* variances of x_1 and x_2 we have then to take into account the covariance of x_1 and x_2 and use Formula (3.93).

Standard error of a linear function

3.71 When the quantities x_1, x_2 ..., vary independently and

$$X = a_1 x_1 \pm a_2 x_2 \pm a_3 x_3 \pm ... \pm a_n x_n$$

then: $\qquad V(X) = a_1^2 V(x_1) + a_2^2 V(x_2) + ... + a_n^2 V(x_n)$ $\qquad$ (3.94)

Note that the variance of X is the sum of the variances of x_1, etc. (with the appropriate multipliers), whether the coefficients in the expression for X are positive or negative.

An interesting particular case arises if X is equal to the mean of a random sample $x_1, ..., x_n$ drawn from a Population with standard deviation σ. Then:

$$X = (x_1 + x_2 + ... + x_n)/n$$

Since x_1, x_2, ... are observations of the same quantity x:

$$V(x_1) = V(x_2) = ... = V(x)$$

Also: $\qquad\qquad\qquad a_1 = a_2 = ... = 1/n$

Therefore: $\qquad V(X) = V(\bar{x}) = \left(\frac{1}{n}\right)^2 V(x) + \left(\frac{1}{n}\right)^2 V(x) + ...$

$$= V(x)/n$$

and $\sigma_m = \sigma/\sqrt{n}$, as stated in (3.8).

Standard error of a general function

3.72 If $X = f(x_1, x_2, ..., x_n)$, where f is any function, then:

$$V(X) = \left(\frac{\partial X}{\partial x_1}\right)^2 V(x_1) + \left(\frac{\partial X}{\partial x_2}\right)^2 V(x_2) + ...$$

$$+ 2\frac{\partial X}{\partial x_1} \cdot \frac{\partial X}{\partial x_2} C(x_1, x_2) + 2\frac{\partial X}{\partial x_1} \cdot \frac{\partial X}{\partial x_3} C(x_1, x_3) + ...$$

$$+ \text{ terms involving higher differentials} \qquad (3.95)$$

The terms involving higher differentials can be ignored if the extent of the variation is so small that they are negligible compared with the terms involving only first differentials. The terms involving covariances vanish if the co-variances are zero, i.e. if x_1, x_2, etc., are mutually independent. If these two conditions are fulfilled, Formula (3.95) reduces to the first line. This form is exact for linear functions of mutually independent variates; for other functions of mutually independent variates it is sufficiently close if the standard deviations are small. A working rule is that the coefficient of variation of each variate should be less than 20%.

Two functions which occur quite frequently are the product and the quotient of two variates; their variances are given in Table 3.6.

If the coefficient of variation of x_1 or x_2 exceeds 20%, the ratio x_1/x_2 is best treated by Fieller's Theorem (Appendix 7A).

TABLE 3.6.   MEAN AND VARIANCE OF A PRODUCT AND A QUOTIENT

Derived Variate, X	Approximate Mean of X	Approximate Variance of X
$x_1 x_2$	$\bar{x}_1 \bar{x}_2$	$\bar{x}_2^2 V(x_1) + \bar{x}_1^2 V(x_2) + 2\bar{x}_1 \bar{x}_2 C(x_1,\ x_2)$
x_1/x_2	$\bar{x}_1/\bar{x}_2$	$\{V(x_1) + \bar{X}^2 V(x_2) - 2\bar{X} C(x_1,\ x_2)\}/\bar{x}_2^2$

Example: Ammonia oxidation efficiency

3.721 In estimating the efficiency of an ammonia oxidation (nitric acid) plant converter the following formula is used:

$$Y = \frac{N(100 - 1\cdot25A)}{A} = \frac{100N}{A} - 1\cdot25N$$

where $A = \%$ NH$_3$ in inlet gas

$N = \%$ NO in exit gas

$Y = \%$ of the NH$_3$ converted to NO = "oxidation efficiency"

Let $V(A) =$ variance of NH$_3$ analysis (experimental error)

$V(N) =$ variance of NO analysis (experimental error)

$V(Y) =$ variance of efficiency determination (experimental error)

The differentials are given by:

$$\frac{\partial Y}{\partial A} = -\frac{100N}{A^2}; \quad \frac{\partial Y}{\partial N} = \frac{100}{A} - 1\cdot25$$

Then from Formula (*3.95*):

$$V(Y) = \frac{10^4 N^2}{A^4} V(A) + \left(\frac{100}{A} - 1\cdot25\right)^2 V(N)$$

For this equation to be valid it is necessary that the errors in A and N should be independent of each other. It would not be valid, for example, to take a number of separate efficiency determinations and use the set of values of A and N to calculate $V(A)$ and $V(N)$. As A increases, N increases (on average), so that the changes in A and N are highly correlated, and far from independent.

Hence the method employed in the present example was to take a number of duplicate samples of inlet and exit gas, the duplicates being taken more or less simultaneously, and to estimate $V(A)$ and $V(N)$ from the differences between the duplicates. This eliminates the effect of variations in A (and hence in N) between the separate tests, and the experimental errors in the determination of A and N are independent. The following values for the means and variances were found experimentally:

$$\bar{A} = 12 \cdot 0 \qquad\qquad \bar{N} = 13 \cdot 5$$
$$V(A) = 59 \cdot 3 \times 10^{-4} \qquad V(N) = 92 \cdot 2 \times 10^{-4}$$
$$\partial Y / \partial A = -(100)(13 \cdot 5)/12^2 = -1{,}350/144 = -9 \cdot 38$$
$$\partial Y / \partial N = 100/12 - 1 \cdot 25 = 8 \cdot 33 - 1 \cdot 25 = 7 \cdot 08$$

(Note that the differentials are evaluated at the mean values of A and N.)

Hence $V(Y) = (-9 \cdot 38)^2 \times 59 \cdot 3 \times 10^{-4} + (7 \cdot 08)^2 \times 92 \cdot 2 \times 10^{-4}$
$$= 87 \cdot 98 \times 59 \cdot 3 \times 10^{-4} + 50 \cdot 13 \times 92 \cdot 2 \times 10^{-4}$$
$$= 0 \cdot 522 + 0 \cdot 462$$
$$= 0 \cdot 984$$

and due to experimental errors in the determination of A and N:

$$S.D.(Y) = \sqrt{0 \cdot 984} = 0 \cdot 992$$

The mean value of Y was about 96, so that the coefficient of variation is:

$$C.V.(Y) = 0 \cdot 992 \times 100/96$$
$$= 1 \cdot 03\%$$

Note that if the "percentage error" in A and N, i.e. their coefficients of variation, had simply been added, we should have found the coefficient of variation of Y as follows:

$$S.D.(A) = \sqrt{(59 \cdot 3 \times 10^{-4})} = 7 \cdot 7 \times 10^{-2} \qquad C.V.(A) = 7 \cdot 7/12 \cdot 0 = 0 \cdot 64\%$$
$$S.D.(N) = \sqrt{(92 \cdot 2 \times 10^{-4})} = 9 \cdot 6 \times 10^{-2} \qquad C.V.(N) = 9 \cdot 6/13 \cdot 5 = \underline{0 \cdot 71\%}$$
$$C.V.(Y) \qquad\qquad = 1 \cdot 35\%$$

This is a serious overestimate of the variation in Y and illustrates the importance of using the correct method.

References

[3.1] Buist, J. M., and Davies, O. L. "Methods of Averaging Physical Test Results. Use of Median", *Rubber Journal*, CXII, 447–454 (1947).
[3.2] Bainbridge, J. R. "Sums of Squares and Products of Deviations from the Mean", *Technometrics*, 5, 292–293 (1963).
[3.3] Davies, O. L., and Pearson, E. S. "Methods of Estimating from Samples the Population Standard Deviation", *J. R. Statist. Soc. Supplement*, 1, 76–93 (1934).
[3.4] Lord, E. "Power of the Modified *t*-Test (*u*-Test) Based on Range", *Biometrika*, 37, 64–77 (1950).
[3.5] Pearson, E. S., and Hartley, H. O. *Biometrika Tables for Statisticians*, Vol. 1 (third edition). Charles Griffin (High Wycombe, 1976).
[3.6] Grubbs, F. E. "Procedures for Detecting Outlying Observations in Samples", *Technometrics*, 11, 1–21 (1969).

Appendix 3A

Proof of Equation (*3.42*)

Let the Population mean be μ, the sample mean be $\bar{x}$, and the Population variance σ^2. Then if x_i represents an observation in a sample of n:

$$(x_i - \mu) \equiv (x_i - \bar{x}) + (\bar{x} - \mu)$$

$$(x_i - \mu)^2 = (x_i - \bar{x})^2 + (\bar{x} - \mu)^2 + 2(x_i - \bar{x})(\bar{x} - \mu)$$

Summing this equation over all values of i from 1 to n:

$$\Sigma(x_i - \mu)^2 = \Sigma(x_i - \bar{x})^2 + n(\bar{x} - \mu)^2 + 2(\bar{x} - \mu)\Sigma(x_i - \bar{x})$$

But $\qquad\qquad\qquad \Sigma(x_i - \bar{x}) = 0$, by definition of $\bar{x}$

Therefore: $\qquad\qquad \Sigma(x_i - \mu)^2 = \Sigma(x_i - \bar{x})^2 + n(\bar{x} - \mu)^2$

i.e. $\qquad\qquad\qquad \Sigma(x_i - \bar{x})^2 = \Sigma(x_i - \mu)^2 - n(\bar{x} - \mu)^2$

If this calculation is repeated for a large number of samples, the mean value of $\Sigma(x_i - \mu)^2$ will tend to $n\sigma^2$ (by definition of σ^2) and the mean value of $n(\bar{x} - \mu)^2$ will similarly tend to n times the variance of $\bar{x}$, i.e. to $n(\sigma^2/n)$ [**§ 3.71**], whence:

$$\Sigma(x_i - \bar{x})^2 \rightarrow n\sigma^2 - n(\sigma^2/n)$$

i.e. $\qquad \Sigma(x_i - \bar{x})^2 \rightarrow (n-1)\sigma^2$

and $\qquad V = \dfrac{\Sigma(x_i - \bar{x})^2}{n-1} \rightarrow \sigma^2$

Thus V is an unbiased estimate of σ^2, i.e. its mean value over a large number of random samples tends to σ^2, being equal to σ^2 for an infinite number of samples.

Chapter 4

Statistical Inference

When a quantity such as the average yield is calculated from a limited sample of variable data the result is subject to error. The degree of uncertainty in the result can conveniently be described by a "confidence interval" within which the true value is almost certainly contained. This chapter describes the calculation of confidence intervals for the mean, the difference between two means, the standard deviation and the ratio of two standard deviations. Also considered is the use of uncertain results as a basis for decision, whether by a "significance test" or by a more explicit evaluation of expected gains and losses.

CONFIDENCE LIMITS

Introduction

4.1 When an estimate of some quantity has been made, for example an estimate of the composition of a chemical based on one or more analyses, it is desirable to know not only the estimated value, which is usually the most likely value, but also how precise this estimate is. A convenient way of expressing the precision is to state limits within which it may reasonably be asserted that the true value lies. Such statements may take the form, for example, that the true value is unlikely to exceed some upper limit, or that it is unlikely to be less than some lower limit, or that it is unlikely to lie outside a pair of limits. Often, such information is as important as the estimate itself. The limits quoted are known as Confidence Limits, i.e. limits within which we are prepared to assert with given confidence (given expectation of being correct) that the true value lies. The degree of confidence to be associated with the limits may be set very high, say at 99% certainty, or more modestly—say at 80% or 90%—depending on the problem in hand. It is possible to calculate such limits for any statistic such as the mean or standard deviation, provided the distribution of the observations is known, and they represent both a more revealing and a more flexible guide to the uncertainties of estimation than the standard errors described in the previous chapter.

In this chapter methods are described for calculating the limits where the original data are Normally distributed. For mean values, for the reasons given in § **3.6**, a close approximation is obtained even if the data are not Normally distributed. For standard deviations, however, a departure from Normality may affect the confidence limits appreciably; an alternative method of calculation is thus outlined for use in those instances in which Normality of the parent distribution cannot be assumed.

Confidence limits for a mean value

4.2 If the form of the distribution is known, and if the true mean and standard deviation are also known, it is possible to make straightforward probability statements about the value of the mean of a number of observations. For a Normal Population with mean μ and standard deviation σ, the probability that the mean of n observations will lie within the range $\mu - 3\sigma/\sqrt{n}$ to $\mu + 3\sigma/\sqrt{n}$ is 0·997 [§ **2.44**]. This is so whatever the values of μ, σ and n. Hence, of every thousand samples randomly drawn from Normal Populations only three, on average, will fail to satisfy the inequality:

$$\mu - 3\sigma/\sqrt{n} < \bar{x} < \mu + 3\sigma/\sqrt{n}$$

Now, if $\bar{x}$ is greater than $\mu - 3\sigma/\sqrt{n}$, then μ is less than $\bar{x} + 3\sigma/\sqrt{n}$; similarly, if $\bar{x}$ is less than $\mu + 3\sigma/\sqrt{n}$, then μ is greater than $\bar{x} - 3\sigma/\sqrt{n}$. Hence when μ is unknown but $\bar{x}$ is known, we may logically assert, with 99·7% confidence, that:

$$\bar{x} - 3\sigma/\sqrt{n} < \mu < \bar{x} + 3\sigma/\sqrt{n}$$

The limits $\bar{x} \pm 3\sigma/\sqrt{n}$ are known as the 99·7% Confidence Limits for μ, and the Confidence Coefficient of 99·7% reflects the fact that, of every thousand such assertions that we make, only three, on average, will be incorrect. For a smaller confidence coefficient the limits are closer and the interval between them—the Confidence Interval—is shorter, for example for 95% confidence the limits are $\bar{x} \pm 1·96\sigma/\sqrt{n}$, and so on. If it is necessary to be virtually certain that our statement is correct, i.e. that our limits do include the true mean, a high confidence coefficient and hence relatively wide limits are needed.

Single- and double-sided limits

4.21 In some situations only one limit may be of interest, say an upper limit if all that is required is a sufficient assurance that μ does not exceed a stated value. Now, the probability that $\bar{x}$ falls short of $\mu - 3\sigma/\sqrt{n}$ is only 0·00135; we may thus assert with some 99·9% confidence that μ is not greater than $\bar{x} + 3\sigma/\sqrt{n}$. In the same way, we may assert with 99·9% confidence that μ is not less than $\bar{x} - 3\sigma/\sqrt{n}$. Generally, the proportion of sample means $\bar{x}$ which will exceed $\mu + u_\alpha\sigma/\sqrt{n}$ (and correspondingly the proportion which will fall short of $\mu - u_\alpha\sigma/\sqrt{n}$) is equal to α, where u_α is the value of u given in

Table A for the probability $P = \alpha$. Hence we may assert with $100(1-\alpha)\%$ confidence *either* that μ is not less than $\bar{x} - u_\alpha \sigma/\sqrt{n}$ *or* that μ is not greater than $\bar{x} + u_\alpha \sigma/\sqrt{n}$. We may also combine the two statements and assert with $100(1-2\alpha)\%$ confidence that μ lies between $\bar{x} - u_\alpha \sigma/\sqrt{n}$ and $\bar{x} + u_\alpha \sigma/\sqrt{n}$.

The situation is illustrated diagrammatically in Figure 4.1, from which we see that we have in effect divided the whole range of possible values of μ into three parts, with two of which (to the right and to the left) we associate confidence coefficients of only $100\alpha\%$ and with the third of which (the central portion) we associate a confidence coefficient of $100(1-2\alpha)\%$. There are thus three statements open to us:

(i) $\mu \geqslant \bar{x} - u_\alpha \sigma/\sqrt{n}$ —confidence coefficient $100(1-\alpha)\%$
(ii) $\mu \leqslant \bar{x} + u_\alpha \sigma/\sqrt{n}$ —confidence coefficient $100(1-\alpha)\%$
(iii) $\bar{x} - u_\alpha \sigma/\sqrt{n} \leqslant \mu \leqslant \bar{x} + u_\alpha \sigma/\sqrt{n}$ —confidence coefficient $100(1-2\alpha)\%$

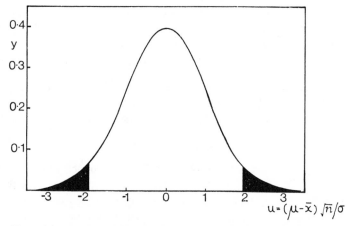

FIG. 4.1. DIAGRAMMATIC REPRESENTATION OF CONFIDENCE LIMITS (95% LIMITS SHOWN)

It is important to note that, while u_α corresponds to probability α, the confidence coefficient of $100(1-\alpha)\%$ applies only to a Single-Sided statement. If we quote a Double-Sided confidence interval (statement (iii)) the appropriate confidence coefficient is $100(1-2\alpha)\%$. It is also important to note that, while the curve of Figure 4.1 has the form of a conventional probability distribution, it is not to be interpreted as a probability distribution for the unknown parameter μ in the usual (relative frequency) sense; although μ is unknown, it is fixed and only the sample estimates vary. The curve is primarily a convenient way of representing the multiplicity of confidence statements which might be made about μ and it would be described, by some schools, as a Fiducial Distribution for μ. The distinction between "confidence" and frequency ratio probabilities is one to which we shall return in § **4.9**.

Confidence limits for a mean value: Standard deviation estimated

4.22 It was assumed above that σ was known exactly. Usually only an estimate, s, of σ is available, based on a limited number (ϕ) of degrees of freedom, and it is necessary to use this estimate in calculating the confidence limits. Since s itself is subject to some uncertainty, the confidence limits for μ are further apart than they would be if σ were known exactly. This uncertainty is allowed for by using, in place of u, the quantity t, which is tabulated in Table C. For a large number of degrees of freedom, say more than 20, the uncertainty in s is comparatively small, and t is practically identical with u, but as the number of degrees of freedom becomes smaller, t becomes progressively larger than u. The following table, for $\alpha = 0.025$, illustrates the point:

ϕ				t
∞	...	...	...	$1.96 (= u)$
20	...	...	...	2.09
10	...	...	...	2.23
5	...	...	...	2.57
2	...	...	...	4.30
1	...	...	...	12.71

The $100(1-2\alpha)\%$ confidence limits are:

$$\text{Lower limit} = \bar{x} - t_\alpha s/\sqrt{n}$$
$$\text{Upper limit} = \bar{x} + t_\alpha s/\sqrt{n}$$

where t_α is found from Table C, using the appropriate number of degrees of freedom.

Difference between two mean values

4.23 A problem that frequently arises is to assess the magnitude of the difference between two mean values, for example the yields of two alternative processes of manufacture or the compositions of two chemicals. The method of comparing the two is to calculate the difference between two observed means, $\bar{x}_1 - \bar{x}_2$, and the standard error of the difference. These values are used to calculate the confidence limits for the true difference, $\mu_1 - \mu_2$. If the lower limit is greater than zero, it can be assumed that μ_1 is greater than μ_2; or if the upper limit is less than zero, that μ_2 is greater than μ_1. If the limits are too wide to lead to sufficiently reliable conclusions more observations must be taken. Note that these conclusions cannot be drawn directly from the confidence limits for the individual means.

Let the group means $\bar{x}_1$ and $\bar{x}_2$ be based on n_1 and n_2 independent observations from Populations with standard deviations σ_1 and σ_2 respectively. The standard error of $\bar{x}_1 - \bar{x}_2$ is given by [§ **3.7**]:

$$S.E. (\bar{x}_1 - \bar{x}_2) = (\sigma_1^2/n_1 + \sigma_2^2/n_2)^{\frac{1}{2}} \tag{4.1}$$

When σ_1 and σ_2 are equal this reduces to:

$$S.E. \ (\bar{x}_1 - \bar{x}_2) = \sigma(1/n_1 + 1/n_2)^{\frac{1}{2}} \qquad (4.11)$$

The confidence limits are therefore:

$$(\bar{x}_1 - \bar{x}_2) \pm u_\alpha \sigma(1/n_1 + 1/n_2)^{\frac{1}{2}} \qquad (4.12)$$

If $n_1 = n_2$, this reduces to:

$$(\bar{x}_1 - \bar{x}_2) \pm u_\alpha \sigma \sqrt{(2/n)} \qquad (4.13)$$

4.231 If σ_1 and σ_2 are not known, estimates s_1 and s_2 are calculated in the usual way; if these do not differ greatly, and if it is reasonable to suppose that σ_1 and σ_2 are the same or nearly so, a common value is calculated as in § **3.342**. The confidence limits are then found as in Equation (*4.12*) or (*4.13*), substituting s for σ and t for u; s is based on $\phi = n_1 + n_2 - 2$ degrees of freedom, and this value is used in finding the appropriate value of t from Table C.

A difficulty arises when σ_1 and σ_2 cannot be assumed equal. An approximate solution is to calculate the standard error from:

$$S.E. \ (\bar{x}_1 - \bar{x}_2) = (s_1^2/n_1 + s_2^2/n_2)^{\frac{1}{2}} \qquad (4.2)$$

where s_1^2 and s_2^2 are the observed variances of the two sets of data. The approximate confidence limits are $(\bar{x}_1 - \bar{x}_2) \pm t_\alpha \{S.E. \ (\bar{x}_1 - \bar{x}_2)\}$, but in finding t from Table C the value of ϕ is not $(n_1 + n_2 - 2)$ but is given [4.1] by the formula:

$$\frac{1}{\phi} = \frac{1}{\phi_1} \left\{ \frac{s_1^2/n_1}{s_1^2/n_1 + s_2^2/n_2} \right\}^2 + \frac{1}{\phi_2} \left\{ \frac{s_2^2/n_2}{s_1^2/n_1 + s_2^2/n_2} \right\}^2 \qquad (4.3)$$

where the suffixes 1 and 2 refer to the two samples. Note that $(s_1^2/n_1 + s_2^2/n_2)$ will already have been calculated. If s_1^2 is much greater than s_2^2, and n_1 and n_2 are not widely different, ϕ approximates to ϕ_1. If it is suspected that s_1^2 will be considerably greater than s_2^2 it is usually best to make n_1/n_2 about equal to s_1^2/s_2^2, so that the variances of the two sample means are approximately equal.

Confidence limits for the *ratio* of two mean values are mentioned in Appendix 7A.

Example 4.1. Laboratory test for ease of filtration

4.24 In a plant manufacturing a nitrogenous fertiliser a main limitation to increased output was the rate of filtration, on rotary filters, to separate the fertiliser solution from insoluble by-products. Laboratory experiments were carried out in an attempt to make the magma more easily filtered, and a standard procedure was established for estimating ease of filtration, consisting simply of filtering a given volume through a standard filter paper under a standard suction and observing the time for filtration. It was first necessary to confirm that this test gave a reasonable indication of the ease of filtration on the plant filters. Samples of plant magma were taken at a time when filtering was judged to be "fair" and also at a time when it was judged to be "fairly good", this being part of a more extensive investigation covering other conditions also. The results of the laboratory tests were variable, but the average for the samples taken during the "fair" period was higher (i.e. poorer filtration) than for those in the "fairly good" period. It was necessary to estimate the magnitude of this difference, and also its limits of uncertainty, not only in assessing the results of the particular experiment but also in planning future trials.

The individual results are given to the nearest one-fifth of a second in Table 4.1, and the necessary calculations beneath.

TABLE 4.1. LABORATORY FILTRATION TEST

"Fair" Period		"Fairly Good" Period	
x_1	x_1^2	x_2	x_2^2
8·4	70·56	9·0	81·00
9·8	96·04	10·2	104·04
12·2	148·84	9·6	92·16
12·6	158·76	4·4	19·36
13·0	169·00	7·0	49·00
9·2	84·64	9·2	84·64
13·6	184·96		
78·8	912·80	49·4	430·20

$\bar{x}_1 = 78\cdot8/7 = 11\cdot26$ sec.　　　　$\bar{x}_2 = 49\cdot4/6 = 8\cdot23$ sec.

$s_1^2 = \{912\cdot80 - (78\cdot8)^2/7\}/6 = 4\cdot29$　　$s_2^2 = \{430\cdot20 - (49\cdot4)^2/6\}/5 = 4\cdot69$

$\phi_1 = 6$　　　　　　　　　　　　　$\phi_2 = 5$

$s_1 = \sqrt{4\cdot29} = 2\cdot07$ sec.　　　$s_2 = \sqrt{4\cdot69} = 2\cdot17$ sec.

Since s_1 and s_2 are almost identical, and since the standard deviations were not expected to be different, we use the combined estimate $s = 2\cdot12$ sec., based on $6+5 = 11$ degrees of freedom [§ **3.342**]. Then:

$$S.E. (\bar{x}_1 - \bar{x}_2) = 2\cdot12\sqrt{(\tfrac{1}{7}+\tfrac{1}{6})} = 1\cdot18$$

$$\phi = 11; \ t_{0\cdot025} = 2\cdot20 \ [\text{Table C}]$$

$$\bar{x}_1 - \bar{x}_2 = 11\cdot26 - 8\cdot23 = 3\cdot03 \text{ sec.}$$

The 95% confidence limits are:

$$3\cdot03 \pm 2\cdot20 \times 1\cdot18 = 0\cdot4 \text{ and } 5\cdot6 \text{ sec.}$$

It is thus reasonably certain that there is a real difference, in the expected direction. The estimate is, however, rather uncertain, and it would be desirable to increase the precision by carrying out more tests. It is clear, too, that several tests would be necessary before any particular magma could be classified with any reasonable degree of confidence.

The following considerations give a guide to the number of tests required. The standard error of the mean of n results carried out under uniform conditions is $2 \cdot 12/\sqrt{n}$, and the 95% confidence limits are $\bar{x} \pm t_{0.025} \times 2 \cdot 12/\sqrt{n}$. If we assume the estimate of $2 \cdot 12$ for the standard deviation to be free from error the confidence limits become:

$$\bar{x} \pm u_{0.025} \times 2 \cdot 12/\sqrt{n} = \bar{x} \pm 1 \cdot 96 \times 2 \cdot 12/\sqrt{n} = \bar{x} \pm 4 \cdot 16/\sqrt{n}$$

If for example it is desired to determine the time with 95% confidence correct to $\pm 1 \cdot 5$ seconds, we have:

$$4 \cdot 16/\sqrt{n} = 1 \cdot 5$$
$$\sqrt{n} = 2 \cdot 77$$
$$n = 7 \cdot 67$$

So eight tests would be required.

The problem of determining the number of trials required is dealt with in more detail in Chapter 5.

Note that whenever the $\pm$ notation is used, it is imperative to support it by a statement giving the meaning of the limits; for example, if they are confidence limits, state the degree of confidence.

Paired comparisons

4.25 When two processes are being compared it is desirable to keep all conditions constant, e.g. to use the same raw materials for both, so that the difference between the results is due only to the difference between the processes. It may be necessary to carry out several trials on each process in order to detect the difference with sufficient assurance, and it may not be possible to prepare sufficient raw material of uniform composition for the entire series of trials. The results of the trials on one process will vary because of experimental error, even if the raw material is constant; but if the raw material varies, the variation will be greater. Thus the standard error of each mean, and hence of the difference between the means, is increased by the use of several batches of raw material. The confidence limits for the true difference are made wider, and the comparison is less sensitive.

In suitable circumstances this difficulty can be overcome. For example, when comparing two plants, provided one batch of raw material is sufficient for at least one trial on each plant, a pair of trials may be carried out with each batch of raw material and the differences between the efficiencies noted. These differences, as distinct from the actual efficiencies, are the quantities studied, because there is no reason to believe that they are affected by variations between batches of raw material. Their standard deviation is the same as if the raw material were constant. This procedure may greatly reduce the number of trials required for a given degree of precision.

Example 4.2. Percentage ammonia in plant gas

4.26 Two analysts carried out simultaneous measurements of the percentage of ammonia in a plant gas on nine successive days to find the extent of the bias, if any, between their results. The gas composition varied somewhat, the variations being large compared with the analytical errors. In this case the differences between the simultaneous measurements are likely to have a smaller standard error than differences between measurements by the two operators carried out at different times. For this reason the tests were designed to permit "paired comparisons" and the statistical analysis is carried out on the differences between simultaneous measurements.

The results are shown in Table 4.2, the figures given being $100(N-12)$ where $N = \%$ ammonia.

TABLE 4.2. PERCENTAGE AMMONIA IN PLANT GAS (TRANSFORMED DATA)

| Sample No. | Operator | | $(x_2-x_1) = y$ | y^2 |
| | A | B | | |
	x_1	x_2		
1	4	18	14	196
2	37	37	0	0
3	35	38	3	9
4	43	36	−7	49
5	34	47	13	169
6	36	48	12	144
7	48	57	9	81
8	33	28	−5	25
9	33	42	9	81
Total	303	351	48	754
Mean	33·7	39·0	5·3	—

The mean value $\bar{y}$ is 5·3, and the standard deviation of y, calculated in the usual way, is 7·9. The standard error of $\bar{y}$ is thus $7·9/\sqrt{9} = 2·6$. To calculate the 95% double-sided confidence limits for $\bar{y}$, we require $\alpha = 0·025$, and for $\phi = 8$ this gives $t = 2·31$; hence the limits are $5·3 \pm 2·31 \times 2·6 = -0·7$ and 11·4. Thus the results do not indicate conclusively that a bias exists. They do however show that at worst B's results are unlikely to be more than 11·4 units higher than A's, i.e. about 0·11% NH_3. Such a bias would not be considered serious in this application.

The conditions under which the "difference" method is used must be carefully noted. The two members of each pair must be naturally associated; it is not valid to pair the results arbitrarily, for example by arranging the two series in order of magnitude and pairing the results in that order. It should also be noted that when the experiment is designed for paired comparisons it should be analysed as such; the results must not be treated as if they were from two independent sets, even if it transpires that the variation between samples is small.

Confidence limits for the standard deviation

4.3 The standard deviation estimated from a sample is, like any other statistic, subject to uncertainty; repeat estimates using different samples taken under the same conditions will neither agree among themselves nor—except coincidentally—agree with the Population value. It is desirable to know how reliable any sample estimate is. One index is the standard error of the sample standard deviation, but since the sampling distribution of the standard deviation is neither Normal nor symmetric it is better to calculate confidence limits. As in the case of the mean, the confidence limits are closer the larger the sample size, i.e. the greater the number of degrees of freedom on which the sample estimate is based. For reasons which will become apparent later, it is convenient to consider first the calculation of confidence limits for the ratio of two standard deviations; those for a single standard deviation are considered in § **4.33**.

Confidence limits for the ratio of two standard deviations

4.31 To compare two standard deviations, σ_1 and σ_2, we calculate the ratio s_1/s_2 of two estimates based on the available observations. The confidence limits for σ_1/σ_2 are obtained by the use of the pair of multipliers L_1 and L_2 given in Table H. Multiplying the s_1/s_2 ratio by L_1 gives the lower confidence limit, and by L_2 the upper confidence limit. Table H gives values of L_1 and L_2 for $\alpha = 0·01$, 0·025, 0·05, 0·10, i.e. for confidence coefficients of 98%, 95%, 90% and 80%. If only one limit is of interest the appropriate limit is calculated and its confidence coefficient is $100(1-\alpha)\%$. Note that ϕ_N in Table H is the number of degrees of freedom for the numerator of the ratio s_1/s_2; ϕ_D is for the denominator. If the same procedure is carried out for s_2/s_1, the limits will be the exact reciprocals of those for s_1/s_2; the results are thus equivalent.

Suppose for example that $s_1 = 1·5$, based on 10 degrees of freedom, and $s_2 = 1$, based on 20 degrees of freedom, and it is required to find the upper and lower 95% confidence limits. The ratio $s_1/s_2 = 1·5$. For $\phi_N = 10$, $\phi_D = 20$ and $\alpha = 0·025$, $L_1 = 0·60$ and $L_2 = 1·85$.

$\therefore$ Lower limit for $\sigma_1/\sigma_2 = 1\cdot5 \times 0\cdot60 = 0\cdot90$

Upper limit for $\sigma_1/\sigma_2 = 1\cdot5 \times 1\cdot85 = 2\cdot78$

We can thus state, with 95% confidence, that σ_1 is between 0·90 times and 2·78 times σ_2.

Alternatively:

$$s_2/s_1 = 0\cdot67 \qquad \phi_N = 20 \qquad \phi_D = 10$$
$$L_1 = 0\cdot54 \text{ and } L_2 = 1\cdot67$$

Lower limit for $\sigma_2/\sigma_1 = 0\cdot67 \times 0\cdot54 = 0\cdot36$

Upper limit for $\sigma_2/\sigma_1 = 0\cdot67 \times 1\cdot67 = 1\cdot11$

These are the reciprocals of the limits of 2·78 and 0·90 found earlier.

Example 4.3. Comparison of two analysts

4.32 A relatively inexperienced analyst had been trained to carry out a routine test, and it was decided to compare him with an experienced man to find whether his results were as reproducible or whether he required a further period of training. It was thought that if the standard deviation of his results was not more than twice that of the experienced man he would be satisfactory. This can be decided by calculating the confidence limits for the ratio of the standard deviations, using a suitable confidence coefficient, say 95%, and concluding that the new analyst is suitable if the upper limit is less than 2.

The results of the experiment were as follows. A is the new analyst.

Analyst	A	B
Number of tests (n)	20	13
Variance (s^2)	294·7	139·0
Standard deviation (s)	17·2	11·8
Degrees of freedom $(\phi = n-1)$ ...	19	12

$$s_1/s_2 = 17\cdot2/11\cdot8 = 1\cdot46$$
$$\phi_N = 19 \qquad \phi_D = 12$$

From Table H ($\alpha = 0\cdot025$), interpolating by inspection, $L_1 = 0\cdot57$ and $L_2 = 1\cdot66$. The 95% confidence limits for σ_1/σ_2 are:

$$1\cdot46 \times 0\cdot57 = 0\cdot83$$
$$1\cdot46 \times 1\cdot66 = 2\cdot42$$

Since the upper limit is greater than 2, analyst A would not be considered suitable from these tests alone. However, the lower limit indicates that there is a reasonable probability that A is as good as B, and the only fair conclusion is that insufficient tests have been carried out to assess analyst A. Further tests are therefore necessary before a sufficiently reliable conclusion can be drawn. The number of tests required is dealt with more fully in Chapter 5.

Confidence limits for a single standard deviation

4.33 In Table H the values for ϕ_N or ϕ_D equal to infinity correspond to the case where one standard deviation is known exactly. Suppose $\phi_D = \infty$, then σ_2 is known exactly, and the ratio is uncertain only to the extent that σ_1 is uncertain; the confidence limits for σ_1 are clearly obtained by multiplying those for the ratio by the known value of σ_2. If $\sigma_2 = 1$, the confidence limits of the ratio are those of σ_1. To find the limits for a single standard deviation σ, therefore, we simply use the line of Table H for $\phi_D = \infty$, with ϕ_N equal to the number of degrees of freedom on which the estimate of σ is based. It will be seen that for a small number of degrees of freedom the limits are wide; for example, if $\phi = 4$ the upper 98% limit is seven times the lower; for $\phi = 2$ it is 21 times as great. Table H shows immediately how many observations are required to obtain a given precision. For example, suppose that the upper limit should be not more than twice the estimate. For the 98% limit the number of observations is 11, since for $\phi = 10, L_2 = 1·98$, and for ϕ less than 10, L_2 is greater than 2.

The 95% confidence limits for the individual standard deviations in Example 4.3 are as follows.

	Analyst A	Analyst B
Estimate s	17·2	11·8
ϕ	19	12
L_1	0·76	0·72
L_2	1·46	1·65
Lower confidence limit	13·1	8·5
Upper confidence limit	25·1	19·5

The effect of non-Normality

4.4 Tables A, B, C, D and H have been calculated on the assumption that the data are Normally distributed. So far as mean values are concerned this assumption does not lead to serious error, since, as shown in § **2.42**, the means of even small samples are very nearly Normally distributed, even if the original data are not. But for the standard deviation, or the ratio of two standard deviations, the values given in Tables B, D and H may be appreciably in error if the data are not Normal; even so, the values when used with caution are a useful guide. If sufficient information about the distribution is available, it may be possible to effect a transformation which gives a more nearly Normal distribution. Alternatively the method outlined below may be used.

Confidence limits for standard deviation: Distribution not assumed Normal

4.41 If the original data are not Normally distributed the confidence limits calculated by the methods of §§ **4.31** and **4.33** may be very much too wide or too narrow, depending on the actual distribution. This question is discussed in [4.2], where a method is given which supplies approximate confidence limits whatever the nature of the distribution, i.e. it is a Distribution-Free method. This procedure has the disadvantage that the number of degrees of freedom available is very much less than the number of observations, so that unless a reasonably large number of observations are available the confidence limits are likely to be wide. Preferably, there should be 30 or more observations, but in the case of a ratio of two standard deviations the method can safely be used with somewhat fewer than 30 in each set. If these conditions are met it is worthwhile to use the distribution-free method unless it is known that the data are Normally distributed or very nearly so.

The method for the ratio of two standard deviations is as follows, that for a single standard deviation being treated as a special case.

4.411 (i) Divide each set of observations at random into small equal groups of three to ten observations, using the method described in § 3.351.

(ii) Calculate the variance V for each of these groups, and take the logarithms. Let $y = \log V$.

(iii) Treat y as a new variable. Calculate its mean and standard deviation separately for each set, and find the confidence limits for $(\bar{y}_1 - \bar{y}_2)$ exactly as in § **4.23**. The number of degrees of freedom used in finding the value of t is the total number of y's less two. Thus if groups of four are used, the number of degrees of freedom is approximately quartered, and so on.

(iv) Take the antilogs of $(\bar{y}_1 - \bar{y}_2)$ and of the confidence limits. Then since $y = \log V$, $(\bar{y}_1 - \bar{y}_2)$ is an estimate of $\log (\sigma_1^2/\sigma_2^2)$, and the confidence limits are those for $\log (\sigma_1^2/\sigma_2^2)$. Thus the antilogs are the estimate and confidence limits of σ_1^2/σ_2^2, and their square roots are those of σ_1/σ_2.

The application to determine the confidence limits for a single standard deviation is obvious. An example is given in Appendix 4A, utilizing the data of Table 2.1. The two methods give similar results, because the data are approximately Normally distributed.

The choice of the number in each sub-group is to some extent a matter of convenience. When a large number of observations are available larger sub-groups can be taken, provided this does not lead to too few sub-groups—say less than 10 for each set—or to too great a wastage of observations. For sets with fewer than 30 observations, it is advisable to take sub-groups of not more than four observations.

TESTS OF SIGNIFICANCE

4.5 A common procedure in present-day statistical analysis is to carry out a Test of Significance as an aid to the interpretation of experimental data. The reasoning behind such a test is best illustrated by a simple example.

Suppose that a process carried out by a standard method has a mean yield of μ_0, the standard deviation of batch yields being σ. A modification is suggested which it is thought might increase the yield, and in order to test whether this is so a series of n batches is prepared by the modified method, giving an average yield of $\bar{x}$. It can be assumed that σ will be the same with the modified as with the normal process. Suppose that in reality the modification has no effect on yield, in other words, that the true yield remains at μ_0. The observed mean may be greater than μ_0 simply because of sampling variations, and it is desirable to avoid the conclusion that the true mean has increased if in fact it has not. It is possible to calculate how large $\bar{x}$ is likely to be if the true mean is μ_0; for example, it is unlikely (1 chance in 40) to exceed $\mu_0 + 1 \cdot 96\sigma/\sqrt{n}$. If it does exceed this value, it is reasonable to conclude that the true yield of the modified process, which we denote by μ_1, is greater than μ_0; if it does not, the evidence is insufficient to conclude that an increase has occurred, and the modified method should not be adopted at this stage, though further tests might well be carried out before the modified process is finally accepted or rejected.

In the above illustration we are, in effect, testing the truth of a hypothesis, namely that the modified process has not produced an improvement in yield. Such a hypothesis is called a Null Hypothesis. In more general terms, the Null Hypothesis takes the form that a parameter does not differ from a particular value. The procedure in a test of significance is to calculate the probability of finding a deviation as extreme as or more extreme than the observed deviation on the assumption that the Null Hypothesis is true. If this probability is sufficiently small, the Null Hypothesis is discredited.

The illustration shows that tests of significance and confidence limits are closely related. The Null Hypothesis is that $\mu_1 - \mu_0 = 0$; from this we deduce that $\bar{x}$ has a probability α of exceeding $\mu_0 + u_\alpha(\sigma/\sqrt{n})$, where u_α is obtained as before from Table A. But $\bar{x} > \mu_0 + u_\alpha(\sigma/\sqrt{n})$ is equivalent to $\mu_0 < \bar{x} - u_\alpha(\sigma/\sqrt{n})$, the right-hand side of which is the lower $100(1-\alpha)\%$ confidence limit for the unknown mean μ_1 [§ **4.21**]. Thus the test of significance of the difference $\mu_1 - \mu_0$ at the level α is exactly the same as determining whether the lower $100(1-\alpha)\%$ confidence limit for μ_1 exceeds μ_0.

In the example, we are interested only in whether μ_1 is greater than μ_0, so that only the lower confidence limit for μ_1 is relevant and the test of significance is single-sided. In other instances, however, we may wish to detect any change, whether increase or decrease. Just as there is a probability α that $\bar{x}$ will exceed $\mu_0 + u_\alpha(\sigma/\sqrt{n})$, so there is a probability α that $\bar{x}$ will fall short of $\mu_0 - u_\alpha(\sigma/\sqrt{n})$. It may readily be shown that a double-sided test of significance at the 2α level corresponds exactly to determining whether μ_0 falls outside the $100(1-2\alpha)\%$ confidence interval for μ_1. Thus if μ_0 falls within the limits, the Null Hypothesis is not discredited and we cannot conclude, at the chosen level of significance, that μ_1 differs from μ_0.

Similar considerations apply in the testing for significance of differences involving other statistics, and no new principle is involved. Confidence limits are usually more useful than a significance test because they give the complete range of values or hypotheses which can be regarded as consistent with the observations. There are however situations where it is sufficient to test a Null Hypothesis, i.e. to test whether or not a particular given Hypothesis is tenable, and in these situations, a significance test may be used.

Since the calculations are similar to those involved in calculating confidence limits, it is not necessary to give the methods of tests of significance in any great detail. It should be noted that a test of significance controls the risk of rejecting the Null Hypothesis when it is true, but has no provision for controlling the risk of not rejecting the Null Hypothesis when it is false.

Significance levels

4.51 When applying a test of significance we calculate the probability P that a given result would occur if the Null Hypothesis were true. If this probability is equal to or less than a given value α, the result is said to be significant at the level α. The appropriate level of significance will depend on the particular problem under consideration. Usually the value $P = 0.05$ gives sufficient assurance, but in certain circumstances a higher degree of assurance such as 0.01 may be required, while in others a lower degree of assurance corresponding to 0.10 may be sufficient. When $P = 0.05$ the result is usually referred to as "significant" and when $P = 0.01$ as "highly significant". The determination of the appropriate significance level to use rests on the same considerations as those for determining the appropriate confidence coefficient, outlined in § **4.2**.

It must be borne in mind that the statistical significance of any difference refers only to the probability that such a difference could have arisen by chance. There is no necessary implication of practical importance, and in particular it is wrong to associate different levels of statistical significance with different degrees of practical importance. Any real difference can be made as significant as we choose by taking sufficient observations, but its practical importance is clearly not thereby altered.

Significance of means

4.6 Two types of tests are involved in the significance of means, these being the Normal Test and the t-Test. The former applies when the standard deviation is known exactly, or is based on a large number of degrees of freedom (in practice > 30), and the latter when the standard deviation has to be estimated from the data and has a limited number ϕ of degrees of freedom. The Normal test is thus a particular case of the t-test when ϕ is large, and need not be considered separately.

The test is applied to the difference between two means μ_0 and μ_1, and there are two cases to consider, (i) μ_0 known and μ_1 estimated from a sample of n observations, (ii) μ_0 and μ_1 estimated from separate samples.

4.61 To test whether μ_1 differs significantly from a given value μ_0 when μ_1 is estimated from a sample of n observations, first calculate the sample mean $\bar{x}$ and form the difference $(\bar{x}-\mu_0)$. Let s be the estimate of the standard deviation based on ϕ degrees of freedom; when s is calculated from the sample, $\phi = (n-1)$. Dividing $(\bar{x}-\mu_0)$ by its standard error $s/\sqrt{n}$, we have:

$$t = (\bar{x}-\mu_0)/(s/\sqrt{n})$$

the significance of which can then be assessed by referring to Table C, using the appropriate number of degrees of freedom.

4.62 In testing whether μ_1 differs significantly from μ_0 when both μ_1 and μ_0 have to be estimated there are several possible situations, and these are discussed fully in § **4.23**. The only case we shall consider here is that in which it is reasonable to assume that the two Populations from which the samples are drawn have equal variances. The combined estimate s^2 of the true variance σ^2 will then be based on $\phi = (n_1 + n_0 - 2)$ degrees of freedom, n_1 and n_0 being the respective sample sizes.

The ratio of the difference between the two sample means, $\bar{x}_1 - \bar{x}_0$, to the standard error of this difference, $s\sqrt{(1/n_1 + 1/n_0)}$, is

$$t = \frac{(\bar{x}_1 - \bar{x}_0)}{s\sqrt{(1/n_1 + 1/n_0)}}$$

the numerical value of which is then referred to Table C.

4.621 The following results were obtained in § **4.24** for the data of Table 4.1:

$$\bar{x}_1 - \bar{x}_0 = 11 \cdot 26 - 8 \cdot 23 = 3 \cdot 03, \qquad s = 2 \cdot 12$$

$$s\sqrt{(1/n_1 + 1/n_0)} = 2 \cdot 12\sqrt{(\tfrac{1}{7} + \tfrac{1}{6})} = 1 \cdot 18$$

$$\phi = 7 + 6 - 2 = 11$$

Therefore $t = 3 \cdot 03/1 \cdot 18 = 2 \cdot 57$

Referring to Table C, we find that for $\phi = 11$ values of t lying outside the interval $(-2 \cdot 20, +2 \cdot 20)$ are significant at the $0 \cdot 05$ level, $2 \cdot 20$ being the tabulated value of t corresponding to $P = 0 \cdot 025$. The difference is therefore significant, and we conclude that the true means, μ_1 and μ_0, are probably different. To be adjudged highly significant the calculated t would have to exceed $3 \cdot 11$, the tabulated value corresponding to $P = 0 \cdot 005$ and $\phi = 11$.

If required, a more exact assessment of the level of significance corresponding to the calculated value of $t = 2 \cdot 57$ can be made with the aid of the Nomogram at the end of the book. The differences between $2 \cdot 57$ and the tabulated values of t for $P = 0.025$ and 0.01 are found thus:

P	0.025		0.01
t for $\phi = 11$	2.20	2.57	2.72
Differences		0.37	0.15
Differences $\times 25$		$9\frac{1}{4}$	$3\frac{3}{4}$

Then on the 0.025 and 0.01 arms of the nomogram we can mark off $-9\frac{1}{4}$ and $+3\frac{3}{4}$ respectively and join with a straight edge. The intersection gives $P = 0.013$. Hence the probability that t will lie outside the interval $(-2.57, 2.57)$ is approximately 0.026.

Single-sided and double-sided tests

4.63 In the above example the alternative to the Null Hypothesis $\mu_1 - \mu_0 = 0$ is that $\mu_1 \neq \mu_0$, irrespective of whether μ_1 is greater than or less than μ_0. Since the Null Hypothesis may be rejected in favour of either of these possibilities, the test is said to be Double-Sided, and it is necessary to double the probabilities in Table C for the critical values of t.

If on the other hand the hypothesis to be tested is $\mu_1 \leqslant \mu_0$, the Null Hypothesis is rejected only if it appears that $\mu_1 > \mu_0$, and the test is said to be Single-Sided. The value of t required for significance must then exceed $t(P = 0.05, \phi)$. Referring to the above example, the calculated t is 2.57, and from Table C we find $t(P = 0.05, \phi = 11) = 1.80$ and $t(P = 0.01, \phi = 11) = 2.72$. The difference in means is therefore significant, though not highly significant, and we conclude that $\mu_1 > \mu_0$.

Distribution-free tests

4.7 The tests just described for comparison of mean values rest on the assumption that the underlying distribution is Normal or nearly so. It is possible to devise tests which make no such assumption and which are applicable to samples from any population. Such Distribution-Free Tests are inevitably somewhat less efficient in particular instances than tests making full use of our assumptions (or beliefs) about the form of the Population distribution; they provide, however, a quick and on the whole conservative means of data evaluation. Two such tests are described below.

Comparison of a mean value with a standard—the sign test

4.71 Given a sample of observations $x_1 \dots x_n$, we may wish to test whether the mean value differs significantly from some standard value μ_0. The procedure is as follows:

(i) For every x_i less than μ_0 record a minus sign $(-)$.

(ii) For every x_i greater than μ_0 record a plus sign $(+)$.

(iii) Discard any x_i exactly equal to μ_0, and reduce the value of n accordingly. (If more than 20% of the observations must be so discarded, it is better not to use this test.)

(iv) Count the number of occurrences of the less frequent sign.
(v) Look up the critical value in Table L at the end of the book for the chosen level of significance.
(vi) Conclude that a significant difference from μ_0 exists if and only if the observed number is less than or equal to the critical value.

Example 4.4.   Fertiliser composition

4.711 The following results were obtained for the percentage of K_2O in 10 samples of a fertiliser:

$$14{\cdot}7,\ 14{\cdot}3,\ 15{\cdot}1,\ 14{\cdot}6,\ 14{\cdot}9,\ 15{\cdot}4,\ 15{\cdot}0,\ 14{\cdot}7,\ 14{\cdot}4,\ 14{\cdot}8$$

Do they indicate a mean analysis significantly different from the declared figure of $15{\cdot}0\%$ K_2O?

There are seven results below $15{\cdot}0\%$, two above, and one of exactly $15{\cdot}0\%$. Taking $\alpha = 0{\cdot}05$ we find the critical value for $n = 9$ is 1. Hence we cannot conclude that the mean differs significantly from $15{\cdot}0\%$.

4.712 Paired comparisons can also be analysed by the Sign Test. Thus for the data of Table 4.2 we count the number of gas samples for which Operator A obtained a higher ammonia content than Operator B and vice versa. We find:

$$A > B \quad \text{for 2 samples}$$
$$A = B \quad \text{for 1 sample}$$
$$A < B \quad \text{for 6 samples}$$

In Table L the critical value for $n = 8$ and $\alpha = 0{\cdot}05$ (double-sided) is zero. Since the number of samples for which $A > B$ is greater than this, we conclude that the two analysts are not obtaining significantly different ammonia percentages.

Comparison of two sample means by the rank-sum test

4.72 One distribution-free test to compare the mean values of two separate samples employs the rank order of the data. To fix ideas, consider again the fertiliser filtration results of Example 4.1, which are reproduced in Table 4.3. To carry out the test we proceed as follows:

(i) Combine the observations from the two samples and rank them in order of increasing size from smallest to largest. Assign rank 1 to the lowest, rank 2 to the next lowest, etc. In the event of ties, assign to each tied member the average of the ranks which would have applied if the tied members had been different. (If more than 20% of observations are tied, do not use this test.)

(ii) Let $n_1 = $ smaller sample size $(= 6)$
$\qquad n_2 = $ larger sample size $\ (= 7)$
$\qquad n\ = n_1 + n_2 \qquad\qquad (= 13)$

(iii) Calculate R, the sum of the ranks for the *smaller* sample, viz. $R = 28.5$.
(If the two samples are equal compute R for either sample.)
Compute also
$$R' = n_1(n+1) - R = 55.5$$

(iv) Look up the critical value in Table M at the end of the book for the chosen level of significance.

(v) If either R or R' is smaller than or equal to the critical value conclude that a significant difference exists. In the example, taking $\alpha = 0.05$ (double-sided), a value of 27 is critical, so that the test just fails to establish significance at the 5% level.

TABLE 4.3. LABORATORY FILTRATION TEST

"Fair" Period		"Fairly Good" Period	
Time	Rank	Time	Rank
8·4	3	9·0	4
9·8	8	10·2	9
12·2	10	9·6	7
12·6	11	4·4	1
13·0	12	7·0	2
9·2	5·5	9·2	5·5
13·6	13		

We may compare this outcome with that of the earlier t-test in § **4.24** which did in fact establish significance. The loss of discrimination in using a distribution-free test procedure is highlighted by this marginal example. In general, differences are fairly small, and such as could frequently be tolerated in return for the gain in arithmetic convenience. For further information on distribution-free tests the reader is referred to [4.3].

The F-test

4.8 The F-test is used for comparing two variances. Suppose it is required to compare the values of two variances σ_1^2 and σ_2^2 from estimates s_1^2 and s_2^2 based on ϕ_1 and ϕ_2 degrees of freedom respectively. If the alternative to the Null Hypothesis is $\sigma_1^2 > \sigma_2^2$, we calculate the ratio $F = s_1^2/s_2^2$ and refer to Table D for the critical values of F with $\phi_N = \phi_1$, and $\phi_D = \phi_2$. This represents a single-sided test. If on the other hand the alternative to the Null Hypothesis is simply $\sigma_1^2 \neq \sigma_2^2$ the test is double-sided, and we then calculate the ratio of the larger estimate to the smaller one, the probabilities in Table D being doubled to give the critical values for this ratio.

To determine whether a single variance estimate s^2 differs significantly from an assumed value σ^2 we follow the same procedure, treating σ^2 as if it were an estimate based on an infinite number of degrees of freedom, i.e. taking $\phi_D = \infty$ or $\phi_N = \infty$ as appropriate.

Examples of the application of the F-test appear in later chapters, notably in Chapter 6 on the Analysis of Variance.

STATISTICAL DECISIONS

4.9 The significance test performs a valuable role in preventing hasty conclusions. An observed difference is not taken to represent a real difference unless the probability is suitably small that it could have arisen purely by chance; failing such assurance, the Null Hypothesis is assumed to be true. However, as a basis for decision this approach may be over-cautious, particularly in instances where the Null Hypothesis is essentially artificial as, for example, in development work searching for improved processes. Thus, while it may be reasonable to set up a Null Hypothesis that two analysts with similar training and experience will perform a given analysis equally well (and we may require a good deal of evidence to make us change our belief), it may make a good deal less sense to postulate that some change in a manufacturing process has had no effect on the yield or quality of the product, particularly if the change was made in the hope or expectation of benefit. In such circumstances the relevant question is not: "Has the change had an effect?" so much as: "Does the effect of the change appear sufficient to justify its cost (or insufficient to offset any savings)?"

Were the effect of the change precisely known, there would be no difficulty in reaching a decision. Commonly, however, the effect will be more or less uncertain, and the decision must then depend on an appraisal of the gains and losses associated with the conceivable outcomes, qualified by our assessment of the relative likelihood of each. But as the instances quoted above make clear, the relative likelihoods are determined not only by any observational evidence, but also by any prior beliefs we may hold concerning the true state of the system. Indeed, if our beliefs are strong enough, we may have no need for the reassurance or clarification that observation can bring, and may be prepared to reach a decision on prior beliefs alone. More generally, however, we will want to take simultaneous account both of our prior beliefs and of subsequent observational evidence. To do this, we make use of what is known as Bayes' Theorem.

Subjective Probabilities and Bayes' Theorem

4.91 The strength of a belief or of an opinion is often described in probability terms, for example "three times as likely as not" or "a 60:40 chance". These probabilities are *subjective*, since they describe only our own opinions, which may not be shared by others, and which are not usually verifiable in terms of the frequency-ratios introduced in § **2.16**. (Where the beliefs concern the outcome of a single event, the whole idea of frequency-ratio verification is incongruous.) They are nevertheless not meaningless, since they summarise —albeit imprecisely and in no very well-defined manner—such cognate information and experience as we possess on the issue in doubt.

Where the belief is about some quantity which may take a range of values, it may be represented by a Subjective Probability Distribution, which summarises the varying degrees of belief we are willing to accord to different parts of the range. For example, we may feel 95% certain that a planned process change will effect some improvement, think there is an even chance that the

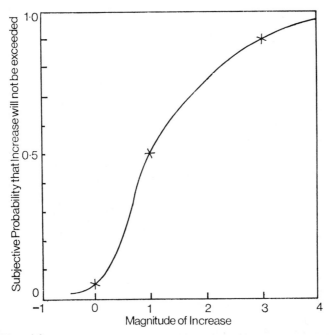

FIG. 4.2. CUMULATIVE SUBJECTIVE PROBABILITY DISTRIBUTION

improvement will exceed 1 unit and put the chance only at one in ten that the improvement will exceed 3 units. From such considerations we may construct a cumulative subjective probability distribution like Figure 4.2 (which is comparable with the cumulative frequency curve in Figure 2.3). Then by drawing tangents to the cumulative curve we can derive a probability density function like Figure 4.3. More commonly, since our beliefs will usually be rather hazy, we may find it more convenient to represent them approximately by some standard distribution such as the rectangular or the Normal distribution. We may in addition wish to associate a finite probability with some specific value, e.g. zero. Thus a full description of our prior beliefs may include both point and distributed probabilities.

The effect of additional information, for example from observation, is to cause us to modify our beliefs. Specifically, the strength of our resultant (posterior) belief in a particular true state is proportional both to our prior belief and to the Likelihood that, if such were indeed the true state, the additional evidence would have been what it was. Let us denote by $p_1(\theta)$ the prior distribution of our beliefs about some quantity θ, and by $L(x \mid \theta)$ the likelihood of the additional information x given that the true value of the

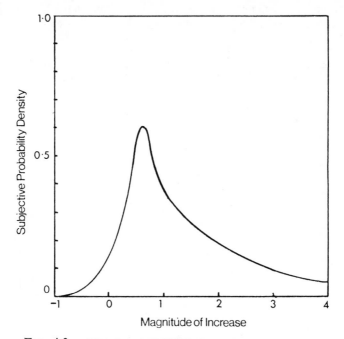

FIG. 4.3.   SUBJECTIVE PROBABILITY DENSITY FUNCTION

unknown is indeed θ. Then by a theorem due originally to the Rev. T. Bayes our posterior distribution of belief is given by

$$p_2(\theta) = \frac{p_1(\theta) . L(x \mid \theta)}{\int p_1(\theta) . L(x \mid \theta) \mathrm{d}\theta} \qquad (4.4)$$

where the integration is taken over the full range over which our prior beliefs extend. The use of this denominator is said to *normalize* the posterior distribution, because it makes the area under the $p_2(\theta)$ curve equal to unity.

Example 4.5

4.911 Suppose that our prior probability distribution is that shown in Figures 4.2 and 4.3. We now carry out an experiment and observe a mean improvement of 1·3 units with standard error 0·5 units. (We assume for purposes of illustration that this latter figure is known exactly.) Then, for a true mean μ, the likelihood of observing 1·3 units is:

$$\frac{1}{\sigma_m\sqrt{(2\pi)}} \exp\left\{-(\bar{x}-\mu)^2/2\sigma_m^2\right\} = \sqrt{(2/\pi)} \exp\left\{-2(1\cdot3-\mu)^2\right\} \qquad (4.41)$$

that is, the ordinate of the corresponding Normal distribution viewed as a function of μ [Appendix 4B].

Values of the likelihood for any assigned μ may now be read from tables of the Normal density function [4.4], and values of the prior probability density from Figure 4.3. Hence we may construct Table 4.4.

TABLE 4.4. CALCULATION OF POSTERIOR PROBABILITIES

(1)	(2)	(3)	(4)
μ	Prior Probability Density	Likelihood of Observed Result	Product (2)×(3)
−1·0	0·005	0·000	0·000
−0·5	0·030	0·001	0·000
0	0·150	0·014	0·002
0·5	0·525	0·111	0·058
1·0	0·395	0·333	0·132
1·5	0·260	0·368	0·096
2·0	0·190	0·150	0·028
2·5	0·130	0·022	0·003
3·0	0·090	0·001	0·000
3·5	0·065	0·000	0·000
4·0	0·050	0·000	0·000

Column 4 of the table gives a relative indication of posterior probability densities; to derive absolute probabilities we must divide by the integral of the function. This must be obtained either numerically or by plotting and counting squares, and is found to be approximately 0·156. Hence the posterior probability distribution is as shown in Figure 4.4 or, in cumulative form, in Figure 4.5. From the latter we see that we can now be virtually certain (subjective probability greater than 99·9%) that there is an increase, and can be almost 95% certain that this increase will exceed 0·5 unit. On the other hand we can now be 95% certain that the increase will not exceed 1·9 units; the greater improvements to which we felt obliged to concede some prior probability have been largely ruled out by the observed result. The effect of the additional information is thus both to sharpen our probability judgments and (as comparison of Figures 4.3 and 4.4 will show) to modify the shape of our subjective probability density function towards that of the likelihood function (4.41).

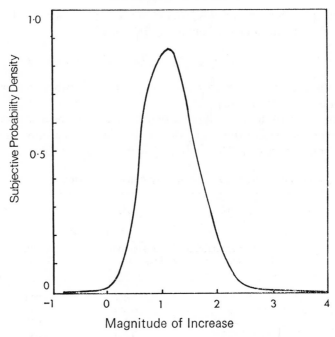

FIG. 4.4.　POSTERIOR PROBABILITY DENSITY FUNCTION

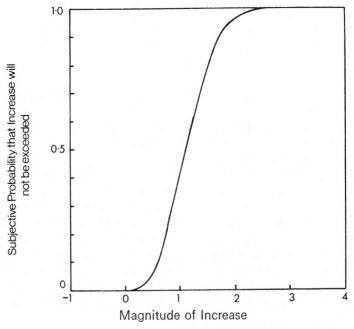

FIG. 4.5.　CUMULATIVE POSTERIOR PROBABILITY DISTRIBUTION

Weak prior beliefs and objective probabilities

4.92 The posterior probability density is proportional both to the prior probability density and to the likelihood of the observed results; it tends to zero as either of these factors tends to zero. If now one factor is relatively constant over the whole range for which the other factor is appreciably different from zero, the shape of the posterior function will owe little to the former and will be almost wholly determined by the latter.

For example, a single very imprecise observation, giving rise to a flat likelihood function, will scarcely modify reasonably firm prior beliefs (nor should we expect it to do so). Similarly, when the prior probability density function is fairly flat over the whole range for which the likelihood function is appreciable, the shape of the posterior probability density function will closely approximate that of the likelihood function, and will not depend appreciably on our prior beliefs.

A flat prior probability density can arise in two ways:

(i) We may believe initially that the true value is equally likely to lie anywhere within a range which, in the event, encompasses the effective range of the likelihood function.

(ii) The observed results may simply be so much more precise than our prior beliefs that, however we have chosen to represent these, they are sensibly uniform over the range of posterior interest.

In either case, there is a reflection of vagueness or lack of definition in our prior beliefs, which in many instances cannot be avoided. In such circumstances the likelihood function (suitably normalised) provides as it were an "objective" posterior probability density function, free of subjective considerations and hence the same for all observers. In the limit, as the precision of observation is increased (for example by replication of results), all posterior distributions will converge to the likelihood function, irrespective of the prior distribution of belief. In this sense also the distribution defined by the likelihood function may be considered objective, but note that it is not objective in the sense that it is capable of experimental verification. Indeed, as was pointed out earlier, there is an essential incongruity in attempting to apply frequency-ratio concepts of probability to the outcome of unique events; any probability measure in such circumstances can only describe the strength of a belief, or the confidence with which we are prepared to make a particular assertion.*

---

* We may note that the setting of confidence limits for a Normal mean [§4.2] is in fact indistinguishable from the use of a posterior probability distribution based on the likelihood function. In this instance, however, it was also possible to attach a frequency ratio interpretation to the confidence coefficient by considering as the "event" the making of an assertion and not the occurrence of a particular value in connection with any one assertion. Such a dual interpretation is not always possible.

A flat, or effectively flat, prior distribution may be termed "non-informative"; it contributes nothing to the form of the posterior distribution and so imparts no information over the range of practical interest. When our prior beliefs are weak, or notably imprecise, there is both practical convenience and computational advantage in specifying a "non-informative" prior distribution, i.e. no distribution, because the posterior distribution is then defined directly by the likelihood function. But when there is substantial prior information (as in the example quoted in the next chapter), we should of course make full use of it in setting up an appropriate prior distribution.

Gains and losses

4.93 Let the posterior distribution of belief about some unknown parameter θ be represented by the probability density function $p_2(\theta)$, and let the Gain which would accrue to us for any particular θ be denoted by $g(\theta)$. For example, θ might be the level of sales of a particular product, and $g(\theta)$ the profit associated with this level of sales. The value of $g(\theta)$ for any particular θ may be either positive or negative; in the latter case it represents a Loss.

The degree of belief in a particular gain is of course precisely the same as that in the corresponding parameter value θ, and is represented by $p_2(\theta)$. The conceptual average gain, called the Expected Gain, is obtained by weighting each conceivable gain by the strength of our belief in it and is given by:

$$\bar{g} = \int g(\theta) \cdot p_2(\theta) \mathrm{d}\theta \qquad (4.5)$$

where as before the integration extends over the whole range spanned by our beliefs. The expected gain $\bar{g}$ may also be positive or negative, but any decision based on θ should clearly be such as to yield a positive expected gain. Thus, if θ represents the improvement due to some change in process conditions and $g(\theta)$ represents the net gain due to such an improvement, the change should be made or perpetuated only if $\bar{g} > 0$. Where a decision must be made among a number of alternatives, the one which yields maximum expected gain should be chosen.

Where $g(\theta)$ is a linear function of θ, say $a + b\theta$, then:

$$\bar{g} = a \int p_2(\theta) \, \mathrm{d}\theta + b \int \theta \cdot p_2(\theta) \, \mathrm{d}\theta$$

$$= a + b\bar{\theta}$$

$$= g(\bar{\theta}) \qquad (4.51)$$

where $\bar{\theta}$ is the expected value of θ for the posterior distribution (equivalent to the mean value for a frequency-ratio probability distribution). Thus the expected gain reduces to the gain corresponding to the expected value. In particular instances—particularly when a non-informative prior distribution is assumed—this results in considerable simplification. For example, if θ is a Population mean value, its expected value for the posterior distribution represented by the likelihood function is simply the observed mean result $\bar{x}$, and the decision is taken precisely as if the observed mean were indeed the true mean.

Utility

4.94 The expected gain is not, of course, a tangible quantity; if subjective probabilities had objective reality, then all gamblers would be rich! Rather, it is the best single estimate which can be made of the consequences of a particular decision, and as such is an appropriate basis for decision. Nevertheless, the actual gain resulting from our decision may be more or less than expected, and indeed may on occasion turn out to be a loss. Numerically the loss of £1,000 is only the opposite of gaining £1,000, but in practical terms the one may be much more damaging than the other is beneficial —particularly if our capital is less than £1,000. If we express $g(\theta)$ and $\bar{g}$ strictly in monetary terms, we cannot allow for the fact that equal deviations above and below expectation may not be of equal consequence. To do this, we need to recognise that the Utility to us of any given sum of money is not a constant, but depends on the amount of money we already have. Thus a complete theory of decision must be based on utilities, rather than on money values, and will lead to decisions which depend not only on the facts, but also on our circumstances.

The utility to be ascribed to any given sum is inherently arbitrary, since for effective decision we need to know only relative utilities. However, to avoid unnecessary arbitrariness, it is helpful to express utilities in quasi-monetary terms, such that zero gain or loss from "present wealth" corresponds to zero utility and such that the initial slope of the curve of utility against monetary worth is unity, i.e. that the utility of the first £1 gained or lost is written as $\pm$£1. A typical utility function might then be of the form shown in Figure 4.6, and would reflect the fact that each succeeding pound gained was of progressively less utility, while each succeeding pound lost was viewed increasingly seriously.

To construct such a plot, we effectively need to consider for what we would sell our chances in, or pay to get out of, particular notional lotteries. Suppose, for example, we held a ticket in a lottery in which there was a one in ten chance of gaining £1000. On some such basis as "a bird in the hand is worth two in the bush", we might be willing to sell our ticket for £60. We thus imply that the utility to us of the £1000 we might win is only ten times that of the £60 we are prepared to take for the ticket; if this is not so, we are acting illogically. Conversely, if we were in a situation where there was one chance in ten that we might have to pay out £1000, we might be willing to pay as much as £200 to be relieved of our responsibilities. We would then imply that a loss of £1000 is viewed by us ten times as seriously as a loss of £200.

Proceeding in this way, it is possible to build up a utility function covering the whole range of possible outcomes associated with any decision. (Some further considerations are discussed in Appendix 4C.) In basing our decision on the expected utility of each possible course of action we are introducing a further element of subjectivity into the decision process, but it is a very necessary one, since the essence of a sound decision is that it be right for us, the decision makers. For example, a bachelor and a man with family responsibilities will take different decisions concerning life insurance primarily because their utility functions are unequal, and not in the belief

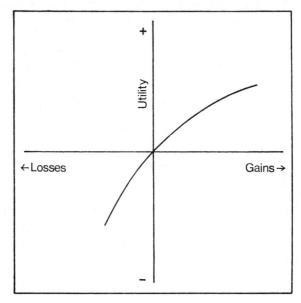

FIG. 4.6.   TYPICAL FORM OF UTILITY FUNCTION

that their life expectancies are different. Similarly, an investment decision appropriate to a company with a capital of £100m. might be quite inappropriate to one with a capital of £1m., even though the potential benefits are the same and the outlay is within the capacity of either company, because the smaller company could accept less readily than the larger the uncertainties inherent in the project. Only a treatment in terms of utilities would highlight such a difference and lead to the appropriate decisions for the two companies.

The examples make clear that the concept of utility must always underlie the decision process. Nevertheless, many cost decisions encountered in industrial firms are relatively small in relation to the total size of the firm, and in such cases a simple linear utility function, i.e. a treatment in terms of money values, is adequate. This incidentally avoids the difficulty of having to establish a corporate as distinct from a personal utility function. But where a decision is one which could possibly make or break the firm, explicit consideration must clearly be given to the utilities of different outcomes.

Example 4.6. Decision on a pilot-plant programme

4.95 To illustrate these ideas in some detail, consider the following problem. A short programme of laboratory work has thrown up an apparently promising new catalyst for an established reaction. Small-scale tests have indicated an average increase in activity of 50% per unit volume compared with the standard catalyst, though at a manufacturing cost estimated to be some 10% greater. It proved difficult to obtain reproducible results on the small scale, and the 95% confidence limits associated with the mean increase were ±30% per unit volume. To establish more clearly the performance of the new catalyst and to finalise the manufacturing recipe it is proposed that a programme of pilot-plant work be undertaken at a cost of £100,000. What should be management's attitude to the proposal?

To decide this, we need to evaluate:
(i) The prior distribution of belief concerning the activity of the catalyst.
(ii) The posterior distribution of belief, in the light of the evidence to hand.
(iii) The gains and losses associated with particular true activities.
(iv) The utilities to be ascribed to such gains or losses.

Item (i) is readily established; catalysis is still so imperfectly understood that almost any result, within reason, must be considered equally likely, and it is thus appropriate to take a flat (non-informative) prior distribution. This means that (ii), the posterior distribution, may be represented for practical purposes by a Normal distribution with mean 50% and standard deviation 15%.

Item (iii), the gains and losses associated with particular true outcomes, are less easily established. It is of course clear that, if the new catalyst is in reality no better than the conventional catalyst, the £100,000 expended on pilot-plant work will bring no return. Similarly, if the increase in activity is below 10%—the increase in cost—any advantages of the new catalyst can only be long-term ones—the possibility of marginal increase in the output of the existing full-scale plant, or capital savings in the construction of new plant. For present purposes we may assume that vessel design, and the capacities of other units, are such that no increase in output is possible in the existing plant, and that future capital savings are too nebulous and uncertain to be accorded any weight. Thus an increase in activity exceeding 10% is necessary for gain to be possible, and the only direct return from the availability of a more active catalyst is the reduced cost of catalyst replacement.

Taking the present cost of catalyst for the plant as £50,000 and its life as 2 years, a 20% more active catalyst at 10% greater cost would permit a saving of £4,167 per replacement, and a 50% more active catalyst would permit a saving of £13,333 per replacement. Assuming a 15% rate of return on capital and a first replacement one year hence, these figures are equivalent to capital sums of £14,800 and £47,500 respectively. Even at the upper limit of the confidence interval (80% greater activity), the equivalent capital sum is only £69,000. Clearly, since the company has only one plant, the direct

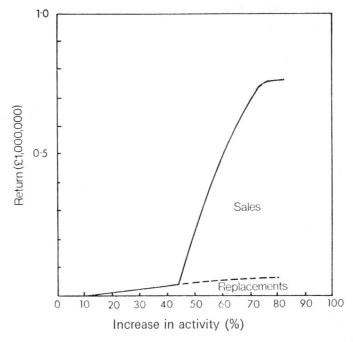

FIG. 4.7. GAIN AS A FUNCTION OF INCREASE IN CATALYST ACTIVITY

savings will in no way justify the cost of the pilot-plant work, though the position would be changed if a number (say 10) of plants existed.

The direct savings do not, however, represent all possible benefits. Given a sufficiently active catalyst, there are good hopes of external sales in a world market estimated conservatively to be worth £2,000,000 p.a. Significant market penetration is believed to be possible with a catalyst 50% more active than the current standard, and effective market domination for a few years is thought to be achievable with any catalyst over 75% more active than the current standard. Preliminary costings indicate net present worths of such a venture ranging from £340,000 at the lower figure to £1,390,000 at the higher. These figures may, however, be eroded by competitive research, and are somewhat arbitrarily halved to allow for this. Hence the monetary gain, as a function of the improvement in activity, is as shown in Figure 4.7, which displays both the rather low returns obtainable from the reduced cost of catalyst replacement and the rapidly rising returns obtainable if a marketable catalyst is achieved. In every instance, the return presupposes a prior expenditure of £100,000 on pilot plant work.

Having established monetary worths, there remains item (iv)—the utility of any gain. The company is large enough for a linear utility function to be

TABLE 4.5. CALCULATION OF DISTRIBUTION OF GAINS

(1)	(2)	(3)	(4)	(5)
Increase (%)	u	Normal Ordinate	Gain (£1,000)	Product (3) × (4)
10	−2·67	0·011	0	0
15	−2·33	0·026	8	0·2
20	−2·00	0·054	15	0·8
25	−1·67	0·099	21	2·1
30	−1·33	0·164	26	4·3
35	−1·00	0·242	31	7·5
40	−0·67	0·319	36	11·5
45	−0·33	0·377	69	26·0
50	0	0·399	217	86·6
55	0·33	0·377	356	134·2
60	0·67	0·319	480	153·1
65	1·00	0·242	593	143·5
70	1·33	0·164	700	114·8
75	1·67	0·099	760	75·2
80	2·00	0·054	765	41·3
85	2·33	0·026	765	20·1
90	2·67	0·011	765	8·7
95	3·00	0·004	765	3·4
100	3·33	0·002	765	1·2
105	3·67	0·0005	765	0·4
110	4·00	0·0001	765	0·1

applicable, but the cost of the pilot-plant work has to be borne by a limited research budget, and acceptance of the proposal would lead to the rejection of other proposals. In the light of past experience, a threefold return of research expenditure is required. Hence, the proposal will only be accepted if the expected gain is in excess of £300,000.

We can now proceed to calculate the expected utility of the project. Since the posterior distribution is taken to be Normal with mean 50% and standard deviation 15%, we record against each increase in Table 4.5 the equivalent

Normal deviate $u = (\text{Increase} - 50\%)/15\%$. The corresponding ordinates are read from the Normal density tables [4.4], and the gains from Figure 4.7. The product of Gain × Ordinate is listed in column (5). The expected gain is obtained by summing the entries in this column and—since the tabulation interval is one-third of a standard deviation—dividing the answer by three. The result is $835\cdot0/3 = 278\cdot3$, i.e. £278,333 on conversion back to true units.

This falls short of the £300,000 target; hence the proposal would be rejected by management at this stage. The difference is however small, and in view of the commercial prospects which would be opened up by a really substantial increase in activity, they might well be prepared to support a limited extension of the laboratory work to see if further increases in catalyst activity could be achieved.

References

[4.1] WELCH, B. L. "The Generalisation of 'Student's' Problems when Several Different Population Variances are Involved", *Biometrika*, **34**, 28–35 (1947).
[4.2] Box, G. E. P. "Non-Normality and Tests on Variances", *Biometrika*, **40**, 318–335 (1953).
[4.3] TATE, M. W., and CLELLAND, R. C. *Non-parametric and Short-cut Statistics.* Interstate (Danville, Illinois, 1957).
[4.4] PEARSON, E. S., and HARTLEY, H. O. *Biometrika Tables for Statisticians*, Vol. 1 (third edition). Charles Griffin (High Wycombe, 1976).

Appendix 4A

Confidence limits for a standard deviation: Data not assumed Normal

Table 4A.1 is a reproduction of Table 2.1, with the working units indicated. The 178 observations arranged in random groups of eight (two observations having been discarded) are shown in Table 4A.2 with the group variances and their logarithms to base 10.

The mean and standard deviation of y are calculated in the usual way:

$$\Sigma y = 11\cdot82 \qquad \bar{y} = 0\cdot5373$$

$$\Sigma y^2 = 7\cdot4754 \qquad (\Sigma y)^2/22 = 6\cdot3506$$

$$\therefore \quad \Sigma(y-\bar{y})^2 = \Sigma y^2 - (\Sigma y)^2/22 = 1\cdot1248$$

$$\therefore \quad V(y) = 1\cdot1248/21 = 0\cdot05356$$

$$V(\bar{y}) = 0\cdot05356/22 = 0\cdot002435$$

$$\therefore \quad S.E.(\bar{y}) = 0\cdot0493$$

$$\phi = 21 \qquad t_{0\cdot025} = 2\cdot08$$

$$t \times S.E.(\bar{y}) = 0\cdot1025$$

The 95% confidence limits for $\log \sigma^2$ are:

$$0\cdot5373 \pm 0\cdot1025 = 0\cdot4348 \text{ and } 0\cdot6398$$

TABLE 4A.1. PERCENTAGE CARBON IN A
MIXED POWDER

% Carbon	x'	Number of Results
4·10–4·19	−5	1
4·20–4·29	−4	2
4·30–4·39	−3	7
4·40–4·49	−2	20
4·50–4·59	−1	24
4·60–4·69	0	31
4·70–4·79	1	38
4·80–4·89	2	24
4·90–4·99	3	21
5·00–5·09	4	7
5·10–5·19	5	3
Total		178

TABLE 4A.2. WORKING DATA ARRANGED IN RANDOM GROUPS OF EIGHT

Group	x'								$V(x')$	$y = \log V$
1	3,	2,	0,	2,	−2,	3,	−1,	0	3·55	0·55
2	0,	1,	1,	−1,	3,	0,	−1,	−1	1·93	0·29
3	−2,	0,	−1,	1,	0,	1,	2,	4	3·41	0·53
4	1,	−1,	−1,	1,	−3,	0,	−2,	−2	2·12	0·33
5	−2,	3,	4,	−2,	1,	−2,	−2,	0	6·00	0·78
6	−2,	2,	−1,	−3,	1,	−2,	0,	3	4·50	0·65
7	1,	3,	1,	1,	2,	1,	−3,	1	2·98	0·47
8	3,	0,	−2,	2,	3,	0,	4,	2	4·00	0·60
9	2,	−1,	2,	−1,	1,	−4,	3,	0	5·07	0·71
10	1,	0,	1,	1,	1,	1,	3,	1	0·70	−0·15
11	2,	2,	5,	−2,	0,	2,	1,	2	4·00	0·60
12	1,	0,	−3,	2,	−4,	−1,	2,	2	5·55	0·74
13	3,	−1,	2,	0,	1,	−3,	1,	3	4·21	0·62
14	2,	−1,	−1,	0,	5,	3,	2,	−1	4·98	0·70
15	−2,	1,	−2,	1,	−1,	−3,	1,	0	2·55	0·41
16	−1,	5,	1,	1,	0,	−1,	2,	2	3·84	0·58
17	0,	0,	−1,	0,	1,	−2,	0,	3	2·12	0·33
18	−5,	−2,	2,	0,	1,	3,	4,	1	8·29	0·92
19	4,	3,	3,	3,	−3,	4,	3,	0	5·84	0·77
20	0,	3,	3,	1,	−2,	−2,	0,	−1	3·93	0·59
21	2,	−1,	−2,	4,	1,	0,	0,	−1	3·70	0·57
22	1,	0,	2,	1,	−1,	−2,	1,	1	1·70	0·23

Taking antilogs:

$$\text{Estimate of } \sigma^2 = 3\cdot44$$

$$\text{Lower confidence limit} = 2\cdot72$$

$$\text{Upper confidence limit} = 4\cdot36$$

For σ we have:

$$\text{Estimate} = 1\cdot86$$

$$\text{Lower confidence limit} = 1\cdot65$$

$$\text{Upper confidence limit} = 2\cdot09$$

These may be compared with the direct estimate of σ [§ **3.341**] of 1·94 working units and 95% confidence limits of 1·76 and 2·19 calculated as in § **4.33**. The ratios of either limit to the estimate are in good agreement for the two methods, since the data in this instance are not far from Normal (see Figure 2.4). There is, however, some underestimation of σ by the present method, which is more noticeable because of the narrowness of the limits based on 178 results. Such underestimation is to be expected, since we are in effect taking a geometric rather than an arithmetic mean of the separate variance estimates. It can be approximately overcome by adding to $\bar{y}$ the quantity $\frac{1}{2}(\log_e 10)V(y)$, i.e. $1\cdot151V(y)$, in those instances where the bias is thought to be of much consequence. In the present instance, this gives a revised estimate of $\sigma^2 = \text{antilog}\,(0\cdot5989) = 3\cdot97$, and a corresponding estimate of $\sigma = 1\cdot99$. The limits correspondingly become 1·77 and 2·24— only marginally wider than those calculated by the method of § **4.33**.

Appendix 4B

The likelihood function

4B.1 If data are Normally distributed, the probability that an observed result x will fall between $(x_1 - \frac{1}{2}dx)$ and $(x_1 + \frac{1}{2}dx)$ is given by:

$$dP = \frac{1}{\sigma\sqrt{(2\pi)}} \exp\{-(x_1-\mu)^2/2\sigma^2\}dx$$

where μ is the Population mean and σ the Population standard deviation. The quantity:

$$\frac{1}{\sigma\sqrt{(2\pi)}} \exp\{-(x-\mu)^2/2\sigma^2\}$$

is termed the probability density function for the distribution characterised by the mean μ and standard deviation σ; alternatively, it is termed the Likelihood of x given μ and σ. In the former sense it is thought of as a function of x only, μ and σ being considered as constants; in the latter it is regarded as a function of all three variables. The term Likelihood was introduced by R. A. Fisher to distinguish the use of the function as a means of establishing an order of preference among different possible values of

μ and σ, given x, from its use in conventional probability calculations in which μ and σ are assumed fixed. A well-established technique for estimation from sample data in fact makes use of the principle of Maximum Likelihood, i.e. estimates of the unknown parameters are chosen so as to maximize the likelihood of the observed data.

4B.2 When several results x_1, ..., x_n are drawn from the same Normal Population, the joint likelihood of all results is given by the product of the separate likelihoods for each individual result, viz.:

$$L = \prod_{i=1}^{n} \left[\frac{1}{\sigma\sqrt{(2\pi)}} \exp\left\{ -(x_i - \mu)^2/2\sigma^2 \right\} \right]$$

$$= \left(\frac{1}{\sigma\sqrt{(2\pi)}} \right)^n \exp\left\{ - \sum_{i=1}^{n} (x_i - \mu)^2/2\sigma^2 \right\} \qquad (4B.1)$$

More generally, if the likelihood of a result x_i given some parameter θ is $l(x_i \mid \theta)$, the likelihood of a series of results x_1, ..., x_n is:

$$L = \prod_{i=1}^{n} l(x_i \mid \theta)$$

Here L is a function of x_1, ..., x_n and θ.

In many practical applications it is found that the expression for L can be simplified. For example, in Equation (4B.1), the sum of squares $\Sigma(x_i - \mu)^2$ breaks down into $\Sigma(x_i - \bar{x})^2 + n(\bar{x} - \mu)^2$, in which only the second term involves μ [Appendix 3A]. Similarly, constant factors such as the $1/\{\sigma\sqrt{(2\pi)}\}^n$ may be disregarded when they do not affect a preference ranking or when, as in the application of Bayes' Theorem, a subsequent normalization is necessary. Hence we may deduce, *inter alia*, that the maximum likelihood estimator of the mean μ of a Normal Population is the sample mean $\bar{x}$, since the L of (4B.1) is maximized when $(\bar{x} - \mu)^2$ is minimized.

For a more detailed discussion of likelihood see [4B.1].

Reference

[4B.1] LINDLEY, D. V. *Introduction to Probability and Statistics*. Cambridge University Press (1965).

Appendix 4C

Utility functions

The main shape of a utility function over the typical range of practical interest is as shown in Figure 4.6, viz. decreasing marginal utility with increase in the amount received and, conversely, increasing marginal (negative) utility with increase in the amount paid out. In detail, however, a number of other features may arise. For example, there may be an upper limit to the utility which can accrue; as private individuals, for instance, we might be completely unable to find a meaningful difference between (say) £10m. and £20m., since either sum exceeds what we could possibly utilise. In the same way, there may be a limit to the utility which can be lost, since ruin is ruin, and the bankruptcy courts will limit our liability.

At a much lower level, there may be sums so trivial that their utility is negligible. For example, we might be willing to spend five new pence on a one-in-a-thousand chance of winning five pounds, even though the expected return is far less than our outlay. If there were not, at least on a personal basis, such a widespread gambling propensity, lotteries and football pools would not exist, since it is well known that the total returns fall far short of the total staked.

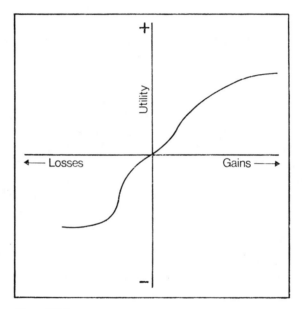

FIG. 4C.1. GENERAL FORM OF UTILITY FUNCTION

A full representation of a utility function is thus notionally as depicted in Figure 4C.1. For a small distance above the origin ("present wealth") there is a region in which "gambling" is preferred, i.e. where a trivial but certain gain would willingly be surrendered for a less than fair chance of a much larger gain. This is succeeded, after a point of inflexion, by the main body of the curve showing diminishing marginal utility with increase in the amount gained, and by a final levelling out. Below the origin, there is an extended "anti-gambling" region in which a certain but limited loss would be preferable to a proportionate risk of much greater loss. As the amounts involved increase, this may eventually be succeeded by a region of desperate gambling in which one would sooner retain a small chance of solvency than opt for certain ruin. The positive and negative parts of the curve may be markedly dissimilar, but overall the curve is likely to be such that symmetrical wagers will be accorded a negative utility.

Whilst all the features enumerated above may be assumed to be present in any utility function, their relative importance may vary widely. In particular, for business decisions, the "gambling" region in the positive part of the curve may generally be assumed to be absent, and the upper and lower bounds to the utility function may be so remote as to be of no consequence. Hence we may well be left with the "typical" utility function shown in Figure 4.6 which may often, as noted in the text, be replaced by a straight line over the limited range of current interest. Nevertheless, it is both instructive and illuminating to construct a utility function appropriate to any decision which must be taken, particularly where the amounts involved are reasonably substantial.

Many books and papers have been published on the subject of utility functions and their ascertainment. The formulation above is basically that of Friedman and Savage [4C.1] as modified by Markowitz [4C.2]. More general reading is available in [4C.3] and [4C.4].

References

[4C.1] FRIEDMAN, M., and SAVAGE, L. J. "The Utility Analysis of Choices Involving Risk", *J. Polit. Econ.*, **56**, 279–304 (1948).

[4C.2] MARKOWITZ, H. "The Utility of Wealth", *J. Polit. Econ.*, **60**, 151–158 (1952).

[4C.3] BLACKWELL, D., and GIRSCHICK, M. A. *Theory of Games and Statistical Decisions.* John Wiley (New York, 1954).

[4C.4] RAIFFA, H., and SCHLAIFER, R. *Applied Statistical Decision Theory.* Harvard University Press (1961).

Chapter 5

Statistical Tests: Choosing the Number of Observations

Decisions may often be taken in the light of evidence
from a limited number of variable observations. This
chapter is concerned with the problem of determining
how many observations should be made so that the
risk associated with any decision is acceptably small.
When certain prior information is available it is
possible to base this determination directly on the
average cost of making wrong decisions. An alternative approach is given for use when, as frequently
happens, this prior information is not available.

Introduction

5.1 The research worker in industry is frequently faced with the problem:
How many observations should be made, given that from these observations
a decision is to be taken which will result in financial loss if it is wrong?

Particular examples of great practical importance arise in designing
sampling inspection schemes and in planning plant trials in process development work. In sampling inspection the problem is to choose the number of
tests that should be carried out in order to decide whether to accept or reject
a consignment of material; this will be discussed in more detail in Chapter 11.
In planning plant trials—that is, experiments on the production scale—the
problem is to choose the number of trials that should be made to decide
whether some modified method of manufacture should be accepted or
rejected.

These problems are particular instances of the decision function problems
which were introduced in the last chapter. A full discussion of these examples
would involve not only the question of the optimum number of observations
but also of the optimum procedure for making the test. A detailed discussion
of this wider problem is not possible in the space available and at the elementary level of this textbook. In this account, therefore, it will be assumed that
the test procedure is given and only the number of observations has to be
decided.

Because of random fluctuations in results there is always some risk that a wrong decision will be taken, however many observations are made. This risk becomes smaller as the number of observations is increased, but since observations cost money, taking too many observations can be as costly as taking too few. In this chapter it will be shown how to calculate the optimum number of observations when certain background information is available and how to arrive at a reasonable procedure when such background information is lacking.

5.11 The background information concerns two things:

(i) The so-called prior distribution, which embodies knowledge available prior to making a series of tests or experiments.

(ii) The costs of making wrong decisions and of experimentation.

We shall first consider the case where both sorts of information are available, illustrating the general principles which then apply in relation to the planning of a sampling inspection (acceptance testing) scheme. We next suppose, as frequently happens, that no information about the prior distribution is available but that information is available on costs. In this situation a procedure that may be regarded as a two-stage modification of the first procedure may be used; this is described and illustrated in relation to the planning of plant trials.

Finally the situation is considered in which neither information on the prior distribution nor information on costs is available. By what may be regarded as a further modification of the first procedure we arrive at methods which, although basically less satisfactory, have at least the advantage that they have been more thoroughly explored. With the first two methods we will consider only the single, but important, case in which the value of a mean is in question; the last procedure, however, is applied also to the problem of comparing variances.

PRIOR DISTRIBUTION AND COSTS KNOWN

Example 5.1. A testing scheme

5.2 The general ideas that may be applied to the economic choice of numbers of observations, when the prior distribution and the costs are known, can best be illustrated by considering the particular example of an acceptance testing scheme.

The quality of a batch of a particular chemical delivered from a supplier is defined as the amount μ of an essential ingredient in the batch. According to the specification, μ, measured in suitable units, should be equal to 90, and in practice, if the mean $\bar{x}$ of chemical analyses of n samples from a given batch is greater than 90, the batch is accepted, and if it is less it is rejected. Assuming that the standard deviation of the error in the sampling and testing method is 1·0, the problem is to decide how many samples from each batch should be tested, i.e. how large n should be.

Outline of the method

5.21 The behaviour of any given scheme can be determined from the curves shown in Figure 5.1 (i) and Figure 5.1 (ii).

The first is the Incoming Quality Curve* or Prior Distribution. This curve

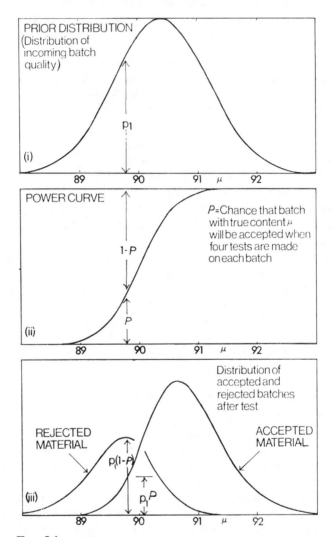

FIG. 5.1.   OPERATION OF A SAMPLING INSPECTION SCHEME

---

* The expressions *incoming quality curve* and *operating characteristic curve* are terms used specifically when these methods are applied to sampling inspection schemes.   The curves are particular instances of *prior distributions* and *power curves* respectively.   These latter terms are used in a wider context.

characterises the material to be tested; it shows how the true batch quality is distributed before the test is applied.

The second is the Operating Characteristic Curve* or Power Curve. This curve characterises the testing scheme; it shows the chance that material of any given quality will be accepted by the testing scheme as satisfactory.

These curves will be described more fully in the next two sections. Having obtained them we can calculate the distribution of quality of the accepted batches and the distribution of quality of the rejected batches (see Figure 5.1 (iii)). Once these two distributions have been determined, the most economic number of observations is easily derived.

The process described may be likened to the sieving of particles where the sieve corresponds to the testing procedure. We have an aggregate of material and we want to divide it into two heaps, one heap containing material which is smaller than a certain size and the other containing material which is larger. The prior distribution corresponds to the size distribution of material before sieving. The sieve does not achieve a perfectly sharp "cut"; its "discrimination" is shown by the power curve. Knowing these two characteristic curves we can easily calculate the size distributions of the particles passing through the sieve and of those remaining in it.

The discrimination effected by the testing scheme, which is measured by the steepness of the power curve, can be made as great as we please by increasing the number of analytical tests. We have to find the point at which the monetary advantage of increasing the discriminating power is just offset by the increased cost of testing. In the method we are discussing this is done numerically by making the calculation for a series of trial values of n and choosing that which gives the lowest costs. Mathematical solutions are possible in particular instances; the numerical method which we give here, however, has the advantage that it can be applied generally.

The prior distribution

5.22 In sampling inspection problems the prior distribution is simply the distribution of the true quality μ of incoming batches. In practice this has to be determined from past records, on the assumption that future experience will resemble past experience. Small changes in the form of the prior distribution are found to have only a minor effect on the optimal values of n, so it is not necessary to know the prior distribution exactly.

We cannot use for the prior distribution the observed distribution of past test results *as it stands*, since this does not give the distribution of the true quality μ but rather the distribution of test observations which are subject to error. The variance of the observed distribution is the sum of the variance of the true quality and the error variance. We have therefore to apply an adjustment to the observed distribution of test results. This may be done as follows.

Suppose that in the past k tests have normally been carried out on each batch and that the best estimate of the batch quality has been taken to be the mean of these k test results (k may be less or greater than the value of n which is to be determined for sampling inspection in the future). Suppose also that a large number, preferably several hundred, of these mean estimates of batch quality are available covering a typical period. A histogram is plotted from these results, and the mean and variance of the observed distribution, which we shall call the unadjusted distribution, are determined in the usual way. From these we can calculate:

Mean of adjusted distribution = Mean of unadjusted distribution

Variance of adjusted distribution = Variance of unadjusted distribution

— variance due to sampling and testing

Since k tests were performed, the variance due to sampling and testing was σ_T^2/k, where σ_T^2 is the variance of an individual test result.

A sufficient approximation to the distribution of μ will now be provided by a curve with the same mean as the adjusted distribution and of the same general shape, but with dispersion reduced by a factor

$$\sqrt{\frac{\text{Variance of adjusted distribution}}{\text{Variance of unadjusted distribution}}}$$

In practice we can proceed by fitting a smooth curve by eye to the histogram and graphically constructing an adjusted curve of the same shape but with suitably reduced dispersion. Alternatively, in some cases the unadjusted curve may happen to be closely similar to some theoretical curve (e.g. the Normal curve), and the prior distribution can then be represented by the same type of curve with suitably reduced dispersion.

In the present instance we will assume that the testing procedure prior to the investigation has been to make one test on each batch and to accept or reject on this result. Past records of these results are available over a considerable period. A histogram plotted for 250 such values shows an approximately Normal distribution with mean 90·4 and variance 1·64. Since we are assuming that the sampling and testing error variance $\sigma_T^2 = 1·0$ and that $k = 1$, the adjusted curve has mean 90·4 and variance $1·64 - 1·00 = 0·64$. We therefore take the prior distribution to be a Normal curve with mean 90·4 and standard deviation 0·8 as drawn in Figure 5.1 (i). A more precise method of adjustment of the observed distribution is given in [5.1].

We denote the ordinate of the prior distribution by $p_1(\mu)$. Only the relative frequencies at various levels of μ are required in the calculations which follow. It is not essential, therefore, that the curve should be drawn on such a scale that its area is unity.

The power curve

5.23 The power curve shows, for a given test procedure, the chance P that batches with true quality μ will be accepted. We assume that our future test procedure will consist of performing n tests on each batch of material and accepting or rejecting the batch depending on whether the observed mean result $\bar{x}$ is higher or lower than the critical level, $\bar{x}_0 = 90$. The derivation of the power curve for such a situation is as follows.

For the reasons given in § **2.42** we may assume $\bar{x}$ to be Normally distributed. Consequently from a table of the Normal curve we can obtain the probability P that, for any given value of μ, $\bar{x}$ will exceed 90 and the batch will be accepted. We are given that in repeated sampling from a particular batch a single analytical result varies because of sampling and testing errors

TABLE 5.1. POWER CURVE: CALCULATION OF PROBABILITY OF ACCEPTING A BATCH WITH MEAN QUALITY μ FOR $\sigma_T = 1$ AND $n = 4$

μ	u	P	μ	u	P
88·0	4·0	0·0000	90·2	−0·4	0·6554
88·2	3·6	0·0002	90·4	−0·8	0·7881
88·4	3·2	0·0007	90·6	−1·2	0·8849
88·6	2·8	0·0026	90·8	−1·6	0·9452
88·8	2·4	0·0082	91·0	−2·0	0·9772
89·0	2·0	0·0228	91·2	−2·4	0·9918
89·2	1·6	0·0548	91·4	−2·8	0·9974
89·4	1·2	0·1151	91·6	−3·2	0·9993
89·6	0·8	0·2119	91·8	−3·6	0·9998
89·8	0·4	0·3446	92·0	−4·0	1·0000
90·0	0·0	0·5000			

about the true batch value μ with standard deviation $\sigma_T = 1·0$. The standard error of the mean of n analyses is therefore $\sigma_T/\sqrt{n} = 1/\sqrt{n}$. To find P we calculate $u = (90 - \mu)\sqrt{n}/\sigma_T$ and find from Table A the probability that u will attain this value or more. If u is positive, P is less than 0·5 and is the entry in Table A; if it is negative, P is greater than 0·5 and is found by subtracting the entry in Table A from unity, i.e. the entry is equal to $(1 - P)$. For instance, if 4 analyses are carried out and $\sigma_T = 1·0$, then $u = 2(90 - \mu)$. The values of P obtained from Table A for values of μ between 88 and 92 are shown in Table 5.1 and are plotted in Figure 5.1 (ii).

The power curves corresponding to $n = 1$, 4 and 16 are shown in Figure 5.2, from which we see the comparative discriminating power of tests based on different numbers of observations. The steeper the curve, the greater is the discriminating power of the test. The "ideal" power curve would be a vertical line at $\mu = 90$; this would give certain acceptance for $\mu > 90$ and certain rejection for $\mu < 90$. Such a curve could only occur, however, if there were no testing error or if an infinite number of analyses were done.

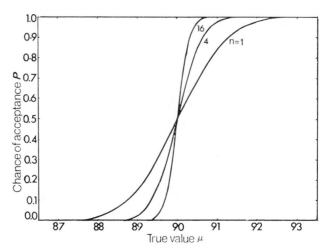

FIG. 5.2.   POWER CURVES FOR VARIOUS VALUES OF n WITH
$\bar{x}_0 = 90$ AND $\sigma_T = 1$

The distribution of outgoing quality

5.24 The relative frequency with which batches of quality μ are offered *and* accepted is obtained by multiplying the ordinate p_1 of the prior distribution, which is the relative frequency with which a batch of quality μ is offered, by the ordinate P of the power curve which is the probability that, having been offered, a batch with quality μ will be accepted; plotting the product $p_1 P$ against μ then gives the distribution of the quality of accepted batches. For the present example this is shown by the curve on the right-hand side of Figure 5.1 (iii). Similarly the distribution of rejected batches is obtained by plotting $p_1(1-P)$ against μ; this is shown on the left-hand side of Figure 5.1 (iii). The complete diagram shows how the incoming Population of batches is split up by the test into two portions. The areas under the distribution curves for accepted and rejected material show the proportions which fall into these two categories.

For the example considered in which 4 repeat tests are performed on each batch these areas are $0.664A$ and $0.336A$ respectively, where A is the area under the prior distribution curve. The mean of the distribution of accepted material is 90.772 and the mean of the distribution of rejected material is 89.661. The method of calculating these numbers will be described in § **5.26.**

Because the test is subject to error we do not get a perfect "cut-off" between batches with a content of less than 90 and batches with a content of more than 90. In fact the 66.4% of accepted batches is made up of 59.8% with a content greater than 90 and 6·6% with a content of less than 90, and the 33·6% of rejected batches is made up of 24·2% of batches with a content of less than 90 and 9·4% of batches with a content greater than 90. These percentages are obtained from the areas of the distributions lying on each side of the value $\mu = 90$.

Economics of the scheme

5.25 To assess the economic implications of the scheme we must consider the value of the accepted material and the cost of testing it. Only accepted batches have to be paid for, and their price is independent of the quality level μ. Suppose the utility of a batch of material with content μ is £100μ (e.g. the value of a consignment with content 90 is £9,000), and the cost of testing is £1 per test. Then, using 4 tests per batch, the average value of the accepted material is £90·772 × 100 = £9,077·2 per batch. Now, the cost of testing is £4 per batch, and since only 66·4% of batches are accepted the cost of testing per accepted batch is £4/0·664 = £6. Thus if we debit the cost of testing, the net value per accepted batch using this scheme is £9071·2. Proceeding in this way for various numbers of tests, Table 5.2 is constructed.

TABLE 5.2. ECONOMICS OF TESTING SCHEME

Number of tests n	0	1	2	3	4	5	6
Percentage accepted	100	62·3	64·6	65·7	66·4	66·9	67·2
Average value in £'s of accepted batches	9040·0	9070·4	9074·5	9076·2	9077·2	9077·8	9078·3
Cost of testing per accepted batch (£'s)	0	1·6	3·1	4·6	6·0	7·5	8·9
Net value in £'s	9040·0	9068·8	9071·4	9071·6	9071·2	9070·3	9069·4

We see that in this particular case there is a considerable monetary advantage (about £30 per batch) in testing the batches rather than in not testing them, and that there is a gain of almost £3 per batch by using three tests per batch rather than one, but no advantage is obtained by using more than three tests.

Practical calculation of an optimal scheme

5.26 The exact evaluation of all the quantities considered above could be carried out on an electronic computer. However, sufficient accuracy can usually be achieved by simple arithmetic on a desk calculator. We shall illustrate this by showing in the case $n = 4$ detailed calculations for the distribution of accepted material, the proportion of accepted batches, and the average value of accepted batches.

TABLE 5.3. QUANTITIES USED IN CALCULATING THE OPTIMAL SCHEME

(1)	(2)	(3)	(4) = (2)×(3)	(5) = 100×(1)
μ	p_1	P	p_1P	Q
87·5	0·0006	0·000	0·0000	8,750
87·7	0·0013	0·000	0·0000	8,770
87·9	0·0030	0·000	0·0000	8,790
88·1	0·0064	0·000	0·0000	8,810
88·3	0·0127	0·000	0·0000	8,830
88·5	0·0238	0·001	0·0000	8,850
88·7	0·0417	0·005	0·0002	8,870
88·9	0·0688	0·014	0·0010	8,890
89·1	0·1065	0·036	0·0038	8,910
89·3	0·1550	0·081	0·0126	8,930
89·5	0·2119	0·159	0·0337	8,950
89·7	0·2721	0·274	0·0746	8,970
89·9	0·3282	0·421	0·1382	8,990
90·1	0·3719	0·579	0·2153	9,010
90·3	0·3958	0·726	0·2874	9,030
90·5	0·3958	0·841	0·3329	9,050
90·7	0·3719	0·919	0·3418	9,070
90·9	0·3282	0·964	0·3164	9,090
91·1	0·2721	0·986	0·2683	9,110
91·3	0·2119	0·995	0·2108	9,130
91·5	0·1550	0·999	0·1548	9,150
91·7	0·1065	1·000	0·1065	9,170
91·9	0·0688	1·000	0·0688	9,190
92·1	0·0417	1·000	0·0417	9,210
92·3	0·0238	1·000	0·0238	9,230
92·5	0·0127	1·000	0·0127	9,250
92·7	0·0064	1·000	0·0064	9,270
92·9	0·0030	1·000	0·0030	9,290
93·1	0·0013	1·000	0·0013	9,310
93·3	0·0006	1·000	0·0006	9,330

$\Sigma p_1 = 3·9994$ $\Sigma p_1 P = 2·6566$ $\Sigma p_1 PQ = 24,114·5$

The quantities required are set out in Table 5.3. In Column 1 of this table the mean value for the batch is given from $\mu = 87·5$ to $\mu = 93·3$ at intervals of 0·2. This allows between 20 and 30 ordinates to be calculated over the range in which the distribution curves have appreciable frequencies. Col. 2 shows the ordinate of the prior distribution curve. In the present example this is simply the ordinate of a Normal curve with mean 90·4 and standard deviation 0·8, and is readily derived from Table 1 of [5.2]. As we have explained, an empirical curve could have been used equally well. Since we are concerned only with relative frequency it is not necessary to scale this curve so that the area beneath it is unity. Col. 3 shows the ordinate of the power curve for each value of μ. This is obtained in the manner already discussed in § **5.23**. The ordinates of the distribution of accepted batches are obtained in Col. 4 by multiplying together the entries in Cols. 2 and 3. Col. 5 shows the value Q in £'s per batch of material with content μ.

The proportion of batches accepted is given by:

$$\int p_1(\mu)P\mathrm{d}\mu \Big/ \int p_1(\mu)\mathrm{d}\mu$$

If we regard the values in Col. 1 as the mid-points of groups of width 0·2, we obtain the following approximation to the proportion of batches accepted:

$$\Sigma p_1 P / \Sigma p_1 = 2\cdot6566/3\cdot994 = 0\cdot6642$$

Similarly, the average value of accepted material is given approximately by:

$$\Sigma p_1 PQ / \Sigma p_1 P = 24{,}114\cdot5/2\cdot6566 = 9{,}077\cdot2$$

We have supposed in this example that the testing procedure was decided in advance and only the effect of changing n was to be studied. If the buyer were free to set the critical rejection level $\bar{x}_0$ above 90, it would obviously be to his advantage, since the average value of the accepted batches would then be higher. However the supplier would be reluctant to agree to such a change, because a larger proportion of his manufacture would then be rejected (assuming the same prior distribution). So in the example we assumed that the critical level of 90 had been set by mutual agreement between supplier and buyer, and only the amount of testing had to be decided.

5.27 In the example the utility is a simple linear function of μ. This will not always be so. Thus in the situation described it could have happened that a batch was worth a certain fixed amount if μ was greater than 90 and nothing otherwise. Another possibility would be that if μ was less than 90 the loss was proportional to $(90-\mu)$ except when μ fell below some limit μ_0, when the loss was much greater. This case might arise if material with $\mu < \mu_0$ could seriously upset the process.

Whatever the nature of the utility function, the method described above can be immediately applied. The average value of the accepted material can always be evaluated once the value of the material for each value of μ and the distribution of the accepted material are known.

The type of approach discussed is applicable to a number of parallel situations. It will be seen that the method not only allows the most economic scheme to be found, but in the course of the calculations all the characteristics of any proposed scheme are revealed. We can calculate, for example, what proportion of the material we reject will be rightly rejected and what proportion will be wrongly rejected. Knowledge of these values may be of considerable use in negotiations with the producer. It should not be forgotten that the introduction of an inspection scheme will probably have the effect of encouraging the producer to improve his material. He may, for example, introduce a testing scheme himself which reduces the lower tail of the incoming quality curve, or he may raise the mean level of his whole production. The prior distribution, therefore, may change, and the consumer should always be conscious of this possibility and of the advantages to be gained by changing his scheme if a revised calculation shows this to be desirable.

PRIOR DISTRIBUTION UNKNOWN

Example 5.2. Planning of plant trials

5.3 In principle the method discussed above for the sampling inspection scheme could be used in a variety of situations, some of which would not at first sight appear to be related to it. For example, suppose that an experiment on a plant was to be conducted to try out some suggested modification, and the problem was that of deciding how many trials should be done before finally accepting or rejecting the modification.

Suppose, to correspond to the previous example, that:
 (i) the normal plant efficiency is 90 units;
 (ii) the true level of process efficiency when the modification is used is μ;
 (iii) the result x of a single trial varies about μ with standard deviation $\sigma = 1$ unit;
 (iv) the capital value of an improvement to some value μ is £$100(\mu - 90)$;
 (v) the cost of an experiment is £1 per trial.

Arithmetically we then have the same problem as before, and in principle it could be solved in the same way.

If we try to apply this technique, however, we are immediately faced with the practical question "What should we take for the prior distribution?" In the sampling inspection example the prior distribution was an observable entity—the distribution of quality of incoming batches. In the present problem, however, it is much more nebulous. We might proceed by asking those most directly concerned with the process modification what increase in yield they thought was most likely (thus fixing the mode of the prior distribution) and what were the largest and smallest increases they thought possible (thus getting some idea of the range). Although such an approach might be thought to be too indefinite, it can in fact often be usefully applied. At least we know that this approach makes full use of expert opinion and that we have worked out the logical consequences of that opinion.

A modification to this approach which does not specifically involve the prior distribution and which has considerable practical appeal is due to Yates [5.3] and Grundy, Healy and Rees [5.4] and [5.5]. They use a two-stage procedure. A first sample of n_1 observations is taken and a calculation is made. This calculation tells us either:

(i) to accept or reject the modification without further test; or

(ii) to take n_2 further observations and then to accept the modification if the mean of all results shows an apparent positive effect (however small) and to reject if the mean of all the results shows an apparent negative effect.

The method can be arrived at by assuming that prior to the first sample nothing is known of the magnitude of the effect which might be found, and we therefore assume the same initial probability for all values of μ. The "effective" prior distribution for the second sample utilises the information gained from the first sample. We shall describe here the important case where the capital value of an improvement is given by $£k'(\mu - \mu_0)$, where μ is the mean value obtained using the modification and μ_0 is the mean value obtained using the normal process. Suppose:

(i) there are n_1 observations in the first sample with mean $\bar{x}_1$;

(ii) there are n_2 observations in the second sample with mean $\bar{x}_2$;

(iii) the mean of all the $n_1 + n_2$ observations is $\bar{x}$;

(iv) the standard deviation is known and equal to σ;

(v) the cost of experimentation is $£k$ per observation;

(vi) the capital value of an improvement from μ_0 to μ is $£k'(\mu - \mu_0)$.

It is shown [5.5] that the course of action to be adopted depends on a quantity:

$$\lambda = \frac{k'}{kn_1} \frac{\sigma}{\sqrt{n_1}}$$

It is seen that the ratio k'/k compares the value of the expected gain from the experiment with the cost of experimentation. We would expect, therefore, that when the value of λ is large, and much is therefore at stake, larger numbers of observations would be justified, and in fact this is found to be the case. The procedure is as follows. The value of λ appropriate to the particular experiment is first calculated. The quantity $u = \dfrac{\bar{x}_1 - \mu_0}{\sigma/\sqrt{n_1}}$ is then referred to Table 5.4 overleaf. If $|u|$ exceeds the critical value given there for the appropriate value of λ, the modification is accepted without further test if $\bar{x}_1 - \mu_0$ is positive, or rejected if $\bar{x}_1 - \mu_0$ is negative.

If $|u|$ is not as large as its critical value, further evidence is needed. A further sample of size $n_2 = cn_1$ should therefore be taken. The size of c depends on u as well as on λ and is given by the entries in Table 5.5.

5.31 To return to the example quoted at the beginning of § **5.3**; $\sigma = 1$, $k' = 100$ and $k = 1$. If one trial only was performed in the first sample $n_1 = 1$, and $\lambda = 100$. From Table 5.4, therefore, if the observed value was greater than $90 + 1·30$ we should accept without further testing, and if it was less than $90 - 1·30$ we should reject without further testing. If an intermediate value was observed we should consult Table 5.5. It will be seen

TABLE 5.4.   CRITICAL VALUE OF $|u|$

Value of λ	100	200	500	1,000	2,000
Critical value of $|u|$	1·30	1·52	1·81	2·02	2·22

that we should then have to take either 3 or 4 further observations and base our decisions on the sign of $\bar{x} - 90$ where $\bar{x}$ was the mean of the whole group.

The above example is somewhat unrealistic; the values of σ and λ are unusually small for a plant trial, and it is most unlikely that a first sample would contain only one observation. The following is a more typical example. Suppose that a 1% increase in yield would save £10,800 per annum

TABLE 5.5.   PROPORTIONATE SIZE OF SECOND SAMPLE

Value of $|u|$	Value of λ				
	100	200	500	1,000	2,000
	Value of $c = n_2/n_1$				
0	3·9	5·3	9·1	13·0	19·0
0·5	3·7	5·0	8·3	12·0	18·0
1·0	3·0	3·8	6·5	10·0	14·0
1·5	—	2·0	4·3	6·7	10·0
2·0	—	—	—	3·0	5·5

and such an improvement would be written off over a 5-year period, thus representing £54,000 in capital value per 1%. This makes $k' = 54,000$. Suppose that each experiment costs £20 to perform, i.e. $k = 20$, and further that $\sigma = 5$. A first sample of $n_1 = 9$ trials is performed, so that:

$$\lambda = \frac{54,000}{20 \times 9} \times \frac{5}{3} = 500$$

Suppose that an average improvement of 1% was actually observed in these trials. We then have $u = 0·6$, and from Table 5.4 we see that no immediate decision can be taken. Entering Table 5.5 with $u = 0.6$ and $\lambda = 500$, we find that c is about 8.0, and we should then recommend that $8.0 \times 9 = 72$ further observations be made. The modification would be judged on the sign of $\bar{x} - \mu_0$, where $\bar{x}$ is the mean for the entire group of 81 observations.

PRIOR DISTRIBUTION AND COSTS UNKNOWN

Many problems occur in which not only can no firm information be obtained about the prior distribution but the costs also are unknown or difficult to assess. These can be dealt with in the manner illustrated below.

Example 5.3. Planning laboratory experiments on dye receptivity

5.4 Suppose laboratory experiments were being conducted to study possible methods for increasing the dye receptivity of a synthetic fibre. It is reasonably certain that if such experiments were successful they would be of monetary value; frequently, however, so much would depend on unforeseeable factors that it might be extremely difficult to guess even the order of the quantities involved. Furthermore, the real cost of experimenting would involve not only direct expenses but also the cost of denying facilities to other projects which might yield even greater gain. In these circumstances the approach outlined above would become difficult or even impossible. We can, however, still make some attempt to choose the size of experiment intelligently by selecting the scheme that gives an acceptable power curve.

To use Figure 5.2 once more, let us suppose that the known standard deviation of observations on dye receptivity is $\sigma = 1$ unit, and that it is decided to use a procedure whereby n tests are performed and the method accepted if $\bar{x}$, the mean of the tests, is greater than $\bar{x}_0 = 90$ and rejected otherwise. Then the graphs marked $n = 1$, $n = 4$, $n = 16$ are the appropriate power curves when 1, 4 and 16 tests respectively are made. These curves show the "sensitivity" of each test procedure. As would be expected, the larger the number of observations, the steeper the power curve and the more sensitive the test.

The effect of changing the critical level $\bar{x}_0$ is to shift the power curve an equal amount along the horizontal axis; for example, if we decided to put the critical level $\bar{x}_0$ equal to 91 instead of to 90 the effect would be to shift the power curve along the horizontal axis by one unit. It will be seen, therefore, that by suitably adjusting the values of n and $\bar{x}_0$ the characteristics of the test as measured by the slope and position of the power curve may be chosen at will. The experimenter can therefore proceed by first deciding on the sort of power curve he wants and then finding values of n and $\bar{x}_0$ which approximately satisfy these requirements.

In the example we are considering here, the power curve is simply a cumulative Normal distribution. This curve is most conveniently plotted on Normal Probability Paper, the vertical axis of which scales the probability so that any cumulative Normal curve becomes a straight line instead of the usual S-shaped curve [§ 2.45]. In practice, therefore, the experimenter proceeds by drawing the straight line on Normal probability paper which he feels will best meet his requirements. If μ_P is used to denote the value of μ corresponding to the probability of acceptance P, the standard error specified for the mean of n tests is given by:

$$\sigma/\sqrt{n} = \tfrac{1}{2}(\mu_{0\cdot84} - \mu_{0\cdot16})$$

Hence the required values of $\bar{x}_0$ and n are

$$\bar{x}_0 = \mu_{0\cdot50} \qquad\qquad (5.1)$$

$$n = 4\sigma^2/(\mu_{0\cdot84} - \mu_{0\cdot16})^2 \qquad\qquad (5.11)$$

Numerical example

5.41 Suppose that the experimental error standard deviation σ is equal to 5 and that after some consideration the experimenter decides that a power curve like that in Figure 5.3 would be acceptable. Reading off some typical values it will be seen that, if a test is employed which has such a power curve, then when the true dye receptivity is as low as 88 there will only be a chance of about 0·3% of being led to believe that a real improvement has occurred. When the true dye receptivity is 89 this chance rises to about 2%; at 90 (when the modified method is of the same effectiveness as the standard method) it is 10%. At the value 91·75 the chances of accepting or rejecting the modification are equal. When the modified method gives a true dye receptivity of 94 the chance of accepting the modified method will be 95%, and this chance rises to 99% and 99·9% respectively for dye receptivities of 95 and 96.

To determine the value of $\bar{x}_0$ and n we read off the values $\mu_{0\cdot16} = 90\cdot40$, $\mu_{0\cdot50} = 91\cdot75$ and $\mu_{0\cdot84} = 93\cdot10$, whence:

$$\bar{x}_0 = 91\cdot75$$

$$n = 4(5/2\cdot7)^2 = 13\cdot7$$

A test with about the required characteristics is therefore provided by making 14 repeat tests. The modification should be accepted if the mean result exceeds 91·75 and rejected otherwise.

Relation to prior distribution and costs

5.42 In practice, if the power curve which the experimenter draws leads to a value of n which he feels is too large, he may decide to modify the requirements expressed by the power curve. In this way by trial and error he will eventually reach the compromise which he thinks is most reasonable. In this process of choice he will of course to some extent be subjectively taking into account such factors as the prior distribution of the amount of improvement to be expected, the possible monetary gains and losses, and the cost of experiments.

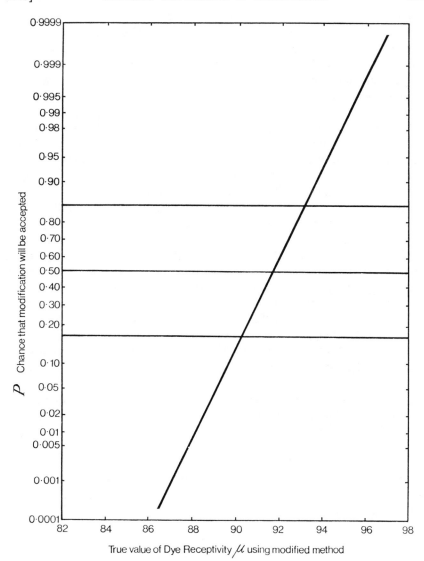

FIG. 5.3. POWER CURVE PLOTTED ON NORMAL PROBABILITY PAPER

The procedure from the point of view of hypothesis testing

5.43 An alternative—and more customary—approach to this problem is via significance tests [§ **4.5**].   In the above example, for instance, we are in effect performing an ordinary significance test, having first taken the precaution of ensuring that the sample size is large enough for important improvements nearly always to be detected.

To see that this is so it should first be noted that the power curve is fully defined as soon as we specify any two points on it. This means that although it is always wisest to plot the complete power curve, it is sufficient from the point of view of specifying the scheme for the experimenter to state the chance of accepting the proposed method for two specific values of μ. The values of $\bar{x}_0$ and n which produce a power curve passing through these two points can then be calculated.

In an example such as that now being considered it is customary to choose the two specific values of μ to be the "standard value" μ_0 and some larger value $\mu_1 = \mu_0 + \delta$ where δ is the difference that it is "important to detect". The hypothesis $\mu = \mu_0$ is then called the Null Hypothesis, i.e. the hypothesis that the modification has no effect, and the hypothesis $\mu = \mu_1 \, (= \mu_0 + \delta)$ is called an Alternative Hypothesis. The probabilities, i.e. the ordinates on the power curve associated with these two values of μ, are denoted by α and $1 - \beta$ respectively.

The probability α is the chance that we shall conclude that an improvement on the standard value has occurred when it has not, i.e. that we shall wrongly reject the Null Hypothesis. It is in fact the familiar "significance level" and is sometimes called the chance of an Error of the First Kind. The probability β is the chance that we shall conclude that no improvement has occurred when in fact a change as large as δ has occurred, i.e. the chance that we shall wrongly accept the Null Hypothesis when in fact an important difference has occurred. This value β is sometimes called the chance of an Error of the Second Kind.

To ensure that the power curve passes through the points (μ_0, α) and $(\mu_0 + \delta, 1 - \beta)$, the values of $\bar{x}_0$ and n must be such that:

$$\mu_0 = \bar{x}_0 - u_\alpha \sigma / \sqrt{n} \qquad (5.2)$$

and
$$\mu_0 + \delta = \bar{x}_0 + u_\beta \sigma / \sqrt{n} \qquad (5.21)$$

where σ is the standard deviation of a single experimental result, and u_α and u_β are the standard Normal deviates associated with the probabilities α and β. Solving these two equations, we thus require:

$$n = \{(u_\alpha + u_\beta)\sigma/\delta\}^2 \qquad (5.22)$$

and
$$\bar{x}_0 = \mu_0 + \delta u_\alpha/(u_\alpha + u_\beta) \qquad (5.23)$$

The quantity $\bar{x}_0 = \mu_0 + u_\alpha \sigma / \sqrt{n}$ in Equation (5.2) is the familiar value with which $\bar{x}$ must be compared in a simple test of significance at the α level of probability. Thus what we are doing is to make an ordinary test of significance at the probability level α, having chosen n large enough to ensure that there is only a small probability β of an improvement of size δ being overlooked.

Notice that for given α and β risks, the required value of n in (5.22) is inversely proportional to δ^2. Thus if we halve the size of improvement we are concerned about we shall have to make four times the number of test measurements.

Numerical example

5.44 Suppose as before that $\mu_0 = 90$, $\sigma = 5$, and that it has been decided that the probability of wrongly asserting that an improvement has occurred should be not more than $\alpha = 0.10$. Suppose further that an increase of $\delta = 4$ units would be of such importance that we want the probability of failing to detect it to be not more than $\beta = 0.05$. From Table A we find $u_\alpha = u_{0.10} = 1.28$ and $u_\beta = u_{0.05} = 1.64$. Thus we have, on substituting in Equations (*5.22*) and (*5.23*):

$$n = \{(1.28 + 1.64)1.25\}^2 = 3.65^2 = 13.3$$
$$\bar{x}_0 = 90 + 4 \times 1.28/2.92 = 91.75$$

It will be noted that, since the points (90, 0.10), (94, 0.95) are both on the power curve of Figure 5.3, this scheme and the previous one discussed in § **5.41** are in fact identical. The slight departure from the value of n found before arises from the slight uncertainty in reading the graph. Our scheme is therefore equivalent to testing whether the mean of a sample of 14 observations is significantly greater than 90; if it is, the modification is accepted; if not, the modification is rejected.

Single- and double-sided tests

5.45 In the above example we were concerned only with whether μ had or had not increased. A decrease in the mean was of no special interest, since the action required (that of rejecting the modification) would be the same whether a harmful negative effect or no effect whatever were produced by the modification. Such a test is called a Single-Sided test. In other circumstances we might be interested in a change in either direction. For example, the analysis of a chemical may have to lie within given limits, an excessive deviation in either direction being undesirable. This is called a Double-Sided test.

Example 5.4. Specification of an organic product

5.46 The specification for a particular organic product states that its true hydroxyl value μ should lie within the range $13.4 \pm 0.2\%$ by weight. As a result of many sets of repeat determinations the standard deviation of variations due to sampling and testing was known to be 0.085%. A procedure for testing the product was required which would ensure with high probability that off-grade material would not be despatched and that good material would not be rejected.

For the single-sided situation in § **5.43** the test procedure was specified by two values, n and $\bar{x}_0$. In the double-sided test, on the other hand, we clearly need three values, these being the number n of repeat tests to be performed on each consignment, and *two* critical values which we may denote by $\bar{x}_{0-}$ and $\bar{x}_{0+}$. So long as $\bar{x}$ lies between the values $\bar{x}_{0-}$ and $\bar{x}_{0+}$ we conclude that no important change has occurred from the value μ_0; when $\bar{x}$ exceeds $\bar{x}_{0+}$ we conclude that it is out of specification on the *high* side, while if it falls short of $\bar{x}_{0-}$ we conclude that it is out of specification on the *low* side.

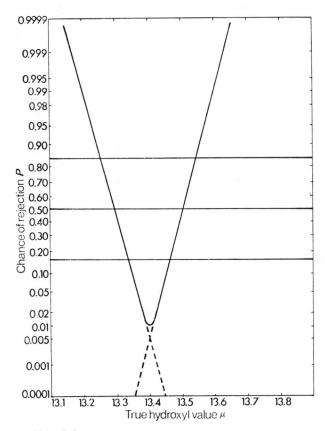

FIG. 5.4.   POWER CURVE FOR DOUBLE-SIDED TEST

The procedure we adopt is equivalent to applying two separate tests, one for an increase and one for a decrease. The power curve for the joint test has the appearance shown in Figure 5.4. The full line representing the joint power curve simply shows the sum of the two probabilities read from the two straight-line graphs. These straight-line graphs are the power curves for the separate tests and are shown by dotted lines where they are distinguishable from the full line. For the low values of μ almost the whole contribution to the probability of rejection comes from the chance that the product will not be accepted because the mean $\bar{x}$ has *too low* a value. This chance is represented by the straight line on the left of the diagram. Conversely for high values of μ almost the whole contribution to the probability of rejection arises from the chance that the product will not be accepted because the mean $\bar{x}$ of the results has too high a value. This chance is represented by the straight line on the right-hand side of the diagram. Only in the neighbourhood of the minimum value is the influence of both probabilities felt, and the value at the minimum corresponding to μ_0, where the straight lines cross, is in fact twice the value read from either straight-line graph. This corresponds to the fact that when $\mu = \mu_0$ the probability α that the consignment will be (wrongly) rejected is made up of two parts—a chance $\frac{1}{2}\alpha$ that it will be rejected because $\bar{x}$ exceeds $\bar{x}_{0+}$ and a chance $\frac{1}{2}\alpha$ that it will be rejected because $\bar{x}$ falls short of $\bar{x}_{0-}$.

To devise a suitable scheme, therefore, the experimenter should as before first draw the power curve that he feels will be suitable. This is done by drawing two straight lines in a V shape on Normal probability paper, equally inclined to the vertical and crossing at the value $\mu = \mu_0$. He does this bearing in mind the fact that the probability of rejection at $\mu = \mu_0$ will be twice the value shown at the intersection of the straight lines, and apart from the region around μ_0 the power curve will closely follow the straight lines. The values of n and $\bar{x}_0$ for a test corresponding to each separate straight line power curve are then determined. The value of n is of course the same for each test, and is the value to be adopted.

For example, suppose the experimenter decided that a power curve like that in Figure 5.4 would meet his requirements. To determine the appropriate test we note from the straight line on the right-hand side of the figure that $\mu_{0.16} = 13.465$, $\mu_{0.50} = 13.505$ and $\mu_{0.84} = 13.545$. Since we are told that $\sigma = 0.085$ we have from Formulae (*5.1*) and (*5.11*):

$$\bar{x}_{0+} = 13.505$$

$$n = 4(0.085/0.080)^2 = 2.125^2 = 4.5$$

Similarly from the left-hand line we have $\mu_{0.16} = 13.335$, $\mu_{0.50} = 13.295$ and $\mu_{0.84} = 13.255$, whence:

$$\bar{x}_{0-} = 13.295$$

$$n = 4(0.085/0.080)^2 = 4.5 \text{ (as before)}$$

The value $\bar{x}_{0-}$ could alternatively be obtained by symmetry.

Five analyses are therefore required. Since n has been rounded up to 5, the risks will be rather less than those specified.

Consider this example from the alternative viewpoint of the significance test. Three points are now needed to define the power curve, and these are taken to be $(\mu_0 - \delta, 1 - \beta)$, (μ_0, α) and $(\mu_0 + \delta, 1 - \beta)$. The value β is dependent on the risk we are prepared to run of failing to detect an important difference

$\pm \delta$ from the value μ_0. As before, α is the risk we are prepared to run of wrongly asserting that the Null Hypothesis $\mu = \mu_0$ is untrue. It is composed of a part $\frac{1}{2}\alpha$, which is the risk of rejecting the Null Hypothesis and saying that an *increase* in μ has occurred, and of a second part $\frac{1}{2}\alpha$ which is the risk of rejecting the Null Hypothesis and saying that a *decrease* in μ has occurred.

The necessary values of n and $\bar{x}_{0+}$ can then be calculated from Formulae (5.22) and (5.23) using $\frac{1}{2}\alpha$ in place of α:

$$n = \{(u_{\frac{1}{2}\alpha} + u_\beta)\sigma/\delta\}^2 \qquad (5.3)$$

$$\bar{x}_{0+} = \mu_0 + \delta u_{\frac{1}{2}\alpha}/(u_{\frac{1}{2}\alpha} + u_\beta) \qquad (5.31)$$

Also $\qquad \bar{x}_{0-} = \mu_0 - \delta u_{\frac{1}{2}\alpha}/(u_{\frac{1}{2}\alpha} + u_\beta) \qquad (5.32)$

The scheme we have given above would have been arrived at, for example, if it had been decided that when the true hydroxyl value μ_0 is 13·4 the risk of rejection should not exceed $\alpha = 0·01$, and when it deviates from 13·4 by more than $\delta = 0·2$ (i.e. if the product is outside specification), the risk of acceptance (and hence passing the batch for despatch) should not exceed $\beta = 0·01$.

Since $u_{\frac{1}{2}\alpha} = u_{0·005} = 2·58$ and $u_\beta = u_{0·01} = 2·33$, we have:

$$n = \{(2·58 + 2·33) \times 0·425\}^2 = (2·087)^2 = 4·4$$

$$\bar{x}_{0+} = 13·4 + 0·2 \times 2·58/4·91 = 13·505$$

$$\bar{x}_{0-} = 13·4 - 0·2 \times 2·58/4·91 = 13·295$$

Again, apart from a small difference in n due to uncertainty in reading the graph, the results are identical with those obtained before.

Comparison of two means

5.5 So far we have discussed only the situation where a single method giving a true mean value μ was to be evaluated. This was done by comparing the observed mean $\bar{x}$ with a critical level $\bar{x}_0$, or in the case of a double-sided situation with two critical levels $\bar{x}_{0-}$ and $\bar{x}_{0+}$. Frequently we have two methods to compare. Suppose μ_1 is the true mean for the first method and μ_2 the true mean for the second. Then, since the quantity of interest is the difference $\mu = \mu_2 - \mu_1$, the problem becomes of the same form as discussed before. In fact we have to consider the observed difference $\bar{x} = \bar{x}_2 - \bar{x}_1$ in relation to a critical value of the difference $\bar{x}_0$ (or in the case of a double-sided situation to two critical values $\bar{x}_{0-}$ and $\bar{x}_{0+}$). The experimenter proceeds as before by first plotting on Normal probability paper the power curve which he feels the test should have. From this the required test specification is readily derived.

Single-sided test situation

5.51 Consider the single-sided situation first. Here the question asked is of the kind "Is the mean value obtained from method 2 greater by an important amount than the mean obtained from method 1?" To resolve this question predetermined numbers of observations n_1 and n_2 would be made with the two methods, and the mean difference $\bar{x} = \bar{x}_2 - \bar{x}_1$ in the results would be compared with a critical value $\bar{x}_0$. If $\bar{x}$ was greater than $\bar{x}_0$ the conclusion would be drawn that method 2 was better, otherwise it would be concluded that no real difference of importance existed. The power curve would show the chance of reaching the conclusion that method 2 was better than method 1 for any value of μ, the true difference.

As before, the critical value $\bar{x}_0$ of the difference would be given by reading off the value $\mu_{0.50}$ from the chosen power curve. Also, since the standard error of the observed mean difference $\bar{x}_2 - \bar{x}_1$ is $(\sigma_1^2/n_1 + \sigma_2^2/n_2)^{\frac{1}{2}}$, the values of n_1 and n_2 necessary to give the chosen power curve would be such that:

$$(\sigma_1^2/n_1 + \sigma_2^2/n_2)^{\frac{1}{2}} = \tfrac{1}{2}(\mu_{0.84} - \mu_{0.16}) \qquad (5.4)$$

We have assumed that σ_1 and σ_2, the standard deviations of observations arising from the two methods, are known in advance. Clearly a variety of values of n_1 and n_2 can be found which satisfy (5.4). It is readily shown that the smallest total $(n_1 + n_2)$ is obtained by arranging that $n_1/n_2 = \sigma_1/\sigma_2$. In this case:

$$n_1 = 4\sigma_1(\sigma_1 + \sigma_2)/(\mu_{0.84} - \mu_{0.16})^2 \qquad (5.41)$$
$$n_2 = 4\sigma_2(\sigma_1 + \sigma_2)/(\mu_{0.84} - \mu_{0.16})^2 \qquad (5.42)$$

Often σ_1 and σ_2 would be expected to be equal; if their common value is denoted by σ the number of observations is then given by:

$$n_1 = n_2 = 8\sigma^2/(\mu_{0.84} - \mu_{0.16})^2 \qquad (5.43)$$

Note that according to this last formula, the number needed in each group is twice that required to compare a single mean with a standard value [§ **5.4**].

Double-sided test situation

5.52 When the question is: "Is there evidence that method 2 *differs* from method 1 by an important amount?" we have the double-sided situation where we are equally interested in a negative and a positive difference μ. We deal with this situation exactly as before. A V-shaped power curve built up from two straight lines equally inclined to the vertical is drawn on Normal probability paper. The critical values $\bar{x}_{0-}$ and $\bar{x}_{0+}$ may be read off immediately as the points on the component straight lines corresponding to the 0·50 probability ordinate. Appropriate values for n_1 and n_2 are obtained as before, using Equation (5.4) and substituting values of $\mu_{0.84}$ and $\mu_{0.16}$ from either of the straight-line components of the V-shaped power curve.

Procedure from the viewpoint of significance testing

5.53 Consideration from the viewpoint of significance testing closely follows that discussed above. In this approach the test procedure is chosen to give a power curve passing through the points $(0, \alpha)$ and $(\delta, 1-\beta)$ in the single-sided test situation, or $(-\delta, 1-\beta)$, $(0, \alpha)$ and $(\delta, 1-\beta)$ in the double-sided test situation. This is done as usual by choosing first the difference δ which it is important to detect, and then assigning suitable values to α and β.

In the single-sided situation the appropriate formulae from which n_1, n_2 and $\bar{x}_0$ can be calculated are:

$$\sigma_1^2/n_1 + \sigma_2^2/n_2 = \delta^2/(u_\alpha + u_\beta)^2 \qquad (5.5)$$

$$\bar{x}_0 = \delta u_\alpha/(u_\alpha + u_\beta) \qquad (5.51)$$

For the double-sided situation they are:

$$\sigma_1^2/n_1 + \sigma_2^2/n_2 = \delta^2/(u_{\frac{1}{2}\alpha} + u_\beta)^2 \qquad (5.52)$$

$$\bar{x}_0 = \pm \delta u_{\frac{1}{2}\alpha}/(u_{\frac{1}{2}\alpha} + u_\beta) \qquad (5.53)$$

As before, important special cases of the formulae occur when n_1/n_2 is put equal to σ_1/σ_2 and when $\sigma_1 = \sigma_2$.

Standard deviation not precisely known

5.6 It is obvious that if nothing is known of the standard deviation we can make no statement whatever about the number of observations needed. Conversely when σ, or in the case of two groups σ_1 and σ_2, are precisely known the number of observations to provide a test with a given power curve can be calculated exactly. We have supposed that values of the standard deviation postulated in advance were entirely reliable and were used not only in planning the experiment but also in making the test itself. In practice the information concerning σ prior to the experiment is seldom exact and in some cases amounts to little more than a guess. For this reason, having obtained the number of observations by the method described, the experimenter may choose to make the actual significance test using the estimates of the standard deviation obtained from the results themselves. To test the Null Hypothesis it will then be appropriate to use the t-test rather than the Normal Curve test. For example, if this were done in comparing a mean $\bar{x}$ with the standard value μ_0 the actual critical value used would not be $\bar{x}_0 = \mu_0 + u_\alpha\sigma/\sqrt{n}$ as in Equation (5.2) but $\bar{x}_0 = \mu_0 + t_\alpha s/\sqrt{n}$, where t_α is the deviate of the t-distribution corresponding to the probability α as given by Table C. By proceeding in this way the experimenter ensures that the risk is precisely α that the Null Hypothesis will be wrongly rejected and therefore ensures that the ordinate of the power curve is correct at the point $\mu = \mu_0$. Whether or not the true power curve is close to that desired at other values of μ will entirely depend on the accuracy of the initial estimate of σ. When it is clear from an unexpectedly large size of the estimate s that σ has been underestimated a further sample should be taken using the estimate s to decide its magnitude. The two sets of observations should then be combined in making the final test. This use of the ordinary significance tables in making a two-stage test is not strictly valid but provides a fair approximation in most cases.

Choosing the number of observations in comparing standard deviations

5.7 Not infrequently the quantity of interest is not the mean but the variability or spread associated with a given measurement or method. For example, we may wish to compare the precisions of two analytical methods, or the variabilities in strength of yarn produced by two different methods.

When the original observations are approximately Normally distributed an appropriate criterion for the comparison of variability is the variance ratio $F = s_1^2/s_2^2$, or equivalently the ratio of the standard deviations $L = s_1/s_2 = \sqrt{F}$, or again equivalently the logarithm of this latter ratio $z = \ln(s_1/s_2) = \frac{1}{2}\ln F$, where ln denotes the Napierian logarithm. Since these three criteria are all functions of s_1/s_2, exactly the same results will be obtained whichever is used. However, it is most convenient to use the quantity z for the purpose of deciding the number of observations because, unlike the other criteria, z is approximately Normally distributed with variance independent of the ratio σ_1/σ_2 of the true values. This approximation is satisfactory provided the numbers of degrees of freedom ϕ_1 and ϕ_2 on which the estimates s_1 and s_2 are based are not too small, say each not less than 10. Since also the mean value of z is linearly related to $\ln(\sigma_1/\sigma_2)$, the power curve for the variance test can be approximately represented by a straight line on Logarithmic Probability Paper as shown in Figure 5.5 overleaf. Using this fact we can apply the same method as before to determine the appropriate numbers of observations.

Example 5.5. Comparison of analytical methods

5.71 Suppose for example that two methods of analysis are to be compared. Suppose method 1 is simpler to perform than method 2, and consequently it is felt that method 1 should be chosen unless it can be demonstrated that it is considerably less accurate than method 2. The subjective feelings contained in this statement may be expressed by the experimenter in quantitative form by drawing a line on logarithmic probability paper like that shown in Figure 5.5. This incorporates the experimenter's ideas on what he feels the chances of accepting method 2 should be for various values of the ratio $\rho = \sigma_1/\sigma_2$.

From such a graph a suitable scheme can be determined in much the same way as before. The variance of z can be shown to be:

$$\sigma_z^2 = \tfrac{1}{2}\{1/(\phi_1-1)+1/(\phi_2-1)\}$$

Also the required variance of z implied by the choice of power curve is:

$$\sigma_z^2 = \tfrac{1}{4}(\ln\rho_{0.84}-\ln\rho_{0.16})^2$$

where $\rho_{0.84}$ and $\rho_{0.16}$ are the values of ρ corresponding to the probabilities 0·84 and 0·16. By equating these two expressions we obtain the following equation, from which appropriate values of ϕ_1 and ϕ_2 can be obtained:

$$\tfrac{1}{2}\{1/(\phi_1-1)+1/(\phi_2-1)\} = \tfrac{1}{4}\{\ln(\rho_{0.84}/\rho_{0.16})\}^2 \qquad (5.6)$$

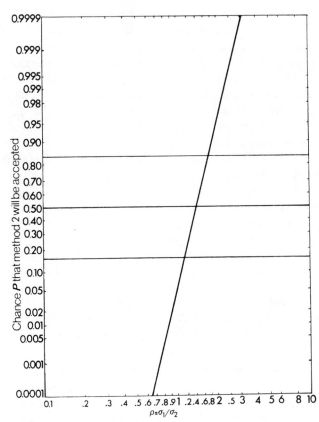

Fɪɢ. 5.5. POWER CURVE FOR COMPARISON OF
STANDARD DEVIATIONS

and the critical value for the observed z is:

$$z_0 = \ln \rho_{0.50} \qquad (5.61)$$

The smallest total number of observations is obtained when $\phi_1 = \phi_2$. Denoting this common value by ϕ, it follows from Equation (5.6) that:

$$\phi = 1 + 4/\{\ln (\rho_{0.84}/\rho_{0.16})\}^2 \qquad (5.62)$$

For this specific example $\rho_{0.84} = 1.73$, $\rho_{0.50} = 1.40$, $\rho_{0.16} = 1.14$, whence $\ln (\rho_{0.84}/\rho_{0.16}) = 0.42$. Thus we need to make sufficient observations to obtain variance estimates each having ϕ degrees of freedom where:

$$\phi = 1 + 4/0.42^2 = 23.7$$

The required number n of observations in each group is therefore 24.7, or 25 to the nearest whole number, and the critical value for making the test is:

$$z_0 = \ln 1.40 = 0.34$$

Thus 25 observations should be made with each method of analysis. The quantity $z = \ln(s_1/s_2)$ should then be calculated and method 2 accepted if the value of z exceeds 0·34, otherwise rejected. Equivalently we may refer the ratio s_1^2/s_2^2 to tables of the F-distribution.

The procedure for the double-sided situation simply involves the combination of two single-sided test procedures. When a single estimate s_1 is to be compared with a "precisely known" value σ_2, the approximate value of ϕ_1 is obtained simply by putting $\phi_2 = \infty$ in Equation (5.6).

Procedure from the viewpoint of significance testing

5.72 In this approach the test procedure is chosen to give a power curve passing through the points $(1, \alpha)$ and $(R, 1-\beta)$, where R is a ratio of the standard deviations which is important in the sense that we wish to run only a small risk β of failing to detect it.

Appropriate values of ϕ_1 and ϕ_2 are then related by:

$$\tfrac{1}{2}\{1/(\phi_1-1)+1/(\phi_2-1)\} = \{(\ln R)/(u_\alpha+u_\beta)\}^2 \qquad (5.7)$$

and the critical value of z is:

$$z_0 = u_\alpha\{1/(\phi_1-1)+1/(\phi_2-1)\}^{\frac{1}{2}}/\sqrt{2} \qquad (5.71)$$

If it is decided to make the groups of equal size, the common value ϕ is given by:

$$\phi = 1 + \{(u_\alpha+u_\beta)/\ln R\}^2$$

and

$$z_0 = u_\alpha/(\phi-1)^{\frac{1}{2}}$$

For the double-sided situation we require the power curve to pass through the three points $(1/R, 1-\beta)$, $(1, \alpha)$ and $(R, 1-\beta)$ and the appropriate formulae are obtained by replacing α by $\tfrac{1}{2}\alpha$ in Equations (5.7) and (5.71).

To illustrate the use of these formulae, suppose that in the previous example the experimenter had decided that if the standard deviations σ_1 and σ_2 were equal (that is, ρ was unity) he would wish to run only a small risk $\alpha = 0·05$ of choosing method 2. If on the other hand σ_1 was twice as large as σ_2, so that $\rho = 2$, he would wish to take a risk of only $\beta = 0·05$ of not choosing method 2. Substituting these values, we have:

$$\phi = 1 + (3·29/0·693)^2 = 1+4·75^2$$
$$= 23·6$$

whence

$$n = 24·6$$

The conclusion is as before that we should include 25 observations in each group. This is to be expected, since the power curve of Figure 5.5 passes through the points $(1, 0·05)$ and $(2, 0·95)$.

The value z_0 is 0·346, and the final comparison of z with z_0 is of course equivalent to a test of significance at the level α. It can be carried out without approximation either by employing the exact z tables as given, for example, in [5.6] or by using either of the equivalent tests based on $F = s_1^2/s_2^2$ or $L = s_1/s_2$.

Sequential testing

5.8 The methods described in §§ **5.4–5.7** indicate the basis on which the experimenter can select the number of observations needed to make the comparison decisive, on the assumption that this number must be fixed before the experiment is performed. It will often be found on a critical examination of an experiment planned on these lines that either:

(i) a statistically significant difference would have been reached with appreciably fewer observations; or

(ii) before all the observations had been obtained the difference was so small that no reasonably likely values in the remaining observations could have resulted in the significance level being reached.

These outcomes are to be expected, since the values of the risks α and β have been chosen to exclude the less likely contingencies.

However, there are many types of chemical and physical research in which the observations are obtained one after another. The time needed to obtain test results, for example chemical analyses, may be short compared with the length of a single experimental run. In these cases it is not necessary to fix the total size of the experiment in advance. Instead, after each observation is made, a simple statistical test is applied to determine whether the results obtained so far indicate a definite conclusion from the experiment, or whether more observations are needed to make it decisive. Thus the experiment ends as soon as a definite conclusion can be drawn, and the average number of observations required in experiments carried out in this manner tends to be smaller than when the number has to be predetermined.

These procedures are known as Sequential Methods, and they are fully described in Chapter 3 of *Design and Analysis*. A reduction in the number of experimental runs is especially important when the observations are expensive or time-consuming. In particular, if the true difference is distinctly greater than the quantity δ, which it is regarded as essential to detect if it exists, then we can establish this quickly.

References

[5.1] BRUNT, D. *The Combination of Observations*. Cambridge University Press (1931).

[5.2] PEARSON, E. S., and HARTLEY, H. O. *Biometrika Tables for Statisticians*, Vol. 1 (third edition). Charles Griffin (High Wycombe, 1976).

[5.3] YATES, F. "Principles Governing the Amount of Experimentation in Developmental Work", *Nature*, **170**, 138–140 (1952).

[5.4] GRUNDY, P. M., HEALY, M. J. R., and REES, D. H. "Decision between Two Alternatives—How Many Experiments?" *Biometrics*, **10**, 317–323 (1954).

[5.5] GRUNDY, P. M., HEALY, M. J. R., and REES, D. H. "Economic Choice of Amount of Experimentation", *J. R. Statist. Soc.*, B, **18**, 32–55 (1956).

[5.6] FISHER, R. A., and YATES, F. *Statistical Tables for Biological, Agricultural and Medical Research* (sixth edition). Longman (London and New York, 1974).

[5.7] BARNARD, G. A. "Sampling Inspection and Statistical Decisions", *J. R. Statist. Soc.*, B, **16**, 151–174 (1954).

Chapter 6

Analysis of Variance

In this chapter the Analysis of Variance is described.
This is a technique by which the variations associated
with defined sources may be isolated and estimated.
For example, a group of chemical test results from
different samples of the same material may contain
differences associated with sampling as well as differences associated with the test itself. The Analysis of
Variance technique allows the sampling variation and
the testing variation to be separated and their
magnitudes estimated.

Introduction

6.1 In this chapter we consider those situations in which a result or an
observation is subject to a number of sources of variation, and the problem is
to separate and estimate these sources of variation. For instance, to assess
the average purity of a chemical product in bulk, a chemist will carry out
analyses on one or more samples taken from the bulk. The method of
analysis may not be perfectly reproducible, and so repeat analyses will vary;
the material, also, may not be perfectly homogeneous, and so repeat samples
will differ among themselves. If in such a case one analysis on one sample is
used as an estimate of the average purity of the bulk, this estimate will be
subject to two types of errors, those arising in the analysis and those arising
in the sampling.

Similar problems may arise in all types of processes—testing processes or
manufacturing processes—which involve more than one stage. For instance,
the test for the effectiveness of a waterproofing treatment involves three
distinct stages, each of which contributes to the overall error of the test.
These sources of error are:

(i) Variation in the fabric which has to be treated with the waterproofing
agent.
(ii) Variation in applying the treatment.
(iii) Error in the assessment of the water repellency.

In manufacturing processes there are usually three types of variation
which contribute to the variation in yield or quality of the final product.

These are:

(i) Variation in the quality of one or more of the raw materials.
(ii) Variation in operating conditions of the process at one or more stages.
(iii) Error arising in sampling and testing the finished product or in estimating the yield.

Additive property of variances

6.11 When two or more independent sources of variation operate, the resulting variance is the sum of the separate variances [§ 3.71]. The two types of errors which arise when estimating the property of a bulk chemical are:

(i) Errors of sampling, with variance denoted by σ_1^2.
(ii) Errors of analysis, with variance denoted by σ_0^2

These sources of error operate independently, and the total variation may be obtained by simple addition of the two. This means that, when the result of one analysis on one random sample is used as an estimate of the quality of the bulk, this estimate will have an error variance of:

$$\sigma^2 = \sigma_1^2 + \sigma_0^2 \tag{6.1}$$

To determine the variances exactly would require an infinite number of observations; in practice the variances can only be estimated from a finite number of observations, and it is these estimates which have to be used in Formula (*6.1*) to derive an estimate of the combined variance.

Estimates of variances are denoted by the symbol s^2 with the appropriate suffix. The estimate of the variance for one analysis on one sample is then:

$$s^2 = s_1^2 + s_0^2 \tag{6.11}$$

When these variances are based on only a few measurements they are not very precise. Errors of estimation can be considerable for small samples.

When n analyses are carried out on the sample and the results averaged, the variance from this source is reduced to σ_0^2/n, and the variance of the mean result when used as an estimate of the average value of the bulk is:

$$\sigma_1^2 + \sigma_0^2/n$$

When m samples are taken from the bulk and n analyses carried out on each, the variance of the mean is:

$$(\sigma_1^2 + \sigma_0^2/n)/m = \sigma_1^2/m + \sigma_0^2/nm$$

It will be noted that the divisor of σ_1^2 is the total number of samples, and the divisor of σ_0^2 is the total number of analyses. To use these formulae in practice we have to substitute estimated values for the σ's. When different numbers of analyses are made on the samples the result is more complicated, and the appropriate treatment is considered in Appendix 6A. It is worth while mentioning here that the precision of a result from any sampling and testing scheme can be assessed and the best scheme derived. This also is considered in Appendix 6A and in § **6.54**.

Types of classification

6.2 The Analysis of Variance is essentially a method of separating the total variance of a response into its various components, corresponding to the sources of variation which can be identified. The data must clearly contain information on any given source of variation before its contribution can be estimated, and as a rule the components are best estimated from experiments which have been designed for this purpose. The procedure to be used in the application of the Analysis of Variance will depend on the number and nature of the independent causes of variation which can be identified. It is always possible to classify the data with respect to each such source of variation, and a complete classification is a necessary first step in the analysis.

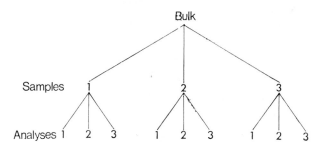

FIG. 6.1.   HIERARCHIC CLASSIFICATION

There are two basic types of classification with which we shall deal in this book. One type has already been illustrated in the sampling and testing example discussed in § **6.11**, in which a bulk chemical is sampled a number of times and repeat analyses are carried out on each. This scheme can be represented by a series of branching lines as in Figure 6.1. This is usually called the Hierarchic Classification, but some statisticians use the alternative term, Nested Classification.

Note in this example that Analyses 1, 2 and 3 of Sample 1 are connected by the fact that they are from the same sample, but there is *no* connection between Analysis 1 of Samples 1, 2 and 3. This relationship is characteristic of a hierarchic classification. The pattern is readily extended to three or more sources of variation.

Hierarchic classifications will be discussed in detail in §§ **6.3–6.5**.

6.21 The second basic arrangement, called the Cross-Classification, is one in which each observation can be classified independently with respect to all sources of variation. For example, in comparing three analytical methods for assessing the amount of a particular impurity in a product, samples from the same material were analysed by the three methods in five laboratories. There are here two sources of variation—laboratory and method of analysis—and these can be displayed by means of a two-way table. It is clear that the laboratory and the analytical method enter the Analysis of Variance on an equal footing and that, unlike a hierarchic classification, no nesting of sources is involved.

The results obtained from each laboratory can be entered in the appropriate cells of Table 6.1; they are then classified with respect to both methods and laboratories. Classification can still be carried out when some laboratories do not use all three methods; in such a case some cells in the table will be blank. This type of classification can also be extended to three or

TABLE 6.1.   TWO-WAY CROSS-CLASSIFICATION

Method of Analysis	Laboratory				
	1	2	3	4	5
1					
2					
3					

more sources of variation. For example, if more than one sample or material is analysed, a table such as the above can be constructed for each sample and a three-way table results. The data are then classified with respect to three sources of variation.

Cross-classifications will be discussed in detail in § 6.6.

6.22 More complicated classifications exist which are combinations of these two basic types, for example a two-way table in which some cells contain two or more observations.

The most satisfactory way to acquire a working knowledge and understanding of the methods used in Analysis of Variance is to work out fully a number of typical examples. The treatment in this chapter is of limited scope, intended mainly to illustrate principles and basic methods of analysis; the examples, however, are of the kind most frequently encountered in the chemical industry. Two of these are of the hierarchic type, the first with one criterion of classification and the second with two. The third is the simplest form of a cross-classification, and the final example is also a cross-classification but with several results in each cell; it is the simplest form combining both methods of classification. Further examples involving a larger number of sources of variation are treated in *Design and Analysis*.

Nature of variation

6.23 Irrespective of the data classification, there are two kinds of problems to which the Analysis of Variance is applied, and these are distinguished by the nature of the separate sources of variation. In one, the individual items examined can be regarded as a random sample from an infinite Population of such items, and the purpose of the analysis is to estimate the variance of the Population. The first illustration given above is of this type, since the repeat analytical tests on each sample of material can be regarded as random elements from a very large Population of analytical tests whose variation is completely specified by the variance σ_0^2. Similarly, the repeat samples from the bulk can be regarded as random elements from a Population of such samples whose variation is completely specified by the variance σ_1^2. The purpose of the experiment is to estimate σ_0^2 and σ_1^2.*

In the other kind of problem we are interested in specific comparisons between the items tested. For instance, suppose it is necessary to compare the purity of batches of a chemical product manufactured by three different processes; one batch is available from each route and it is sampled and analysed several times. Then the "between batches" source of variation, which may be assumed to be due mainly to changes in the method of manufacture, is of a different kind from that due to sampling and analytical errors. Even if a Population of batches can be postulated it is of no interest, because we only want to compare the three specific batches with each other; that is, we require to compare means and not to estimate variances. This is simply an extension of the problem of comparing two group means which was considered in Chapter 4 and it will be discussed in detail in § **6.7**.†

The two problems dealt with by the Analysis of Variance are referred to respectively as Estimation of Variances and Comparison of Mean Values. The arithmetical procedures are similar for the two cases; they differ only in the final stage and in the interpretation.

SEPARATION AND ESTIMATION OF VARIANCES

Analysis of Variance of hierarchic data

6.3 The simplest form of hierarchic data involves only two sources of variation, and an illustration has already been given [§ **6.11**] in which several

---

* When the Population is of infinite size, this case is described in some statistical literature as a Model II situation.

† When the Population is restricted to the groups which have been examined, this case is described in some statistical literature as a Model I situation.

samples are taken from a bulk suspected to be variable, and a number of analytical tests carried out on each sample. The purpose of the analysis is to separate and estimate the variances due to testing and to sampling. Only the general principles involved in the derivation of the methods of analysis will be dealt with here; the full step-by-step details of the recommended methods are given in Appendix 6B.

Estimation of analytical error variance

6.31 Suppose there are k samples and n repeat analyses on each, giving a total number of analyses $N = kn$. The analytical error is responsible for the variation in the repeat analyses on each sample; denote its variance by σ_0^2. The usual method for estimating the size of σ_0^2 is as follows:

Denote the observations on the ith sample by $(x_{i1}, x_{i2}, ..., x_{in})$ with mean $\bar{x}_i$. The sum of squares about the mean is $\sum_{j=1}^{n} (x_{ij} - \bar{x}_i)^2$, which has $n - 1$ degrees of freedom. The estimate of the variance of analytical error based on this sample is then $\sum_{j=1}^{n} (x_{ij} - \bar{x}_i)^2/(n-1)$. A similar expression can be obtained for each sample, and on the reasonable assumption that the analytical error does not vary from sample to sample, the variances may be combined as shown in § 3.342. Accordingly σ_0^2 is estimated by:

$$\frac{\text{Total of the sums of squares about the sample means}}{\text{Total of the degrees of freedom}}$$

$$= \sum_{i=1}^{k} \sum_{j=1}^{n} (x_{ij} - \bar{x}_i)^2/k(n-1) \qquad (6.2)$$

The numerator is referred to as the sum of squares within samples.

An alternative expression for the numerator which is more convenient for computation is:

$$\sum_{i=1}^{k} \sum_{j=1}^{n} x_{ij}^2 - \sum_{i=1}^{k} \left(\sum_{j=1}^{n} x_{ij} \right)^2 \Big/ n \qquad (6.21)$$

The first term is the sum of squares of each of the nk observations and the second is the sum of the squares of each sample total divided by n. The arithmetical computation can be further simplified by the use of the devices described in § 3.22. In particular, the use of a "working origin" avoids Expression (6.21) becoming a relatively small difference between two large numbers; the subtracted constant does not have to be restored at the end of the calculation when we are only concerned with a variance.

In combining the sums of squares within each sample to obtain an estimate of the experimental error, the tacit assumption is made that the system of chance causes resulting in the experimental error is the same for all the samples. There is usually no reason for supposing that this is not the case in sampling and testing problems. Even if there is reason for expecting the variances to differ for the various groups, the estimate for σ_0^2 will still be an estimate of the average experimental error variance.

Such situations could arise when different analysts or different methods of analysis are used. It may then be required to compare the variances for the different analysts or methods; this is a separate problem which has already been considered for two sets of results in §§ **4.31** and **4.8**. An extension of the method for comparing the variances from more than two sets is described in Appendix 6E.

Estimation of the sampling error variance

6.32 Let the means of the samples be $\bar{x}_1, \bar{x}_2, \ldots, \bar{x}_k$ and the grand mean $\bar{x}$. The variance between the sample means is simply:

$$\sum_{i=1}^{k} (\bar{x}_i - \bar{x})^2/(k-1) \tag{6.3}$$

Denoting the variance due to sampling error by σ_1^2, the variance of the mean of n tests on any one sample is:

$$\sigma_1^2 + \sigma_0^2/n \tag{6.4}$$

Expression (6.3) is therefore an estimate of (6.4). If we multiply (6.3) and (6.4) by n we obtain:

$$n \sum_{i=1}^{k} (\bar{x}_i - \bar{x})^2/(k-1) \rightarrow n\sigma_1^2 + \sigma_0^2 \tag{6.5}$$

where the arrow denotes "is an estimate of".

The expression $n \sum_{i=1}^{k} (\bar{x}_i - \bar{x})^2$ is referred to as "the sum of squares between samples". An alternative expression for this sum of squares which is more suitable for computing is:

$$\sum_{i=1}^{k} \left(\sum_{j=1}^{n} x_{ij} \right)^2 \bigg/ n - \left(\sum_{i=1}^{k} \sum_{j=1}^{n} x_{ij} \right)^2 \bigg/ nk \tag{6.51}$$

Previously, we obtained in Equation (6.2):

$$\sum_{i=1}^{k} \sum_{j=1}^{n} (x_{ij} - \bar{x}_i)^2/k(n-1) \rightarrow \sigma_0^2 \tag{6.6}$$

An estimate of σ_1^2 can therefore be obtained by solving (6.5) and (6.6). This will be dealt with more fully in § **6.35**.

6.33 Another expression required is "the total sum of squares", which is the sum of squares of all the nk observations about the grand mean $\bar{x}$; it has $(nk-1)$ degrees of freedom:

$$\text{Total sum of squares} = \sum_{i=1}^{k} \sum_{j=1}^{n} (x_{ij} - \bar{x})^2 \tag{6.7}$$

An alternative expression more suitable for computing is:

$$\sum_{i=1}^{k} \sum_{j=1}^{n} x_{ij}^{2} - \left(\sum_{i=1}^{k} \sum_{j=1}^{n} x_{ij} \right)^{2} \Big/ nk \qquad (6.71)$$

In the Expressions (6.21), (6.51) and (6.71) for the sums of squares, the first terms are referred to as "the crude sum of squares" and the second terms as "the correction for the mean". It is seen that the divisor in each of the second terms is the number of observations making up the total which is squared. This is true in all hierarchic data, whether the number of observations in each group is the same or not.

The step-by-step details of the method of analysis most suitable for numerical calculation are given in Appendix 6B.1.

Analysis of Variance table

6.34 The sums of squares and degrees of freedom "between samples", "within samples" and "total" may be set out in tabular form, called the Analysis of Variance Table, as in Table 6.2.

TABLE 6.2.   ANALYSIS OF VARIANCE TABLE

Source of Variation	Sum of Squares	Degrees of Freedom	Mean Square	Quantity estimated by Mean Square
Between Samples	$n \sum_{i=1}^{k} (\bar{x}_i - \bar{x})^2 = S_1$	$k-1$ $(= \phi_1)$	$S_1/(k-1)$ $(= M_1)$	$\sigma_0^2 + n\sigma_1^2$
Within Samples	$\sum_{i=1}^{k} \sum_{j=1}^{n} (x_{ij} - \bar{x}_i)^2 = S_0$	$k(n-1)$ $(= \phi_0)$	$S_0/k(n-1)$ $(= M_0)$	σ_0^2
Total	$\sum_{i=1}^{k} \sum_{j=1}^{n} (x_{ij} - \bar{x})^2$	$nk-1$		

All three sums of squares and degrees of freedom have been derived independently. It is seen that the degrees of freedom for "Total" is the sum of the other two. When the alternative Expressions (6.21), (6.51) and (6.71) are substituted for the sums of squares, it is also seen that the sum of squares for "Total" is the sum of the other two sums of squares. This is an important result: it is an expression of the additive property of the sum of squares and of the degrees of freedom, and is of considerable importance in the Analysis of Variance technique, as will be seen in the numerical examples which follow. This additive property arises from the particular arrangement of the experiment, and cases exist where it does not hold [*Design and Analysis,*

Chapters 6 and 11]. However, in all the examples dealt with in this chapter the sums of squares are additive.

The quotients obtained when each sum of squares is divided by its degrees of freedom are known as Mean Squares, and they are entered in the fourth column of the Analysis of Variance Table. Arranging the computation in such a table makes the additive property of the sums of squares clear and also enables a useful comparison to be made between the mean squares. The implications of this will be considered more fully in a later section.

Estimates of variances and their confidence limits

6.35 The main purpose of the Analysis of Variance is to estimate σ_0 and σ_1. It is clear from the Analysis of Variance table how these estimates may be derived. Denote by M_1 the mean square between samples, and by M_0 the mean square within samples. Then:

$$M_0 \rightarrow \sigma_0^2$$
$$(M_1 - M_0)/n \rightarrow \sigma_1^2$$

or in terms of the standard deviations:

$$\sqrt{M_0} \rightarrow \sigma_0$$
$$\sqrt{\{(M_1 - M_0)/n\}} \rightarrow \sigma_1$$

One of the most important functions of statistical methods is to provide estimates and their confidence limits. Estimates are of little value unless we can attach at least approximate confidence limits to them. The confidence limits for the estimate of σ_0 can readily be derived from Table H at the end of the volume by entering with $k(n-1)$ degrees of freedom [§ **4.33**]. If the multipliers for probability α are denoted by $L_1(\alpha)$ and $L_2(\alpha)$, then the $100(1-2\alpha)\%$ confidence limits are:

$$L_1(\alpha)\sqrt{M_0} \text{ and } L_2(\alpha)\sqrt{M_0}$$

The estimate of σ_1^2 is derived from a difference between two mean squares, and its exact confidence limits are not easy to calculate. However, exact confidence limits can be calculated for the ratio σ_1^2/σ_0^2, and multiplying these limits by the estimate of σ_0^2 will give approximate confidence limits for σ_1^2 which are sufficiently accurate for most practical purposes when, as is usually the case, the estimate of σ_0^2 is based on 10 or more degrees of freedom.

Let M_1/M_0 = ratio of the two mean squares.

ϕ_1 = degrees of freedom of M_1 ($= k-1$ in Table 6.2).

ϕ_0 = degrees of freedom of M_0 ($= k(n-1)$ in Table 6.2).

$L_1(\alpha)$ = the multiplier for the lower confidence limit of the ratio of two standard deviations based on ϕ_1 & ϕ_0 degrees of freedom for probability α.

$L_2(\alpha)$ = the corresponding multiplier for the upper confidence limit.

Then the $100(1-2\alpha)\%$ confidence limits (i.e. the lower $100(1-\alpha)\%$ and the upper $100(1-\alpha)\%$ limits) for σ_1^2/σ_0^2 are given by [Appendix 6C.12]:

$$(M_1 L_1^2 - M_0)/n M_0 \text{ and } (M_1 L_2^2 - M_0)/n M_0$$

Now M_0 is the estimate of σ_0^2, and therefore the lower and upper confidence limits for σ_1 may be approximated by:

$$\sqrt{\{(M_1 L_1^2 - M_0)/n\}} \quad \text{and} \quad \sqrt{\{(M_1 L_2^2 - M_0)/n\}}$$

the estimate for σ_1 being:

$$\sqrt{\{(M_1 - M_0)/n\}}$$

When $M_1 L_1^2 \leqslant M_0$ the lower limit must be taken as zero.

Although these confidence limits are a little too narrow, because full allowance has not been made for the errors in using M_0 as an estimate of σ_0^2, they are sufficiently good for most practical purposes, but caution must be exercised in their interpretation when M_0 is based on fewer than 10 degrees of freedom. Numerical applications are given in the examples later in the chapter.

Unequal numbers of repeat analyses

6.36 Sometimes the numbers of repeat analyses carried out on each sample are not the same. More generally, it often happens in hierarchic data with two sources of variation that the numbers of results in the groups are unequal. The experiment is then described as Unbalanced.

Fortunately the additive property of the sums of squares still holds good in these cases, and the Analysis of Variance is only a little more complicated than for balanced experiments. If n_i denotes the number of repeat analyses on the ith sample, then the crude sum of squares between samples becomes $\sum_{i=1}^{k} S_i^2/n_i$, where S_i is the sum of the observations on the ith sample. The mean square between samples now provides an estimate of $\sigma_0^2 + \bar{n}\sigma_1^2$, where $\bar{n}$ is a function of the numbers of repeat analyses; $\bar{n}$ is approximately equal to the mean number of analyses per sample.

Full details of the computation are given in Appendix 6B.1.

Example 6.1. Variation of dyestuff yields

6.4 Before examining the further uses of the Analysis of Variance it will be helpful to illustrate the derivation of the variance components by means of a numerical example.

This example deals with an investigation to find out how much the variation from batch to batch in the quality of an intermediate product (H-acid) contributes to the variation in the yield of a dyestuff (Naphthalene Black 12B) made from it. In the experiment six samples of the intermediate, representing different batches of works manufacture, were obtained, and five preparations of the dyestuff were made in the laboratory from each sample. The equivalent yield of each preparation as grams of standard colour was determined by dye-trial, and the results are recorded in Table 6.3.

TABLE 6.3.   YIELDS OF NAPHTHALENE BLACK 12B

Sample No. of H-acid	1	2	3	4	5	6
Individual yields in grams of standard colour	1,545	1,540	1,595	1,445	1,595	1,520
	1,440	1,555	1,550	1,440	1,630	1,455
	1,440	1,490	1,605	1,595	1,515	1,450
	1,520	1,560	1,510	1,465	1,635	1,480
	1,580	1,495	1,560	1,545	1,625	1,445
Mean	1,505	1,528	1,564	1,498	1,600	1,470

Grand mean $= 1,527 \cdot 5$ g.

The experimental error, or simply "error", is measured by the variation between the five preparations from each sample of H-acid.

The arithmetical computations are simplified by subtracting a constant quantity from each yield so that we deal with smaller numbers. This does not affect the estimates of the variances, since these are functions of the

TABLE 6.31.   COMPUTATION OF SUMS OF SQUARES   (BASED ON TABLE 6.3)

Sample No. of H-acid	1	2	3	4	5	6	
	145	140	195	45	195	120	
	40	155	150	40	230	55	
Yield $-1,400$	40	90	205	195	115	50	Total
	120	160	110	65	235	80	
	180	95	160	145	225	45	
(1) Sum $= S$	525	640	820	490	1,000	350	3,825
(2) Crude Sum of Squares	71,025	86,350	140,250	66,900	210,000	28,350	602,875
(3) Correction $= S^2/5$	55,125	81,920	134,480	48,020	200,000	24,500	544,045
(4) (2)$-$(3) $=$ Sum of Squares about Sample Mean	15,900	4,430	5,770	18,880	10,000	3,850	58,830
(5) Degrees of freedom	4	4	4	4	4	4	24

deviations from the means. The calculation of the sum of squares for each sample is shown in Table 6.31; the initial figures in this table are derived by subtracting 1,400 from each observation.

We derive:

Sum of squares of the deviations of the observations from the sample means (called sum of squares within samples) $= 58,830$

Degrees of freedom within samples $= 24$ (i.e. 4 for each sample)

Variance within samples (i.e. experimental error variance) $= 58,830/24 = 2,451$

To derive the other sums of squares required for the Analysis of Variance we make use of Formulae (*6.51*) and (*6.71*). The calculations are:

Correction for the grand mean $= (3,825)^2/30 = 487,687·5$

Sum of squares between samples = total of row (**3**) less the correction for the grand mean $= 544,045 - 487,687·5 = 56,357·5$

Finally:

Sum of squares of all 30 observations about the grand mean (called the Total Sum of Squares) = total of row (**2**) less the correction for the mean $= 602,875 - 487,687·5 = 115,187·5$.

When applying the Analysis of Variance for the first few times it is instructive to record the results in detail as in Table 6.31. However, the details given for each sample in rows (**2**), (**3**), (**4**), and (**5**) are not always required, and it is usually sufficient to quote only the quantities in the Total column.

The above calculations lead to the Analysis of Variance Table 6.32.

TABLE 6.32.   ANALYSIS OF VARIANCE TABLE   (BASED ON TABLE 6.31)

Source of Variation	Sum of Squares	Degrees of Freedom	Mean Square	Quantity estimated by the Mean Square*
(1) Between Samples	56,357·5	5	11,272	$\sigma_0^2 + 5\sigma_1^2$
(2) Within Samples	58,830·0	24	2,451	σ_0^2
Total	115,187·5	29	(3,972)	

* The multiplier for σ_1^2 is the number of observations for each sample.

Note that in this table the sum of the "sums of squares" and the sum of the "degrees of freedom" of rows (**1**) and (**2**) are respectively equal to the corresponding values in the row marked Total and that these quantities have been calculated independently.

This additive property of the sums of squares may be used to simplify the arithmetical computations. Thus it is not necessary to calculate all three sums of squares: two are sufficient, and the third may be derived using the additive property. The easiest to calculate are the total sum of squares and the sum of squares between samples.

The additive property of the sums of squares is very useful in the Analysis of Variance and becomes increasingly so for more complex examples. It should not be confused with the additive law of the combined effect of several independent variances. The former is governed by an algebraic identity and must apply to any arithmetically correct calculations, while the latter is a property of "true" variances of infinite Populations, and is never completely realised by practical estimates of variances, which are necessarily based on a finite number of observations.

Estimates of variances

6.41 From Table 6.32 we readily deduce:

$$\sigma_0^2 \rightarrow 2{,}451$$
$$\sigma_1^2 \rightarrow (11{,}272 - 2{,}451)/5 = 1{,}764$$

where the arrow now denotes "is estimated as".

Expressed as standard deviations:

$$\sigma_0 \rightarrow 49{\cdot}5$$
$$\sigma_1 \rightarrow 42{\cdot}0$$

These are respectively the estimates of the standard deviations of the experimental error and of the variation between batches of H-acid.

Confidence limits

6.42 For 24 degrees of freedom the factors for the 95% confidence limits for the estimate of σ_0 derived from Table H are $0{\cdot}78$ and $1{\cdot}39$. These give $38{\cdot}6$ and $68{\cdot}8$ for the 95% confidence limits of σ_0.

The approximate 95% confidence limits for σ_1 are derived using the formulae of § **6.35**. From Table H we find that for $\phi_1 = 5$, $\phi_0 = 24$, $\alpha = 0{\cdot}025$:

$$L_1^2(0{\cdot}025) = (0{\cdot}56)^2 = 0{\cdot}317$$
$$L_2^2(0{\cdot}025) = (2{\cdot}51)^2 = 6{\cdot}28$$

Whence:

Lower limit $= \sqrt{\{(11{,}272 \times 0{\cdot}317 - 2{,}451)/5\}} = \sqrt{224} \quad = 15{\cdot}0$

Upper limit $= \sqrt{\{(11{,}272 \times 6{\cdot}28 \ -2{,}451)/5\}} = \sqrt{13{,}667} = 117$

Thus the confidence limits for σ_1 are $15{\cdot}0$ and 117. These limits are much wider than those for the estimate of σ_0. This was to be expected, because (i) the number of degrees of freedom, 5, for the mean square between samples is much smaller than that for the experimental error, and (ii) the estimate of σ_1 is derived from a difference between two mean squares, between samples and within samples, and this increases the variance of the estimate.

Test of significance

6.43 In some problems the possibility that a given source of variation has had no effect has to be considered. For instance, in Example 6.1 one purpose of the experiment might have been to find out whether or not the variation in the observed yield from batch to batch could be accounted for by experimental error. Setting up the Null Hypothesis that there is no batch-to-batch variation, i.e. that $\sigma_1^2 = 0$, it is evident from Table 6.32 that on this hypothesis the mean square between samples is also an estimate of σ_0^2. There are then two independent estimates of σ_0^2, one from the mean square within samples and the other from the mean square between samples.

We may now test whether these two estimates differ significantly, i.e. whether they differ by more than can be reasonably explained on the grounds of errors in the estimates. This is done by taking the ratio of the mean square between samples to the mean square within samples—called the F-ratio [§ **4.8**]—and referring to the F-tables (see Table D at the end of the volume). A significant value of F discredits the Null Hypothesis.

When the observations are distributed Normally, the ratio of the two mean squares is distributed exactly as F and depends only on the degrees of freedom of the mean squares. This is also sufficiently close to the truth even when the distribution of the observations departs markedly from the Normal form, provided the numbers of analyses on each sample do not differ widely. There is an important distinction between the application of the F-test to the comparison of mean squares in an Analysis of Variance table and the application to the comparison of two variances calculated from different sets of observations, e.g. when comparing the variability of two analytical methods. The latter is highly dependent on the form of the distribution of the observations, and when this departs from Normality, even to a relatively small extent, the F-test has to be interpreted with caution [§ **4.4**].

Applying the F-test to the two mean squares of Table 6.32 we have:

$$F = \frac{\text{Mean square between samples}}{\text{Mean square within samples}} = \frac{11{,}272}{2{,}451} = 4.60$$

For 5 & 24 degrees of freedom the 0·05 and 0·01 values of F are respectively 2·62 and 3·90. The observed ratio of 4·60 is greater than either of these. Hence the hypothesis that $\sigma_1^2 = 0$ is discredited, and it is concluded that real variations do exist in the quality of H-acid from batch to batch.

Note on application of tests of significance

6.44 It should be clearly understood that the F-test applied to Table 6.32 tests the hypothesis that σ_1^2 is zero, and such a test will have a meaning only when it is reasonable to make this hypothesis. In most applications of the Analysis of Variance to sampling and testing problems, repeat samples are taken only when the material is expected or suspected to be variable, and in such circumstances a Null Hypothesis that the sampling error variance is zero is illogical. We know that a sampling error exists, though it may be small, and the purpose of the analysis is to estimate its magnitude and determine the confidence limits.

There are some problems, however, involving an Analysis of Variance for which it is reasonable to set up a Null Hypothesis. For instance, when a blender is used to obtain a bulk quantity of uniform material from a number of batches which are liable to vary, the material from such a blend is expected to be uniform, and only on rare occasions when something has gone wrong would the material not be homogeneous. When testing a number of samples from such a material it would be reasonable to set up the Null Hypothesis that $\sigma_1{}^2$ is zero and to interpret a significant value of F as indicating that something had gone wrong in the blending operation.

Hierarchic classification with three sources of variation

6.5 A sampling and testing example containing three sources of variation will now be considered. The example refers to deliveries of a chemical paste product contained in casks where, in addition to sampling and testing errors, there are variations in quality between deliveries which require to be estimated.

Example 6.2. Variation of paste strengths

6.51 As a routine, three casks selected at random from each delivery were sampled, and the samples were kept for reference. It was desired to estimate the variability in the paste strength from cask to cask and from one delivery to another. Ten of the delivery batches were chosen at random and two analytical tests carried out on each of the 30 samples. In order to ensure that the tests were independent, all 60 strength determinations were carried out in a random order. The resulting data are given in Table 6.4 overleaf.

To illustrate the method of analysis of the data we will first consider the more general case of k batches, n casks per batch, and q tests per cask. Denote the analytical error variance by $\sigma_0{}^2$, the cask-to-cask variance by $\sigma_1{}^2$, and the variance between batches by $\sigma_2{}^2$. For each cask the average of the q strength determinations will have a variance of $\sigma_1{}^2 + \sigma_0{}^2/q$ due to the two lower sources of variation in the hierarchy.

Now consider the table of data formed by the averages of the tests on each cask: it involves k batches with n averages per batch and forms a classification identical with that of § **6.2**. Analysing these means will give an Analysis of Variance table similar to Table 6.2 but with $\sigma_1{}^2$ replaced by $\sigma_2{}^2$, and $\sigma_0{}^2$ replaced by $\sigma_1{}^2 + \sigma_0{}^2/q$ in the column for the "quantity estimated by the mean squares". In order to make the coefficient of σ_0^2 unity all sums of squares have to be multiplied by q, and we obtain the expressions listed in Table 6.41 for the total variation *between* the cask averages.

TABLE 6.4.   % PASTE STRENGTH OF SAMPLES

Batch	Cask 1		Cask 2		Cask 3	
1	62·8	62·6	60·1	62·3	62·7	63·1
2	60·0	61·4	57·5	56·9	61·1	58·9
3	58·7	57·5	63·9	63·1	65·4	63·7
4	57·1	56·4	56·9	58·6	64·7	64·5
5	55·1	55·1	54·7	54·2	58·8	57·5
6	63·4	64·9	59·3	58·1	60·5	60·0
7	62·5	62·6	61·0	58·7	56·9	57·7
8	59·2	59·4	65·2	66·0	64·8	64·1
9	54·8	54·8	64·0	64·0	57·7	56·8
10	58·3	59·3	59·2	59·2	58·9	56·6

TABLE 6.41. ANALYSIS OF VARIANCE FOR TWO HIGHEST SOURCES IN A HIERARCHY

Source of Variation	Sum of Squares	Degrees of Freedom	Mean Square	Quantity estimated by Mean Square
Between Batches	$qn \sum_{i=1}^{k} (\bar{x}_i - \bar{x})^2$	$k-1$	M_2	$\sigma_0^2 + q\sigma_1^2 + nq\sigma_2^2$
Between Casks within Batches	$q \sum_{i=1}^{k} \sum_{j=1}^{n} (\bar{x}_{ij} - \bar{x}_i)^2$	$k(n-1)$	M_1	$\sigma_0^2 + q\sigma_1^2$
Total Between Casks	$q \sum_{i=1}^{k} \sum_{j=1}^{n} (\bar{x}_{ij} - \bar{x})^2$	$nk-1$		

The information in this table is sufficient to estimate σ_2^2. To complete the analysis we note that there are q tests on each of the nk casks, and the variance within these tests estimates σ_0^2 directly. This estimate is based on $nk(q-1)$ degrees of freedom. The completed Analysis of Variance is shown in Table 6.42.

For the purpose of computation the following alternative expressions for the sums of squares are preferable:

(1) $$qn \sum_{i=1}^{k} (\bar{x}_i - \bar{x})^2 = \sum_{i=1}^{k} S_i^2/qn - S^2/qnk \qquad (6.81)$$

where S_i is the total of the qn tests on the ith batch and S the grand total of the qnk tests. $\sum_{i=1}^{k} S_i^2/qn$ is frequently referred to as the crude sum of squares between batches and S^2/qnk the correction for the grand mean.

(2)
$$q \sum_{i=1}^{k} \sum_{j=1}^{n} (\bar{x}_{ij} - \bar{x}_i)^2 = \sum_{i=1}^{k} \sum_{j=1}^{n} S_{ij}^2/q - \sum_{i=1}^{k} S_i^2/qn \qquad (6.82)$$

where S_{ij} is the sum of the q observations for the jth cask in the ith batch.

(3)
$$q \sum_{i=1}^{k} \sum_{j=1}^{n} (\bar{x}_{ij} - \bar{x})^2 = \sum_{i=1}^{k} \sum_{j=1}^{n} S_{ij}^2/q - S^2/qnk \qquad (6.83)$$

(4)
$$\sum_{i=1}^{k} \sum_{j=1}^{n} \sum_{t=1}^{q} (x_{ijt} - \bar{x}_{ij})^2 = \sum_{i=1}^{k} \sum_{j=1}^{n} \sum_{t=1}^{q} x_{ijt}^2 - \sum_{i=1}^{k} \sum_{j=1}^{n} S_{ij}^2/q \qquad (6.84)$$

where $\sum_{i=1}^{k} \sum_{j=1}^{n} \sum_{t=1}^{q} x_{ijt}^2$ is the actual (crude) sum of squares of the qnk observations.

(5)
$$\sum_{i=1}^{k} \sum_{j=1}^{n} \sum_{t=1}^{q} (x_{ijt} - \bar{x})^2 = \sum_{i=1}^{k} \sum_{j=1}^{n} \sum_{t=1}^{q} x_{ijt}^2 - S^2/qnk \qquad (6.85)$$

TABLE 6.42. ANALYSIS OF VARIANCE OF HIERARCHIC DATA WITH THREE SOURCES OF VARIATION

Source of Variation	Sum of Squares	Degrees of Freedom	Mean Square	Quantity estimated by Mean Square
(1) Between Batches	$qn \sum_{i=1}^{k} (\bar{x}_i - \bar{x})^2$	$k-1$ $(= \phi_2)$	M_2	$\sigma_0^2 + q\sigma_1^2 + nq\sigma_2^2$
(2) Between Casks within Batches	$q \sum_{i=1}^{k} \sum_{j=1}^{n} (\bar{x}_{ij} - \bar{x}_i)^2$	$k(n-1)$ $(= \phi_1)$	M_1	$\sigma_0^2 + q\sigma_1^2$
(3) Total between Casks	$q \sum_{i=1}^{k} \sum_{j=1}^{n} (\bar{x}_{ij} - \bar{x})^2$	$nk-1$		
(4) Analytical Error	$\sum_{i=1}^{k} \sum_{j=1}^{n} \sum_{t=1}^{q} (x_{ijt} - \bar{x}_{ij})^2$	$nk(q-1)$ $(= \phi_0)$	M_0	σ_0^2
(5) Total	$\sum_{i=1}^{k} \sum_{j=1}^{n} \sum_{t=1}^{q} (x_{ijt} - \bar{x})^2$	$nkq-1$		

Expressed in these alternative forms, the additive property of the sums of squares is apparent. It is sufficient, therefore, to calculate (6.81), (6.83) and (6.85) and derive (6.82) and (6.84) by subtraction. In presenting the final Analysis of Variance table row (3) is usually omitted. The general method of analysis, set out in a form most convenient for numerical calculation, is given in Appendix 6B.2.

It is clear from the quantities estimated by the mean squares that to test the hypothesis that σ_2^2 is zero we compare the mean squares of (1) and (2), and to test the hypothesis that σ_1^2 is zero we compare the mean squares of (2) and (4). Usually the problem is not one of testing these hypotheses but of estimating the variances σ_0^2, σ_1^2 and σ_2^2. σ_0^2 is given directly by (4), σ_1^2 is derived from the difference of the mean squares (2) and (4), and σ_2^2 is derived from the difference of the mean squares (1) and (2).

Analysis of data of Example 6.2

6.52 To simplify the arithmetic, subtract 50 from the observations of Table 6.4. This gives the working data listed in Table 6.5.

TABLE 6.5. % STRENGTH LESS 50

Batch	Cask 1		Cask 2		Cask 3		Total per Batch
	Observations	Total	Observations	Total	Observations	Total	
1	12·8 12·6	25·4	10·1 12·3	22·4	12·7 13·1	25·8	73·6
2	10·0 11·4	21·4	7·5 6·9	14·4	11·1 8·9	20·0	55·8
3	8·7 7·5	16·2	13·9 13·1	27·0	15·4 13·7	29·1	72·3
4	7·1 6·4	13·5	6·9 8·6	15·5	14·7 14·5	29·2	58·2
5	5·1 5·1	10·2	4·7 4·2	8·9	8·8 7·5	16·3	35·4
6	13·4 14·9	28·3	9·3 8·1	17·4	10·5 10·0	20·5	66·2
7	12·5 12·6	25·1	11·0 8·7	19·7	6·9 7·7	14·6	59·4
8	9·2 9·4	18·6	15·2 16·0	31·2	14·8 14·1	28·9	78·7
9	4·8 4·8	9·6	14·0 14·0	28·0	7·7 6·8	14·5	52·1
10	8·3 9·3	17·6	9·2 9·2	18·4	8·9 6·6	15·5	51·5

Total number of observations = 60   603·2

Putting $k = 10$, $n = 3$ and $q = 2$, the various sums of squares are calculated using the Formulae (6.81), (6.83) and (6.85) as shown in Table 6.51.

(6.83) is the sum of squares between all 30 casks. The sum of squares and degrees of freedom for the variation between casks within batches are derived by subtracting (6.81) from (6.83), and the sum of squares and degrees of freedom within casks are derived by subtracting (6.83) from (6.85). This leads to the Analysis of Variance Table 6.52.

A simple check on the calculations is available in this case because each cask has only two readings; the sum of squares of the differences for each cask divided by 2 thus provides an alternative expression for the sum of squares within casks. It is $(0\cdot2^2 + 1\cdot4^2 + \ldots + 0\cdot9^2 + 2\cdot3^2)/2 = 40\cdot68/2 = 20\cdot34$, in agreement with the entry in Table 6.52.

TABLE 6.51. COMPUTATION OF SUMS OF SQUARES

Source of Variation	Crude Sum of Squares	Correction for the Mean	Corrected Sum of Squares	Degrees of Freedom
(*6.81*) Between Batches	$(73\cdot6^2 + 55\cdot8^2 + \ldots$ $+ 51\cdot5^2)/6$ $= 6,311\cdot57$	$(603\cdot2)^2/60$ $= 6,064\cdot17$	247·40	$10 - 1$ $= 9$
(*6.83*) Total between Casks	$(25\cdot4^2 + 21\cdot4^2 + \ldots$ $+ 14\cdot5^2 + 15\cdot5^2)/2$ $= 6,662\cdot48$	6,064·17	598·31	$30 - 1$ $= 29$
(*6.85*) Total	$12\cdot8^2 + 12\cdot6^2 + \ldots$ $+ 6\cdot6^2 = 6,682\cdot82$	6,064·17	618·65	$60 - 1$ $= 59$

TABLE 6.52. ANALYSIS OF VARIANCE TABLE

Source of Variation	Sum of Squares	Degrees of Freedom	Mean Square	Quantity estimated by Mean Square
Between Batches	247·40	9	27·49	$\sigma_0^2 + 2\sigma_1^2 + 6\sigma_2^2$
Between Casks within Batches	350·91	20	17·55	$\sigma_0^2 + 2\sigma_1^2$
Within Casks	20·34	30	0·68	σ_0^2
Total	618·65	59		

Estimates of σ_0^2, σ_1^2 and σ_2^2 are obtained by solving the following "equations":

$$\sigma_0^2 + 2\sigma_1^2 + 6\sigma_2^2 \to 27\cdot49$$
$$\sigma_0^2 + 2\sigma_1^2 \to 17\cdot55$$
$$\sigma_0^2 \to 0\cdot68$$

The estimates are:

Variance between batches $= \sigma_2^2 \to 1\cdot66$; $\sigma_2 \to 1\cdot29$
Variance between casks $= \sigma_1^2 \to 8\cdot44$; $\sigma_1 \to 2\cdot91$
Testing error variance $= \sigma_0^2 \to 0\cdot68$; $\sigma_0 \to 0\cdot82$

It is seen that the cask-to-cask variation is by far the largest. The ratio of the mean square between casks within batches, based on 20 degrees of freedom, to the mean square within casks, based on 30 degrees of freedom, is $17\cdot55/0\cdot68 = 25\cdot9$ and it far exceeds the upper probability points of the F-ratio listed in Table D. If it is required to test the Null Hypothesis that the apparent variation between batches is due to the variation between casks and to analytical error, then the mean square of $27\cdot49$ based on 9 degrees of freedom must be compared with the mean square of $17\cdot55$ based on 20 degrees of freedom. The F-ratio is $1\cdot57$, and from Table D it is seen that a ratio of $2\cdot39$ is required for significance at the $0\cdot05$ level and $1\cdot96$ at the $0\cdot10$ level. Hence there is insufficient evidence to enable us to reject this Null Hypothesis.

Confidence limits

6.53 Since σ_0 is based on 30 degrees of freedom, the factors for the 95% confidence limits obtained from Table H for $\alpha = 0\cdot025$ are $0\cdot80$ and $1\cdot34$. These give $0\cdot66$ and $1\cdot10$ for the 95% confidence limits of σ_0. To calculate the confidence limits for σ_1/σ_0 we require the following quantities from Table H:

$$L_1{}^2(0\cdot025, 20, 30) = 0\cdot67^2 = 0\cdot45; \quad L_2{}^2(0\cdot025, 20, 30) = 1\cdot53^2 = 2\cdot35$$

Referring to the Analysis of Variance Table 6.52 for the values of the mean squares and to the formulae of § **6.35**, we obtain the following approximations for the 95% confidence limits for σ_1:

$$\sqrt{\{(17\cdot55 \times 0\cdot45 - 0\cdot68)/2\}} = \sqrt{3\cdot60} = 1\cdot89$$

and
$$\sqrt{\{(17\cdot55 \times 2\cdot35 - 0\cdot68)/2\}} = \sqrt{20\cdot28} = 4\cdot50$$

the estimate for σ_1 being $2\cdot91$.

The confidence limits for σ_2 are derived from the confidence limits for $\sigma_2{}^2/\sigma_1{}^2$ in much the same way as the confidence limits for σ_1 are derived from the confidence limits for $\sigma_1{}^2/\sigma_0{}^2$. The results are:

0 (Since the quantity under the square root sign is negative, this limit is taken to be zero)

and
$$\sqrt{\{(27\cdot49 \times 3\cdot67 - 17\cdot55)/6\}} = \sqrt{13\cdot89} = 3\cdot73,$$

the estimate for σ_2 being $1\cdot29$. As expected, σ_2 is subject to larger errors of estimation than σ_1 or σ_0.

Precision of sampling and testing schemes

6.54 The precision of a sampling and testing scheme can readily be determined from the above estimates of the sampling and testing errors. For example:

Scheme 1

Sample 4 casks and test each sample once:

Variance of mean $= (\sigma_1{}^2 + \sigma_0{}^2)/4 = (8\cdot44 + 0\cdot68)/4 = 2\cdot28$

S.E. of mean $= \sqrt{2\cdot28} = 1\cdot51$

Scheme 2

Sample 10 casks, blend the samples and test twice:

$$\text{Variance of mean} = \sigma_1^2/10 + \sigma_0^2/2 = (8.44/10) + (0.68/2) = 1.18$$
$$\text{S.E. of mean} = \sqrt{1.18} = 1.09$$

If sampling is cheaper than analytical testing, the second scheme, in addition to giving a smaller standard error, may be cheaper. The cost of any scheme can readily be calculated, given the costs of sampling and testing. For further consideration of the economics of sampling and testing see Chapter 4 of *Design and Analysis*.

Unequal numbers of observations in the groups

6.55 The analysis is simplest when the experiment is balanced, i.e. when there are the same number of tests on each sample, the same number of samples in each batch, etc. It is not always possible or even desirable to use a balanced experiment, and in such a case some modifications have to be introduced into the analysis. The calculations of the sums of squares are similar, provided the alternative expressions introduced to simplify the computations, (6.81), (6.82), etc., are used, and that each total squared is divided by the number of observations involved therein.

The quantities estimated by the mean squares in an unbalanced experiment are a little more complicated:

Mean square between batches $= M_2 \rightarrow \sigma_0^2 + \bar{n}_2\sigma_1^2 + \bar{n}_3\sigma_2^2$ d.f. $= \phi_2$

Mean square between samples
within batches $= M_1 \rightarrow \sigma_0^2 + \bar{n}_1\sigma_1^2$ d.f. $= \phi_1$

Mean square within samples $= M_0 \rightarrow \sigma_0^2$ d.f. $= \phi_0$

The coefficients $\bar{n}_1$, $\bar{n}_2$ and $\bar{n}_3$ are functions of the numbers of samples in the batches and the numbers of tests on the samples. Frequently $\bar{n}_1$ and $\bar{n}_2$ may be approximated by the mean number of tests per sample and $\bar{n}_3$ by the mean number of tests per batch.

Full details of the computation for an unbalanced experiment with three sources of variation are given in Appendix 6B.2. Confidence limits for the components of variance are discussed in Appendix 6C.

Design of sampling and testing schemes

6.56 One disadvantage of a balanced hierarchic design is that σ_1^2 is estimated with less precision than σ_0^2, and σ_2^2 with less precision than σ_1^2, etc. A balanced design will have to involve at least twice as many samples as batches and at least twice as many tests as samples; in order to obtain a reasonably accurate estimate of σ_2^2 this may mean multiplying the number of tests to a prohibitive number. One way of getting over the difficulty is to use an unbalanced design such as that shown in Figure 6.2 overleaf.

In this scheme two samples are taken from each batch; one of the samples is tested once only and the other twice. For this arrangement, if there are n batches, there will be $(n-1)$ degrees of freedom for batches, n for samples and n for analytical testing. This is not necessarily the most efficient arrangement, but it is better for some purposes than the balanced arrangement with the same number $3n$ of analytical tests.

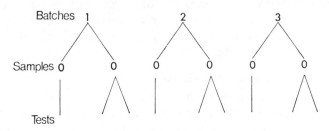

FIG. 6.2.   UNBALANCED HIERARCHIC DESIGN

A method is required which will enable the efficiencies of various arrangements to be compared. Unfortunately it is not practicable to lay down precise rules for doing this, but the following considerations will assist.

We may take the ratio of the upper confidence limit to the actual estimate as a reasonable measure of the precision of an estimated variance. Denote this ratio by Q. Clearly we will prefer designs which result in low values of Q. For σ_0^2 the ratio depends only on the degrees of freedom and may be obtained directly from Table H at the end of the volume. For σ_1^2 the ratio is:

$$(M_1 L_2^2 - M_0)/(M_1 - M_0)$$

where L_2 is the factor for the upper confidence limit for ϕ_1 & ϕ_0 degrees of freedom and the appropriate level of α. Substituting for M_0 and M_1 their expected values σ_0^2 and $\sigma_0^2 + \bar{n}_1 \sigma_1^2$ respectively, the ratio may be written as:

$$Q(\sigma_1^2) = \{(1 + \bar{n}_1 R_1)L_2^2 - 1\}/\bar{n}_1 R_1 = L_2^2 + (L_2^2 - 1)/\bar{n}_1 R_1$$

where R_1 is the ratio σ_1^2/σ_0^2. This formula gives the exact ratio of the upper confidence limit of σ_1^2/σ_0^2 to the estimated value of this ratio. Frequently $(L_2^2 - 1)/\bar{n}_1 R_1$ will be found to be small compared with L_2^2, and it can then be ignored; in such cases the ratio is given directly by the value of L_2^2 corresponding to the probability level chosen, and will depend only on the degrees of freedom of the mean squares.

The corresponding ratio for σ_2^2 is more complicated, but for the designs usually used (such as those considered in Appendix 6D) a sufficiently accurate measure is obtained by using the above formula with R_1 replaced by R_2, where $R_2 = \sigma_2^2/(\sigma_0^2 + \bar{n}_1 \sigma_1^2)$, $\bar{n}_1$ replaced by $\bar{n}_3$, and L_2 by the value for ϕ_2 & ϕ_1 degrees of freedom. For a completely balanced design $\bar{n}_2 = \bar{n}_1$, and the formula gives the exact measure. The designs generally used in practice are balanced or nearly so and the error involved in taking $\bar{n}_2 = \bar{n}_1$ is very small, so that the above formula may be used as an approximation to the ratio.

In Appendix 6D five different arrangements for estimating $\sigma_2{}^2$, $\sigma_1{}^2$ and $\sigma_0{}^2$ are compared, each design being based on a total of about 48 analyses but with different numbers of batches, numbers of samples per batch and numbers of tests per sample. The conclusion reached is that no design is best for estimating all three variances; the design which is best for estimating $\sigma_2{}^2$ is not the best for estimating $\sigma_1{}^2$, etc. Design 3, in which two samples are taken from each batch, one sample is analysed twice and the other once, is a reasonable compromise. However, the final choice in any particular case will have to take other considerations into account, e.g. the cost of the design and whether or not some variances are required to be estimated more accurately than others.

General hierarchic classification

6.57 The analysis for a larger number of sources of variation can be carried out in stages in a similar way, and no new principle is involved. Once the method of analysis is understood for three sources of variation, its extension to any larger number presents no difficulty.

Cross-classifications

6.6 We now turn to the second basic type of data arrangement, the cross-classification, in which each source of variation enters the analysis on an equal footing. In a two-way cross-classification the data can be arranged in a rectangular table with r rows and c columns.

Example 6.3. Two-way cross-classification in penicillin assay variations

6.61 The results of an experiment of this type are shown in Table 6.6. The investigation which it represents is concerned with penicillin testing. The particular problem was to assess the variability between samples of penicillin by the *B. subtilis* plate method. In this test method a bulk-inoculated nutrient agar medium is poured into a Petri dish of approximately 90 mm. diameter, known as a plate. When the medium has set, six small hollow cylinders or pots (about 4 mm. in diameter) are cemented on to the surface at equally spaced intervals. A few drops of the penicillin solutions to be compared are placed in the respective cylinders, and the whole plate is placed in an incubator for a given time. Penicillin diffuses from the pots into the agar, and this produces a clear circular zone of inhibition of growth of the organisms, which can readily be measured. The diameter of the zone is related in a known way to the concentration of penicillin in the solution [§ **7.613**].

In this example the six samples were compared on twenty-four replicate plates. The circle diameters (in millimetres) of the zones of inhibition are given in Table 6.6. A constant quantity (20 mm.) has been subtracted from each observation in order to simplify the arithmetic; this, of course, does not affect the variance.

When analysing such a table of results we have to consider three sources of variation: that between columns (samples of penicillin), that between rows (plates), and that due to testing error. It is assumed that there is a large Population of samples of penicillin and a large Population of plates, and one

TABLE 6.6.   24-PLATE COMPARATIVE TEST ON SAMPLES OF PENICILLIN
(CIRCLE DIAMETER MINUS 20 MM.)

Plate No.	Penicillin Samples						Total	Mean
	1	2	3	4	5	6		
1	7	3	6	3	3	1	23	3·8
2	7	3	6	3	3	1	23	3·8
3	5	1	5	4	4	0	19	3·2
4	6	3	5	3	3	0	20	3·3
5	5	2	6	2	3	0	18	3·0
6	4	2	5	3	2	−1	15	2·5
7	4	0	3	1	2	−1	9	1·5
8	6	2	6	4	4	1	23	3·8
9	4	1	4	2	2	0	13	2·2
10	4	1	4	3	2	−1	13	2·2
11	6	3	6	4	4	1	24	4·0
12	5	2	6	4	4	0	21	3·5
13	6	4	6	4	5	2	27	4·5
14	6	3	6	3	3	0	21	3·5
15	6	3	5	4	4	2	24	4·0
16	5	2	5	3	3	0	18	3·0
17	5	1	4	3	3	0	16	2·7
18	5	2	4	3	3	−1	16	2·7
19	4	1	3	1	1	−1	9	1·5
20	6	3	6	4	4	1	24	4·0
21	5	1	4	2	2	−2	12	2·0
22	5	2	5	2	2	0	16	2·7
23	4	1	4	2	4	−1	14	2·3
24	4	1	4	2	1	−2	10	1·7
Total	124	47	118	69	71	−1	428	
Mean	5·17	1·96	4·92	2·88	2·96	−0·04	Grand mean = 2·97	

purpose of the experiment is to estimate their variances. Denote these variances respectively by σ_c^2, σ_r^2 and σ_0^2. The variance σ_c^2 may be regarded as a sampling variance, σ_r^2 as the component of error due to variation between plates and σ_0^2 as the remainder of the error. Apart from experimental error, the differences between any two samples are expected to be the same for all the plates. This implies that the deviation δ of any single result from the grand mean can be regarded as a sum of the three separate components corresponding respectively to the operation of σ_c^2, σ_r^2 and σ_0^2:

$$\delta = \delta_c + \delta_r + \delta_0$$

Each plate in Table 6.6 can be regarded as having been used for 6 penicillin samples, and if for the moment we ignore the fact that the same samples were associated with every plate, we can construct a hierarchic Analysis of Variance between and within plates. If the grand mean is denoted by $\bar{x}..$ and a plate (row) mean by $\bar{x}_{i.}$, then the sum of squares between plates is $c\sum_{i=1}^{r}(\bar{x}_{i.}-\bar{x}..)^2$ based on $r-1$ degrees of freedom, where $r = 24$ and $c = 6$ in this example. For Table 6.6 this sum of squares comes to 106, based on 23 degrees of freedom.

In a similar way the table can be regarded as consisting of 6 samples all tested 24 times, and we may calculate the sum of squares between samples (columns). If a column mean is denoted by $\bar{x}_{.j}$, this sum of squares is $r\sum_{j=1}^{c}(\bar{x}_{.j}-\bar{x}..)^2$ based on $c-1$ degrees of freedom. For the data of Table 6.6 this comes to 449, based on 5 degrees of freedom.

The sum of squares for all individual results is 590. From these figures we now construct the actual Analysis of Variance Table 6.61 for the cross-classification. The remainder sum of squares and degrees of freedom are obtained by subtraction from the total.

TABLE 6.61. ANALYSIS OF VARIANCE TABLE (BASED ON TABLE 6.6)

Source of Variation	Sum of Squares	Degrees of Freedom	Mean Square
Between Plates	106	23	4·6
Between Samples	449	5	89·8
Remainder (= Error)	35	115	0·30
Total	590	143	

The calculations can be summarized as follows:

Correction for the mean $= (428)^2/144 = 1,272$

Sum of squares between plates
$$= (23^2 + 23^2 + \ldots + 14^2 + 10^2)/6 - 1,272 = 106 \qquad \text{d.f.} = 23$$

Sum of squares between samples
$$= \{124^2 + 47^2 + \ldots + 71^2 + (-1)^2\}/24 - 1,272 = 449 \qquad \text{d.f.} = 5$$

Total sum of squares
$$= \{7^2 + 3^2 + \ldots + 1^2 + (-2)^2\} - 1,272 = 590 \qquad \text{d.f.} = 143$$

Estimates of variances

6.62 Since there are 24 plates and 6 samples per plate, the mean squares estimate the following quantities:

$$\text{Between plates:} \quad \sigma_0^2 + 6\sigma_r^2$$
$$\text{Between samples:} \ \sigma_0^2 + 24\sigma_c^2$$
$$\text{Remainder:} \qquad \sigma_0^2$$

We note that the variance between samples does not enter into the quantity estimated by the mean square between plates, and *vice versa*; this is because all the samples are common to every plate, so that the differences between the means of the plates are unaffected by the differences between the samples, and *vice versa*.

Using the mean squares of Table 6.61 and the expressions of the quantities estimated by them, we readily derive the following estimates:

$$\sigma_0^2 \to 0.30 \qquad\qquad\qquad \sigma_0 \to 0.55$$
$$\sigma_r^2 \to (4.6 - 0.30)/6 \quad = 0.72 \qquad \sigma_r \to 0.85$$
$$\sigma_c^2 \to (89.8 - 0.30)/24 = 3.73 \qquad \sigma_c \to 1.93$$

The confidence limits may be calculated, using the same formulae as those used in the simple hierarchic classification. Referring to Table H, we find that for 115 degrees of freedom the factors for $\alpha = 0.025$ are 0.89 and 1.14, whence the 95% confidence limits for σ_0 are 0.49 and 0.63.

Following the procedure of § **6.35**, to find approximate 95% confidence limits for σ_r we look up the values of L_1 and L_2 in Table H for 23 & 115 degrees of freedom and obtain $L_1 = 0.75$ and $L_2 = 1.43$, giving $L_1^2 = 0.562$ and $L_2^2 = 2.04$. The 95% confidence limits for σ_r are therefore:

$$\sqrt{\{(4.6 \times 0.562 - 0.30)/6\}} \quad \text{and} \quad \sqrt{\{(4.6 \times 2.04 - 0.30)/6\}}$$

i.e. 0.62 and 1.23. Similarly the 95% confidence limits for σ_c are:

$$\sqrt{\{(89.8 \times 0.373 - 0.30)/24\}} \quad \text{and} \quad \sqrt{\{(89.8 \times 6.07 - 0.30)/24\}}$$

i.e. 1.18 and 4.76.

Because of the larger number of degrees of freedom available for the error variance these limits are narrower than those obtained in previous examples.

Test of significance

6.621 If it is required to test the hypothesis $\sigma_r^2 = 0$ we compare the mean square between plates with the remainder mean square, and for the hypothesis $\sigma_c^2 = 0$ we compare the mean square between samples with the remainder mean square. Usually, however, the problem is not one of testing hypotheses but of separating and estimating the variances.

Sums of squares

6.622 The identity which expresses the additive property of the sums of squares is derived and proved in Appendix 6F. This identity is of interest because it shows that the remainder sum of squares is the sum of squares of the deviations of the observations from the values expected on the assumption that the plate "effect" and the sample "effect" are additive. A plate effect is defined as the difference between the plate mean and the grand mean. Thus, for example, the mean of Plate 1 is 0·86 mm. greater than the grand mean and represents an "effect" of +0·86 mm. Further, the mean of the first sample is 2·20 mm. above the grand mean and represents an "effect" of +2·20 mm. On the assumption of an additive law for the effects of plates and samples we would expect Sample 1 on Plate 1 to be 0·86 + 2·20 = 3·06 above the grand mean of 22·97. This gives an expected value of 26·03 mm. The actual value is 27·00 mm., which deviates from the expected by 0·97 mm. The residual quantities remaining after the effect of rows and columns have been allowed for are used to estimate the experimental error variance σ_0^2. This will be explained more fully in § **6.65**.

The additive property of the sums of squares shows that:

(i) Sum of squares within plates = Sum of squares between samples + Remainder sum of squares

(ii) Sum of squares within samples = Sum of squares between plates + Remainder sum of squares

General two-way classification

6.63 The analysis of the general unreplicated two-way cross-classification with r rows and c columns is carried out similarly; the step-by-step details are set out in a form convenient for numerical computation in Appendix 6F.

Missing values

6.64 In Example 6.3 the data comprised a complete 24×6 cross-classification with 144 measurements. Sometimes, however, one or more results in an unreplicated cross-classification may be unreliable or missing, due to some abnormal cause or owing to an accident. If only one result is missing its value can be estimated and the Analysis of Variance completed by the method given in Appendix 6G. When more than one result is missing the analysis is more complicated; this problem is discussed in Appendix 6B of *Design and Analysis*.

Interaction

6.65 In the analysis of a two-way table the remainder sum of squares is the sum of squares of the deviations of the observations from the values expected on the assumption that the effects of the variations due to rows and to columns are additive. The remainder mean square is generally referred to as the Interaction Mean Square. When, as in the previous example, the residuals after allowing for the effect of rows and columns are due to experimental error, the interaction mean square is an estimate of the experimental error variance.

The hypothesis that the effects are additive is discredited when the remainder mean square is significantly greater than that expected from the experimental error variance; the sources of variation represented by the rows and columns are then said to interact. In order that this hypothesis can be tested a separate estimate of the error variance is needed; this is usually obtained from previous experiments or by replication.

The simplest example of an interaction is found in a 2×2 table. Consider, for example, two samples of penicillin compared by each of two different methods. The true results by each method on each sample are depicted as follows:

Method	Penicillin Sample	
	A	B
1	a_1	b_1
2	a_2	b_2

An interaction is a consistent effect such as would arise if Sample A is more effective than B with one method and not with the other. The difference between A and B by Method 1 is $d_1 = (a_1 - b_1)$, and by Method 2 is $d_2 = (a_2 - b_2)$. If d_1 and d_2 differ, then the samples and methods are said to interact and the extent of the difference $(d_1 - d_2)$ is a measure of the interaction. The reader may care to substitute numbers in place of the symbols to appreciate the interaction feature more clearly.

The true values are, of course, unknown and have to be estimated experimentally. Suppose n repeat observations have been carried out on each sample by each method, and let the means be $(\bar{x}_1, \bar{x}_2)$ for Sample A and $(\bar{y}_1, \bar{y}_2)$ for Sample B. The interaction is then estimated by $d = (\bar{x}_1 - \bar{y}_1) - (\bar{x}_2 - \bar{y}_2)$. The error variance $\sigma_0{}^2$ can be estimated from the replicate tests and will be based on $4(n-1)$ degrees of freedom.

The table is symmetrical, and we could equally well consider the differential effect of the methods for the two samples; thus for A the estimate of the effect of methods is $(\bar{x}_1 - \bar{x}_2)$, and for B it is $(\bar{y}_1 - \bar{y}_2)$. The difference between these is $(\bar{x}_1 - \bar{x}_2) - (\bar{y}_1 - \bar{y}_2)$, which is identical with the expression obtained previously.

The significance of the differential effect can readily be ascertained in the following manner. The variance of each mean is σ_0^2/n and that of the difference $(\bar{x}_1 - \bar{y}_1)$ is $2\sigma_0^2/n$. The variance of $(\bar{x}_2 - \bar{y}_2)$ is also $2\sigma_0^2/n$, and therefore the variance of the difference between them is $4\sigma_0^2/n$. We must substitute for σ_0^2 its estimate s_0^2 based on $4(n-1)$ degrees of freedom. The standard error of the interaction d is then $2s_0/\sqrt{n}$, and we compute the ratio:

$$t = d/(2s_0/\sqrt{n})$$

which is based on $4(n-1)$ degrees of freedom. If the value of t is significant [Table C], the Null Hypothesis of no interaction is discredited.

When there are more than two columns or rows the interaction cannot be estimated in the above simple way. In general it is necessary to calculate an interaction mean square and to compare it with the error mean square. An example of this will be considered in § **6.75**. When there is no replication ($n = 1$), it is not possible to test the significance of an interaction unless a prior estimate of the error variance is available.

COMPARISON OF MEANS

Specific values model

6.7 It was pointed out in § **6.23** that there were two basic types of problem which could be dealt with by the technique of Analysis of Variance. Up to now we have considered only the one in which the items in each classification are random elements from a large Population of such items, and the only measures of interest are the variances of these Populations. The method analyses the variance into its separate components and enables these components to be estimated and the confidence limits calculated.

The other type of problem is that of comparing the specific values of certain items as opposed to the estimation of the variance of a large Population of such items. This is not strictly an Analysis of Variance problem, since we are not concerned with variances except that of experimental error, but nevertheless the same technique of analysis up to the point of deriving the mean squares may usefully be employed. This second class of problem is called Comparison of Means. An illustration will now be considered.

In the example of Table 6.3 the six samples of H-acid represented batches from bulk manufacture (i.e. a Population of batches) and we required to estimate the variance of this Population. This estimate, together with that of experimental error (the variance within batches) and their confidence limits give a complete analysis of the data.

Suppose now that our interest lay in specific comparisons among samples of H-acid; this would arise if the six samples represented material prepared by different processes or obtained from different suppliers. The main interest would then lie in comparing one material with another and ultimately in choosing the most suitable one. There would be little point in estimating the variance of a Population of different materials, because it would have little meaning in relation to the investigation. It is possible, of course, to follow the normal procedure of the Analysis of Variance, but the quantity estimated by the mean square between samples would not then contain σ_1^2, because such a quantity does not exist. The σ_1^2 of a normal Analysis of Variance would have to be replaced by $\sum_{i=1}^{k} (M_i - \overline{M})^2/(k-1)$, where M_i is the true mean of the ith sample and $\overline{M}$ is the mean of the M_i's. This expression is a measure of the overall variation between the given samples, and if an overall test of significance is required for this variation it is provided by a comparison of the mean squares. The main purpose of such an Analysis of Variance is, however, to supply a valid estimate of error for the comparisons between the means of the samples of H-acid. This estimate of error is given by the mean square within samples.

The problem is seen to be an extension of the one considered in § **4.23**, in which two group means were compared; the only difference here is that there are more than two means to compare. It will be recalled that for two group means, each based on n observations, the $100(1 - 2\alpha)\%$ confidence limits for the difference are:

$$(\bar{x}_1 - \bar{x}_2) \pm t_\alpha s \sqrt{(2/n)}$$

where s is the estimate of σ_0 based on $2(n-1)$ degrees of freedom, and t_α is the corresponding value of t for the probability α.

Comparison of k group means

6.71 An obvious extension when there are more than two groups is to compare the groups two at a time, using the combined variance within all the groups as the estimate of error—in other words, using the estimate of σ_0^2 derived in the Analysis of Variance Table. Suppose there are k group means $\bar{x}_1, \bar{x}_2, ..., \bar{x}_k$, each based on n observations; the $100(1 - 2\alpha)\%$ confidence limits for the comparisons are then:

$$(\bar{x}_i - \bar{x}_j) \pm t_\alpha s \sqrt{(2/n)}$$

where s is the estimate of σ_0 based on $k(n-1)$ degrees of freedom and t_α is the corresponding value of t for the probability α.

If the purpose of the experiment is to make specific comparisons of this sort, i.e. if the experimenter can specify beforehand what comparisons he is going to make, then the above confidence limits may be applied to each comparison as though they were independent. This means that there is a risk 2α of any one set of confidence limits not containing the true difference, and in the long run only in $100(2\alpha)\%$ of the total number of such comparisons made will the confidence limits not contain the true difference. If we work with 95% confidence limits, then in the long run only in one case out of 20 will the confidence interval for the difference between a pair of means not contain the true difference; in one case in 40 the true difference will exceed the upper limit, and in one case in 40 the true difference will be below the lower limit.

Now in an experiment in which k groups are compared there are $k(k-1)/2$ possible pairs, so that the chance that the confidence limits for one or more of these differences will not contain the true value may be quite high. There is in principle no difference between making m specific comparisons between several samples in one experiment and making these comparisons one at a time in each of m experiments on pairs of samples. What must be guarded against is selection of only those comparisons which are suggested by the data. For instance, if only the highest and the lowest results are compared there is a high chance of a misleading conclusion being reached, and the proportion of wrong conclusions due to the error of the first kind for such comparisons may be very high.

Comparison with a control

6.72 One application of these multiple comparisons is to chemotherapeutic research, in which a biological test is used as a screen for a large number of compounds. Usually several compounds are compared with a control in each application of the test. An effect is defined as the difference between the mean of a treated group and that of the control. Those compounds which show an effect above a certain predetermined value will be retained for further examination, but those showing a smaller effect will be discarded. The only comparisons normally made are between a treatment mean and the control.

To be of technical importance a compound should have a true effect greater than a given amount d. A critical value c for an observed effect is chosen so that a compound is classified as effective or ineffective according to whether the observed difference is greater or less than c. In such screening tests there is a risk of observing an effect greater than c from a compound which really has no effect. This is the error of the first kind which gives rise to "false positives". There is also a risk of a "false negative", i.e. of observing an effect less than c when the compound has a true effect greater than d and so is of potential interest. This corresponds to the error of the second kind. Confidence limits can be used to control these risks in the following manner [Chapter 5]. Suppose we are prepared to take a risk α of a false negative and a similar risk of a false positive, then the distance between the $100(1-2\alpha)\%$ confidence limits should be equal to d, and since the limits will be symmetrical about the observed value the critical value c should be taken at $d/2$. The number of observations on each group must then be chosen to give a $100(1-2\alpha)\%$ confidence interval of d. This can readily be determined when the standard error is known or when an estimate is available from previous trials.

Global comparisons

6.73 There are other kinds of experiments in which the experimenter, although requiring to compare the sample means in pairs, may require to choose the confidence limits so that the chance of *any one or more* of these limits in each experiment being wrong is 2α. The experiment is then considered as an entity and not as a set of smaller ones, and the experimenter wishes to ensure that in only $100(2\alpha)\%$ *of such experiments* will any wrong conclusions be reached.

The confidence limits must then be wider than those used previously. Suitable limits have been derived by J. W. Tukey; for any difference $\bar{x}_i - \bar{x}_j$ he advocates the calculation of:

$$(\bar{x}_i - \bar{x}_j) \pm q_{2\alpha}s/\sqrt{n}$$

where $q_{2\alpha}$ is the Studentised Range given in Table 29 of [6.1]; $q_{2\alpha}$ is in fact the upper $100(2\alpha)\%$ probability point for the range of k observations from a Normal Population divided by an independent estimate of the standard deviation based on ϕ degrees of freedom. These limits tend to be rather wide when they are applied to every difference, since they are essentially based on the largest difference. Other tests exist which have advantages in certain circumstances, and a full discussion will be found in [6.2].

Illustrations of multiple confidence intervals and Tukey's test

6.74 The following illustrations are intended to show the circumstances in which the various types of confidence interval discussed in the preceding sections should be used.

6.741 Assume that in Example 6.1 the six samples represented six different qualities of H-acid, e.g. six different suppliers, and that Sample 1 represented the one in current use, e.g. own manufacture. It was required to find out if higher yields could be obtained from any of the other qualities.

Here the correct procedure is to compare each mean with that of the control (Sample 1), using confidence limits derived from the t-table. The standard deviation of each observation is $49\cdot5$ [§ **6.41**], whence the standard error of a difference between sample means is $49\cdot5\sqrt{(2/5)} = 31\cdot3$. The $0\cdot025$ value of t for 24 degrees of freedom is $2\cdot06$. The 95% confidence limits to be attached to any observed difference are then $\pm64\cdot5$; in other words, there is a chance of 1 in 40 (or less) of being wrong if it is concluded that a real improvement has occurred when the observed difference in means is $64\cdot5$ (or more). One sample, No. 5, is thus seen to be significantly better than the control.

6.742 Now assume that in Example 6.1 the six samples represented six consecutive batches of current manufacture. Is there justification for carrying out an analytical investigation on some of the batches for the purpose of discovering causes of variation in yield?

Here we are not interested in specific comparisons. When the experiment was planned there was no reason for carrying out particular comparisons between particular batches. Here we want a global comparison to ascertain which differences between batches are too large to be explained as having arisen from experimental error.

Tukey's test may therefore be applied. For the 5% significance level, $k = 6$ and $\phi = k(n-1) = 24$, the value of q is 4·37 (see Table 29 of [6.1]). The confidence limits for the largest difference are then:

$$\pm 4{\cdot}37 \times 49{\cdot}5/\sqrt{5} = \pm 97$$

The differences between the means of Samples 4 and 5 and of Samples 5 and 6 exceed this value. Hence there is sufficient evidence for carrying out an analytical investigation.

6.743 In Example 6.3 the quantity estimated by the mean square for penicillin samples was obtained on the assumption that the samples could be regarded as representative of a Population of samples, and this gave meaning to σ_c^2. If, however, the samples cannot be so regarded, e.g. when the samples represent specific grades of the material, then σ_c^2 no longer has a meaning and it has to be replaced by the expression $\sum\limits_{j=1}^{c} (M_j - \overline{M})^2/(c-1)$ where M_j is the true mean of the jth sample and $\overline{M}$ is the mean of the M_j's. The main purpose of the experiment is then to compare the six particular samples.

In this example there is a meaning to the variance between plates, because the plates used in the test are a sample from a much larger number of such plates. However, this source of error has been eliminated in the comparisons of the samples.

The residual standard deviation was estimated as 0·55 mm. in § **6.62**, based on 115 degrees of freedom. The standard error of each sample mean over the twenty-four plates is therefore $0{\cdot}55/\sqrt{24} = 0{\cdot}11$ mm. This is the appropriate standard error in comparisons between the sample means:

A = 25·17	B = 21·96	C = 24·92
D = 22·88	E = 22·96	F = 19·96

Assume that one of the samples, A, is a standard material and we require to compare each of the other samples with it. For this purpose we calculate the differences between the mean of A and each of B to F and their confidence limits. The 95% confidence limits of each difference are:

$$\pm 1{\cdot}98 \times 0{\cdot}55\sqrt{(2/24)} = \pm 0{\cdot}31$$

All the samples except C are thus seen to differ significantly from A.

Example 6.4. Testing of cement mortar

6.75 This example is of a type which combines a simple hierarchic classification with a two-way cross-classification. It is the commonly occurring design of a two-way cross-classification with repeat tests in each cell.

A sample of Portland cement was mixed and reduced in a sampling machine to small samples for testing. The cement mortar was "gauged", i.e. mixed with water and worked for a standard time, by each of three gaugers and cast into $\frac{1}{2}$ in. cube moulds. The cubes were removed from the moulds after twenty-four hours, stored at 18°C for seven days, and then tested for compressive strength. Three testers or "breakers" were employed. Each gauger gauged twelve cubes which were divided into three sets of four, and each breaker tested one such set from each gauger. The experimental design was thus a 3×3 cross-classification with 4 results in each cell.

TABLE 6.7. TESTS ON SAMPLES OF PORTLAND CEMENT (WORKING DATA)

	Breaker 1		Breaker 2		Breaker 3		Means
Gauger 1	28	52	−66	−60	−84	18	4·8
	−24	80	2	120	32	−40	
Gauger 2	−58	28	34	−12	−82	−20	−3·2
	58	−10	−4	120	−40	−52	
Gauger 3	36	116	72	−24	−54	−7	33·6
	68	50	62	56	−32	60	
Means	35·3		25·0		−25·1		11·75

The purpose of the investigation was to examine four sources of variation:

(*a*) Between the three gaugers.

(*b*) Between the three breakers.

(*c*) Interaction of gaugers and breakers.

(*d*) Between repeat tests, i.e. testing error.

Some variation might be expected between gaugers, since the personal factor is strong unless the operation is carefully standardized. Between breakers (i.e. the assistants operating the testing machine) there should normally be no variation, since the machine is automatic. In this case an old machine was in use, and one object of the test was to find whether personal factors in the preliminary adjustment of the machine could cause variations in the results.

The results were about 5,000 lb./sq. in. To save arithmetic 5,000 has been subtracted from each, and since the figures were recorded to the nearest 10 lb./sq. in. the differences were divided by 10. The adjusted data are shown in Table 6.7.

Analysis

6.751 The first stage in the analysis is to construct Table 6.71, in which each entry is the sum of the corresponding repeat tests in Table 6.7.

TABLE 6.71. TABLE OF SUMS DERIVED FROM TABLE 6.7

	Breaker 1	Breaker 2	Breaker 3	Total
Gauger 1	136	−4	−74	58
Gauger 2	18	138	−194	−38
Gauger 3	270	166	−33	403
Total	424	300	−301	423

The estimate of the error variance $\sigma_0{}^2$ is obtained directly from the variance within the groups of four repeat tests.

The variance associated with the entries in Table 6.71 is $4\sigma_0{}^2$, since each entry is a sum of four independent observations. If, therefore, the two-way table of totals (Table 6.71) is analysed in the usual way, the quantities estimated by the mean squares will include the item $4\sigma_0{}^2$. Each sum of squares must therefore be divided by 4, in order to make the coefficient of $\sigma_0{}^2$ unity.

The correct computation is almost automatic if we follow the rule for all calculations involved in Analysis of Variance, namely that when determining the crude sum of squares of similar quantities which are sums of individual observations, each square has to be divided by the number of observations contained in it.

The details of the calculations of the sums of squares are:

(i) Correction for the mean $= (423)^2/36 = 4{,}970$

(ii) Sum of squares between gaugers
$= \{58^2 + (-38)^2 + 403^2\}/12 - 4{,}970 = 8{,}965$ d.f. $= 2$

(iii) Sum of squares between breakers
$= \{424^2 + 300^2 + (-301)^2\}/12 - 4{,}970 = 25{,}061$ d.f. $= 2$

(iv) Total sum of squares between the 9 sub-groups
$$= \{136^2 + (-4)^2 + \ldots + 166^2 + (-33)^2\}/4 - 4,970$$
$$= 40,664 \qquad \text{d.f.} = 8$$

(v) Total sum of squares of all 36 observations
$$= \{28^2 + 52^2 + \ldots + 60^2\} - 4,970 = 114,819 \qquad \text{d.f.} = 35$$

TABLE 6.72.   ANALYSIS OF VARIANCE TABLE (BASED ON TABLE 6.71)

Source of Variation	Sum of Squares	Degrees of Freedom	Mean Square	Quantity estimated by the Mean Square
Between Gaugers	8,965	2	4,482	$\sigma_0^2 + 6\Sigma(G_i - \mu)^2$
Between Breakers	25,061	2	12,530	$\sigma_0^2 + 6\Sigma(B_j - \mu)^2$
Interaction of Gaugers and Breakers	6,638	4	1,659	$\sigma_0^2 + \Sigma\Sigma(\mu_{ij} - G_i - B_j + \mu)^2$
(Total between Sub-groups)	(40,664)	(8)		
Within Sub-groups	74,155	27	2,746	σ_0^2
Total	114,819	35		

The Analysis of Variance table may then be constructed as in Table 6.72. The sum of squares and degrees of freedom within sub-groups were found by subtracting (iv) from (v) and the interaction of gaugers and breakers by subtracting the sum of (ii) and (iii) from (iv). The row in brackets is usually omitted.

The quantities estimated by the mean squares given in Table 6.72 can readily be determined from the following considerations. Assume for the moment that there is no error and that the true values for each cell of Table 6.7 are as follows:

			Means
μ_{11}	μ_{12}	μ_{13}	G_1
μ_{21}	μ_{22}	μ_{23}	G_2
μ_{31}	μ_{32}	μ_{33}	G_3
Means B_1	B_2	B_3	μ

Each entry must then be multiplied by 4 to obtain the results corresponding to Table 6.71; the row and column means have to be multiplied by 12 and the grand mean by 36 to give the corresponding totals. Carrying out the Analysis of Variance on these true values would give the mean squares shown in Table 6.72, but with σ_0^2 omitted. σ_0^2 has now to be added to allow for the errors in the determinations of the breaking strengths.

More generally, when the classification consists of r rows, c columns and n repeat tests in each cell, the multipliers for the summation terms in the expected mean squares are $nc/(r-1)$, $nr/(c-1)$ and $n/(r-1)(c-1)$ for the row, column and interaction effects respectively.

Interpretation

6.752 The interaction mean square does not differ significantly from the error mean square; in fact it is smaller, but not significantly so. All the apparent interaction between breakers and gaugers can therefore be explained by the experimental error, and it is reasonable to conclude that no real interaction exists. An interaction could arise if a breaker obtained better results with cubes made by a certain gauger than with the others, not due to any real difference between the cubes but because he carried out the test in such a way as to show better results.

We are interested in assessing whether any real variation exists between the breakers and between the gaugers, and therefore it is appropriate to assess the significance of the mean squares. The computed F-ratios are 4,482/2,746 $= 1.63$ for gaugers and 12,530/2,746 $= 4.56$ for breakers. Each is based on 2 & 27 degrees of freedom and from Table D the critical levels of the F-ratio are 3.35 and 5.49 for the 5% and 1% significance levels respectively. Hence the variation between gaugers is judged not significant, but the variation between breakers is significant at the 5% level.

The application of Tukey's test is, however, more informative because it gives directly the confidence limits of the differences between the means. To apply this test we refer to Table 29 of [6.1] to obtain the value of $q_{2\alpha}$ for $\phi = 27$, $k = 3$ and $\alpha = 0.025$. This is found to be 3.51. The standard error of each breaker or gauger mean is $\sqrt{(2,746/12)} = 15.1$, and therefore the 95% confidence limits of any difference between two means are ± 53.0. The differences in the mean values obtained by the three gaugers are all less than this value, which confirms the conclusion already arrived at that there is insufficient evidence to presume that the gaugers were different. On the other hand, Breaker 3 is seen to differ from Breaker 1 because the difference in their means exceeds 53.0.

Testing significance in two-way tables with repeats

6.753 In this problem we are concerned with comparisons between the three gaugers and three breakers. If, however, the problem was one in which we were interested in a whole Population of gaugers and breakers, then the summation terms in the quantities estimated by the mean squares in Table 6.72 would be replaced by expressions involving σ_G^2 (for gaugers), σ_B^2 (for breakers) and $\sigma_I{}^2$ (for interaction). The variance $\sigma_I{}^2$ represents a variable bias, and if this is appreciable it would be an additional testing error which would have to be taken into account in assessing the significance of the variation between breakers and between gaugers. It would then be necessary to assess the significance of the first two mean squares of Table 6.72 against the interaction and not against the error variance. Had the interaction been significant it would have been necessary to investigate the experimental conditions to identify the possible causes and eliminate them in future tests. Even when the interaction is not significant the first two mean squares can still be assessed against the interaction mean square, but this involves a lower level of precision since there are fewer degrees of freedom for interaction than for error.

Many textbooks recommend that any mean square which is not significant be combined with the error mean square to give a new error mean square based on a larger number of degrees of freedom. The combination of non-significant mean squares should be carried out with discretion and not as a general rule; it should be restricted to those mean squares corresponding to effects which from prior considerations are not expected to be appreciable, but whenever possible the experiment should be carried out on a sufficient scale to supply an adequate number of degrees of freedom for error. Indiscriminate combination of non-significant mean squares can sometimes lead to absurd conclusions. In this example the interaction and error mean squares could be combined to give $(6{,}638 + 74{,}155)/(4 + 27) = 2{,}606$ based on 31 degrees of freedom. The significance of the mean squares between gaugers and between breakers could be assessed against this new error. There is no serious objection to this procedure in this particular example because the interaction was not expected, but it offers no advantage.

Further aspects of the Analysis of Variance

6.8 The examples already given cover a substantial proportion of the types of applications of the Analysis of Variance likely to be encountered in chemical work. More complex cases usually represent extensions and combinations of these examples. Example 6.4, for instance, is an extension of Example 6.3. The method of analysis in the complex cases follows the pattern set by the examples given above. The main point to remember is that when calculating crude sums of squares, the square of any figure which is the sum of a given number of observations must be divided by the given number of observations constituting the sum.

Most electronic computers have had Analysis of Variance programs written for them. The user should consult his own installation for specifications of the order and the format in which the data must be assembled for machine processing. Often a suite of programs is available, one for each of the commonly occurring classification combinations. Alternatively there may be a universal program which caters for all standard data arrangements. Use of a computer speeds up the calculations, but the output must be carefully interpreted by the analyst according to the principles described in this chapter.

When the comparison of means is required for several sets of data but no computer is available, short-cut methods based on sums of ranges can reduce the labour of calculation with only a slight loss in efficiency [6.3].

Many problems in the Analysis of Variance can also be treated by the methods of Multiple Regression [Chapter 8]. However, the procedures described in the present chapter are usually more appropriate. One exception to this general rule occurs when the factors, instead of being *qualitative* (e.g. 6 penicillin samples and 24 plates in Example 6.3), are levels of *quantitative* variables. For example, suppose we determine the yield of product from a chemical reaction at each of the 12 combinations of 4 temperature levels and 3 pressure levels. Then it is usually better to postulate an equation

$$y = f(t, p; \theta_1, \theta_2, \ldots)$$

relating yield to temperature and pressure, and to estimate the unknown parameters $\theta_1, \theta_2, \ldots$ in the equation, than to treat the data by Analysis of Variance methods as a simple 4×3 cross-classification. Also, when most of the factors are quantitative but one or two are qualitative, the latter can be incorporated in a multiple regression equation by the use of "dummy variables" [§ **8.55**].

Different numbers of observations in the cells

6.81 The method of analysis used for Example 6.4 applies only when there are the same number of repeat observations in each cell. In the general cross-classification the number of repeat observations will not be constant. The simpler methods of the Analysis of Variance will then usually be inapplicable because the sums of squares for the sources of variation are no longer additive. The design is said to be Non-Orthogonal in these cases.

However, there is one instance where the numbers of repeat tests are unequal but the simpler techniques do still apply. This is when the ratios of the numbers of observations in corresponding cells of the rows, or of the columns, are the same throughout the table, i.e. when:

$$n_{ij} = n_{i.} n_{.j} / N \qquad \text{for all } i, j \tag{6.9}$$

where n_{ij} is the number of observations in the (i, j) cell,

   $n_{i.}$ is the total number of observations in the ith row,

   $n_{.j}$ is the total number of observations in the jth column,

and N is the overall total number of observations.

Such Proportional Frequencies are a feature of the two designs:

					Row Numbers					Row Numbers
	6	6	6	6	24		9	12	15	36
	5	5	5	5	20		6	8	10	24
	3	3	3	3	12		3	4	5	12
Column Numbers	14	14	14	14	56	Column Numbers	18	24	30	72

In these cases the analysis is carried out in a similar manner to that of Example 6.4, applying the above rule for the divisor in the calculation of each crude sum of squares and each correction.

For the general case in which the respective numbers in the rows and columns are not proportional, the non-orthogonal analysis is complicated and beyond the scope of this book. A comprehensive treatment is given in Chapters 35 and 36 of [6.4]. Frequently, however, it is possible to obtain good approximate analyses by adding dummy observations to some of the cells or by assuming that the means of others are based on fewer results in order to satisfy the proportional frequencies condition. A dummy observation is set equal to the mean for the cell to which it is added. The variance within cells is calculated from the original observations in the usual way, since this particular variance does not require the proportionality condition.

Finite Populations

6.82 In § **6.23** we distinguished between cases in which items in a classification are a random sample from an infinite Population of such items and cases where the analysis is concerned with just the particular items measured. We usually want to estimate components of variance in the first instance, but to compare mean values in the second.

There is also an intermediate case where the total Population is larger than the number of items tested but far from infinite. For example, a set of similar machines may be used for a certain operation but measurements are obtained from only a limited number of them. Although the Analysis of Variance is computed in the standard way in this intermediate case, the quantities estimated by the mean squares are different.

Mixed models also occur in which one classification, say rows, comprises the whole Population of items while the other, the columns, is a small random sample from an infinite Population. A general method is available to determine the expected mean squares in all these cases. It involves applying a rather complex sequence of rules for each source of variation and the reader should refer to § 7.63 of [6.5] for full details of the procedure. Knowing the expected mean squares, it is then possible to decide which of their ratios should be computed to test for the statistical significance of the sources of variation.

References

[6.1] PEARSON, E. S., and HARTLEY, H. O. *Biometrika Tables for Statisticians*, Vol. 1 (third edition). Charles Griffin (High Wycombe, 1976).

[6.2] O'NEILL, R., and WETHERILL, G. B. "The Present State of Multiple Comparison Methods", *J. R. Statist. Soc.* B, 33, 218–241 (1971).

[6.3] KURTZ, T. E., LINK, R. F., TUKEY, J. W., and WALLACE, D. L. "Short-cut Multiple Comparisons for Balanced Single and Double Classifications: Part 1, Results", *Technometrics*, 7, 95–161 (1965).

[6.4] KENDALL, M. G., and STUART, A. *The Advanced Theory of Statistics*, Vol. III (fourth edition). Charles Griffin (High Wycombe, 1983).

[6.5] BENNETT, C. A., and FRANKLIN, N. L. *Statistical Analysis in Chemistry and the Chemical Industry*. John Wiley (New York, 1954).

Appendix 6A

Combination of results from several samples

Suppose there are k independent unbiased estimates of a mean μ denoted by $\bar{x}_1, \bar{x}_2, ..., \bar{x}_k$ with variances $V_1, V_2, ..., V_k$. The problem is to combine these means in order to derive the best estimate of μ.

The expression $X = (w_1\bar{x}_1 + w_2\bar{x}_2 + ... + w_k\bar{x}_k)/W$, where $W = \Sigma w_i$, gives a weighted mean of the $\bar{x}$'s, which is a linear estimate of μ. It can be shown that the best linear combination of the $\bar{x}$'s, i.e. the one which has minimum variance, is obtained when the weights are equal (or proportional) to the inverse of the variances of the corresponding $\bar{x}_i$'s, i.e. when $w_i = 1/V_i$. The variance of the weighted mean is then:

$$V(X) = 1/W = 1 \bigg/ \left(\frac{1}{V_1} + \frac{1}{V_2} + ... + \frac{1}{V_k} \right)$$

This result can be applied to the analyses on a number of samples taken from a batch. Suppose there are k samples whose means are $\bar{x}_1, \bar{x}_2, ..., \bar{x}_k$ respectively and let n_i denote the number of analyses on the ith sample. Denote the variance of sampling by σ_1^2 and the variance of analytical error by σ_0^2, then the variance of $\bar{x}_i$ is $V_i = \sigma_1^2 + \sigma_0^2/n_i$. The weight to be applied to $\bar{x}_i$ in order to obtain the best estimate for the batch mean is then $w_i = 1/(\sigma_1^2 + \sigma_0^2/n_i)$.

When estimates are available for σ_1^2 and σ_0^2, the precision of any sampling and testing scheme can be assessed. Given the costs of sampling and testing, the cost efficiency can also be assessed. Thus, if $Q_1 = $ cost of one sample, $Q_0 = $ cost of one analysis, the total cost of the above scheme is $kQ_1 + \Sigma n_i Q_0$. Its cost efficiency is $W/(kQ_1 + \Sigma n_i Q_0)$, where $W = \Sigma w_i = \Sigma 1/(\sigma_1^2 + \sigma_0^2/n_i)$.

The weight w_i, which is the inverse of the variance of $\bar{x}_i$, is sometimes referred to as the amount of information contributed by $\bar{x}_i$. The smaller the variance, the more precise the estimate and the greater the amount of information it contains. The sum of the weights $W = \Sigma w_i$ is the total amount of information contained in the n sample means. Since the variance of X is $1/W$, it follows there is no loss of information when the means are combined in this way.

Appendix 6B

Analysis of hierarchic data

Hierarchic data with two sources of variation

6B.1 Hierarchic data with two sources of variation can be set out as in Table 6B.1.

TABLE 6B.1. DATA WITH TWO SOURCES OF VARIATION

Group No.	Number of Observations in the Group	Observations	Group Total	Group Mean
1	n_1	$x_{11}, x_{12}, x_{13}, \dots, x_{1n_1}$	S_1	$\bar{x}_1$
2	n_2	$x_{21}, x_{22}, x_{23}, \dots, x_{2n_2}$	S_2	$\bar{x}_2$
3	n_3	$x_{31}, x_{32}, x_{33}, \dots, x_{3n_3}$	S_3	$\bar{x}_3$
.	.	. . . .	.	.
.	.	. . . .	.	.
.	.	. . . .	.	.
k	n_k	$x_{k1}, x_{k2}, x_{k3}, \dots, x_{kn_k}$	S_k	$\bar{x}_k$

The numbers of observations in each group are not necessarily the same.

Total number of observations $= N = n_1 + n_2 + n_3 + \dots + n_k$

Grand total $= S = S_1 + S_2 + S_3 + \dots + S_k$

Grand mean $= S/N = \bar{x}$

The quantities required for the construction of the Analysis of Variance table are:

(i) *Correction for the mean.* Square the grand total and divide by the total number of observations, i.e. S^2/N.

(ii) *Total sum of squares.* Square each observation and add, then subtract the correction for the mean:

$$\text{Total sum of squares} = (x_{11}^2 + x_{12}^2 + \dots + x_{kn_k}^2) - S^2/N$$

$$= \sum_{i=1}^{k} \sum_{j=1}^{n_i} x_{ij}^2 - S^2/N$$

This is based on $(N-1)$ degrees of freedom.

The quantity $\sum\limits_{i=1}^{k} \sum\limits_{j=1}^{n_i} x_{ij}^2$ is called the crude total sum of squares.

(iii) *Sum of squares between groups.* Square each group total, divide each by the number of observations in the group, and add. This gives the crude sum of squares between groups. Subtract the correction for the mean:

Sum of squares
between groups $= (S_1^2/n_1 + S_2^2/n_2 + \ldots + S_k^2/n_k) - S^2/N$

$$= \sum_{i=1}^{k} (S_i^2/n_i) - S^2/N$$

When all groups contain the same number $n = N/k$ of observations, the crude sum of squares between groups is equal to $1/n$ times the sum of squares of the individual group totals.

The degrees of freedom between groups are one less than the number of groups, i.e. $\phi_1 = k-1$.

(iv) *Sum of squares within groups.* Subtract (iii) from (ii). When each group has just two observations, the sum of squares within groups can also be obtained directly by accumulating one-half of the sum of squares of the within-group differences (see § **3.342** for details).

TABLE 6B.2.   ANALYSIS OF VARIANCE TABLE (BASED ON TABLE 6B.1)

Source of Variation	Sum of Squares	Degrees of Freedom	Mean Square
Between Groups	$\sum\limits_{i=1}^{k} (S_i^2/n_i) - S^2/N$	$k-1$ $(= \phi_1)$	M_1
Within Groups	$\sum\limits_{i=1}^{k} \sum\limits_{j=1}^{n_i} x_{ij}^2 - \sum\limits_{i=1}^{k} (S_i^2/n_i)$	$N-k$ $(= \phi_0)$	M_0
Total	$\sum\limits_{i=1}^{k} \sum\limits_{j=1}^{n_i} x_{ij}^2 - S^2/N$	$N-1$	

The Analysis of Variance may then be set out as in Table 6B.2. The mean squares M_1 and M_0, are the respective sums of squares divided by their degrees of freedom. The quantities estimated by the mean squares are as follows:

(i) *For equal numbers in the groups:*

$$M_1 \rightarrow \sigma_0^2 + n\sigma_1^2$$
$$M_0 \rightarrow \sigma_0^2$$

where σ_0^2 is the true variance within groups and σ_1^2 the true variance between groups.

(ii) *For unequal numbers in the groups:*

$$M_1 \rightarrow \sigma_0^2 + \bar{n}\sigma_1^2$$
$$M_0 \rightarrow \sigma_0^2$$

where $\bar{n} = \left(N^2 - \sum_{i=1}^{k} n_i^2\right) \Big/ (k-1)N$. ($\bar{n}$ reduces to N/k $(= n)$ when all the n_i are equal).

Hierarchic data with three sources of variation

6B.2 Suppose there are k groups, each consisting of a number of sub-groups. Let the ith group contain r_i sub-groups, and denote the number of observations in these sub-groups by n_{i1}, n_{i2}, etc. Denote the totals of the observations in each sub-group by S_{ij} and the group totals obtained by summing the totals in each of the component sub-groups by S_i. The numbers of observations and the corresponding totals may be set out as in Table 6B.3.

TABLE 6B.3. DATA WITH THREE SOURCES OF VARIATION

Group No.	Number of Observations in Group	Group Total	Number of Observations in Sub-groups	Sub-group Totals
1	n_1	S_1	$n_{11}, n_{12}, \ldots, n_{1r_1}$	$S_{11}, S_{12}, \ldots, S_{1r_1}$
2	n_2	S_2	$n_{21}, n_{22}, \ldots, n_{2r_2}$	$S_{21}, S_{22}, \ldots, S_{2r_2}$
.	.	.	. . .	. . .
.	.	.	. . .	. . .
k	n_k	S_k	$n_{k1}, n_{k2}, \ldots, n_{kr_k}$	$S_{k1}, S_{k2}, \ldots, S_{kr_k}$

Total number of sub-groups $= R = r_1 + r_2 + \ldots + r_k$
Total number of observations $= N = n_1 + n_2 + \ldots + n_k$
Grand total of observations $= S = S_1 + S_2 + \ldots + S_k$

Then the steps in the computation are as follows:

(i) *Correction for the mean.* Square the grand total and divide by the total number of observations, i.e. S^2/N.

(ii) *Total sum of squares.* This is the sum of squares of all N observations minus the correction for mean. It is based on $(N-1)$ degrees of freedom.

(iii) *Sum of squares between groups.* This is derived from the group totals as follows:

$$(S_1^2/n_1 + S_2^2/n_2 + \ldots + S_k^2/n_k) - S^2/N$$

It is based on $(k-1)$ degrees of freedom.

(iv) *Total sum of squares between sub-groups.* This is derived from the totals of the R sub-groups. The crude sum of squares is obtained by squaring each sub-group total, dividing by the corresponding numbers of observations, and adding.

Total sum of squares between sub-groups

$$= (S_{11}^2/n_{11} + S_{12}^2/n_{12} + \ldots + S_{k1}^2/n_{k1} + \ldots) - S^2/N$$

This is based on $(R-1)$ degrees of freedom.

(v) From the above quantities we derive:

Sum of squares between sub-groups within groups = (iv) − (iii)

Sum of squares within sub-groups   = (ii) − (iv)

TABLE 6B.4.   ANALYSIS OF VARIANCE TABLE (BASED ON TABLE 6B.3)

Source of Variation	Sum of Squares	Degrees of Freedom	Mean Square
Between Groups	$\sum\limits_{i=1}^{k} (S_i^2/n_i) - S^2/N$	$k-1$ $(= \phi_2)$	M_2
Between Sub-groups within Groups	$\sum\limits_{i=1}^{k} \sum\limits_{j=1}^{r_i} (S_{ij}^2/n_{ij}) - \sum\limits_{i=1}^{k} (S_i^2/n_i)$	$R-k$ $(= \phi_1)$	M_1
Within Sub-groups	$\sum\limits_{i=1}^{k} \sum\limits_{j=1}^{r_i} \sum\limits_{t=1}^{n_{ij}} x_{ijt}^2$		
	$- \sum\limits_{i=1}^{k} \sum\limits_{j=1}^{r_i} (S_{ij}^2/n_{ij})$	$N-R$ $(= \phi_0)$	M_0
Total	$\sum\limits_{i=1}^{k} \sum\limits_{j=1}^{r_i} \sum\limits_{t=1}^{n_{ij}} x_{ijt}^2 - S^2/N$	$N-1$	

The Analysis of Variance may then be set out as in Table 6B.4. Mean squares are obtained by dividing the sums of squares by the corresponding degrees of freedom. The quantities estimated by the mean squares are as follows:

(i) *All groups containing the same number* r *of sub-groups and all sub-groups containing the same number* n *of observations*:

$$M_2 \rightarrow \sigma_0^2 + n\sigma_1^2 + rn\sigma_2^2$$
$$M_1 \rightarrow \sigma_0^2 + n\sigma_1^2$$
$$M_0 \rightarrow \sigma_0^2$$

where σ_0^2 = true variance within sub-groups = experimental error variance

σ_1^2 = true variance between means of sub-groups

σ_2^2 = true variance between means of groups

(ii) *Sub-group numbers and group numbers not all equal, as in Table 6B.4:*

$$M_2 \rightarrow \sigma_0^2 + \bar{n}_2\sigma_1^2 + \bar{n}_3\sigma_2^2$$

$$M_1 \rightarrow \sigma_0^2 + \bar{n}_1\sigma_1^2$$

$$M_0 \rightarrow \sigma_0^2$$

where $\bar{n}_1 = \left\{ N - \sum_{i=1}^{k} \left(\sum_{j=1}^{r_i} n_{ij}^2/n_i \right) \right\} \bigg/ (R-k)$

$$\bar{n}_2 = \left\{ \sum_{i=1}^{k} \left(\sum_{j=1}^{r_i} n_{ij}^2/n_i \right) - \sum_{i=1}^{k} \sum_{j=1}^{r_i} n_{ij}^2/N \right\} \bigg/ (k-1)$$

$$\bar{n}_3 = \left(N^2 - \sum_{i=1}^{k} n_i^2 \right) \bigg/ \{(k-1)N\}$$

$\bar{n}_1$ and $\bar{n}_2$ are kinds of average of the numbers of observations in the sub-groups n_{ij}; they both reduce to n when all the r_i and all the n_{ij} are equal. $\bar{n}_3$ is a kind of average of the number of observations per group; it reduces to rn when all the r_i and all the n_{ij} are equal.

Appendix 6C

Confidence limits for the components of variance

6C.1 Consider the Analysis of Variance of an unbalanced hierarchic classification with three sources of variation. The quantities estimated by the mean squares are given in Appendix 6B.2.

Estimates of the variances are:

$$\sigma_0^2 \rightarrow M_0$$

$$\sigma_1^2 \rightarrow (M_1 - M_0)/\bar{n}_1$$

$$\sigma_2^2 \rightarrow \frac{1}{\bar{n}_3} \left\{ (M_2 - M_1) + \frac{(\bar{n}_1 - \bar{n}_2)}{\bar{n}_1}(M_1 - M_0) \right\}$$

The expression $(\bar{n}_1 - \bar{n}_2)/\bar{n}_1$ is usually negligible, in which case the expression for the estimate of σ_2^2 reduces to $(M_2 - M_1)/\bar{n}_3$.

Confidence limits for σ_0

6C.11 σ_0 is estimated by $\sqrt{M_0}$ based on ϕ_0 degrees of freedom. The factors for the lower and upper confidence limits are given in Table H, from which the confidence limits for σ_0, are readily calculated [§4.33].

Confidence limits for σ_1

6C.12 In the first place we note that:

$$(\sigma_0^2 + \bar{n}_1\sigma_1^2)/\sigma_0^2 = \{1 + \bar{n}_1(\sigma_1/\sigma_0)^2\} \rightarrow M_1/M_0$$

M_1/M_0 is the ratio of two independent mean squares and is exactly comparable with the ratio s_1^2/s_0^2 of estimates of two separate variances. The corresponding confidence limits have been considered in § **4.31**. The $100(1 - 2\alpha)\%$ confidence limits for the ratio of the true variances are:

$$(s_1^2/s_0^2)L_1^2 \quad \text{and} \quad (s_1^2/s_0^2)L_2^2$$

where L_1 and L_2 are the multipliers for the lower and upper confidence limits for the ratio of two standard deviations based respectively on ϕ_1 and ϕ_0 degrees of freedom, and the probability level α (Table H). These factors have to be squared to apply to the ratio of variances.

The confidence limits for $\{1 + \bar{n}_1(\sigma_1/\sigma_0)^2\}$ are thus:

$$(M_1/M_0)L_1^2 \quad \text{and} \quad (M_1/M_0)L_2^2$$

Accordingly, the confidence limits for σ_1/σ_0 are:

$$\sqrt{\{(M_1L_1^2 - M_0)/\bar{n}_1 M_0\}} \text{ and } \sqrt{\{(M_1L_2^2 - M_0)/\bar{n}_1 M_0\}}$$

These confidence limits are exact. Multiplying these by σ_0 will give the exact confidence limits for σ_1. However σ_0 is not known exactly and has to be replaced by its estimate, $\sqrt{M_0}$. Approximate confidence limits for σ_1 are therefore:

$$\sqrt{\{(M_1L_1^2 - M_0)/\bar{n}_1\}} \text{ and } \sqrt{\{(M_1L_2^2 - M_0)/\bar{n}_1\}}$$

These limits are good approximations which can safely be applied when ϕ_0 is more than 10, but some caution must be exercised when ϕ_0 is less than 10. Situations in which ϕ_0 is so small should occur only very rarely.

When $M_1L_1^2 \leqslant M_0$, the lower confidence limit must be taken to be zero. This arises when M_1 is not significantly larger than M_0.

Confidence limits for σ_2

6C.13 These limits are given only for those cases in which $\bar{n}_1$ and $\bar{n}_2$ can be assumed equal. Only in very unbalanced designs, which should rarely occur, will this assumption be seriously in error.

Equating $\bar{n}_2$ to $\bar{n}_1$ we have:

$$\sigma_2^2 \rightarrow (M_2 - M_1)/\bar{n}_3$$

Approximate confidence limits for σ_2 are then derived in exactly the same way as for σ_1, substituting M_1 for M_0 and M_2 for M_1. The confidence limits are:

$$\sqrt{\{(M_2L_1^2 - M_1)/\bar{n}_3\}} \quad \text{and} \quad \sqrt{\{(M_2L_2^2 - M_1)/\bar{n}_3\}}$$

where L_1 and L_2 are the appropriate multipliers for ϕ_2 & ϕ_1 degrees of freedom for the chosen value of α.

Appendix 6D

Comparison of the efficiencies of certain hierarchic designs

6D.1 There is no single measure for the efficiency of a hierarchic design when the total number of measurements is fixed. However the efficiency obviously depends on the precisions of the estimates of the separate variances, so the latter may be used for comparing alternative designs. We use the following expression as a convenient measure of the precision of an estimate of a component of variance:

$$Q = \frac{\text{Upper confidence limit for the estimate}}{\text{Estimated value}}$$

Thus we will prefer designs which give low values of Q.

For experiments consisting of the analysis of samples drawn from batches of material, let σ_0^2 denote the analytical error, σ_1^2 the sampling error, and σ_2^2 the variance between batches. The Analysis of Variance table will be of the form shown in Table 6B.4.

$Q(\sigma_0^2)$, the precision of the estimate M_0 of the analytical error, depends only on the number of degrees of freedom available for the estimate, and its value is given by $[L_2(\alpha)]^2$, where $L_2(\alpha)$ is the entry in Table H for the upper confidence limit corresponding to degrees of freedom of ϕ_0 & ∞ and the chosen probability level α.

Using the expression obtained in Appendix 6C.12 for the upper confidence limit of the estimate of σ_1^2, the precision of the estimate of the sampling error is approximately:

$$Q(\sigma_1^2) = \frac{M_1 L_2^2 - M_0}{M_1 - M_0} = L_2^2 + \frac{L_2^2 - 1}{\bar{n}_1 R_1} \qquad (6D.1)$$

where $R_1 = \sigma_1^2/\sigma_0^2$, L_2 is the value of $L_2(\alpha, \phi_1, \phi_0)$ in Table H and $\bar{n}_1$ is as defined in Appendix 6B.2.

The precision of the estimate of the variance between batches is more complicated, but provided the design is balanced (i.e. $\bar{n}_1 = \bar{n}_2$) or nearly so, the following formula is a good approximation:

$$Q(\sigma_2^2) = \frac{M_2 L_2'^2 - M_1}{M_2 - M_1} = L_2'^2 + \frac{L_2'^2 - 1}{\bar{n}_3 R_2} \qquad (6D.2)$$

where $R_2 = \sigma_2^2/(\sigma_0^2 + \bar{n}_1 \sigma_1^2)$, L_2' is the value of $L_2(\alpha, \phi_2, \phi_1)$ in Table H and $\bar{n}_3$ is as defined in Appendix 6B.2.

Example

6D.2 Suppose it is required to determine the most efficient arrangement of sampling and analysing a number of batches subject to the condition that the same number of samples (not exceeding 3) is taken from each batch, the same number of analyses is carried out on each batch and the total number of analyses made is roughly 48. The five designs possible under these conditions are considered below, the precisions of the variance estimates being calculated using the upper limit of the 95% confidence interval, i.e. $\alpha = 0.025$, in the formula for Q.

Design 1

Two samples from each of 12 batches, each sample analysed twice. For this design, which is a balanced one, i.e. the same number of analyses in each sample, we have:

$N =$ total number of analyses $= 48$

$n_{ij} =$ number of analyses on the jth sample from the ith
       batch $= 2$ for all i, j

$n_i = \sum_{j=1}^{r_i} n_{ij} =$ number of analyses on the ith batch $= 4$ for all i

$\bar{n}_1 = \bar{n}_2$ $= 2$

$\bar{n}_3 =$ $= 4$

$k =$ number of batches $= 12$

$R =$ total number of samples $= 24$

We also require the degrees of freedom for the three sources of variation. These are:

$$\phi_2 = \text{degrees of freedom for batches} \quad = 11$$

$$\phi_1 = \text{degrees of freedom for sampling error} = 12$$

$$\phi_0 = \text{degrees of freedom for analytical error} = 24$$

From Table H we find, interpolating where necessary:

$$L_2(\alpha) = L_2(0.025, 24, \infty) = 1.39$$

$$L_2 \quad = L_2(0.025, 12, 24) = 1.74$$

$$L'_2 \quad = L_2(0.025, 11, 12) = 1.85$$

Hence $Q(\sigma_0^2) = 1.39^2 = 1.94$

$$Q(\sigma_1^2) = 1.74^2 + (1.74^2 - 1)/2R_1 = 3.02 + 1.01/R_1$$

$$Q(\sigma_2^2) = 1.85^2 + (1.85^2 - 1)/4R_2 = 3.43 + 0.61/R_2$$

Design 2

Three samples from each of 8 batches, each sample analysed twice. This design is also balanced.

The following results are obtained as in Design 1:

$$N = 48 \quad n_{ij} = 2 \quad n_i = 6 \quad k = 8 \quad R = 24$$

$$\bar{n}_1 = 2 \quad \bar{n}_2 = 2 \quad \bar{n}_3 = 6$$

$$\phi_2 = 7 \quad \phi_1 = 16 \quad \phi_0 = 24$$

$$L_2(\alpha) \quad = L_2(0.025, 24, \infty) = 1.39$$

$$L_2 \quad = L_2(0.025, 16, 24) = 1.62$$

$$L'_2 \quad = L_2(0.025, 7, 16) \quad = 2.13$$

$$Q(\sigma_0^2) = 1.39^2 \approx 1.94$$

$$Q(\sigma_1^2) = 1.62^2 + (1.62^2 - 1)/2R_1 = 2.63 + 0.81/R_1$$

$$Q(\sigma_2^2) = 2.13^2 + (2.13^2 - 1)/6R_2 = 4.54 + 0.59/R_2$$

Design 3

Two samples from each of 16 batches, one sample analysed once and the other twice.

Here: $\quad N = 48 \quad n_{ij} = 1, 2 \quad n_i = 3 \quad k = 16 \quad R = 32$

From Appendix 6B.2:

$$\bar{n}_1 = \{48 - 16(1^2 + 2^2)/3\}/16 = 4/3 \qquad = 1.33$$

$$\bar{n}_2 = \{16(1^2 + 2^2)/3 - 16(1^2 + 2^2)/48\}/15 = 5/3 = 1.67$$

$$\bar{n}_3 = (48^2 - 16 \times 3^2)/(15 \times 48) \qquad = 3$$

$$\phi_2 = 15 \quad \phi_1 = 16 \quad \phi_0 = 16$$

$$L_2(\alpha) = L_2(0.025, 16, \infty) = 1.52$$

$$L_2 \quad = L_2(0.025, 16, 16) = 1.66$$

$$L'_2 \quad = L_2(0.025, 15, 16) = 1.68$$

Therefore $\quad Q(\sigma_0^2) = 2.32$

$$Q(\sigma_1^2) = 1.66^2 + (1.66^2 - 1)/1.33R_1 = 2.76 + 1.32/R_1$$

Since $\bar{n}_1$ and $\bar{n}_2$ do not differ appreciably we may assume them to be equal, and it follows from Equation (6D.2) that:

$$Q(\sigma_2^2) = 1.68^2 + (1.68^2 - 1)/3R_2 = 2.84 + 0.61/R_2$$

Design 4

Three samples from each of 12 batches, two samples analysed once and the other twice.

$$N = 48 \qquad n_{ij} = 1, 1, 2 \qquad n_i = 4 \qquad k = 12 \qquad R = 36$$

$$\bar{n}_1 = \{48 - 12(1^2 + 1^2 + 2^2)/4\}/24 = 5/4 \qquad\qquad = 1 \cdot 25$$

$$\bar{n}_2 = \{12(1^2 + 1^2 + 2^2)/4 - 12(1^2 + 1^2 + 2^2)/48\}/11 = 1 \cdot 5$$

$$\bar{n}_3 = (48^2 - 12 \times 16)/(11 \times 48) \qquad\qquad\qquad = 4$$

$$\phi_2 = 11 \qquad \phi_1 = 24 \qquad \phi_0 = 12$$

$$L_2(\alpha) = L_2(0 \cdot 025, 12, \infty) = 1 \cdot 65$$

$$L_2 \qquad = L_2(0 \cdot 025, 24, 12) = 1 \cdot 59$$

$$L'_2 \qquad = L_2(0 \cdot 025, 11, 24) = 1 \cdot 78$$

$$Q(\sigma_0{}^2) = 2 \cdot 72$$

$$Q(\sigma_1{}^2) = 1 \cdot 59^2 + (1 \cdot 59^2 - 1)/1 \cdot 25R_1 = 2 \cdot 54 + 1 \cdot 23/R_1$$

and since $\bar{n}_1$ and $\bar{n}_2$ do not differ appreciably we have approximately:

$$Q(\sigma_2{}^2) = 1 \cdot 78^2 + (1 \cdot 78^2 - 1)/4R_2 = 3 \cdot 17 + 0 \cdot 54/R_2$$

Design 5

Three samples from each of 10 batches, two samples analysed twice and the third once. (Note that in this design there are 50 analyses altogether as opposed to 48 in the previous designs.)

$$N = 50 \qquad n_{ij} = 2, 2, 1 \qquad n_i = 5 \qquad k = 10 \qquad R = 30$$

$$\bar{n}_1 = \{50 - 10(2^2 + 2^2 + 1^2)/5\}/20 = 8/5 \qquad\qquad = 1 \cdot 6$$

$$\bar{n}_2 = \{10(2^2 + 2^2 + 1^2)/5 - 10(2^2 + 2^2 + 1^2)/50\}/9 \quad = 1 \cdot 8$$

$$\bar{n}_3 = (50^2 - 10 \times 5^2)/(9 \times 50) \qquad\qquad\qquad = 5$$

$$\phi_2 = 9 \qquad \phi_1 = 20 \qquad \phi_0 = 20$$

$$L_2(\alpha) = L_2(0 \cdot 025, 20, \infty) = 1 \cdot 44$$

$$L_2 \qquad = L_2(0 \cdot 025, 20, 20) = 1 \cdot 57$$

$$L'_2 \qquad = L_2(0 \cdot 025, 9, 20) \quad = 1 \cdot 91$$

$$Q(\sigma_0{}^2) = 2 \cdot 09$$

$$Q(\sigma_1{}^2) = 1 \cdot 57^2 + (1 \cdot 57^2 - 1)/1 \cdot 6R_1 = 2 \cdot 46 + 0 \cdot 92/R_1$$

and the approximate result:

$$Q(\sigma_2{}^2) = 1 \cdot 91^2 + (1 \cdot 91^2 - 1)/5R_2 = 3 \cdot 67 + 0 \cdot 53/R_2$$

These results are conveniently summarized in Table 6D.1 overleaf.

The preferred design is the one which has the lowest values for the Q ratios. However it is evident from the results that no design emerges as the best with respect to all three sources of variation. Consequently the ultimate choice will have to be decided by other considerations, such as the economics of sampling and analysing, an account of which will be found in Chapter 4 of *Design and Analysis.*

If a balanced design is preferred, then we note that whereas the first design gives the more precise estimate of σ_2^2 it gives the less precise estimate of σ_1^2

TABLE 6D.1. COMPARISON OF FIVE DESIGNS FOR ANALYSIS OF SAMPLES

Design Number	1	2	3	4	5
No. of Batches	12	8	16	12	10
No. of Analyses per Batch	4	6	3	4	5
No. of Samples per Batch	2	3	2	3	3
Total No. of Analyses (d.f. for analytical error)	48 (24)	48 (24)	48 (16)	48 (12)	50 (20)
Total No. of Samples (d.f. for sampling error)	24 (12)	24 (16)	32 (16)	36 (24)	30 (20)
d.f. for Batches	11	7	15	11	9
$Q(\sigma_0^2)$	1·94	1·94	2·32	2·72	2·09
$Q(\sigma_1^2)$	3·02 $+1\cdot01/R_1$	2·63 $+0\cdot81/R_1$	2·76 $+1\cdot32/R_1$	2·54 $+1\cdot23/R_1$	2·46 $+0\cdot92/R_1$
$Q(\sigma_2^2)$	3·43 $+0\cdot61/R_2$	4·54 $+0\cdot59/R_2$	2·84 $+0\cdot61/R_2$	3·17 $+0\cdot54/R_2$	3·67 $+0\cdot53/R_2$

The difference, however, is more marked between the precisions of σ_2^2 than those of σ_1^2, particularly when R_1 is greater than unity, as is usually the case; this may be sufficient to favour the first of the balanced designs in these circumstances.

Design 3 has the largest number of batches and, as expected, gives the most precise estimate of σ_2^2. The precisions of σ_0^2 and σ_1^2 are not much inferior to the other designs. This design in which each batch is sampled twice, one sample analysed once and the other twice, is a good compromise.

The above expressions for Q may be simplified for comparisons of the precisions of the estimates of σ_1^2 and σ_2^2. For if we assume that R_1 and R_2 are greater than unity, it is evident that for comparison purposes the values of $Q(\sigma_1^2)$ and $Q(\sigma_2^2)$ may be fairly accurately replaced by L_2^2 and $L_2'^2$ respectively, so that we need only consider the terms independent of R_1 and R_2.

Appendix 6E

Comparison of several variance estimates

Methods are given in the literature for the comparison of several variance estimates. The best known of these is the Bartlett test [6E.1], which is designed to test the Null Hypothesis that all the variance estimates being compared are estimates of the same variance. This method, however, is very sensitive to departures from the assumption of Normality; indeed, it has been shown that the Bartlett test is a good one for the purpose of testing departures from Normality ([6E.2] and [Appendix 2A of *Design and Analysis*]). Unless there is sufficient evidence to assume that the distributions are at least approximately Normal, the result of the Bartlett test should be interpreted with caution.

In § **4.411** a method is given for the comparison of two variance estimates. This method is insensitive to departures from Normality and may safely be used in all practical situations provided sufficient observations are available to justify a comparison of two variance estimates. This method may readily be extended to the comparison of several variance estimates [6E.2].

Suppose we wish to compare the variations in two or more sets of observations. Begin by dividing each group of observations into sub-groups of equal size, using a suitable process of randomization. Calculate the variance of each sub-group and the logarithms of these variances. The logarithms are then the "observations" to which we apply the usual method of the Analysis of Variance. The result of this analysis will be an Analysis of Variance table giving a mean square between groups and a mean square within groups. A comparison of these two mean squares by the F-test will test the Null Hypothesis that all the variances of the original groups of observations are equal. We may also use the mean square within groups as an error variance to compare any two groups.

With a slight sacrifice of information we may use the logarithms of the ranges of the sub-groups in place of the logarithms of the variances.

The method may also be used when the groups belong to a cross-classification, provided sufficient observations exist in each cell of the table. The observations in each cell of the table are divided into sub-groups as explained above, and the variances of these sub-groups are calculated. A new two-way table is formed, the entries in each cell now being the logarithms of the variances of the sub-groups of the observations in the corresponding cell of the original table.

An Analysis of Variance of these log variances is carried out in exactly the same way as the analysis of Example 6.4, and the resulting Analysis of Variance table enables one to decide whether the variability differs "between breakers" or "between gaugers", in just the same way as the Analysis of Variance of the original observations enables the corresponding *means* to be compared.

References

[6E.1] BARTLETT, M. S. "Properties of Sufficiency and Statistical Tests", *Proc. Roy. Soc.* A, **160**, 268–282 (1937).

[6E.2] Box, G. E. P. "Non-Normality and Tests on Variances", *Biometrika*, **40**, 318–335 (1953).

Appendix 6F

Analysis of two-way cross-classification without repeats

6F.1 The data are arranged in a two-way table as follows:

TABLE 6F.1.   TWO-WAY CLASSIFICATION

Row	Column 1	2	3	...	c	Row Totals
1	x_{11}	x_{12}	x_{13}	...	x_{1c}	R_1
2	x_{21}	x_{22}	x_{23}	...	x_{2c}	R_2
3	x_{31}	x_{32}	x_{33}	...	x_{3c}	R_3
.	.	.	.		.	.
.	.	.	.		.	.
.	.	.	.		.	.
r	x_{r1}	x_{r2}	x_{r3}	...	x_{rc}	R_r
Column totals	C_1	C_2	C_3	...	C_c	S

There are r rows and c columns, giving $N = rc$ observations. Each cell contains one observation, x_{ij}, and the grand total:

$$S = \sum_{i=1}^{r} \sum_{j=1}^{c} x_{ij} = \sum_{i=1}^{r} R_i = \sum_{j=1}^{c} C_j$$

The successive steps in computing the Analysis of Variance are as follows:

(i) Correction for the mean $= S^2/N$

(ii) Sum of squares between rows $= \sum_{i=1}^{r} R_i^2/c - S^2/N$

(iii) Sum of squares between columns $= \sum_{j=1}^{c} C_j^2/r - S^2/N$

(iv) Total sum of squares = sum of squares of all N individual observations

minus the correction for the mean $= \sum\limits_{i=1}^{r} \sum\limits_{j=1}^{c} x_{ij}^2 - S^2/N$

The Analysis of Variance table may now be constructed as in Table 6F.2, where the "remainder" is derived by subtraction from the "total". The mean squares M_R, M_C and M_0 are the sums of squares divided by the appropriate degrees of freedom. σ_R^2, σ_C^2 and σ_0^2 are the true variances of the three sources of variation.

TABLE 6F.2. ANALYSIS OF VARIANCE TABLE

Source of Variation	Sum of Squares	Degrees of Freedom	Mean Square	Quantity estimated by the Mean Square
Between Rows	$\sum\limits_{i=1}^{r} R_i^2/c - S^2/N$	$r-1$	M_R	$\sigma_0^2 + c\sigma_R^2$
Between Columns	$\sum\limits_{j=1}^{c} C_j^2/r - S^2/N$	$c-1$	M_C	$\sigma_0^2 + r\sigma_C^2$
Remainder	$\sum\limits_{i=1}^{r} \sum\limits_{j=1}^{c} x_{ij}^2 + S^2/N$ $- \sum\limits_{i=1}^{r} R_i^2/c - \sum\limits_{j=1}^{c} C_j^2/r$	$(r-1)(c-1)$	M_0	σ_0^2
Total	$\sum\limits_{i=1}^{r} \sum\limits_{j=1}^{c} x_{ij}^2 - S^2/N$	$N-1$		

Interpretation of the remainder sum of squares

6F.2 Let $\bar{x}_{i.}$ represent the mean of the ith row, $\bar{x}_{.j}$ the mean of the jth column and $\bar{x}_{..}$ the grand mean. (It is helpful when averaging over a suffix to replace that suffix with a dot.)

The deviation of x_{ij} from the grand mean may be expressed as:

$$(x_{ij} - \bar{x}_{..}) = (\bar{x}_{i.} - \bar{x}_{..}) + (\bar{x}_{.j} - \bar{x}_{..}) + (x_{ij} - \bar{x}_{i.} - \bar{x}_{.j} + \bar{x}_{..}) \qquad (6F.1)$$

Now $(\bar{x}_{i.} - \bar{x}_{..})$ and $(\bar{x}_{.j} - \bar{x}_{..})$ are simply the deviations of the row and column means from the grand mean; in other words the "row effect" and the "column effect" respectively. If the row and column effects are assumed to be purely additive, the expected value of x_{ij} would be:

$$\bar{x}_{..} + (\bar{x}_{i.} - \bar{x}_{..}) + (\bar{x}_{.j} - \bar{x}_{..}) \qquad (6F.2)$$

The deviation of the observation from its expected value is therefore:

$$x_{ij} - \{\bar{x}_{..} + (\bar{x}_{i.} - \bar{x}_{..}) + (\bar{x}_{.j} - \bar{x}_{..})\} = (x_{ij} - \bar{x}_{i.} - \bar{x}_{.j} + \bar{x}_{..}) \qquad (6F.3)$$

which is precisely the third term on the right-hand side of the identity (6F.1).

Squaring both sides of (6F.1) and summing over all observations, we obtain the following identity for the sum of squares:

$$\sum_{i=1}^{r} \sum_{j=1}^{c} (x_{ij} - \bar{x}_{..})^2 = \sum_{i=1}^{r} c(\bar{x}_{i.} - \bar{x}_{..})^2 + \sum_{j=1}^{c} r(\bar{x}_{.j} - \bar{x}_{..})^2$$

$$+ \sum_{i=1}^{r} \sum_{j=1}^{c} (x_{ij} - \bar{x}_{i.} - \bar{x}_{.j} + \bar{x}_{..})^2 \qquad (6F.4)$$

The left-hand side of this identity is the total sum of squares; the terms on the right-hand side are respectively the sums of squares between rows, between columns and remainder. Hence this identity expresses the additive property of the sums of squares in the Analysis of Variance Table 6F.2.

From Equation (6F.3) we see that the remainder sum of squares is the sum of squares of the deviations of the actual observations from their expected values according to a simple row and column effects additive model.

If we set up the Null Hypothesis that there are no real differences in row means or column means, then each mean square in Table 6F.2 supplies an unbiased estimate of the experimental error variance σ_0^2.

Appendix 6G

Estimation of a missing value in a two-way cross-classification without repeats

It sometimes happens in an experiment of the type considered in Example 6.3, i.e. with data which can be classified independently in two ways, that one or more of the results is unreliable or missing, due to some abnormal cause or owing to an accident. For example, in penicillin testing a pot may be accidentally disturbed, causing the penicillin solution to leak on the surface.

When only one result in the set is missing or abnormal it may be estimated by the following formula:

$$K(r-1)(c-1) = (r+c-1)S - rS_R - cS_C \qquad (6G.1)$$

where K = estimate of missing value,

r = number of rows,

c = number of columns,

S = sum of all known $(rc-1)$ observations,

S_R = sum of row totals, excluding the row from which the result is missing,

S_C = sum of column totals, excluding the column from which the result is missing.

For example, suppose that in Table 6.6 the result on Sample 1 of Plate 1 were missing. We have:

$$r = 24, \quad c = 6, \quad S = 428 - 7 = 421$$

$$S_R = 428 - 23 = 405, \quad S_C = 428 - 124 = 304$$

Therefore, substituting in the formula:

$$K \times 23 \times 5 = 29 \times 421 - 24 \times 405 - 6 \times 304$$

or
$$115K = 665$$

i.e.
$$K = 665/115 = 5 \cdot 8$$

This differs from the observed value, owing to experimental error.

The Analysis of Variance is then computed in the standard manner of Appendix 6F, using the estimate of the missing value in the vacant cell. Subsequently the sum of squares between rows should be adjusted for bias by subtracting the small quantity:

$$\{(S - S_c) - (r - 1)K\}^2 / r(r - 1) \qquad (6G.2)$$

and the sum of squares between columns should similarly be reduced by:

$$\{(S - S_r) - (c - 1)K\}^2 / c(c - 1) \qquad (6G.3)$$

The degrees of freedom for the remainder and for the total sum of squares are each reduced by unity. After applying these adjustments the row and column effects can each be tested for significance against the remainder in the usual manner.

When two or more values are missing from an unreplicated two-way cross-classification the analysis is more complicated, and the reader should refer to Appendix 6B of *Design and Analysis* for a full discussion of the procedure in such cases.

Chapter 7

Linear Relationships between Two Variables

In experimental work we often wish to find out
whether, and by how much, the level of one quantity
changes with changes in the level of some other
quantity. In this chapter a number of problems of
this sort are considered, in which the relationship can
be represented approximately by a straight line graph.

Introduction

7.1 Physical and chemical laws can usually be expressed in mathematical
form. A simple example is Boyle's Law relating volume and pressure of a
perfect gas under constant temperature. Some of these laws have been
derived empirically and others, for example the sedimentation laws of sus-
pensions of particles in a viscous medium, have been derived mathematically.
Each law expresses a relationship between a number of variables and can be
used to estimate the value of any one variable, given values of all the others.
This is the principle used in instruments for measuring temperatures, pressures, flow rates of liquids and gases, etc.

This chapter deals mainly with the problem of estimating empirically a
relation between variables. The problem is frequently encountered throughout industry, as for instance in the calibration of instruments. It also arises
in laboratory testing when the property of main interest can be determined
only by destroying the article, as in tests for breakage strength, or by un-
economical means, as in most service tests on finished materials. It is then
desirable to find more economical tests by measuring some other properties
which can be related to the required property.

There are many other applications, one of particular importance in the
chemical industry being to estimate the dependence of yield and quality of
products on reaction conditions such as temperature, pressure, concentration,
time of reaction, etc. This is required for determining the best operating
conditions of the process and for assessing what latitude can be tolerated in
the reaction conditions. In chemical processes a further advantage of estimating empirically the relation between variables is that it may increase one's
understanding of the chemistry and the kinetics of the reaction.

Functional relationship

7.11 When a unique relationship between the variables exists or can be postulated, then the variables are said to be Functionally Related. As will be shown later, not all related variables are functionally related. If, however, it is known that a variable y and another variable x are functionally related, then an approximation to the form of this relationship over given ranges of the variables can be estimated empirically from measurements of the corresponding values of y and x when these are varied over the given ranges. An example is the calibration of a platinum resistance thermometer, where the resistance is measured at a number of fixed reference points of temperature—such as the boiling points and freezing points of certain liquids—which, under specified conditions, are known not to vary and have been determined with a high degree of accuracy in relation to the absolute temperature scale. For simplicity we shall assume that the resistance and temperature are linearly related, although in practice this is only a rough approximation.

Measurements of resistance are subject to error, and the temperature reference points depend on having exact conditions which may not be experimentally attainable. The plotted points will not therefore fall exactly on a straight line but will vary randomly about one, that is, about the graph of the functional relation. If the determinations are repeated again and again and the average values plotted, then in the limit the points will, if there are no systematic errors, fall exactly on the graph of the functional relation.

The example illustrates the situation in which a unique functional relationship exists and independent repeat determinations on the same sample or on the same reference points vary randomly about it; in other words, the variation of the experimentally determined points about the graph of the functional relationship is due entirely to random experimental errors. The functional relationship describes the relationship between the true values of the variables, and is the one to use when estimating the true value of one variable from an experimentally determined value of the other, even though the latter may be subject to random error. The more precisely the one variable is measured the more precise will be the estimate of the other variable.

Regression

7.12 Relationships between variables are not always unique in the sense that, apart from experimental errors, a particular value of one variable always corresponds to the same value of the other variable. An example is the relation between the weight and height of adult males for a given population. If the weight of each individual is plotted against his height, a diagram such as that given in Figure 7.1 results.

This method of graphical presentation is called a Dot or Scatter Diagram. For any given height there is a wide range of observed weights, and *vice versa*. Some of this variation will be due to errors of measurement of weight and height, but most of it will be due to real variation between individuals. There is thus no unique relationship between true weight and true height. However, the average observed weight for a given observed height increases with increasing height, and the average observed height for a given observed weight increases with increasing weight. The graph of the mean value of one variable for given values of the other variable, when referred to the whole

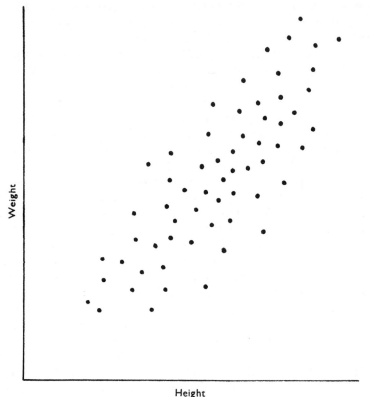

FIG. 7.1. SCATTER DIAGRAM

population, is called a Regression Curve, or simply a Regression. The graph, or in other words the *locus* of the mean weight of all individuals of given height, is called the Regression of Weight upon Height, and the locus of mean height of all individuals of given weight is called the Regression of Height upon Weight. *These two regression lines are generally different*, as shown in Figure 7.2.

The ellipse is intended to depict the distribution of weight and height. The mid-points of the vertical lines depict the mean weights for given heights, and these fall in a straight line. The mid-points of the horizontal lines depict the mean heights for given weights, and these also fall on a straight line, but this line is different from the other. Both pass through the centre of the ellipse. The two lines are closer together for more elongated ellipses.

Another example is given in Figure 7.31. If the two variables in this figure were functionally related and the scatter arose because of errors of measurement, the two regression lines would still exist and a third line could be drawn representing the functional relationship. The three lines are different, but all pass through the grand mean of the data. The line representing the functional relationship lies between the two regression lines.

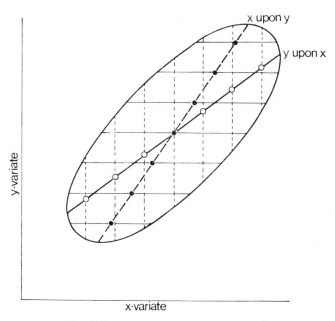

FIG. 7.2.   THE TWO REGRESSION LINES

The possible existence of all three lines may appear a little confusing. Usually, a functional relationship is determined from a planned experiment carried out under carefully controlled conditions. Under these circumstances the variation about the line is small and all three lines are practically identical.

It must be emphasised that a functional relationship exists only in those situations where the variation about the line is due entirely to experimental errors, and it represents the relationship between the "true" values of the variables which need not refer to any given Population. On the other hand, the regression of variate y upon x must refer to a Population of values of y for any given x, and similarly the regression of variate x upon y must refer to a Population of values of x for any given y. Specifically the regression lines represent the relationships between the means of the values of one variable and any specified values of the other. They are used in those situations where we are mainly interested in mean values. The following is an illustration based on the example of Figure 7.1. Suppose we had observed the height of an individual from this population but not his weight, and we required to estimate the latter. For the given observed height there is a whole range of possible observed weights; the mean of these gives, however, an unbiased estimate of the weight we might expect to observe for the individual.

This chapter deals only with linear regression and functional relationships between two variables, more complex relationships being considered in Chapter 8. Regressions are treated first, because the calculations are usually simpler, and because in most experiments designed to estimate functional relationships the independent variables (e.g. temperature, pressure and concentration in a chemical process) are usually determined without appreciable error, and in these circumstances it can be shown that the regression of the dependent variable (yield, quality) on the independent variables is also the best estimate of the functional relationship. It will be appreciated, however, that it is usually a functional relationship which is of most interest in the chemical industry.

Another class of problems involving related variables are those in which we are not so much interested in estimating one variable from another as in studying the joint distribution of the variables and in obtaining a general measure of the extent of association between them. Such problems come under the title of "Correlation", and are dealt with in § **7.8**.

In an experiment comparing the effectiveness of two drug treatments on groups of individuals in a clinical trial, the individuals within each group vary in a number of characteristics, such as weight, age, etc., which may influence the effectiveness of the drug. These unavoidable variations will increase the variation of the response to the drugs within the groups, and if they are not allowed for may seriously reduce the precision of the comparison between the two groups. Methods exist to eliminate the effect of these variations between the individuals, and thus increase the sensitivity of the experiment. These methods come under the title of "Allowance for Concomitant Variation", usually known as "Analysis of Covariance", and are considered in § **7.7** for one "nuisance" factor. More complicated cases of Multiple Regression are dealt with in Chapter 8.

LINEAR REGRESSION

7.2 An example will first of all be described which requires the use of linear regression. This will be followed by an exposition of the methods involved in deriving the equation of the regression in the general case, and in deriving the confidence limits for the equation and for predictions obtained from it. These methods will then be applied to the particular example, and will be followed by a more detailed discussion of the meaning and the uses of regression equations.

Example 7.1. Hardness and abrasion resistance of rubber

7.21 In an investigation of the properties of a synthetic rubber, varying amounts and types of compounding material were used to prepare a number of specimens.. A wide range of physical properties was obtained. Table 7.1

TABLE 7.1. PHYSICAL PROPERTIES OF RUBBER SAMPLES

Specimen No.	Abrasion Loss (g./h.p.-hour)	Hardness (I.R.H. (Shore) units)
1	372	45
2	206	55
3	175	61
4	154	66
5	136	71
6	112	71
7	55	81
8	45	86
9	221	53
10	166	60
11	164	64
12	113	68
13	82	79
14	32	81
15	228	56
16	196	68
17	128	75
18	97	83
19	64	88
20	249	59
21	219	71
22	186	80
23	155	82
24	114	89
25	341	51
26	340	59
27	283	65
28	267	74
29	215	81
30	148	86

lists the abrasion loss and the hardness of thirty different specimens; these results are shown graphically in Figure 7.3 overleaf.

An inspection of these figures shows that high values of abrasion loss tend to coincide with low values of hardness. This is brought out more clearly in the scatter diagram of Figure 7.3.

The errors in measuring the abrasion loss and hardness are small compared with the scatter of the points about any vertical or horizontal line; i.e. repeat measurements on the same samples will give substantially the same results. Different samples may have the same hardness but different abrasion losses, and *vice versa*, these differences being mainly due to real variations between the samples. There is no question, therefore, of a unique functional relationship between these two properties.

The determinations of hardness can be carried out quickly and simply, but the determinations of abrasion loss require an elaborate apparatus and are much more difficult to carry out. It would therefore be a considerable

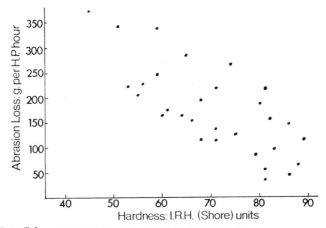

FIG. 7.3.   SCATTER DIAGRAM OF ABRASION LOSS AND HARDNESS

advantage if abrasion loss could be predicted with sufficient accuracy from the measurements of hardness. We require an equation expressing abrasion loss in terms of hardness which in the long run will give the best estimates of abrasion losses on future samples for which only hardness measurements will be carried out. Let y denote abrasion loss and x hardness.

Figure 7.3 represents observed values on 30 specimens; suppose observations were available on an indefinitely large number of specimens. For any given observed value of hardness there would be a distribution of observed values of abrasion loss. This distribution will be assumed to have the same standard deviation, σ, throughout the range of hardness. This assumption may seem a very restrictive one, but in practice it is usually sufficiently closely satisfied, and where it is not, the data can usually be transformed to make it valid, or alternatively a weighted regression can be carried out [§ 7.5]. Let $Y(x)$ be the mean abrasion loss for the given hardness, then the distribution of the possible values that the abrasion loss can take for a future specimen with a hardness of x can be expressed in terms of $Y(x)$ and σ. $Y(x)$, considered as a function of x, is the regression of abrasion loss on hardness, and it is clear from an inspection of Figure 7.3 that the regression can be assumed to be linear. Neither $Y(x)$ nor σ is known, and these have to be estimated from the data. Even in a fairly large sample from a Population in which the regression is linear, the plot of the means of the actual observations of one variable for given values of the other will not usually fall exactly on a straight line, owing to random sampling errors. The regression calculated from a sample will be an *estimate* of the Population regression, and will be subject to the usual random sampling errors. The estimated regression equation of the abrasion loss upon hardness gives the predicted value of abrasion loss for any given value of hardness. The confidence limits of the predicted value can be calculated, and the methods for doing this are considered in later sections.

Estimation of regression lines

7.22 The actual method of estimating a linear regression, usually referred to as fitting a straight line to the data, is straightforward. Denoting abrasion loss by y and hardness by x, we require to estimate the values of the parameters α and β in the equation:

$$Y = \alpha + \beta x \tag{7.1}$$

The capital letter is introduced to denote the "predicted" or mean value of the dependent variable, and the small letter is used to denote an observed value.

There are several possible methods of estimating α and β in Equation (7.1), but the one of most interest is called the Method of Least Squares. This consists of finding the values of α and β that minimize the sum of squares of the deviations of the observed values (in this case the abrasion loss) from the line. If we follow a maximum likelihood approach [Appendix 4B], it happens that this reduces to the method of least squares when the observed values of y are Normally distributed about Y. The estimated values of α and β will be denoted by a and b respectively.

If we denote the observations by

$$(x_1, y_1), (x_2, y_2), \ldots, (x_n, y_n)$$

then for any point x_i, the predicted value Y_i of y_i is $\alpha + \beta x_i$. The vertical distance between the actual y_i and the predicted Y_i is therefore $y_i - (\alpha + \beta x_i)$ and the sum of squares of all of the n such deviations is:

$$S.S. = \Sigma\{y - (\alpha + \beta x)\}^2$$

Minimizing $S.S.$ by partial differentiation with respect to α and β and equating to zero, we obtain two equations for the parameter estimates a and b:

$$na + b\Sigma x = \Sigma y \tag{7.11}$$
$$a\Sigma x + b\Sigma x^2 = \Sigma xy \tag{7.12}$$

which are usually referred to as the Normal Equations. Dividing by n, Equation (7.11) may be written as:

$$a = \bar{y} - b\bar{x} \tag{7.13}$$

indicating that the regression line passes through the centroid of the data, and that the regression Equation (7.1) may be written as:

$$Y = \bar{y} + b(x - \bar{x}) \tag{7.14}$$

From Equations (7.12) and (7.13) we obtain:

$$b = (\Sigma xy - n\bar{x}\bar{y})/(\Sigma x^2 - n\bar{x}^2) \tag{7.15}$$

which may be rewritten as

$$b = \frac{\Sigma xy - (\Sigma x)(\Sigma y)/n}{\Sigma x^2 - (\Sigma x)^2/n} \tag{7.16}$$

or alternatively as

$$b = \Sigma(x - \bar{x})(y - \bar{y})/\Sigma(x - \bar{x})^2 \tag{7.17}$$

The form (7.16) is the simplest to use for numerical calculation.

The quantity b is called the Estimated Regression Coefficient of y upon x. It will be seen that the numerator of b is the sum of products of the deviations of the x's and y's from their respective means, whilst the denominator is the sum of squares of the x's about their mean. The quantity Σxy in Equation (7.16) is called the Crude Sum of Products of the observations, and $(\Sigma x)(\Sigma y)/n$ is the Correction for the Means, in an obvious extension of the univariate nomenclature.

A simpler notation is to let

$$S_{xx} = \Sigma(x - \bar{x})^2 = \Sigma x^2 - (\Sigma x)^2/n$$

and

$$S_{xy} = \Sigma(x - \bar{x})(y - \bar{y}) = \Sigma xy - \Sigma x \Sigma y/n$$

whence:

$$b = S_{xy}/S_{xx} \tag{7.18}$$

In order to simplify the arithmetical computations, a constant quantity may be subtracted from each value of x and similarly from each y; the constant subtracted from the x's need not be the same as that subtracted from the y's. Similarly the data may be multiplied or divided by an appropriate constant factor, remembering in both cases to descale the final equation in order to restore the proper units of measurement.

The arithmetical calculations necessary for performing a regression analysis are rather laborious, particularly if there are a large number of original observations, but appropriate computer programs are now available for virtually every model of electronic computer. If the calculations have to be performed on a desk calculating machine with an array keyboard, there is a considerable saving in time if the y observations are entered on the left-hand side of the keyboard simultaneously with the x's being entered on the right-hand side. The accumulating registers then give Σy, Σx, Σy^2, $2\Sigma xy$ and Σx^2, i.e. all the basic summations.

In the above equations and all subsequent ones, the corresponding regression analysis of variate x upon y is obtained by the straightforward interchange of the symbols x and y. The estimated regression equation of x upon y is:

$$X = \bar{x} + b'(y - \bar{y}) \qquad (7.19)$$

where $b' = S_{xy}/S_{yy}$, and results from minimizing the sum of squares of the x deviations, i.e. the *horizontal* distances from the regression line. Note that b' is *not* simply the reciprocal of b, and in fact $|b'| \leqslant 1/|b|$, the equality only being attained when all points fall exactly on a single line.

Example with step-by-step calculations

7.221 In the data of Table 7.1 let y = abrasion loss and x = hardness.

(i) *Calculate Sum of Squares of Hardness about Mean*

$$\text{Number of observations} = 30$$
$$\Sigma x = 2,108$$
$$\text{Hence } \bar{x} = 2,108/30 = 70 \cdot 27$$
$$\text{Crude sum of squares} = 152,422$$
$$\text{Correction for the mean} = (2,108)^2/30 = 148,122$$
$$\text{Hence sum of squares about mean, } S_{xx} = 152,422 - 148,122$$
$$= 4,300$$

(ii) *Calculate Sums of Squares of Abrasion Loss about Mean*

$$\Sigma y = 5,263$$
$$\text{Hence } \bar{y} = 5,263/30 = 175 \cdot 4$$
$$\text{Crude sum of squares} = 1,148,317$$
$$\text{Correction for the mean} = (5,263)^2/30 = 923,306$$
$$\text{Hence sum of squares about mean, } S_{yy} = 1,148,317 - 923,306$$
$$= 225,011$$

(iii) *Calculate Sum of Products of Abrasion Loss and Hardness about Their Respective Means*

$$\text{Crude sum of products of abrasion}$$
$$\text{loss and hardness, } \Sigma xy = 346,867$$
$$\text{Correction for the means} = (2,108 \times 5,263)/30$$
$$= 369,813$$
$$\text{Sum of products about means, } S_{xy} = 346,867 - 369,813$$
$$= -22,946$$

(iv) *Calculate Regression Coefficient of y upon x*

Regression Coefficient, $b = S_{xy}/S_{xx} = -22{,}946/4{,}300$

$$= -5 \cdot 336$$

(v) *Calculate Regression Equation of y upon x*

This is: $\qquad Y = \bar{y} + b(x - \bar{x})$

i.e. $\qquad Y = 175 \cdot 4 - 5 \cdot 336(x - 70 \cdot 27)$

$$= 550 \cdot 4 - 5 \cdot 336x$$

and is used to predict the expected value of abrasion loss, given the hardness of a rubber specimen.

(vi) *Calculate Regression of x upon y*

Regression coefficient of hardness upon abrasion loss:

$$b' = S_{xy}/S_{yy} = -22{,}946/225{,}011 = -0 \cdot 1020$$

Hence the regression equation is: $\qquad X = 70 \cdot 27 - 0 \cdot 1020(y - 175 \cdot 4)$

$$= 88 \cdot 16 - 0 \cdot 1020y$$

This could be used to predict the expected value of hardness given the abrasion loss of a rubber specimen, but would not in fact be required in the present example since hardness can be measured much more easily than abrasion loss.

The two regression lines have been plotted in Figure 7.31, where it is seen that the two regressions differ quite considerably in this example.

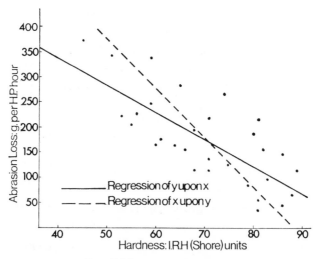

FIG. 7.31.   REGRESSION LINES

Variation about the regression

7.23 The estimated regression equation of y upon x based upon a sample of n observations is:

$$Y = \bar{y} + b(x - \bar{x})$$

where b is given by Equations (*7.15–7.18*). The precision of this regression equation will depend on n, on the extent of scatter about the regression and on the range of values of the x variable. The variation about the regression, represented by the deviations of the observed y's from the line, must be less than the total variation of the observed y's about their mean, and the extent by which it is less represents the amount of variation accounted for by the regression.

Consider any one observation (x_1, y_1) of the sample. The point on the regression line corresponding to x_1, given by substituting x_1 in the regression equation, is $Y_1 = \bar{y} + b(x_1 - \bar{x})$. The deviation of the actual observation y_1 from the regression can be expressed as the algebraic difference between two deviations, that of the deviation of the observation from the average, $y_1 - \bar{y}$, and that of the deviation of the corresponding point on the regression from the average, $Y_1 - \bar{y}$; thus:

$$y_1 - Y_1 = (y_1 - \bar{y}) - (Y_1 - \bar{y}) \qquad (7.2)$$

This is seen to be an algebraic identity. The deviation $(Y_1 - \bar{y})$ is seen from the regression equation to be equal to $b(x_1 - \bar{x})$.

Similar identities can be written down for each of the other $n - 1$ deviations. Squaring each identity and summing the equations, we obtain the following expression for the sum of squares of the deviations from the regression:

$$
\begin{aligned}
\Sigma(y - Y)^2 &= \Sigma(y - \bar{y})^2 - 2\Sigma(y - \bar{y})(Y - \bar{y}) + \Sigma(Y - \bar{y})^2 \\
&= \Sigma(y - \bar{y})^2 - 2b\Sigma(y - \bar{y})(x - \bar{x}) + b^2\Sigma(x - \bar{x})^2 \\
&= \Sigma(y - \bar{y})^2 - 2b^2\Sigma(x - \bar{x})^2 + b^2\Sigma(x - \bar{x})^2 \qquad \text{from (7.17)} \\
&= \Sigma(y - \bar{y})^2 - b^2\Sigma(x - \bar{x})^2 \\
&= \Sigma(y - \bar{y})^2 - \Sigma(Y - \bar{y})^2 \qquad\qquad\qquad (7.21)
\end{aligned}
$$

Now, in Equation (*7.21*), $\Sigma(y - \bar{y})^2$ is the sum of squares of the observations about their mean $\bar{y}$, and $\Sigma(Y - \bar{y})^2$, called the sum of squares due to the regression, is the sum of squares of the deviations of the corresponding points on the regression from the mean. We thus have the identity:

$$\frac{\text{Sum of squares of } y}{\text{about regression}} = \frac{\text{Sum of squares}}{\text{of } y \text{ about mean}} - \frac{\text{Sum of squares due}}{\text{to the regression}} \qquad (7.22)$$

This is equivalent to:

$$\frac{\text{Sum of squares}}{\text{of } y \text{ about mean}} = \frac{\text{Sum of squares}}{\text{about regression}} + \frac{\text{Sum of squares due}}{\text{to the regression}} \qquad (7.23)$$

The sum of squares due to the regression represents the amount by which the total sum of squares of one variable is reduced when allowance is made for variations in the other variable. The alternative expression for $\Sigma(Y-\bar{y})^2$ is $b^2\Sigma(x-\bar{x})^2$ and the latter is the one normally used for the computation. The variations in x are thus responsible for a component $b^2\Sigma(x-\bar{x})^2$ of the total variation in y. When this component is expressed as a percentage of the total variation we obtain a measure called the Percentage Fit—a very simple and useful statistic which indicates how "good" the relationship is between the variables.

The sum of squares about the regression is based on $n-2$ degrees of freedom, since two degrees of freedom have been used up in calculating the regression, one for the mean $\bar{y}$ and one for the regression coefficient b. Dividing the sum of squares about the regression by the degrees of freedom gives an estimate, s^2, of the Variance about the Regression, σ^2.

If the regression equation were known exactly, or estimated from an extremely large number of observations, the variance about the regression would represent the error with which any observed value of y could be predicted from a given value of x. This variance is often referred to as the error variance, and an estimate of it based on $n-2$ degrees of freedom is given by the variance about the regression for the sample. This variance is of fundamental importance because the accuracy of the estimate of the regression equation from a sample depends upon it.

Analysis of variance for regression

7.24 The sums of squares Equation (*7.23*), together with the corresponding degrees of freedom and mean squares, may conveniently be presented in the form of an Analysis of Variance, as shown in Table 7.2.

TABLE 7.2.   ANALYSIS OF VARIANCE FOR REGRESSION

Source of Variation	Sum of Squares	Degrees of Freedom	Mean Square	Quantity estimated by Mean Square
Due to regression	$b^2\Sigma(x-\bar{x})^2$	1	M_1	$\sigma^2 + \beta^2\Sigma(x-\bar{x})^2$
About regression	$\Sigma(y-\bar{y})^2 - b^2\Sigma(x-\bar{x})^2$	$n-2$	M_0	σ^2
Total	$\Sigma(y-\bar{y})^2$	$n-1$		

The sum of squares *about* the regression is usually computed by subtracting the sum of squares *due* to regression from the total sum of squares. The mean squares due to and about regression are denoted by M_1 and M_0 respectively.

It can be shown that, for random samples from Populations in which the observations of y are independently and Normally distributed about $\alpha + \beta x$ with a common variance σ^2, the expected values of M_1 and M_0 are $\sigma^2 + \beta^2 \Sigma(x - \bar{x})^2$ and σ^2 respectively. Further, if in any specific example M_1 is judged significantly greater than M_0 by an F-ratio test, then the regression is statistically significant, i.e. the variates y and x are probably connected by a genuine relationship. Thus, just as in Chapter 6 we used the Analysis of Variance to determine whether certain sources contribute significantly to the variations in a set of data, so now we can test whether the values of x offer meaningful predictors for the values of y.

Example with step-by-step calculations

7.241 Continuing the analysis of the data of Table 7.1, we first calculate the sum of squares due to regression:

$$b^2 \Sigma(x - \bar{x})^2 = (S_{xy}/S_{xx})^2 S_{xx} = S_{xy}^2/S_{xx}$$
$$= (-22,946)^2/4,300$$
$$= 122,446$$

The sum of squares about regression is obtained by subtracting this sum from the total sum of squares of abrasion loss about the mean, which was calculated as 225,011 in § **7.221**. The Analysis of Variance Table may then be constructed as in Table 7.21.

TABLE 7.21.   ANALYSIS OF VARIANCE FOR REGRESSION OF ABRASION
LOSS UPON HARDNESS (BASED ON TABLE 7.1)

Source of Variation	Sum of Squares	Degrees of Freedom	Mean Square
Due to regression	122,446	1	122,446
About regression	102,565	28	3,663
Total	225,011	29	

The percentage fit is $100 \times 122,446/225,011 = 54.4\%$.

To assess the significance of the regression we calculate the ratio of the mean squares due to and about the regression. This gives $122,446/3,663 = 33.4$. For 1 & 28 degrees of freedom the $P = 0.05$ and 0.01 critical values of F are 4.20 and 7.64 respectively [Table D]. The regression is thus highly significant, as might have been expected from a cursory inspection of Figure 7.3.

Similarly the Analysis of Variance for the regression of hardness upon abrasion loss can be constructed as shown in Table 7.22.

TABLE 7.22.   ANALYSIS OF VARIANCE FOR REGRESSION OF HARDNESS
UPON ABRASION LOSS (BASED ON TABLE 7.1)

Source of Variation	Sum of Squares	Degrees of Freedom	Mean Square
Due to regression	2,340	1	2,340
About regression	1,960	28	70
Total	4,300	29	

Note that the percentage fit at 54·4% and the ratio of the mean squares at 33·4 are identical with those for Table 7.21.   From the standpoint of statistical significance both regressions are equivalent.

Standard error of the regression coefficient

7.25 The more scatter there is about the regression line, the less precisely will the slope of the line, b, be known.   Now the regression coefficient of y upon x:

$$b = \frac{\Sigma(x-\bar{x})(y-\bar{y})}{\Sigma(x-\bar{x})^2}$$

$$= \frac{\Sigma(x-\bar{x})y}{\Sigma(x-\bar{x})^2} - \frac{\bar{y}\Sigma(x-\bar{x})}{\Sigma(x-\bar{x})^2} = \frac{\Sigma(x-\bar{x})y}{\Sigma(x-\bar{x})^2}$$

may be expanded as:

$$b = \frac{1}{\Sigma(x-\bar{x})^2} \{(x_1-\bar{x})y_1+(x_2-\bar{x})y_2+\ldots+(x_n-\bar{x})y_n\} \qquad (7.3)$$

The quantities $y_1, y_2, \ldots, y_n$ are the observed values of y for given values of x, which is our independent variable.   The expression b is thus of the form:

$$a_1y_1+a_2y_2+\ldots+a_ny_n \qquad (7.31)$$

where y_1 to y_n are independent observations and the a's are constants.

The variance of b is therefore given by [§**3.71**]:

$$V(b) = \Sigma a_i^2 V(y_i) = \sigma^2 \Sigma a_i^2$$

$$= \sigma^2 \frac{\Sigma(x-\bar{x})^2}{\{\Sigma(x-\bar{x})^2\}^2} = \frac{\sigma^2}{\Sigma(x-\bar{x})^2} \qquad (7.32)$$

Hence: $$S.E. \; (b) = \sigma/\sqrt{\{\Sigma(x - \bar{x})^2\}} \qquad (7.321)$$

We have to substitute for σ its estimated value, s, based on $n-2$ degrees of freedom. If we assume that the variation about the line is Normal, the $100(1 - 2\alpha)\%$ confidence limits for b are:

$$b \pm t_\alpha s/\sqrt{\{\Sigma(x - \bar{x})^2\}} \qquad (7.33)$$

where t_α is the value of t from Table C corresponding to probability α for $n-2$ degrees of freedom. The confidence limits are not sensitive to departures from the assumption of Normality, and may be assumed to be "correct" in most practical situations.

It will be seen that the precision with which the regression coefficient is estimated depends upon the variance about the regression, the range of the values of the x variable and the size of the sample. To test whether b differs from zero we calculate the expression:

$$t = |b|/S.E. \; (b) = |b|\sqrt{\{\Sigma(x - \bar{x})^2\}}/s \qquad (7.331)$$

and refer the value to the t-tables. t_α and t have the same number of degrees of freedom as s^2, in this case $n-2$. This test is exactly equivalent to the F test in § **7.24** for the significance of the regression relationship, since, when $\phi_N = 1$, the value of F for a given probability point is equal to the square of the value of t for the two-sided test.

Example with step-by-step calculation

7.251 We have already shown that for the data of Table 7.1 the estimated variance about the regression is 3,663 based on 28 degrees of freedom, and that the sum of squares of the x's about their mean is 4,300.

Hence the variance of the regression coefficient is $3,663/4,300 = 0.8519$ and its standard error is $\sqrt{0.8519} = 0.923$.

The t factor for $P = 0.025$ and 28 degrees of freedom is 2.05 and therefore the 95% confidence limits for the regression coefficient are

$$-5.336 \pm 2.05 \times 0.923 = -3.444 \text{ and } -7.228$$

These limits show the precision with which the regression coefficient is estimated. Since they do not include zero, the regression relationship is statistically significant. Alternatively we may draw the same conclusion by computing the ratio

$$|b|/S.E. \; (b) = 5.336/0.923 = 5.78$$

which exceeds 2.76, the critical level of t for $P = 0.005$ and 28 degrees of freedom. Note that $5.78^2 = 33.4$, the same as the ratio of mean squares obtained in § **7.241**, so that the two methods of testing the significance of the regression are exactly equivalent.

The difference between two regression coefficients

7.26 When two estimates of a regression coefficient are available, for example coefficients estimated from two sets of experiments or from two subsets of a body of data, it is usually required to compare their values. Let the two regression coefficients be b_1 and b_2 and let σ^2 be the error variance, which is usually estimated by combining the mean squares about the two regressions, i.e.

if $s_1^2 =$ variance about the first regression (slope b_1)
with ϕ_1 degrees of freedom

and $s_2^2 =$ variance about the second regression (slope b_2)
with ϕ_2 degrees of freedom

then σ^2 is estimated by:

$$s^2 = (\phi_1 s_1^2 + \phi_2 s_2^2)/(\phi_1 + \phi_2)$$

From Equation (7.32) and substituting s for σ:

$$V(b_1) = s^2/\Sigma_1(x-\bar{x})^2 \qquad V(b_2) = s^2/\Sigma_2(x-\bar{x})^2$$

where $\Sigma_1(x-\bar{x})^2$ is based on the observations from which b_1 was calculated and similarly for $\Sigma_2(x-\bar{x})^2$. Therefore, since b_1 and b_2 are independent estimates [§ 3.71]:

$$V(b_1 - b_2) = s^2 \left\{ \frac{1}{\Sigma_1(x-\bar{x})^2} + \frac{1}{\Sigma_2(x-\bar{x})^2} \right\} \qquad (7.34)$$

and

$$S.E.\ (b_1 - b_2) = s \sqrt{\left\{ \frac{1}{\Sigma_1(x-\bar{x})^2} + \frac{1}{\Sigma_2(x-\bar{x})^2} \right\}} \qquad (7.341)$$

These enable confidence limits for the difference to be calculated, using the value of t with $(\phi_1 + \phi_2)$ degrees of freedom. If the variances about the two regressions cannot be assumed equal, then the confidence limits are calculated using a method similar to that of § 4.231.

Standard error of the intercept

7.27 It is sometimes desired to test whether the estimate of the intercept a of a regression line, i.e. the value of y when $x = 0$, differs significantly from some specified value (usually zero).

Now from Equation (7.13), $a = \bar{y} - b\bar{x}$. It can be shown that $\bar{y}$ and b are independent of each other, whence it follows from § 3.71 that, since the x's are regarded as constants:

$$V(a) = V(\bar{y} - b\bar{x}) = V(\bar{y}) + \bar{x}^2 V(b)$$

Now $V(\bar{y}) = \sigma^2/n$ and $V(b)$ is given by Formula (7.32). Hence:

$$V(a) = \sigma^2 \left\{ \frac{1}{n} + \frac{\bar{x}^2}{\Sigma(x-\bar{x})^2} \right\} = \frac{\sigma^2 \Sigma x^2}{n \Sigma(x-\bar{x})^2} \qquad (7.35)$$

and

$$S.E.\ (a) = \sigma \sqrt{\left\{ \frac{\Sigma x^2}{n\Sigma(x-\bar{x})^2} \right\}} \qquad (7.351)$$

Therefore the $100(1-2\alpha)\%$ confidence limits for a are given by:

$$a \pm t_\alpha s \sqrt{\left\{ \frac{\Sigma x^2}{n\Sigma(x-\bar{x})^2} \right\}} \tag{7.36}$$

where s is the estimate of σ, the standard error about the regression, and t_α corresponds to a probability α for $n-2$ degrees of freedom.

Standard error of the regression estimate

7.28 It has been shown that both parameters in the regression equation are subject to error because they are estimated from a limited number of observations. An error in the intercept gives rise to a constant error for any point on the line, since a change in this parameter moves the line up or down without changing its slope. An error in the slope of the line gives rise to an error which is zero at the point $(\bar{x}, \bar{y})$ and increases as x departs from $\bar{x}$. Any prediction made from the regression equation is therefore also subject to error.

The regression estimate at a given value x_0 of x is given by:

$$Y_0 = \bar{y} + b(x_0 - \bar{x})$$

As previously stated, $\bar{y}$ and b are independent, so that the variance of the regression estimate is given by:

$$V(Y_0) = V(\bar{y}) + (x_0 - \bar{x})^2 V(b)$$

$$= \frac{\sigma^2}{n} + \frac{(x_0 - \bar{x})^2 \sigma^2}{\Sigma(x-\bar{x})^2} \tag{7.37}$$

and

$$\text{S.E. } (Y_0) = \sigma \sqrt{\left\{ \frac{1}{n} + \frac{(x_0 - \bar{x})^2}{\Sigma(x-\bar{x})^2} \right\}} \tag{7.371}$$

We substitute for σ the estimate s of the standard error about the regression. The error in the regression estimate is thus a minimum at $x_0 = \bar{x}$ and increases as x_0 deviates farther from $\bar{x}$. By substituting $x_0 = 0$ in Equation (7.37) we obtain the variance of the intercept as in Equation (7.35).

Standard error of a prediction of a further observation

7.281 Formula (7.371) is the standard error, due to random sampling variation, of a point on the regression, i.e. it is the standard error of the mean value of y for a given x. The *observed* values of y vary about the mean value with a standard deviation of σ. These two variations are independent; therefore, when Y_0 is used as an estimate of y for a future *individual* observation for which the value of x is x_0, the variance of this estimate is the sum of the variance of the mean value (Formula (7.37)) and the variance about this value (σ^2). Thus $V(Y_0)$ becomes:

$$\sigma^2 \left\{ 1 + \frac{1}{n} + \frac{(x_0 - \bar{x})^2}{\Sigma(x-\bar{x})^2} \right\} \tag{7.38}$$

and $S.E.(Y_0)$ becomes:

$$\sigma \sqrt{\left\{1 + \frac{1}{n} + \frac{(x_0 - \bar{x})^2}{\Sigma(x - \bar{x})^2}\right\}} \tag{7.381}$$

95% confidence limits for a predicted value are thus:

$$Y_0 \pm ts \sqrt{\left\{1 + \frac{1}{n} + \frac{(x_0 - \bar{x})^2}{\Sigma(x - \bar{x})^2}\right\}} \tag{7.382}$$

where s is the estimated value of σ and t is the value obtained from Table C for $n-2$ degrees of freedom corresponding to $P = 0{\cdot}025$.

Application to rubber specimens example

7.282 The variance of the regression estimate, i.e. the variance of the mean value of abrasion loss for a given value x_0 of hardness, is:

$$s^2\{1/n + (x_0 - \bar{x})^2/\Sigma(x - \bar{x})^2\}$$
$$= 3{,}663\{1/30 + (x_0 - \bar{x})^2/4{,}300\}$$
$$= 122{\cdot}1 + 0{\cdot}852(x_0 - \bar{x})^2 \tag{7.39}$$

where $\bar{x} = 70{\cdot}27$. The standard error is the square root of this quantity and the 95% confidence limits are plotted in Figure 7.32.

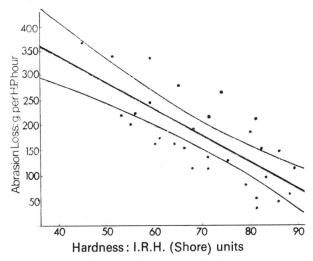

FIG. 7.32.   REGRESSION OF ABRASION LOSS UPON HARDNESS AND
ITS 95% CONFIDENCE LIMITS

The variance of an individual value of the abrasion loss estimated from a value x_0 of hardness is the Expression (7.39) plus the variance about the regression, i.e.

$$\text{Variance} = 3{,}785 + 0{\cdot}852 (x_0 - \bar{x})^2$$

The second part of this expression becomes appreciable only when x_0 deviates by more than 20 units from the mean value of 70·27. The 95% confidence limits are:

$$Y_0 \pm 2 \cdot 05 \sqrt{\{3,785 + 0 \cdot 852 \, (x_0 - \bar{x})^2\}}$$

Conditions for valid estimation of the regression equations

7.3 Regression equations refer to a given Population. For the example of abrasion loss and hardness, the Population is the totality of samples of the given synthetic rubber produced by varying the qualities and amounts of the given compounding materials within specified practical limits. The regression equations may well be different if we change to another medium, such as natural rubber or another type of synthetic rubber. The predictions of abrasion loss from hardness therefore apply only to samples of rubber from the specified Population.

When both regression equations are required, the observations from which the regression equations are estimated should be completely random with respect to the Population. A laboratory research programme cannot always be carried out on the basis of strict randomness, particularly when the question of relating variables is only a part of the programme. In the programme from which this example was drawn, the variations in the amounts of compounding ingredients were made to cover the ranges of interest, and in this respect the samples of rubber were not strictly random. However, no selection was made in relation to the results of the tests obtained on abrasion loss and hardness, and in this respect the results could be considered random.

Because the regression of y upon x is the locus of the mean of the observations of y for given values of x, we can select the values of x. Thus, an unbiased estimate of the regression of abrasion loss on hardness can be obtained using only those results having preselected values of hardness. This is illustrated by Figure 7.4 overleaf.

The mean of each of the arrays gives an unbiased estimate of the regression value for the appropriate hardness, and therefore the calculated regression of abrasion loss upon hardness from these restricted results will be unbiased. It will naturally not be as precise in this case as an estimate based on all the tests. In some situations, however, it may be practicable to select values of the independent variable and to concentrate effort on these points. An estimate of the regression obtained from such results will usually be considerably more precise than one obtained from the same number of points scattered over the same range of the independent variable. This principle is used more extensively in planned experiments to estimate the functional relationship, and will be considered later.

Although selection of results with respect to hardness will not bias the regression of abrasion loss on hardness, such results will be useless for estimating the regression of hardness on abrasion loss.

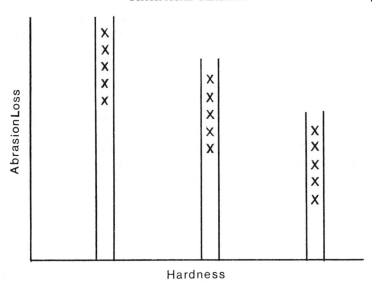

FIG. 7.4.   ARRAY DISTRIBUTIONS (SCHEMATIC)

Cause and effect

7.31 The existence of a statistically significant regression line does not necessarily mean that the changes in x actually *caused* the changes in y, nor that any chosen value of x will, in the future, necessarily result in the predicted value of y. Significance in this sense only implies that the pattern of the y's is significantly non-random with respect to the x's. It is always possible that by some unknown mechanism or by coincidence the values of x are closely associated with some third variable and that this third variable is really causing the changes in y. Consider, for example, a chemical plant where a statistically significant relationship is observed between the dependent variable y and the temperature T_1 at a certain stage of the process. It may be that the real operating variable is the temperature T_2 in some other part of the process, but that T_1 is related to T_2 because of heat transfer between neighbouring pieces of equipment. As long as such a hidden association continues to hold in the future it is reasonable to use T_1 to predict y, but one cannot necessarily be certain that the same pattern will carry on indefinitely.

Whether or not a causative relationship exists between x and y is something that must be inferred *by the experimenter* in the light of his other knowledge concerning the problem in hand. The mere fact that one is attempting to determine the regression line often implies that there are reasonable grounds to suppose that such a causal relationship would be feasible.

The inferential difficulties arise particularly when the data are collected over a period of time and do not come from a carefully planned and controlled experiment. This topic is discussed further in § **7.85** and in Chapter 8.

Replication and lack of fit

7.32 Data on which a regression analysis is to be performed sometimes consist of more than one value of the dependent variable y for each value of x, and the question then arises as to how to make the best use of the additional information. It is *not* necessarily preferable, or even legitimate, simply to regard each value of y as a separate datum and thereby to gain greater sensitivity in significance tests by virtue of the extra degrees of freedom associated with the residual mean square about regression.

The key to this problem is differentiating between cases where replicate values of y are merely *confirmations* of the first observation in each set and cases where replications provide genuinely new information. The former type arise from procedures such as repeat testing of the same sample, repeat samples from the same batch, or several samples taken simultaneously from different parts of a multi-unit machine. These observations are all locally bound with each other, and in these circumstances the regression analysis should be carried out using just the average value of y for each value of x. Theoretically, there should be an equal number of replicates at each x value, since otherwise the errors associated with the mean y's would not be uniform, but one need not be too pedantic in practice.

The second type of replication arises when the different values of y are obtained at essentially independent occurrences of the same value of x, as, for example, in a planned experiment where the same value of the independent variable is used on two separate occasions. In this case the regression parameters should be estimated with *all* the observations regarded as separate data points. However, the statistical significance of the parameters cannot necessarily be judged directly against the residual mean square about the regression, since this potentially consists of two distinct components:

(i) The "pure error" of measurement, sampling and reproducibility of circumstances, and

(ii) The variability associated with the extent to which the data cannot adequately be represented by a straight line, i.e. when some form of non-linear equation is really necessary.

The "pure error" is easily calculated from the combined *within group* sum of squares; it has $\Sigma(n_i - 1)$ degrees of freedom, where n_i is the number of observations at the ith value of x. When this sum of squares is subtracted from the original residual sum of squares about the regression, the remainder is termed the Lack-of-Fit sum of squares. The adequacy of the linear model can then be judged by comparing the lack-of-fit mean square with the pure-error mean square using an F-ratio test. The significance of the regression slope is also judged by reference to the pure-error mean square; in fact the latter quantity now assumes the role of the residual mean square about the regression in the straightforward case of single observations.

Example 7.2. Replicate determinations of chemical yield

7.321 As a simple example, consider the data listed in Table 7.3 where x

TABLE 7.3.   YIELD OF A CHEMICAL PRODUCT

Temperature, x	110	120	130	140
Yield, y	92·8 93·0	94·0 94·3	95·1 94·8	94·9 94·8
Mean yield, $\bar{y}$	92·9	94·15	94·95	94·85

represents the temperature in °C at a certain point in a chemical process and the two values of y are different determinations of the yield of the product at that temperature.

A linear regression analysis of the eight data points gives the slope coefficient $b = 0·0665$ and the primary Analysis of Variance table shown in Table 7.31. (The full details of the calculation have been omitted in this instance.)

TABLE 7.31.   PRIMARY ANALYSIS OF VARIANCE TABLE
(BASED ON TABLE 7.3)

Source of Variation	Sum of Squares	Degrees of Freedom	Mean Square
Due to regression	4·422	1	4·422
About regression	1·047	6	0·174
Total	5·469	7	

The pure-error sum of squares is then calculated by combining the within-temperature sums of squares. Since there are only two replicates, this is most simply achieved from the four ranges [§ **3.342**]:

$$\tfrac{1}{2}\{(0·2)^2 + (0·3)^2 + (0·3)^2 + (0·1)^2\} = 0·115$$

It has four degrees of freedom. After obtaining the lack-of-fit sum of squares and degrees of freedom by differencing, the final Analysis of Variance is as given in Table 7.32.

The significance of the regression is tested by comparing the mean square due to the regression with the pure-error; here, it is clearly highly significant. To test whether there is a significant departure from a straight line, we calculate the variance ratio for the lack-of-fit as $0.466/0.0287 = 16.2$. Comparing this with the values of F listed in Table D for 2 & 4 degrees of freedom, we find that the observed ratio approaches significance at the 1% level, so we may conclude that the data cannot adequately be represented by a straight line. Moreover, the observed values suggest that there may be a maximum yield which is achieved at the two higher temperature levels.

TABLE 7.32.   ANALYSIS OF VARIANCE TABLE SHOWING LACK-OF-FIT

Source of Variation	Sum of Squares	Degrees of Freedom	Mean Square
Due to regression	4.422	1	4.422
Lack-of-fit	0.932	2	0.466
Pure error	0.115	4	0.0287
Total	5.469	7	

Even though in any particular case the lack-of-fit term may not be statistically significant, it is still desirable to apply the techniques which will be described in § **8.54** for the examination of Residuals, i.e. the differences $(y_i - Y_i)$, since the introduction of polynomial terms or transformations of the variables may give a more satisfactory model.

Regression through the origin

7.4 Prior knowledge concerning the relationship sometimes requires that the regression equation should pass through the origin, i.e. $y = 0$ when $x = 0$. The regression equation then reduces to the simple form:

$$Y = \beta x \qquad (7.4)$$

and only the slope parameter β has to be estimated.

Great care must, however, be exercised in the adoption of this simplified model, since any bias in the measurements of x or y would result in a biased regression coefficient and could lead to a false picture. Further, although the relationship may genuinely pass through the origin, it may not be linear over the whole range of x values. Any straight line fitted over a region away from the origin may adequately represent the data in that region, but its extrapolation should not normally be constrained to pass through the origin itself.

The parameter β is estimated by minimizing the sum of squares:

$$S.S. = \Sigma(y-\beta x)^2$$

Differentiating with respect to β:

$$\frac{dS.S.}{d\beta} = -2\Sigma x(y-\beta x)$$

Setting the derivative equal to zero leads to the estimated slope:

$$b = \Sigma xy/\Sigma x^2 \qquad (7.41)$$

This expression is similar to the unconstrained Formula (7.17) except that the sum of products and sum of squares are measured from zero and not from the respective mean values, i.e. there are now no corrections for the means.

A similar simplification applies to all the derived regression formulae:

(i) The sum of squares due to the constrained regression is:

$$b^2\Sigma x^2 \qquad (7.42)$$

(ii) The sum of squares about the regression is:

$$\Sigma(y-bx)^2 = \Sigma y^2 - b^2\Sigma x^2 \qquad (7.43)$$

The best estimate of the variance about the regression s^2 is obtained by dividing this quantity by $n-1$ (not $n-2$ as previously, since now only the one parameter is estimated from the observations).

(iii) The standard error of the regression coefficient is given by:

$$S.E.(b) = s/\sqrt{(\Sigma x^2)} \qquad (7.44)$$

(iv) The standard error of a regression prediction at the point x_0 is given by:

$$S.E.(Y_0) = sx_0/\sqrt{(\Sigma x^2)} \qquad (7.45)$$

(v) The standard error of the predicted value of a further individual observation is given by:

$$S.E.(Y_0) = s\sqrt{(1+x_0^2/\Sigma x^2)} \qquad (7.46)$$

Weighted linear regression

7.5 Standard regression analysis by the method of least squares assumes that the error variance σ^2 is the same for all values of y, but several cases arise in practice where this assumption is not valid. Thus the amount and the reliability of the information about the value of y for each x may differ substantially because of unequal numbers of repeat measurements or samples. Also different instruments may be used to measure the y values even though it is known that the instruments vary in accuracy. A third possibility is the situation in which some underlying physical or chemical law causes the error variance to be directly dependent on x; for example, in measurements of radio-activity the variance decreases with time.

If the variances can be determined, even approximately, it is reasonable to weight each y observation by the inverse of the associated variance, and to estimate the regression parameters by minimizing the weighted sum of squares:

$$S.S. = \Sigma w_i(y_i - \alpha - \beta x_i)^2$$

where $w_i = 1/\sigma_i^2$. Denote the average of the weighting factors by $\bar{w}$ and put $c_i = w_i/\bar{w}$. Then it may be shown that the estimated regression coefficient and intercept are given by:

$$b = \frac{\Sigma c_i x_i y_i - (\Sigma c_i x_i)(\Sigma c_i y_i)/n}{\Sigma c_i x_i^2 - (\Sigma c_i x_i)^2/n} \qquad (7.5)$$

and

$$a = \{\Sigma c_i y_i - b\Sigma c_i x_i\}/n \qquad (7.51)$$

These expressions are similar to Formulae (7.16) and (7.13) for the standard regression analysis, except that each term in each summation has been multiplied by a relative weighting factor $w_i/\bar{w}$.

In the special case when the line is constrained to pass through the origin ($a \equiv 0$), it often happens that the standard deviation associated with each y observation is proportional to its magnitude, i.e. there is a constant coefficient of variation. Apart from a constant factor, the weighted sum of squares then reduces to $\Sigma(y_i/x_i - \beta)^2$, which achieves its minimum when the slope is estimated by the arithmetic mean of the observed ratios y_i/x_i.

LINEAR FUNCTIONAL RELATIONSHIPS

7.6 When the true values of two variables (i.e. the limits to which mean values tend on repetition of the determinations) fall on a straight line, the two variables are said to be linearly functionally related. In practice a linear functional relationship exists between two variables when the variation about a suitably drawn straight line is due, or can be assumed to be due, entirely to experimental errors.

Chemists and physicists are usually interested in functional relationships, but it is not always possible to estimate them from observed results. However, we can always estimate the locus of the mean of the y variable for given values of the x variable, and also the locus of the mean of the x variable for given values of the y variable. Although, as shown earlier in Figure 7.2, these lines are not the same, they will often serve satisfactorily for prediction purposes. There are, however, a number of situations in which the functional relationships can be readily estimated.

Functional relations are usually estimated from the results of a controlled experiment in which one variable is obtained at certain chosen values, and the other variable is then measured at these conditions. An example which has already been considered in the introduction is the calibration of a platinum resistance thermometer, where the resistance is measured at given temperature reference points. Other examples arise in the development of chemical processes when the dependence of yield and quality of the product on such factors as temperature, concentration, time of reaction, etc., have to be determined.

The controlled variable is the independent variable, and the other is the dependent variable. Usually the former can be obtained with negligible error compared with the latter; for example, the temperature in a chemical reaction can usually be controlled fairly precisely, and practically all the experimental error arises in the assessment of the yield.

Linear functional relationships when one variable has negligible error

7.61 When the independent variable can be set or determined with negligible error, the functional relationship is estimated directly by the regression equation of the dependent variable upon the controlled independent variable. The reason for this will be apparent from the following considerations. Let x_0 be one of the values chosen for the independent variable and let $y_1, y_2, ..., y_n$ be repeat values of the other variable determined at the chosen value x_0 of x. The value x_0 of x is assumed free from error, and it therefore represents a "true" value. The corresponding point on the functional relation must then be on the vertical through x_0. The values $y_1, y_2, ..., y_n$ vary, but their mean will be an unbiased estimate of the true value which will become more precise as n increases; in other words, the mean value of y for the given x is an unbiased estimate of a point on the functional relationship. But by definition it is also an estimate of a point on the regression line. The same applies for all other values of x, and therefore the regression of y upon x is also an estimate of the functional relationship.

The calculations are exactly as those given previously in §§ **7.22–7.28**; thus the estimate of the linear functional relation is:

$$Y - \bar{y} = b(x - \bar{x}) \qquad (7.6)$$

where
$$b = S_{xy}/S_{xx} \qquad (7.61)$$

The variance about the line is estimated by:

$$s^2 = \Sigma(y - Y)^2/(n - 2) \qquad (7.62)$$

Corresponding to Formula (*7.321*), the standard error of the slope b is:

$$S.E.(b) = s/\sqrt{S_{xx}} \qquad (7.63)$$

and corresponding to Formula (*7.371*), the standard error of the estimate obtained from the functional relation for a given value $x = x_0$ is:

$$S.E.(Y_0) = s\sqrt{\{1/n + (x_0 - \bar{x})^2/S_{xx}\}} \qquad (7.64)$$

Here we are not interested in the values of y which would be observed, but only in the estimate of the true value. Consequently Formula (*7.381*), which gives the standard error of an individual future observation of y, is not required. Since the variation about the functional relationship is assumed to be due entirely to experimental errors in the observations y, the variance about the functional relationship is an estimate of the error variance of y. If an independent estimate of the latter is available it is then possible to test the hypothesis that the variation about the functional relationship is due entirely to experimental errors in y.

Application to instrument calibration

7.611 In a calibration problem it can usually be assumed that one variable has negligible error; for example, in the calibration of a platinum resistance thermometer referred to earlier it can be assumed that the errors in the values of the temperature are negligible, so that the regression equation:

$$Y = \bar{y} + b(x - \bar{x})$$

where Y is the resistance and x the temperature, estimates the functional relationship. The latter is often referred to as the Standard Curve.

In subsequent use of the resistance thermometer the temperature is estimated from a measurement of the resistance. If this were a regression problem we should use the regression of temperature upon resistance, i.e. x upon y, for this purpose, but since we are interested in estimating true values, and since the functional relation is unique, the same equation should be used for estimating both x from y and y from x. Herein lies an important difference between regression equations and functional relationships.

The estimate of x from a given value y_0 of y is, from Equation (7.6):

$$X_0 = (y_0 - \bar{y})/b + \bar{x} \qquad (7.65)$$

Standard error of a prediction

7.612 It is now required to determine the confidence limits for this value of x corresponding to the observation y_0 of y. An approximate formula will be given in this section; the exact formula, which is more complicated, is obtained from Fieller's Theorem [Appendix 7A].

Since the estimate of x for a given value y_0 of y is given by Equation (7.65), and since $V(\bar{x})$ is zero, the variance of the estimate is:

$$V(X_0) = \frac{V(y_0 - \bar{y})}{b^2} + \left(\frac{y_0 - \bar{y}}{b}\right)^2 \frac{V(b)}{b^2}$$

This expression is derived using only the first two terms of Formula (3.95) and is therefore an approximation. Now:

$$V(b) = \sigma^2/S_{xx}$$

where σ^2 is the variance about the line, and:

$$V(y_0 - \bar{y}) = \left(1 + \frac{1}{n}\right)\sigma^2$$

If, instead of being a single measurement, y_0 is the mean of m separate determinations of the resistance, we would have:

$$V(y_0 - \bar{y}) = \left(\frac{1}{m} + \frac{1}{n}\right)\sigma^2$$

In this general case, therefore:

$$V(X_0) = \left(\frac{1}{m} + \frac{1}{n}\right)\frac{\sigma^2}{b^2} + \left(\frac{y_0 - \bar{y}}{b}\right)^2 \frac{\sigma^2}{b^2 S_{xx}}$$

$$= \frac{\sigma^2}{b^2}\left\{\frac{1}{m} + \frac{1}{n} + \frac{(y_0 - \bar{y})^2}{b^2 S_{xx}}\right\} \qquad (7.651)$$

and

$$S.E.(X_0) = \frac{\sigma}{b}\sqrt{\left\{\frac{1}{m} + \frac{1}{n} + \frac{(y_0 - \bar{y})^2}{b^2 S_{xx}}\right\}} \qquad (7.652)$$

The estimate s of the standard deviation about the regression based on $n-2$ degrees of freedom has to be used for σ; the confidence limits for X_0 can then be derived using the appropriate value of t.

We note that $S.E.(X_0)$ decreases as m increases, i.e. the accuracy of the predicted value X_0 increases with the number of replications for the determination of y. The reason for this is that the effect of experimental errors is reduced by replicating, but for a given estimate of the functional relationship the standard error cannot be reduced below the value given by Expression (7.652) with the term $1/m$ omitted.

Example 7.3. Penicillin assay calibration

7.613 The following example, which is used to illustrate functional relationship with one variable controlled, is in effect a calibration problem and refers to the assay of samples of penicillin. The data are given in Table 7.4, and are to be used for the determination of a standard curve.

Six concentrations of pure penicillin differing in twofold steps from 1 unit to 32 units per ml. are set up on a plate (see § **6.61** for details of technique). Table 7.4 gives the circle diameters of the zones of inhibition in millimetres for each concentration.

In this example the x-variable (i.e. concentration of penicillin) has been obtained without appreciable error at certain predetermined values. However, errors arise in the determination of the circle diameter.

The results of Table 7.4 are plotted in Figure 7.5. A logarithmic scale to the base 2 is taken for the concentration of penicillin, since on this scale, the relation appears linear. Moreover, the series of concentrations 1, 2, 4, 8, ... become equally spaced, twofold increases being represented by unit displacements along the concentration axis.

TABLE 7.4.   PENICILLIN CONCENTRATIONS

Concentration of penicillin solution	1	2	4	8	16	32
$x = \log_2$ (concentration)	0	1	2	3	4	5
Circle diameter (mm.) $= y$	15·87	17·78	19·52	21·35	23·13	24·77

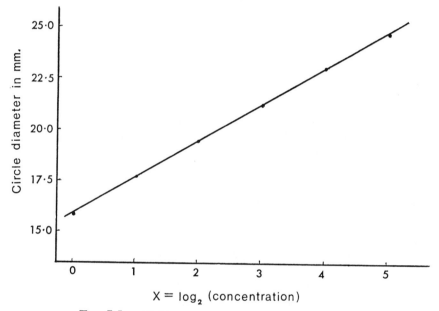

FIG. 7.5.   RELATION BETWEEN CIRCLE DIAMETER AND CONCENTRATION OF PENICILLIN

The calculation of the regression of y upon x for this example is identical with that for Example 7.1. The details are:

$$\Sigma x = 15 \qquad\qquad n = 6 \qquad\qquad \Sigma y = 122\cdot42$$
$$\bar{x} = 2\cdot50 \qquad\qquad\qquad\qquad \bar{y} = 20\cdot40$$
$$\Sigma x^2 = 55 \qquad\qquad \Sigma xy = 337\cdot24 \qquad\qquad \Sigma y^2 = 2{,}553\cdot3880$$
$$(\Sigma x)^2/n = 37\cdot5 \qquad (\Sigma x)(\Sigma y)/n = 306\cdot05 \qquad (\Sigma y)^2/n = 2{,}497\cdot7761$$
$$S_{xx} = 17\cdot5 \qquad\qquad S_{xy} = 31\cdot19 \qquad\qquad S_{yy} = 55\cdot6119$$
$$b = S_{xy}/S_{xx} = 31\cdot19/17\cdot5 = 1\cdot782$$

The regression of y upon x is therefore given by:

$$Y = 20 \cdot 40 + 1 \cdot 782(x - 2 \cdot 50)$$

$$\text{i.e. } Y = 15 \cdot 94 + 1 \cdot 782x$$

This equation is an estimate of the functional relationship, and is therefore the one to use for estimating y from a given x and also for estimating x from a given y. In practice the equation is required for the purpose of estimating the concentration of an unknown sample from an observed value of the circle diameter. We therefore write it in the form:

$$X = (y - 15 \cdot 94)/1 \cdot 782$$

where x, Y have been replaced by X, y respectively, in accordance with our notation of capital letters for predicted values and small letters for observed values.

From (7.22):

$$\text{Sum of squares about the regression} = S_{yy} - b^2 S_{xx} = S_{yy} - (S_{xy})^2/S_{xx}$$

$$= 55 \cdot 6119 - (31 \cdot 19)^2/17 \cdot 5 = 0 \cdot 0224$$

Whence the estimated variance about the regression is:

$$s^2 = 0 \cdot 0224/4 = 0 \cdot 0056$$

Substituting in Formula (7.652) with $m = 1$, the standard error of the estimated value of x corresponding to a single observed circle diameter y is:

$$\frac{\sqrt{0 \cdot 0056}}{1 \cdot 782} \sqrt{\left\{1 + \frac{1}{6} + \frac{(y - 20 \cdot 40)^2}{55 \cdot 59}\right\}} = 0 \cdot 042 \sqrt{\left\{1 \cdot 1667 + \frac{(y - 20 \cdot 40)^2}{55 \cdot 59}\right\}}$$

Linear functional relationship when both variables are subject to error

7.62 Now let us assume that a linear functional relationship exists between y and x, and that the only reason why the observations depart from this is the existence of errors in the sampling and measurement of both y and x. These errors are assumed to be independent; they usually are in practice.

Let the error variance of y be σ_y^2 and that of x be σ_x^2. A separate experiment may be necessary to determine these, but often when obtaining the initial data repeat observations will be made in order to supply estimates of σ_x^2 and σ_y^2.

The functional line passes through the mean of the observations and therefore has the form:

$$Y = \bar{y} + b(x - \bar{x}) \tag{7.7}$$

Suitable estimates for the slope b depend on the information available about the error variances. Thus [7.1]:

(i) If *only* σ_x *is known* use:

$$b = \frac{S_{xy}}{S_{xx} - (n-1)\sigma_x^2} \tag{7.71}$$

unless the denominator is negative, when the slope should be taken as infinite.

(ii) If *only σ_y is known* use:

$$b = \frac{S_{yy} - (n-1)\sigma_y^2}{S_{xy}} \tag{7.72}$$

unless the numerator is negative, when the slope should be taken as zero.

(iii) If *both error variances are known, or if the ratio $\lambda = \sigma_y^2/\sigma_x^2$ is known* use:

$$b = \frac{(S_{yy} - \lambda S_{xx}) + \sqrt{\{(S_{yy} - \lambda S_{xx})^2 + 4\lambda S_{xy}^2\}}}{2S_{xy}} \tag{7.73}$$

In all three cases the sign of the estimated slope is the same as the sign of S_{xy}. Equation (7.73) is used most frequently in practice, because the experimenter often has rather imprecise estimates of σ_x and σ_y individually but can be more certain of the true value of the error variance ratio λ. The third case will therefore be considered in rather more detail.

Formulae for functional relationship when the ratio of error variances is known

7.621 When λ is assumed known and the errors are independently and Normally distributed, Formula (7.73) is the maximum likelihood estimator [Appendix 4B] of the slope of the linear functional relationship. Once it has been calculated it can be used to estimate the individual error variances from the expression:

$$\sigma_y^2 = \lambda \sigma_x^2 = \frac{S_{yy} - bS_{xy}}{n-2} \tag{7.74}$$

Confidence limits β_1, β_2 for the true value β of the slope are given by the rather complex formula [7.2]:

$$\lambda^{\frac{1}{2}} \tan \left(\tan^{-1}\{b\lambda^{-\frac{1}{2}}\} \pm \frac{1}{2} \sin^{-1} \left\{ 2t \left[\frac{\lambda(S_{xx}S_{yy} - S_{xy}^2)}{(n-2)\{(S_{yy} - \lambda S_{xx})^2 + 4\lambda S_{xy}^2\}} \right]^{\frac{1}{2}} \right\} \right) \tag{7.75}$$

where the value of t is obtained from Table C for $n-2$ degrees of freedom and the chosen confidence percentage. If 95% confidence limits are calculated, $(\beta_2 - \beta_1)^2/16$ provides a rough estimate of the variance of the slope b. If the argument of the inverse sine in Formula (7.75) exceeds unity, so that no real angle is possible, we must conclude that no value of the slope can be ruled out at the chosen confidence level.

When $\lambda = 0$, Formula (7.73) reduces to S_{yy}/S_{xy}, which is the slope $(1/b')$ of the regression line of x upon y [§ **7.22**]. At the other extreme, as $\lambda \to \infty$ (corresponding to $\sigma_x^2 \to 0$) Formula (7.73) tends to S_{xy}/S_{xx}, the regression coefficient of y upon x. Hence the linear functional relationship always lies between the two regression lines. When $\lambda S_{xx}/S_{yy}$ is sufficiently small, say less than 0·01, we may in practice decide to disregard the complexity of Formula (7.73) and use S_{yy}/S_{xy} as the slope of the functional line: conversely, when $\lambda S_{xx}/S_{yy}$ is sufficiently large, say greater than 100, S_{xy}/S_{xx} can be used as the estimate of b.

In the case of equal errors ($\lambda = 1$), Formula (7.73) is the same expression as obtained by minimizing the sum of squares of the *perpendicular* distances between the observed values and a straight line.

Linear functional relationship when the error variances are unknown

7.63 A simple method for obtaining an estimate of a linear functional relationship in certain situations when the errors in the variables are un-known is given by M. S. Bartlett [7.3]. This method may also be used as a quick way to obtain an approximate relationship (either functional or regres-sion) even when the variances are known.

Since it is known that the linear functional relation must pass through the mean, one point on the line is readily determined, and all that remains is to estimate the slope. The procedure is to plot the points on a graph and to divide them into three groups, the two end groups having the same number, m, of points chosen to be as near to $n/3$ as possible. The three groups must be non-overlapping when considered in at least one direction, e.g. in the x-direction. It is assumed that the errors in the observations do not interfere with the division made. This condition is usually satisfied in a planned experiment because the intervals between adjacent values of the dependent (or the independent) variable are much larger than the errors in their measure-ment. This is particularly so for the x-variable, with respect to which the division should therefore be made.

The slope is determined as follows. Compute the means of each of the two end groups and denote them by $(\bar{x}_1, \bar{y}_1)$, $(\bar{x}_3, \bar{y}_3)$. The line joining these two points gives the estimate of the slope:

$$b = (\bar{y}_3 - \bar{y}_1)/(\bar{x}_3 - \bar{x}_1)$$

The functional relation is then a line, with this slope, passing through the grand mean.

The confidence limits β_1 and β_2 of the true value of the slope are given by the roots of the quadratic equation with respect to β:

$$\tfrac{1}{2}m(\bar{x}_3 - \bar{x}_1)^2(b - \beta)^2 = t^2(S_{yy} - 2\beta S_{xy} + \beta^2 S_{xx})/(n - 3) \qquad (7.76)$$

where n = number of pairs of observations,

$\qquad m$ = number of pairs of observations in each end group,

$\qquad b$ = estimate of the slope given above,

$\qquad \bar{x}_1, \bar{x}_3$ = means of the two outer groups,

$\qquad S_{xx}$ = corrected sum of squares of x *within* the three groups,

$\qquad S_{xy}$ = corrected sum of products of x and y *within* the three groups,

$\qquad S_{yy}$ = corrected sum of squares of y *within* the three groups,

$\qquad t$ = the value of t for the confidence interval required for the $n-3$ degrees of freedom available within the groups.

The fact that the points can only be grouped according to the observed values, whereas the underlying theory requires ordering according to the *true* x-values, results in the slope estimate b being biased slightly below the true slope β of the functional relationship. However this bias is negligible when σ_x is less than 5% of the range of the x values.

Example 7.4

7.631 We shall illustrate Bartlett's method by applying it to the following data:

$$x = \quad 2 \quad 5 \quad 6 \quad 7 \mid 11 \quad 13 \quad 16 \quad 17 \quad 19 \mid 20 \quad 21 \quad 23 \quad 24$$
$$y = 11 \quad 10 \quad 13 \quad 16 \mid 20 \quad 18 \quad 25 \quad 22 \quad 26 \mid 30 \quad 32 \quad 30 \quad 32$$

Here $n = 13$ and the nearest value to $n/3$ is $m = 4$. Clearly in this example there is no overlapping of the groups in either the x direction or the y direction. From the first 4 pairs of observations:

$$\bar{x}_1 = 20/4 = 5, \qquad \bar{y}_1 = 50/4 = 12\cdot5$$

and from the last 4 pairs of observations:

$$\bar{x}_3 = 88/4 = 22, \qquad \bar{y}_3 = 124/4 = 31\cdot0$$

Therefore: $b = (31\cdot0 - 12\cdot5)/(22 - 5) = 1\cdot088$

The overall means are:

$$\bar{x} = 184/13 = 14\cdot154, \qquad \bar{y} = 285/13 = 21\cdot923$$

The estimated equation of the line is therefore:

$$Y - 21\cdot923 = 1\cdot088 \, (x - 14\cdot154)$$

i.e. $$Y = 6\cdot523 + 1\cdot088x$$

To derive 95% confidence limits for the estimate of the slope we require the following quantities:

$$S_{yy} = \{11^2 + 10^2 + 13^2 + 16^2 - (11 + 10 + 13 + 16)^2/4\}$$
$$+ \{20^2 + \dots + 26^2 - (20 + \dots + 26)^2/5\}$$
$$+ \{30^2 + \dots + 32^2 - (30 + \dots + 32)^2/4\}$$
$$= 7{,}003 - 6{,}933\cdot2 = 69\cdot8$$

and similarly: $$S_{xy} = 4{,}714 - 4{,}665\cdot2 = 48\cdot8$$
$$S_{xx} = 3{,}256 - 3{,}191\cdot2 = 64\cdot8$$

The value of t in Table C for $\alpha = 0\cdot025$, $\phi = n - 3 = 10$ is $2\cdot23$. Substituting in Equation (7.76), the 95% confidence limits for the slope are the solutions of the equation:

$$\tfrac{1}{2} \times 4 \times 17^2 (1\cdot088 - \beta)^2 = 2\cdot23^2 (69\cdot8 - 2\beta \times 48\cdot8 + 64\cdot8\beta^2)/10$$

or $$5{,}458\beta^2 - 2 \times 6{,}046\beta + 6{,}495 = 0$$

i.e. $$\beta = 1\cdot108 \pm 0\cdot193$$

The 95% confidence limits are therefore (0·915, 1·301), the estimate of the slope being 1·088.

Controlled and uncontrolled experiments

7.64 There are in general two ways of obtaining an estimate of the relationship between two variables; one is a controlled experiment in which one variable is held at (or very close to) certain predetermined values, and the other an uncontrolled situation in which the variations largely occur by chance. Consider a chemical process in which variations occur in the yield of product from batch to batch and it is necessary to investigate the causes of this variation. It is found on examination of the batch records that there are variations in the reaction temperature from batch to batch. It is required to find whether temperature variations are the cause or a contributory cause of the yield variation. There are two ways of investigating this:

(a) To carry out a controlled experiment in which the temperature is carefully maintained as closely as possible to certain chosen values for a number of batches; for example, two or more batches may be made at each of four different temperatures covering a range larger than that found in normal manufacture.

(b) To use the actual process records of yield and temperature and examine the relation between them.

In the controlled experiment it is clear that, both for estimating the slope of the relationship and for predicting the yield at any given temperature, the relation of interest is a functional one. Usually, however, the controlled variable, which is considered as the independent variable and is denoted by x, can be set with a high degree of accuracy in relation to the error in the other variable, in which case the regression of y on x provides an estimate of the functional relationship. As an added safeguard, the range of the independent variable can be made very large compared with the error in its determination, in which case both regression equations and the functional relationship, for practical purposes, coincide.

Berkson [7.4] differentiates between the two following cases:

(i) The independent variable is actually measured objectively. In this case the observed values should, in principle, be used rather than the target values (if they differ) and the regression line should then be used as an approximation to the functional relationship because the errors in determination of the x's are negligible.

(ii) The independent variable is merely brought up to some preset value on a scale, e.g. by moving a pointer to a particular place on a dial, adding a substance to a weigh pan until it balances the opposing weights, or automatically activating some instrument when a liquid level reaches a certain height. In this second case there is no error whatsoever in the recorded figure, because it merely records our *intention* and is not a separate observation in the true sense of the word. The only relationship which the experiment could possibly then produce would be a line connecting the observed responses with the *intended values* of the independent variables. We must, therefore, ascribe all the error to y and take x as being accurately known, i.e. the regression of y upon x really is the required functional relationship, not just an approximation to it.

In the uncontrolled situation the batches must be considered as a sample from the Population of batches. Many factors other than variation in temperature and errors in measurement will contribute to the variation in yield. Errors in estimation of yield and temperature may, in fact, account for only a small proportion of the variation. Therefore, although we may be interested in the functional relationship, we cannot estimate it from the data, and must then use the regression lines to assess the association between the two variables. Predictions from these regression lines apply only to the given process operated in the same way as before, on the assumption that the process is stable.

The second method, (*b*), is not very satisfactory, owing to the probable existence of several other uncontrolled variations. If the temperature cannot be kept under control, it is fair to assume that other factors, known or unknown, also cannot be controlled, and variations in these factors will tend to mask the effect of temperature and may be sufficient to mask it altogether. If the temperature varies as a result of variation in some other factor, misleading results as to the effect of temperature would be obtained. If numerical data are available for any of these other factors, the problem is best dealt with by the techniques of Multiple Regression which are described in the next chapter.

Whenever possible the relationship between variables should be determined from controlled experiments. Examination of relationships between variables on the basis of process records should usually be considered as a preliminary to controlled experiments. This topic is referred to again in Chapter 8.

Comparison of several regression lines

7.7 Problems sometimes arise in which it is desired to compare several related regression lines with each other. A simple case of this, where only two lines are concerned, is in assay work as described in the next few sections. The more general case will be considered in § **7.73**.

Parallel line assays: Example 7.5

7.71 Regression equations are frequently used for assay purposes when comparing some property of a sample against a standard curve for that property.

One example of such an assay in biological work has already been described [§ 7.613]. Similar examples arise when assessing the amount of a given dye-stuff present in a sample by means of colorimetric and spectrophotometric methods; such methods are useful for estimating the strength of dyestuffs when there is no simple specific chemical test for this purpose.

We shall illustrate the method by considering a numerical example. The strength of a sample of a dyestuff is generally expressed in terms of that of a standard. Table 7.5 gives the logarithms of the optical densities of varying concentrations of solutions of a sample and of a standard. The problem is to determine the strength of the sample in terms of that of standard. In these circumstances a series of dilutions of both sample and standard should be carried out in the same experiment because, as shown below, the results on the series of dilutions of the sample can be used to obtain an improved estimate

TABLE 7.5.   LOG OPTICAL DENSITIES

Relative Concentrations	Log (10 × Concentration)	Standard	Sample
0·80	0·903	0·521	0·519
0·90	0·954	0·579	0·570
1·00	1·000	0·626	0·622
1·10	1·041	0·672	0·666

of the standard curve. It is not always necessary to take a series of dilutions of the sample to compare with the standard—an estimate can be obtained from one dilution only, but a series of dilutions gives a higher degree of accuracy which may be required in particular circumstances. "Standard curves" are standard only for the conditions appertaining at the time, and may vary slightly from time to time.

Figure 7.6 shows the graphs of the logarithms of the optical densities plotted against the logarithms of the concentrations. The relation between these transformed variables was expected to be linear, but it will be apparent from what follows that there is an additional reason for choosing a logarithmic scale for the concentrations.

Whatever transformation of the optical density is used, a horizontal line drawn through any point on the vertical axis will cut the lines for standard and sample respectively at points corresponding to values x_1 and x_2 of the concentration. Since x_1 and x_2 give the same response, the concentration of pure dyestuff present in each should be the same. The ratio x_1/x_2 then measures the strength of the sample in terms of standard. If the only differ-ence between the sample and the standard material is the amount of diluent present, then the ratio x_1/x_2 will be the same for all values of the optical density. On the logarithmic scale for the concentration, $\log x_1 - \log x_2$ will be constant, and this means that the horizontal distance between the curve for the standard and the curve for the sample will be the same for all values of the optical density, irrespective of the scale in which the latter is expressed. The logarithmic scale is thus the natural one to use for the concentration.

The best transformation to use for the optical density is the one which gives the simplest relationship—linear if possible—between optical density and log (concentration), and in addition makes the variance of the response the same at every point of the range considered. When such a transformation

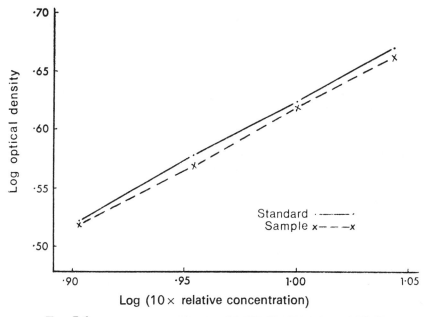

FIG. 7.6.   OPTICAL DENSITY AND CONCENTRATION OF DYESTUFF

does not exist the method of fitting becomes more complicated, and reference should be made to Finney [7.5].

For most applications in the industrial field the ranges of the variables can be so chosen that the above conditions—linearity and equality of variances—are approximately satisfied.

Transformations commonly used for the response are the logarithm, square root, reciprocal, and sometimes $\log(y+a)$, $\sqrt{(y+a)}$, or $1/(y+a)$, where y is the observed response and a a constant to be determined from the data.

The concentration can be set up without appreciable error and can be assumed exact. The regression of log (optical density) upon log (concentration) is then the relationship to use for the purpose of estimating concentration from a measure of the optical density.

In the experiment represented by the data of Table 7.5 and graphed in Figure 7.6 the range of optical density is small and the variance of log (optical density) can be assumed to be constant.

The stages in the estimation of the strength of the sample in terms of the standard are:

(i) Fit separate straight lines to the results for the sample and for the standard.

(ii) Compare the regression coefficients and combine them if they do not differ appreciably, to obtain the slope of the best fitting parallel lines to represent the two sets of results.

(iii) Estimate the horizontal distance between the parallel lines. This gives the logarithm of the ratio of the strength of the sample to that of the standard.

(iv) Calculate the confidence limits for the ratio.

These steps will now be applied to the data of Table 7.5, where we take $x = \log(10 \times \text{relative concentration})$ and $y = \log(\text{optical density})$.

Fitting parallel lines to the data of Example 7.5

7.711 The calculations are given in Table 7.51.

The regression coefficients for the sample and standard clearly do not differ appreciably, and we can proceed to derive the best-fitting pair of parallel lines. If the regression coefficients differed appreciably it would not be possible to give a unique value for the strength of the sample. This could happen if the active ingredients of the two samples were not chemically identical.

The common regression coefficient b for the pair of parallel lines, obtained by combining b_1 and b_2, is given by:

$$b = \frac{\Sigma_1(x-\bar{x})(y-\bar{y})+\Sigma_2(x-\bar{x})(y-\bar{y})}{\Sigma_1(x-\bar{x})^2+\Sigma_2(x-\bar{x})^2} \qquad (7.8)$$

where Σ_1 denotes summation over the observations for the standard and Σ_2 denotes summation over the observations for the sample.

$$\therefore \quad b = \frac{0\cdot011530+0\cdot011356}{0\cdot010605+0\cdot010605} = \frac{0\cdot022886}{0\cdot021210} = 1\cdot0790$$

In this case, since the same values of the concentration were used for both sample and standard, b is simply the average of b_1 and b_2.

The pair of parallel lines are then:

$$\text{Standard:} \quad Y = 0\cdot5995+1\cdot0790(x-0\cdot9745)$$
$$\text{Sample:} \quad Y = 0\cdot5943+1\cdot0790(x-0\cdot9745)$$

since each line must pass through its own centroid.

TABLE 7.51.   CALCULATION INVOLVED IN THE ANALYSIS OF THE
DATA OF TABLE 7.5

	x	y	
		Standard	Sample
n	4	4	4
Sum	3·898	2·398	2·377
Mean	0·9745	0·5995	0·5943
Crude sum of squares	3·809206	1·450142	1·424701
Correction = (sum)²/n	3·798601	1·437601	1·412532
Corrected sum of squares	0·010605	0·012541	0·012169
Crude sum of products of x and y	—	2·348381	2·327743
Correction	—	2·336851	2·316387
Corrected sum of products	—	0·011530	0·011356
Regression coefficient	—	1·0872	1·0708
Sum of squares due to regression	—	0·012536	0·012160
Sum of squares about regression	—	0·000005	0·000009

Combined variance about regression = 0·000014/4 = 0·0000035.   d.f. = 4

Standard error of each $\sqrt{(0·0000035/0·010605)} = \sqrt{0·000330}$
regression coefficient $= 0·018$

Standard error of the difference between the two regression coefficients
$= 0·018\sqrt{2} = 0·025$. Actual difference $= 0·016$.

Estimation of the ratio of the strengths and the confidence limits of the ratio

7.712 The horizontal distance between the two parallel lines:

$$Y = \bar{y}_1 + b(x - \bar{x}_1), \quad \text{i.e. } x = \frac{Y - \bar{y}_1}{b} + \bar{x}_1$$

and

$$Y = \bar{y}_2 + b(x - \bar{x}_2), \quad \text{i.e. } x = \frac{Y - \bar{y}_2}{b} + \bar{x}_2$$

is:

$$\frac{\bar{y}_2 - \bar{y}_1}{b} - (\bar{x}_2 - \bar{x}_1) \tag{7.81}$$

For the present example $\bar{x}_1$ is equal to $\bar{x}_2$ and the horizontal distance is simply $(\bar{y}_2 - \bar{y}_1)/b$, which has the value $-0·00482$.

This is the logarithm of the ratio of the strength of sample to standard, and its antilogarithm then gives the actual ratio. The antilogarithm is 0·9890. In other words, the strength of the sample is found to be 98·9% of that of the standard.

The next step is to find the confidence limits for this result. The confidence limits for a ratio of two parameters have been derived by Fieller, and details will be found in Appendix 7A. For this type of problem, however, the following method gives a sufficiently accurate approximation. We have first to estimate the variance about the regression. The sum of squares of the observations accounted for by the pair of parallel lines is given by:

$$\frac{[\Sigma_1(x-\bar{x})(y-\bar{y})+\Sigma_2(x-\bar{x})(y-\bar{y})]^2}{\Sigma_1(x-\bar{x})^2+\Sigma_2(x-\bar{x})^2} = \frac{(0\cdot022886)^2}{0\cdot021210} = 0\cdot024694$$

The combined corrected sum of squares is $0\cdot012541+0\cdot012169 = 0\cdot024710$. The sum of squares about the parallel regressions is therefore $0\cdot000016$.

Since we have fitted three constants—two means and one regression coefficient—the degrees of freedom are $8-3 = 5$. The variance about the regression is therefore $0\cdot0000032$.

Using the method of § **3.72**, the variance of Expression (*7.81*) is approximately:

$$\frac{V(\bar{y}_2-\bar{y}_1)}{b^2} + \left(\frac{\bar{y}_2-\bar{y}_1}{b}\right)^2 \frac{V(b)}{b^2} \qquad (7.82)$$

since the error variance of the x observations is zero.

$\bar{y}_1$ and $\bar{y}_2$ are each means of four observations, and therefore $V(\bar{y}_2-\bar{y}_1) = \frac{1}{2}\sigma^2$, where σ^2 is the variance about the regression. Also:

$$V(b) = \sigma^2/0\cdot021210$$

By substituting these values in Expression (*7.82*) we have:

Variance of log ratio of strengths $= (\sigma^2/b^2)\{\frac{1}{2}+(0\cdot00482)^2/0\cdot02121\}$

The second expression inside the brackets is clearly small compared with $\frac{1}{2}$ and can be neglected in this particular case. We then obtain:

Variance of log ratio $= \frac{1}{2}\times0\cdot0000032/(1\cdot079)^2 = 0\cdot00000137$

Standard error of log ratio $= 0\cdot00117$

To find the 95% confidence limits we note that the appropriate value of t for 5 degrees of freedom is 2·57, and the confidence limits are $-0\cdot00482 \pm 0\cdot00301 = -0\cdot00783$ and $-0\cdot00181$, the antilogs of which are 0·982 and 0·996. The relative strength of the sample is then 98·9% with 95% confidence limits of 98·2% and 99·6%. The narrowness of the confidence limits indicates that this method of estimating the concentration is fairly accurate.

Adjustment for concomitant variation

7.72 A further use of parallel line regression is in *t*-tests and Analysis of Variance where we need to make adjustments for the concomitant variation of extraneous variables.

Suppose we require to compare the quality of a chemical product made on the plant by two processes, A and B. The usual method is to make a number of batches by each of A and B, measure the quality of each batch, calculate the means of the two groups, their difference, and the confidence limits for the difference [§ **4.231**].

When variations are known to exist between lots of raw materials used in the manufacture of the chemical, it is usual to make a batch of A and a batch of B from each lot [§ **4.25**]; but there are situations where each lot of raw material is insufficient to make both a batch of A and a batch of B, and blending of lots may not be practicable or economical. The quality of the raw material may be well defined, for example by its purity, its crystallising point, or the percentage of a given impurity, and these may be related to the quality of the final product. In these circumstances one batch is made from each lot of raw material and the purity or other relevant property of each lot is also determined.

Example 7.6. Tinctorial strength of dyestuff
7.721 Data obtained from an experiment of this kind are given in Table 7.6.

TABLE 7.6.   QUALITY OF DYESTUFF: COMPARISON OF TWO PROCESSES

	Process A		Process B	
	Tinctorial strength of dyestuff	Log (% impurity) in raw material	Tinctorial strength of dyestuff	Log (% impurity) in raw material
	3	0·84	1	0·73
	−2	0·89	8	0·46
	8	0·58	−3	0·82
	4	0·60	5	0·54
	1	0·95	−3	0·77
	1	0·73	−2	0·84
	8	0·65	0	0·59
	−2	1·00	−6	1·01
	5	0·73	3	0·58
	4	0·68	0	0·70
Means	3·0	0·77	0·3	0·70

The finished product is a dyestuff, and one quality of interest is its tinctorial strength, which is determined by comparing the samples of dyestuff with a standard material by spectrophotometric means; the figures quoted are percentage differences from the standard. This tinctorial strength is affected by an impurity in the raw material. This impurity, an unwanted isomer which is difficult to remove, is present in amounts that vary from about 3% to about 10%. There is a lower limit to the impurity, and the standard deviation is large compared with the mean; in such circumstances the distribution is almost certainly skew, and a transformation is desirable. In this case the logarithmic transformation is satisfactory, and therefore the figures quoted in Table 7.6 for the quality of the raw material are the logarithms of the percentage of the impurity.

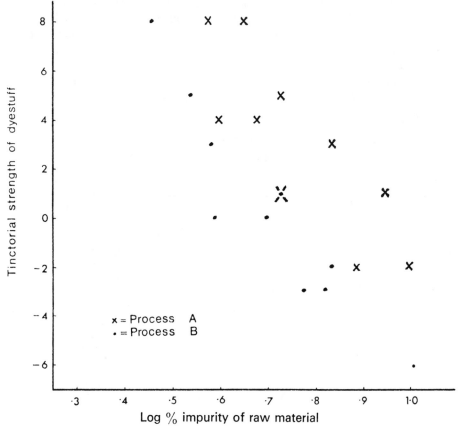

FIG. 7.7.   TINCTORIAL STRENGTH OF DYESTUFF AND IMPURITY OF RAW MATERIAL

The quality of the raw material varies from lot to lot, and an inspection suffices to show that the tinctorial strength of the dyestuff is related to the quality of the raw material. This relationship can be seen more clearly in Figure 7.7, where these two properties are plotted, the two processes A and B being shown separately.

The mean value of log (% impurity) in the raw materials used for A differs a little from that in the materials used for B, and this alone would be expected to give rise to a difference in the mean tinctorial strengths of the products made from them. The effect of the process used on the tinctorial strength of the product is thus confused to a certain extent with that of differences in the quality of the material used. It is desirable to separate these two effects, and this is one of the purposes of the method of analysis developed in this section. Another important purpose of the analysis is to allow for the variation between the lots of raw material used in each group and thereby improve the sensitivity of the comparison between the processes.

The lots of raw material used for processes A and B were selected at random, and if no information were available on the quality of the raw material the data would have to be analysed using the usual method for the comparison of the means of two groups as follows:

Mean A = 3·0   Mean B = 0·3   Difference = 2·7

Sum of squares within groups = 270·1

Variance within groups = 270·1/18 = 15·01

Variance of the difference between the two means = 3·00

Standard error of the difference between the two means = 1·73

95% confidence limits = 2·70 ± 3·63 = −0·93 and 6·33

The confidence limits include the value zero, therefore we cannot conclude with this degree of confidence that there is a real difference between the tinctorial strengths of materials produced by the two processes.

A substantial proportion of the variation in tinctorial strength of the dyestuff from batch to batch is caused by the variation in the raw material, and this contributes to the uncertainty in the estimate of the difference between the two processes. It will be shown in the next section that the variation in the purity of the raw materials can be allowed for, and the result of this is to improve the precision of the experiment to that (or almost that) which would be obtained if the raw material had been kept constant for the whole experiment.

Method of analysis

7.722 The method of analysis resembles very closely the method discussed in § **7.711** for the estimation of two regression lines. To make the explanations clearer we plot in Figure 7.8 an exaggerated situation in which the points obtained by the two systems are separated.

Assume that the relationship between the tinctorial strength of dyestuff and the logarithm of the percent impurity in the raw material is linear and that the regression lines for the two processes A and B are parallel. This means that the relationship is the same for the two processes; this is usually a safe assumption to make.

The distributions of the qualities of the raw materials depicted in Figure 7.8 are different for A and B.   If the two regression lines coincide or do not differ appreciably, then we could say that the apparent difference in mean tinctorial strength by the two processes is due to the variation in the raw materials.   If the two lines are separate, the tinctorial strength of dyestuff produced by the two processes is different, irrespective of whether the quality of raw materials used for each is the same or not.

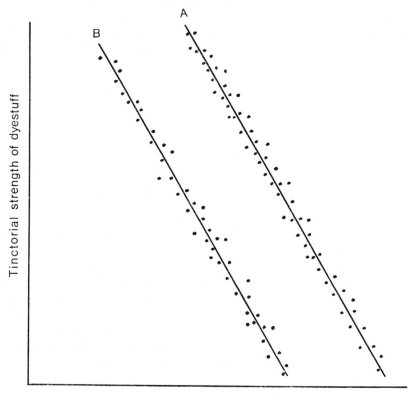

Log % impurity of raw material

FIG. 7.8.   SEPARATE REGRESSIONS (EXAGGERATED)

It is clear that the difference between the two processes, allowing for the effect of the differences in the distribution of impurities of raw materials, is given by the *vertical* distance between the two regression lines.   These lines, assumed to have.the same slope, are:

$$Y_A = \bar{y}_A + b(x - \bar{x}_A)$$
$$Y_B = \bar{y}_B + b(x - \bar{x}_B)$$

where the independent variable x is the logarithm of the percent impurity in the raw material. The vertical displacement is:

$$Y_A - Y_B = (\bar{y}_A - \bar{y}_B) - b(\bar{x}_A - \bar{x}_B) \tag{7.83}$$

The assumption is usually made that the errors in the x-variable are not appreciable, which assumption is reasonable in this case. The variance of $(Y_A - Y_B)$ is therefore:

$$V(\bar{y}_A - \bar{y}_B) + (\bar{x}_A - \bar{x}_B)^2 V(b)$$

$$= \sigma^2 \left\{ \frac{1}{n} + \frac{1}{m} + \frac{(\bar{x}_A - \bar{x}_B)^2}{\Sigma_A(x-\bar{x})^2 + \Sigma_B(x-\bar{x})^2} \right\} \tag{7.84}$$

where n and m are the numbers of samples tested for A and B respectively.

We now require to estimate σ^2. This is the variance about the regression lines and is calculated in the same way as in § 7.712. It is the variance of the tinctorial strength from batch to batch *after allowing for the variations in the percentage of impurity in the raw materials*, and is the same variance that would be obtained had it been possible to use one homogeneous lot of raw material for all 20 batches of product. Following the details given in § 7.712, σ^2 is estimated by:

$$s^2 = \frac{\Sigma_A(y-\bar{y})^2 + \Sigma_B(y-\bar{y})^2 - \dfrac{[\Sigma_A(x-\bar{x})(y-\bar{y}) + \Sigma_B(x-\bar{x})(y-\bar{y})]^2}{\Sigma_A(x-\bar{x})^2 + \Sigma_B(x-\bar{x})^2}}{m+n-3}$$

We have now to consider the situation in which the two regression lines are not parallel. In these circumstances the effect of changing the process is not unique and will depend on the percent impurity in the raw materials. The two lines are:

$$Y_A = \bar{y}_A + b_A(x - \bar{x}_A)$$

$$Y_B = \bar{y}_B + b_B(x - \bar{x}_B)$$

Therefore: $Y_A - Y_B = (\bar{y}_A - \bar{y}_B) + b_A(x - \bar{x}_A) - b_B(x - \bar{x}_B)$

and $V(Y_A - Y_B) = V(\bar{y}_A - \bar{y}_B) + (x - \bar{x}_A)^2 V(b_A) + (x - \bar{x}_B)^2 V(b_B)$

Both the estimate of the difference in tinctorial strength of dyestuff by the two processes and the variance of the difference depend on the value of x considered.

When the slopes are different, the relation between tinctorial strength of dyestuff and percent impurity of raw material differs for the two production processes A and B. Although difficulties of interpretation may arise, this difference in itself shows that the two processes give different results in certain circumstances.

Application to the data of Table 7.6

7.723 It is convenient to summarize the calculations in the form of the following table.

TABLE 7.61. CALCULATION INVOLVED IN ANALYSIS OF DATA OF TABLE 7.6

	Process A		Process B	
Number of batches	10		10	
	Tinctorial strength (y)	Log % impurity (x)	Tinctorial strength (y)	Log % impurity (x)
Mean	3·0	0·765	0·3	0·704
Sum of squares about mean	114·0	0·1950	156·1	0·2454
Sum of products of x and y about means	− 3·920		− 5·772	
Slope	−20·1		−23·5	
Sum of squares about regression	35·20		20·34	

Combining the sums of squares about the two separate regressions gives a variance of 3·471 based on 16 degrees of freedom. Hence the standard errors of the slopes are 4·2 and 3·8 respectively, from which it is clear that the two slopes do not differ appreciably and can be assumed to be equal.

The combined slope is then:
$$-9·692/0·4404 = -22·0$$
Sum of squares due to the pair of parallel regressions is:
$$(9·692)^2/0·4404 = 213·3$$
Sum of squares about the pair of parallel regressions is:
$$270·1 - 213·3 = 56·8$$
This is based on $20-3 = 17$ degrees of freedom.
Therefore variance about regressions is:
$$56·8/17 = 3·341$$
The two regression lines are:
$$Y = 3·0 - 22·0(x - 0·765) = 19·83 - 22·0x$$
$$Y = 0·3 - 22·0(x - 0·704) = 15·79 - 22·0x$$
The vertical distance between the regressions = 4·04. This is the estimate of the difference in tinctorial strengths produced by the two processes, A being superior to B. Using Formula (7.84), the variance of this estimate is:
$$3·341\{1/5 + (0·061)^2/0·4404\} = 0·696$$
so that its standard error is 0·834.

For 17 degrees of freedom the value of t for $P = 0·025$ is 2·11, and therefore the 95% confidence limits for the estimate of the difference in tinctorial strength of dyestuff produced by the two processes, allowing for the variation in percent impurities, are 2·28 and 5·80. There is thus an appreciable difference between the two processes which was not revealed by the analysis ignoring the information on the variation in the raw materials. The error variance of the corrected tinctorial strength per batch is reduced from 15·01 to 3·34, which represents a considerable improvement.

Reference to Equation (7.83) shows that the estimated difference in tinctorial strengths produced by the two processes, 4·04, is composed of two parts. One is the difference in mean observed tinctorial strengths amounting to 2·70, and the other part a correction for the difference in mean percent impurity in the raw materials. This difference is $0·765 - 0·704 = 0·061$, which, when multiplied by the slope of the regression lines, results in the correction of 1·34.

Alternative methods of analysis

7.724 There are other methods of analysing data of the type given in Table 7.6. All are equivalent, but the method given in the previous section shows most clearly the various stages involved and the assumptions which may have to be made. Some of these assumptions are disguised in the more formal method called the Analysis of Covariance. This latter method, which is the one usually given in textbooks, can be more readily generalised for complicated situations.

When more than one "nuisance" factor is operating, the analysis is carried out by the method of multiple regression, which is considered in the next chapter.

The general case of several lines

7.73 Suppose we require the relationship between a variable x and a response y, and that the data comprise several separate sets, e.g. from several machines, pieces of test equipment, lots of raw material, different men, etc. In these circumstances we may wish to know whether all the data points may be considered to lie on a single line common to all sets, or on separate but parallel lines, or on independent skew lines. These questions may be resolved by testing an ordered sequence of hypotheses.

Suppose there are k sets of data points and that a typical set, the ith, contains n_i points, the jth point of the ith set being (x_{ij}, y_{ij}). Let $\bar{x}_{i.}$ denote the mean of the values of x in the ith set and $\bar{x}_{..}$ denote the grand mean over all the sets. The means of the values of y are indicated similarly.

The sequence of alternative hypotheses to be considered is as follows:

H_1: Data in the ith set satisfy the relation:

$$Y = \bar{y}_{i.} + \beta_i(x - \bar{x}_{i.}) \qquad (i = 1, 2, ..., k)$$

where the β_i are different for each set, i.e. independent skew lines.

H_2: Data in the ith set satisfy the relation:

$$Y = \bar{y}_{i.} + \beta(x - \bar{x}_{i.}) \qquad (i = 1, 2, ..., k)$$

where the $\bar{y}_{i.}$ and the $\bar{x}_{i.}$ are different, i.e. separate but parallel lines.

H_3: *All* the data satisfy the relation:

$$Y = \bar{y}_{..} + \beta'(x - \bar{x}_{..})$$

i.e. a single common line.

The least-squares estimates of the k slopes β_i under hypothesis H_1 are given by:

$$b_i = \frac{\sum\limits_{j=1}^{n_i} (x_{ij} - \bar{x}_{i.})(y_{ij} - \bar{y}_{i.})}{\sum\limits_{j=1}^{n_i} (x_{ij} - \bar{x}_{i.})^2} \qquad (i = 1, 2, ..., k) \qquad (7.85)$$

which are the standard regression estimates with each set treated independently. The intercepts with the y-axis are given by $\bar{y}_{i.} - b_i \bar{x}_{i.}$.

The least-squares estimate of the single slope β under hypothesis H_2 is given by:

$$b = \frac{\sum\limits_i \sum\limits_j (x_{ij} - \bar{x}_{i.})(y_{ij} - \bar{y}_{i.})}{\sum\limits_i \sum\limits_j (x_{ij} - \bar{x}_{i.})^2} \qquad (7.86)$$

Thus in this case the k numerators and the k denominators from Equation (7.85) are added together. The intercepts are given by $\bar{y}_{i.} - b\bar{x}_{i.}$.

The least-squares estimate of β' under hypothesis H_3 is given by:

$$b' = \frac{\sum\limits_i \sum\limits_j (x_{ij} - \bar{x}_{..})(y_{ij} - \bar{y}_{..})}{\sum\limits_i \sum\limits_j (x_{ij} - \bar{x}_{..})^2} \qquad (7.87)$$

i.e. the data points are treated as members of one large set. The intercept is given by $\bar{y}_{..} - b'\bar{x}_{..}$.

An overall F-test of significance is first applied to the family of skew lines by comparing the value of the ratio:

$$\frac{\sum\limits_i \sum\limits_j \{b_i^2(x_{ij} - \bar{x}_{i.})^2\} / k}{\sum\limits_i \sum\limits_j \{(y_{ij} - \bar{y}_{i.})^2 - b_i^2(x_{ij} - \bar{x}_{i.})^2\} / (\sum\limits_i n_i - 2k)}$$

with the probability points listed in Table D for k & $(\Sigma n_i - 2k)$ degrees of freedom. In this ratio the numerator is the mean sum of squares due to fitting the lines independently, and the denominator is the mean residual sum of squares. If the ratio is not significant it is concluded that there is no meaningful relation between x and y and the analysis is usually terminated at this stage.

However, if an overall significant relationship is shown to exist, then a separate significance test may be made for each line by comparing the ratio:

$$\frac{\sum_j \{b_i^2(x_{ij}-\bar{x}_{i.})^2\}\Big/1}{\sum_j \{(y_{ij}-\bar{y}_{i.})^2 - b_i^2(x_{ij}-\bar{x}_{i.})^2\}\Big/(n_i-2)} \qquad (i=1, 2, \ldots, k)$$

with the values of F in Table D for 1 & (n_i-2) degrees of freedom, as for a single simple regression.

In order to discriminate between the hypotheses H_1 and H_2, we determine the significance of the increase in the residual sum of squares which results from fitting parallel lines instead of independent lines. This is given by an F-test on the ratio of the mean sum of squares for this increase to the residual mean square for the independent lines:

$$\frac{\sum_i \sum_j \{b_i^2(x_{ij}-\bar{x}_{i.})^2 - b^2(x_{ij}-\bar{x}_{i.})^2\}\Big/(k-1)}{\sum_i \sum_j \{(y_{ij}-\bar{y}_{i.})^2 - b_i^2(x_{ij}-\bar{x}_{i.})^2\}\Big/(\sum_i n_i-2k)}$$

with $(k-1)$ & (Σn_i-2k) degrees of freedom. If this ratio is significant we conclude that an independent relation exists for each set and no further calculations are required. However, if it is not significant we then have to discriminate between hypotheses H_2 and H_3 by determining the significance of the further increase in the residual sum of squares which results from fitting a single common line. This is obtained by an F-test on the ratio of the mean sum of squares for this further increase to the residual mean square for parallel lines:

$$\frac{\sum_i \sum_j \{(y_{ij}-\bar{y}_{..})^2 - b'^2(x_{ij}-\bar{x}_{..})^2 - (y_{ij}-\bar{y}_{i.})^2 + b^2(x_{ij}-\bar{x}_{i.})^2\}\Big/(k-1)}{\sum_i \sum_j \{(y_{ij}-\bar{y}_{i.})^2 - b^2(x_{ij}-\bar{x}_{i.})^2\}\Big/(\sum_i n_i-k-1)}$$

with $(k-1)$ & (Σn_i-k-1) degrees of freedom. We conclude that a set of parallel relationships exist if this ratio is significant; otherwise we conclude that the data are best represented by a single line.

In the latter case the significance level of the common line is calculated in the standard manner from an F-test on the expression:

$$\frac{\sum_i \sum_j \{b'^2(x_{ij}-\bar{x}_{..})^2\}\Big/1}{\sum_i \sum_j \{(y_{ij}-\bar{y}_{..})^2 - b'^2(x_{ij}-\bar{x}_{..})^2\}\Big/(\sum_i n_i-2)}$$

with 1 & (Σn_i-2) degrees of freedom.

Example 7.7

7.731 A very simple example will now be described to illustrate the analysis. Consider two sets of data consisting of the five points:

$$(1, 1), \quad (2, 3), \quad (3, 2), \quad (4, 3), \quad (5, 4)$$

and the six points:

$$(1, 2), \quad (2, 2), \quad (3, 1), \quad (4, 1\cdot5), \quad (5, 1), \quad (6, 0\cdot5)$$

Even in such a simple case the calculations are laborious and it is best to employ an electronic computer. The complete analysis for these data is therefore displayed in Table 7.7 in the form of a typical computer output.

TABLE 7.7. COMPUTER OUTPUT FOR COMPARISON OF REGRESSION LINES

(1)
Independent Regression Lines
Analysis of Variance

Source	DF	SS		MS		F		Sig. Level
Due to	2	5·0286,	0	2·5143,	0	8·7791,	0	1·24 p.c.
About	7	2·0048,	0	2·8639,	−1			
Total	9	7·0333,	0					

(2)

Set	Slope		Intercept		Res. S.D.		DF	Sig. Level
1	6·000,	−1	8·000,	−1	7·303,	−1	3	8·05 p.c.
2	−2·857,	−1	2·333,	0	3·181,	−1	4	1·98 p.c.

(3)
Parallel Regression Lines
Analysis of Variance

Source	DF	SS		MS		F		Sig. Level
Due to	1	3·6364,	−2	3·6364,	−2	4·1576,	−2	84·35 p.c.
About	8	6·9970,	0	8·7462,	−1			
Total	9	7·0333,	0					

(4)
Common Slope = 3·636, −2
Residual Standard Deviation = 9·352, −1 with 8 DF

Set	Intercept		Mean x		Mean y	
1	2·491,	0	3·000,	0	2·600,	0
2	1·206,	0	3·500,	0	1·333,	0

(5)
Improvement of Independent over Parallel Lines

Source	DF	SS		MS		F		Sig. Level
Improvement	1	4·9922,	0	4·9922,	0	1·7431,	1	0·42 p.c.
Ind. Resid.	7	2·0048,	0	2·8639,	−1			

(6)
Common Regression Line
Analysis of Variance

Source	DF	SS		MS		F		Sig. Level
Due to	1	1·8768,	−2	1·8768,	−2	1·4830,	−2	90·58 p.c.
About	9	1·1390,	1	1·2656,	0			
Total	10	1·1409,	1					

(7)

Mean x		Mean y		Slope		Intercept		Res. S.D.		DF
3·273, 0		1·909,	0	−2·581,	−2	1·994,	0	1·125,	0	9

(8)
Improvement of Parallel Lines over Single Line

Source	DF	SS		MS		F		Sig. Level
Improvement	1	4·3934,	0	4·3934,	0	5·0231,	0	5·53 p.c.
Par. Resid.	8	6·9970,	0	8·7462,	−1			

This uses a conventional floating decimal point notation for the levels of the variables; for example 2·8639, −1 is equivalent to 0·28639 in fixed-point notation.

It will be seen at step **5** that the improvement of independent lines over parallel lines is statistically highly significant (0·42% level) and the lines themselves are significant (1·24%). We therefore accept the model consisting of the two skewed lines as printed out at step **2**; in this particular case the first line has a positive slope whilst the second has a negative gradient. Having accepted the two skew lines, the rest of the analysis is really superfluous, since there is no point in continuing the sequence of hypothesis testing to determine whether the data are better represented by a single line than by a set of parallel ones.

CORRELATION

7.8 There is a class of problems in descriptive statistics involving two or more related variates where the main interest is not in estimating one variate from another, but in the joint distribution of the variates and in the extent of the association between them. This is the class of problem in which correlation methods are most useful. The following provides an illustration.

A hat manufacturer would be interested in the distribution of the head measurements of a population, because this must have a bearing on the number of hats of any given shape and size he is likely to sell. Suppose he takes a random sample of individuals in his potential market and measures the head length and breadth of each individual. This supplies him with a mass of data which will imply very little unless it is presented in a condensed form in order to bring out the essential features. As shown in Chapter 2, any one measurement, e.g. the length, can be completely described by means of a distribution, e.g. a histogram, and if this is of the Normal form, two statistics —the mean and standard deviation—give a complete description. If we take length and breadth and make the assumption that both measurements are Normally distributed—for biometric data this will usually be true—then each can be represented by its mean and standard deviation. These four figures, although adequate to describe each property separately, do not give a complete description of the whole data, because no information is supplied on the relationship between the two measurements. As shown previously, a convenient method of exhibiting the relationship is to plot the two measurements, one against the other, for all the individuals in the form of a scatter diagram, as in Figure 7.9 overleaf.

An association or a correlation between the two measurements is indicated by the diagonal tendency. Long heads tend to be associated with broad heads and vice versa, but for any given value of each measurement there is a considerable spread in the other.

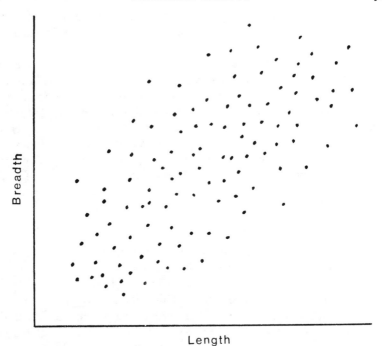

Breadth

Length

FIG. 7.9.   DISTRIBUTION OF HEAD MEASUREMENTS

The Normal surface

7.81 As an alternative to the scatter diagram, the distribution can be represented by a solid histogram formed by first of all dividing the area into
rectangles, and then erecting on each one a block whose volume is proportional to the number of observations covered by the rectangle. In the limit
for an indefinitely large sample and indefinitely small areas we arrive at a
smooth surface.

There exists a Normal distribution for two variates, just as for one; it is
referred to as the Bivariate Normal Distribution. Its probability density
can be represented by the expression:

$$\frac{1}{2\pi\sigma_x\sigma_y\sqrt{(1-\rho^2)}}\exp\left[-\frac{1}{2(1-\rho^2)}\left\{\frac{(x-\mu)^2}{\sigma_x^2}-2\rho\frac{(x-\mu)(y-\eta)}{\sigma_x\sigma_y}+\frac{(y-\eta)^2}{\sigma_y^2}\right\}\right]$$

where σ_x^2 and σ_y^2 are respectively the variances of the two variates x and y,
μ and η their means, and ρ is the additional coefficient which measures the
association between the two variates. These five parameters give a complete
description of the Normal bivariate distribution; the means and variances are
estimated from the sample in the usual way [Chapter 3]. ρ is called the
Correlation Coefficient and is also estimated from the sample. To determine

this value we need to calculate another statistic called the Covariance [§ 3.7]:

$$C(x,y) = \Sigma(x-\bar{x})(y-\bar{y})/(n-1)$$

In words, this is the sum of the products of the deviation of each x from the mean $\bar{x}$ and the deviation of the corresponding y from the mean $\bar{y}$, divided by the number of degrees of freedom, $n-1$. The covariance has thus a similar form to the variance except that product terms are involved. This similarity is brought out more clearly by displaying all these expressions as follows:

$$s_x^2 = \Sigma(x-\bar{x})(x-\bar{x})/(n-1)$$
$$C(x, y) = \Sigma(x-\bar{x})(y-\bar{y})/(n-1)$$
$$s_y^2 = \Sigma(y-\bar{y})(y-\bar{y})/(n-1)$$

The correlation coefficient is then estimated by $C(x, y)/s_x s_y$, and is denoted by the letter r:

$$r = \Sigma(x-\bar{x})(y-\bar{y})/\sqrt{\{\Sigma(x-\bar{x})^2\Sigma(y-\bar{y})^2\}}. \qquad (7.9)$$

All the quantities used in calculating the correlation coefficient have already been encountered in calculating a regression coefficient and its confidence limits. As for regression analysis, if there is a large amount of data then the calculations are best performed on an electronic computer.

Using the data of Table 7.1 as an example and letting x be hardness and y be abrasion loss, we calculated in § **7.221**:

$$\Sigma(x-\bar{x})^2 = 4,300 \qquad \Sigma(x-\bar{x})(y-\bar{y}) = -22,946 \qquad \Sigma(y-\bar{y})^2 = 225,011$$

Hence $r = -22,946/\sqrt{(4,300 \times 225,011)} = -0.738$.

Properties of the correlation coefficient

7.82 The correlation can be negative or positive. When it is positive, one variate tends to increase as the other increases; when it is negative, one variate tends to decrease as the other increases. It can be readily shown that the correlation coefficient cannot take values outside the limits -1 and $+1$. A high absolute value of r indicates a close relationship and a small value, a less definite relationship. When the absolute value of r is unity, the points fall exactly on a straight line and the relationship is perfect. When $r = 0$ the points scatter in all directions, and the variates are linearly independent. The sign of r is the same as the sign of the covariance. Typical scatter diagrams for a low and a high correlation are shown in Figure 7.91 overleaf.

It must be remembered that the correlation coefficient is an estimate (calculated from the sample) of the degree of association between two variates, and this estimate is only valid when the sample is randomly drawn from the Population.

The correlation coefficient is of less value when the bivariate distribution is not Normal, but it can still be of use as an overall measure of *linear* association between the variates.

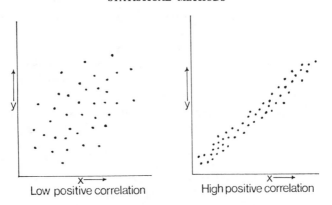

Fig. 7.91. SCATTER DIAGRAMS FOR LOW AND HIGH CORRELATIONS

Significance of the correlation coefficient

7.83 One question usually of interest is whether the value of the correlation coefficient found for a sample could have arisen with a given probability from a Normal Population with $\rho = 0$, i.e. with x and y uncorrelated. A table exists for this purpose (Table E at the end of this volume), which gives the 0·10, 0·05, 0·02 and 0·01 significance points for various numbers of degrees of freedom. In assessing the significance of the correlation coefficient the table is entered with $(n-2)$ for the degrees of freedom, corresponding to the residual sum of squares about the regression as described in earlier sections. For the correlation of hardness and abrasion loss data in Table 7.1, $n-2 = 28$ and the 5% and 1% points obtained by interpolation are $\pm 0·36$ and $\pm 0·46$ respectively. The actual value of $-0·738$ for the correlation coefficient is therefore highly significant, and we conclude that the sample of observations was drawn from a Population in which a negative correlation existed.

This assessment of the significance of the correlation coefficient gives precisely the same result as that of comparing the regression coefficient with its standard error, using the t-table [§ **7.25**].

An approximate method—Tukey's corner test

7.831 A rapid, but approximate, graphical method of determining the significance of the correlation between two variates is the Corner Test, devised by J. W. Tukey. The method is applied as follows.

First plot the n pairs of observations as a scatter diagram. Then draw horizontal and vertical Medial Lines, each dividing the n points into equal-sized groups, thus forming four quadrants. Assign a score of $+1$ to each point in the upper-right and bottom-left quadrants and a score of -1 to each point in the other two quadrants. Starting from the top of the diagram and proceeding in the direction of the y-axis, accumulate the point scores until the *vertical* medial line is crossed for the first time.   Repeat this process from the bottom towards the top of the diagram.   Then count similarly from right to left and from left to right, but stopping at the horizontal medial line in these cases.   If the sum of the four scores (taking signs into account) is ± 11 or more extreme, then the correlation is judged significant at the 0·05 probability level.   If the total score is ± 15 or more extreme, the correlation is significant at the 0·01 level; but when $n > 50$, ± 14 is sufficient for the latter conclusion.   When several points are reached at the same time, a score of $+1/(1 + $ no. of "-1" points) is assigned to each positive quadrant point and vice versa.

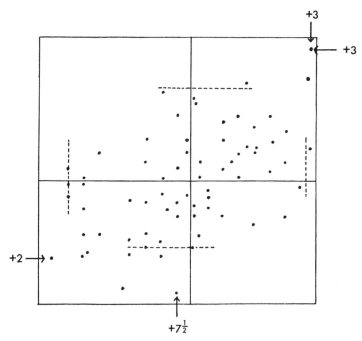

FIG. 7.92.   TUKEY'S CORNER TEST

This test is not as sensitive as the numerical test of the previous section, and therefore failure to detect a significant association does not mean that a significant association could not be detected by the full analysis.   But the detection of a significant association by the corner test often makes the more formal analysis unnecessary.

As an example, consider the data plotted in Figure 7.92.   Starting from the top there are three "$+1$" points before crossing the vertical medial line; from the bottom there are seven "$+1$" points before a "$+1$" point and a "-1" point are met simultaneously, giving a score of $+7\frac{1}{2}$.   From right to left the score is $+3$; from left to right it is $+2$, because there is one clear

"$+1$" point before two "$+1$" points and one "-1" point occur simultaneously, giving a score of $1 + 2(\frac{1}{2}) = 2$. The total quadrant sum is therefore $15\frac{1}{2}$, so that there is strong evidence of an association between the variates ($P < 0.01$).

Confidence limits and the comparison of correlation coefficients

7.84 In order to compare correlation coefficients a useful transformation due to R. A. Fisher may be used. This is:

$$z = \tanh^{-1} r = \tfrac{1}{2}\{\log_e (1+r) - \log_e (1-r)\} \qquad (7.91)$$

which is very nearly Normally distributed with standard deviation $1/\sqrt{(n-3)}$. This transformation enables us to calculate approximate confidence limits for a correlation coefficient and for a difference between two correlation coefficients, and also to combine correlation coefficients from two or more samples and calculate the confidence limits. The factors used to calculate the confidence limits are obtained from the table of Normal deviates (Table A) and not from the t-table.

The correlation coefficients to be compared or combined are first transformed to their z values, using the above expression; these z values, considered as though each were the mean of $(n-3)$ observations from a Population with unit standard deviation, may then be used in the calculation of confidence limits or in any of the tests appropriate for comparisons or combinations of such means.

Correlation between time series

7.85 Difficulties may arise in the interpretation of a correlation coefficient when the observations are taken over a period of time. Consider, for example, the correlation of quality of a product with quality of raw materials taken at intervals over a period, say one year. Let us assume that during this period there has been a gradual tightening up on supervision, which would of itself result in improved yield. Assume also that during the same period a certain small impurity that does not affect the yield has also shown a gradual increase. If the yield is plotted against this impurity a high correlation may result, but it does not follow that the yield increase is due to increase in the impurity. Therefore a significant correlation does not necessarily indicate a causal relationship between the two variables, since this correlation may be due to a common factor, particularly if a time or space factor is involved. This type of correlation is commonly referred to as Spurious Correlation. In a similar context, G. A. Barnard expressed the difficulties very succinctly by saying, "Because the yield rose when the temperature rose, it does not necessarily follow that the yield will rise if the temperature is raised". In certain circumstances the effect of time can be removed by the use of multiple regression methods, which are considered in the next chapter, but generally in a multiple regression analysis the possibility of being misled by spurious correlations is much greater.

Relation between the coefficients of correlation and regression

7.86 The sample estimates of the correlation coefficient and the regression coefficient are:

$$r = \frac{\Sigma(x-\bar{x})(y-\bar{y})}{\sqrt{\{\Sigma(x-\bar{x})^2\Sigma(y-\bar{y})^2\}}} = \frac{C(x, y)}{s_x s_y}$$

and

$$b = \frac{\Sigma(x-\bar{x})(y-\bar{y})}{\Sigma(x-\bar{x})^2} = \frac{C(x, y)}{s_x^2} = r\frac{s_y}{s_x}$$

The regression equation of y upon x can therefore be written as:

$$\frac{Y-\bar{y}}{s_y} = r\frac{x-\bar{x}}{s_x}$$

In other words, the correlation coefficient is numerically the same as the regression coefficient when the variates are expressed in standard measure [§ **2.43**].

The regression equation of x upon y is similarly:

$$\frac{X-\bar{x}}{s_x} = r\frac{y-\bar{y}}{s_y}$$

The sum of squares due to the regression of y upon x is:

$$b^2\Sigma(x-\bar{x})^2 = \{\Sigma(x-\bar{x})(y-\bar{y})\}^2/\Sigma(x-\bar{x})^2 = r^2\Sigma(y-\bar{y})^2$$

In words, r^2 represents the proportion of the total sum of squares in one variable accounted for by the other. The sum of squares about the regression is $(1-r^2)\Sigma(y-\bar{y})^2$, and therefore the ratio of the mean square due to regression to the mean square about the regression is $(n-2)r^2/(1-r^2)$.

These relationships show that "correlation" and "regression" are mathematically equivalent; but on the practical side they represent different types of problems, and it is best to keep them separate. In industrial work the problems classed under regression are by far the more important ones, and for this reason they have received prominence in this chapter.

References

[7.1] WILLIAMS, E. J. *Regression Analysis.* John Wiley (New York, 1959)
[7.2] CREASY, M. A. "Confidence Limits for the Gradient in the Linear Functional Relationship", *J. R. Statist. Soc.*, B, **18**, 65–69 (1956).
[7.3] BARTLETT, M. S. "Fitting a Straight Line when Both Variables are Subject to Error", *Biometrics*, **5**, 207–212 (1949).
[7.4] BERKSON, J. "Are There Two Regressions?" *J. Amer. Statist. Ass.*, **45**, 164–180 (1950).
[7.5] FINNEY, D. J. *Statistical Method in Biological Assay* (third edition). Charles Griffin (High Wycombe, 1978).

Appendix 7A

Fieller's theorem

Confidence limits are sometimes required for the ratio of two estimated parameters, e.g. the ratio of two means, the ratio of two regression coefficients, or the ratio of two potencies such as that used in § **7.71**.

Let a, b be unbiased estimates of two parameters, and assume that a and b are Normally distributed. We require to determine the confidence limits for the true value of the ratio a/b.

Put $a/b = v$, and form the expression $x = a - vb$. This is a linear function of Normally distributed variates, and x is therefore itself Normally distributed with variance $V = V(a) - 2v\,C\,(a, b) + v^2 V(b)$. The expected value of x is zero, so that $x/\sqrt{V}$ is distributed as t, i.e. $(a - vb)^2/V$ is distributed as t^2. The confidence limits for v are therefore given by the values which satisfy the equation:

$$(a - vb)^2/V = t^2$$

using the appropriate value of t. This is a quadratic in v, and the confidence limits of v are:

$$\frac{\dfrac{a}{b} - \dfrac{t^2\,C\,(a, b)}{b^2} \pm \dfrac{t}{b}\sqrt{\left\{ V(a) - \dfrac{2a}{b}C\,(a, b) + \dfrac{a^2}{b^2}V(b) - \dfrac{t^2 V(b)}{b^2}\left(V(a) - \dfrac{[C(a, b)]^2}{V(b)} \right) \right\}}}{1 - \dfrac{t^2 V(b)}{b^2}}$$

The approximate formula, derived in a different manner and used in § **7.712**, is:

$$\frac{a}{b} \pm \frac{t}{b}\sqrt{\left\{ V(a) + \frac{a^2}{b^2}V(b) \right\}}$$

In this situation a and b are independent, so that $C\,(a, b) = 0$. Making this substitution in Fieller's formula and comparing with the approximate formula, we see that in the latter the assumption has been made that $V(b)/b^2$ is small compared with unity. This is generally the case in a well-designed experiment.

Chapter 8

Multiple and Curvilinear Regression

Frequently a quantity of interest such as the yield of
a product will be dependent on the levels of not one
but a number of variables. The situation is often
complicated by the fact that these variables are themselves related. This chapter shows how such multiple
relationships can be elucidated and how some common
pitfalls may be avoided. In the later part of the
chapter the theory is applied to the case in which
quantities are related not by straight lines but by
curves.

Introduction

8.1 In the previous chapter a relationship was obtained [§ **7.221**] between the
abrasion loss and hardness of rubber samples, of the form:

Abrasion loss (g./h.p.-hour) = 550·4 − 5·336 hardness (I.R.H. (Shore) units)

$$\text{i.e.} \quad Y = a + bx_1$$

The value of this equation is that it shows how measurement of a property
which can be fairly readily and rapidly determined—in this instance hardness
—may be used as an indication of a more important but less readily determined property such as abrasion loss. It is, however, by no means an exact
relationship; the residual scatter of points about the line defined by the
relationship still has standard deviation 60·5 g./h.p.-hour compared with a
standard deviation of 88·1 g./h.p.-hour for the overall variation of all abrasion
loss results. Thus it is unlikely that in practice a knowledge of hardness
alone would be a sufficient guide to abrasion loss. It is possible, however,
that other information on a sample in addition to a knowledge of hardness
might enable us to predict abrasion loss more precisely while still avoiding the
elaborate testing called for by a direct determination. We might, in fact,
attempt to relate abrasion loss to more than one other property of a sample
and fit a Multiple Regression Equation showing its dependence on each of the
several factors considered, rather than the simple regression on hardness alone.
Techniques are available for the fitting of such equations analogous to those
already described for the fitting of simple regressions, and these are discussed

in the present chapter. The same techniques may be applied to the fitting of curvilinear (polynomial) relationships, since successive powers of a single independent variable are mathematically equivalent to additional independent variables. Thus, a section of the chapter [§ **8.6**] is devoted to the fitting of curvilinear regressions. Measures of precision and tests of significance for multiple and curvilinear regression are also described, and the use of regression equations for deriving empirical models of a process and for purposes of prediction is discussed.

MULTIPLE LINEAR REGRESSION

Example 8.1. Improved prediction of abrasion loss

8.2 In the example already considered, a further property of a sample of rubber which might reasonably be expected to be related to abrasion loss is its tensile strength. Results for tensile strength are shown alongside those for hardness and abrasion loss in Table 8.1, and Figure 8.1 is a scatter diagram of abrasion loss versus tensile strength.

The diagram shows few signs of any association between abrasion loss and tensile strength, and it does not seem, at first sight, that the additional

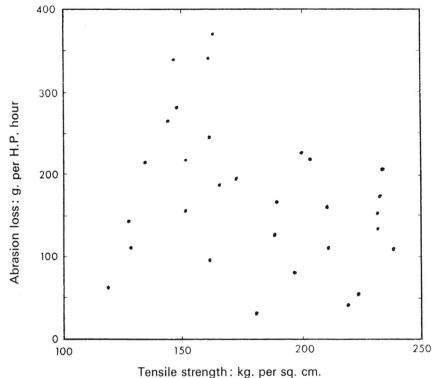

FIG. 8.1. SCATTER DIAGRAM. ABRASION LOSS *v*. TENSILE STRENGTH
FOR RUBBER SAMPLES

information will be of much value for purposes of prediction. Figure 8.1 does not, however, tell the whole story. If we are to use tensile strength to assist in the prediction of abrasion loss, we must relate the observed variations

TABLE 8.1.   PHYSICAL PROPERTIES OF RUBBER SAMPLES

Sample No.	Abrasion Loss (g./h.p.-hour) (y)	Hardness (I.R.H. (Shore) units) (x_1)	Tensile Strength (kg./sq. cm.) (x_2)
1	372	45	162
2	206	55	233
3	175	61	232
4	154	66	231
5	136	71	231
6	112	71	237
7	55	81	224
8	45	86	219
9	221	53	203
10	166	60	189
11	164	64	210
12	113	68	210
13	82	79	196
14	32	81	180
15	228	56	200
16	196	68	173
17	128	75	188
18	97	83	161
19	64	88	119
20	249	59	161
21	219	71	151
22	186	80	165
23	155	82	151
24	114	89	128
25	341	51	161
26	340	59	146
27	283	65	148
28	267	74	144
29	215	81	134
30	148	86	127
Mean	175·4	70·27	180·5

in tensile strength to the as yet unexplained variations in abrasion loss, i.e. to the variations of abrasion loss *about the regression of abrasion loss upon hardness*, rather than to the overall variation. If we do this, then we should also eliminate any variations in tensile strength which can themselves be associated with variations in hardness. Thus we must look for a relationship between the residual variations in abrasion loss after allowing for hardness, and the residual variations in tensile strength, again after allowing for any effects of hardness. The scatter diagram of these residual variations is shown as Figure 8.2.

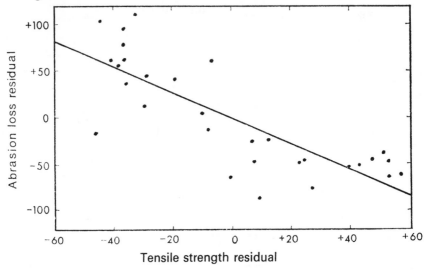

FIG. 8.2. SCATTER DIAGRAM. RESIDUAL VARIATIONS IN ABRASION LOSS AND TENSILE STRENGTH AFTER ALLOWING FOR HARDNESS

There is now a clear indication of a relationship. It follows that a knowledge of tensile strength as well as hardness *will* permit more precise prediction of abrasion loss than a knowledge of hardness alone; the residual scatter will be reduced from that about the regression on hardness only (the overall scatter in Figure 8.2) to that about the relationship which may be fitted to the values in Figure 8.2. The slope of the latter is calculated as -1.374, i.e. a unit increase in tensile strength is associated with a decrease of 1.374 g./h.p.-hour in abrasion loss, and the residual scatter is found to have a standard deviation of 36.5 g./h.p.-hour. The details of the calculations are given later.

Separation of effects

8.21 By eliminating the effects of hardness (x_1) we have effectively studied variations in abrasion loss (y) and tensile strength (x_2) at constant hardness. Because it applies to variations at constant hardness, the relationship revealed by Figure 8.2 will be a true measure of the effect of tensile strength on abrasion loss when both hardness and tensile strength are considered. Conversely, however, the simple relationship between abrasion loss (y) and hardness (x_1) derived earlier will not be applicable when both hardness (x_1)

and tensile strength (x_2) are used for prediction, since it does not relate to variations in abrasion loss and hardness at *constant tensile strength*. The Total Regression Coefficient derived earlier will differ from the more specific Partial Regression Coefficient now applicable, because the former describes the effects both of variations in hardness, and of any associated variations in tensile strength. The difference will be greater or smaller depending on the extent of the association between hardness and tensile strength.

Let b continue to denote the total regression coefficient of abrasion loss upon hardness and let m denote the slope of the regression of tensile strength upon hardness. If b_2 denotes the slope of the regression of residual variations in abrasion loss upon residual variations in tensile strength, then the equation fitted in Figure 8.2 to allow for the dependence on tensile strength (x_2) is:

$$\{(Y-\bar{y})-b(x_1-\bar{x}_1)\} = b_2\{(x_2-\bar{x}_2)-m(x_1-\bar{x}_1)\}$$

The deviations from this regression are inexplicable in terms of either hardness or tensile strength, so that the equation in fact represents the multiple regression of y upon x_1 and x_2. Rearrangement gives:

$$(Y-\bar{y}) = (b-mb_2)(x_1-\bar{x}_1)+b_2(x_2-\bar{x}_2) \qquad (8.1)$$

Thus the partial regression coefficient of y upon x_1 is given by:

$$b_1 = b-mb_2 \qquad (8.11)$$

The value obtained for b_1 is -6.571, so that the partial effect of hardness is greater than the total effect found earlier. The explanation lies in the negative correlation which exists between hardness and tensile strength; hard samples tend to have a low tensile strength, so that the effect of hardness is partially offset by the contrary effect of associated variations in tensile strength. The final multiple regression equation to express the estimated dependence of abrasion loss (y) upon hardness (x_1) and tensile strength (x_2) is:

$$(Y-175.4) = -6.571(x_1-70.27)-1.374(x_2-180.5) \qquad (8.2)$$

i.e.   Abrasion loss $= 885.2-6.571$ hardness -1.374 tensile strength (8.21)

Total and partial regression coefficients

8.22 The distinction between total and partial regression coefficients is an important one, and one which should never be overlooked. It is sometimes the practice, when a dependence on more than one factor is suspected, to plot the values of the "dependent" variable against the corresponding values of each of the "independent" variables in turn, and to assume that a dependence exists wherever this seems indicated by such a plot. This, however, may be completely misleading; the relationships shown by such two-dimensional plots are all total regressions, and these may depend much more on the particular arrangement of the observations with respect to the independent variables than on any real effects of the factors concerned.

To illustrate the distinction it is convenient to make use of a more extreme example than that discussed above on the physical properties of rubber samples. Figure 8.3 is a (schematic) plot of results obtained in the manufacture of an organic chemical. The ordinate (x_1) is the "strength" of the base material, i.e. the proportion of the active ingredient, and the abscissa (x_2) is the time required for completion of the reaction. Each point plotted represents a single batch. The dependent variable y was the purity of the final product, which was related to strength of base material and time of reaction by means of a multiple regression equation. It was found that, in fact,

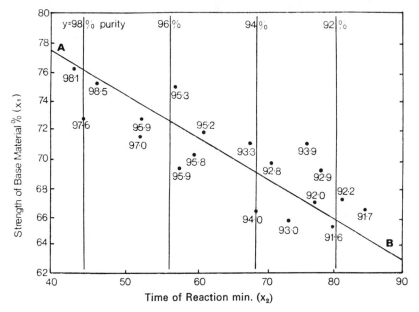

FIG. 8.3. MANUFACTURE OF AN ORGANIC CHEMICAL. PLOT OF RESULTS, SHOWING ESTIMATED CONTOURS OF CONSTANT PRODUCT PURITY

purity depended only on time of reaction—the longer the time of reaction, the lower the purity. Thus contours of constant purity (lines joining all points for which a given purity would be predicted from the regression equation) are the vertical lines shown in the figure. Such a contour diagram is often a useful practical representation of a multiple regression equation with two independent variables (see also § **8.82**).

The effect which must be assigned to either factor when both are considered—i.e. the partial regression coefficient—is the rate of change of y with the factor when the other is held constant. It is thus the rate at which contours are crossed by a point travelling parallel to the appropriate co-ordinate axis. The contours of Figure 8.3 illustrate that, while purity is related to time of reaction (x_2), it is not related to strength of base material (x_1), since no contours are ever crossed in the x_1 direction. The observations available are not, however, scattered evenly over the plane of the paper; they show a general diagonal tendency, lower strengths being associated with longer reaction times. Thus, in the results available, movement in the x_1 direction takes place, on the average, along the line AB—the regression of x_2 upon x_1. This line does cross the purity contours, at a rate determined by the slope of the regression of x_2 upon x_1. If we ignore or lose sight of the fact that, *in the results available*, movement in the x_1 direction is only obtained at the cost of movement in the x_2 direction, we would then conclude that purity

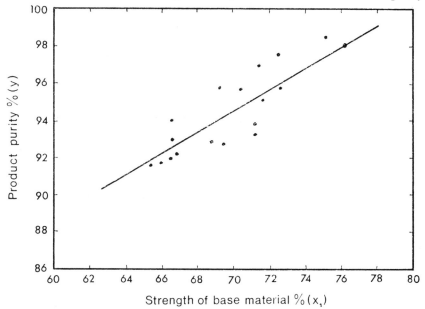

FIG. 8.4.   MANUFACTURE OF AN ORGANIC CHEMICAL.   APPARENT
DEPENDENCE OF PRODUCT PURITY ON STRENGTH OF BASE MATERIAL

of product is affected by strength of base material. Such a conclusion had in fact been drawn, before the multiple regression analysis showed it to be fallacious, from a plot [Figure 8.4] of purity of product against strength of base material; the apparent plausibility of the result—decreasing purity associated with decreasing strength of base material—served to distract attention from the important connection between strength of base material and reaction time.

The rate of change of y with x_1 as we move along the line AB is the total regression coefficient for y upon x_1, and in this instance it gives a completely false impression of the role of x_1 (strength of base material); the apparent dependence of y on x_1, which it shows, results solely from the correlation between x_1 and x_2, i.e. from the arrangement of the observations with respect to x_1 and x_2. If, by accident or design, this arrangement had been different, a different total regression coefficient would necessarily have been obtained. The partial regression coefficients would not, however, have been affected, except by inevitable sampling fluctuations; they do not, as the representation shows, depend on the arrangement of the observations.

It will be observed that, in the present instance, the total regression coefficient for the regression of y (purity) upon x_2 (time of reaction) is the same as the corresponding partial regression coefficient, and is unaffected by the arrangement of the observations. The reason for this is that there is no real dependence of y on x_1 (strength of base material). Usually, however, the contours will be inclined, and both partial regression coefficients will exist. The general effect of correlation between the independent variables is then to throw part of the real effect of x_1 into the apparent effect of x_2, and part of the real effect of x_2 into the apparent effect of x_1, so that both total regression coefficients are affected.

Generalization of the method of least squares

8.3 The discussion above [§§ **8.2, 8.21**] shows how, in principle, a multiple regression equation may be fitted, but the method used is not a particularly practical one—especially when more than two "independent" variables must be considered. Now the separate regressions are fitted by the Method of Least Squares [§ **7.22**], in which the sums of squares of deviations from the relationship are minimized. This same method may be applied directly to the fitting of multiple regression equations, and short-circuits the repeated fitting and adjustment carried out above. For the dependence of one variable on two other variables the relationship to be fitted is of the form:

$$Y = \beta_0 + \beta_1 x_1 + \beta_2 x_2 \qquad (8.3)$$

Just as in Chapter 7 for a single independent variable we obtained the best fitting *line* in two dimensions, so now we require the best fitting *plane* in the three dimensions y, x_1, x_2.

For any observed pair of values x_{1i}, x_{2i} of the "independent" variables x_1, x_2, we may calculate a predicted value Y_i of the "dependent" variable from the relationship:

$$Y_i = \beta_0 + \beta_1 x_{1i} + \beta_2 x_{2i} \qquad (8.31)$$

The difference between the actual observed value y_i and this predicted value is given by:

$$(y_i - Y_i) = (y_i - \beta_0 - \beta_1 x_{1i} - \beta_2 x_{2i}) \qquad (8.32)$$

The values of the parameters β_0, β_1 and β_2 are as yet unknown. The problem is to derive those values for the parameters which will give rise to the least disagreement, overall, between observation and prediction. As a measure of the overall disagreement we take the sum of squares of the deviations:

$$S.S. = \sum_i (y_i - Y_i)^2 \qquad (8.33)$$

In fitting by the method of least squares the estimated parameters b_0, b_1 and b_2 are such as to minimize $S.S.$ This may be done by the methods of differential calculus; the partial derivatives of $S.S.$ with respect to β_0, β_1, β_2 are each equated to zero, giving a set of simultaneous equations for b_0, b_1 and b_2. After a little algebraic manipulation these equations become:

$$
\left.
\begin{aligned}
b_1 \Sigma(x_1 - \bar{x}_1)^2 + b_2 \Sigma(x_1 - \bar{x}_1)(x_2 - \bar{x}_2) &= \Sigma(x_1 - \bar{x}_1)(y - \bar{y}) \\
b_1 \Sigma(x_1 - \bar{x}_1)(x_2 - \bar{x}_2) + b_2 \Sigma(x_2 - \bar{x}_2)^2 &= \Sigma(x_2 - \bar{x}_2)(y - \bar{y}) \\
b_0 &= \bar{y} - b_1 \bar{x}_1 - b_2 \bar{x}_2
\end{aligned}
\right\}
\quad (8.34)
$$

Extending the notation introduced in Chapter 7, we may also write the first two equations more concisely as:

$$
\left.
\begin{aligned}
b_1 S_{11} + b_2 S_{12} &= S_{y1} \\
b_1 S_{12} + b_2 S_{22} &= S_{y2}
\end{aligned}
\right\}
\quad (8.35)
$$

where S_{11} denotes $\Sigma(x_1 - \bar{x}_1)^2$, S_{12} denotes $\Sigma(x_1 - \bar{x}_1)(x_2 - \bar{x}_2)$ and so on.

For the multiple regression of abrasion loss on hardness and tensile strength discussed above, Equations (8.35) become:

$$
\left.
\begin{aligned}
4{,}300 b_1 - 3{,}862 b_2 &= -22{,}946 \\
-3{,}862 b_1 + 38{,}733 b_2 &= -27{,}857
\end{aligned}
\right\}
\quad (8.351)
$$

These may readily be solved to give:

$$
b_1 = -6 \cdot 571 \qquad b_2 = -1 \cdot 374
$$

so that the solution is identical with that quoted earlier.

For the more general problem, in which the dependence of a variable y on a number of other variables $x_1 \dots x_p$ is to be estimated, the relationship may be written as:

$$
Y = \beta_0 + \beta_1 x_1 + \beta_2 x_2 + \dots + \beta_p x_p \quad (8.36)
$$

and the Least-Squares Equations are:

$$
\left.
\begin{aligned}
b_1 S_{11} + b_2 S_{12} + \dots + b_p S_{1p} &= S_{y1} \\
b_1 S_{21} + b_2 S_{22} + \dots + b_p S_{2p} &= S_{y2} \\
\vdots \qquad \vdots \qquad \vdots \qquad \vdots \\
b_1 S_{p1} + b_2 S_{p2} + \dots + b_p S_{pp} &= S_{yp} \\
\text{and } b_0 = \bar{y} - b_1 \bar{x}_1 - b_2 \bar{x}_2 - \dots - b_p \bar{x}_p
\end{aligned}
\right\}
\quad (8.37)
$$

These equations are sometimes also referred to as the Normal Equations (unrelated to the Normal distribution). The number of equations is equal to the number of parameters to be estimated, i.e. $(p+1)$. Since $S_{ij} = S_{ji}$ by definition, the coefficients in the first p equations form a symmetrical pattern, with corrected sums of squares along the principal diagonal and corrected sums of products elsewhere.

It will be observed in Equations (8.34), (8.35) and (8.37) that all the values of the x_i enter the analysis as deviations from their mean values; data in this form are described as Centred.

The final equation in (8.37) shows that the fitted model can also be written in centred form as:

$$Y - \bar{y} = b_1(x_1 - \bar{x}_1) + b_2(x_2 - \bar{x}_2) + \ldots + b_p(x_p - \bar{x}_p) \qquad (8.371)$$

Solution of the equations

8.31 When there are only two independent variables, the solutions can be written out at once. We have:

$$b_1 S_{11} + b_2 S_{12} = S_{y1}$$
$$b_1 S_{12} + b_2 S_{22} = S_{y2}$$

and hence, by appropriate cross-multiplication and subtraction:

$$b_1(S_{11}S_{22} - S_{12}^2) = S_{22}S_{y1} - S_{12}S_{y2}$$
$$b_2(S_{11}S_{22} - S_{12}^2) = S_{11}S_{y2} - S_{12}S_{y1}$$

i.e.
$$\left. \begin{array}{l} b_1 = (S_{22}S_{y1} - S_{12}S_{y2})/(S_{11}S_{22} - S_{12}^2) \\ b_2 = (S_{11}S_{y2} - S_{12}S_{y1})/(S_{11}S_{22} - S_{12}^2) \end{array} \right\} \qquad (8.38)$$

Generally speaking, however, some systematic method of solution, with appropriate arithmetical checks, is required.

Several methods are available, of which the simplest and most readily comprehensible is the schoolroom method of systematic elimination of unknowns. More sophisticated methods for the rapid and accurate solution of the simultaneous least-squares equations are available, but it is outside the scope of this volume to describe them in detail. Computer programs are now readily available for multiple regression calculations and a suitable method of solution will have been built into the program. Indeed, the user does not have to concern himself with this aspect of the numerical technique in normal circumstances. Those readers who do not have access to such programs should refer to Appendix 8B of the Third Edition of this book or to Woolf's method in reference [8.1].

Matrix notation for least-squares estimation

8.32 For readers familiar with matrix notation, the theory and equations of multiple regression analysis can be written in a very concise form.

Measure the dependent variable and each of the independent variables relative to their respective mean values. Denote the set of n centred observed values of the dependent variable y by the $n \times 1$ column vector y and the set of p unknown parameters $\beta_1, \ldots, \beta_p$ by the $p \times 1$ column vector β. Also let the set of np centred values of the independent variables be denoted by the $n \times p$ matrix x. This is often referred to as the Design Matrix, since it describes the combinations of the levels of the independent variables which have been included in the experimental design. Then the centred model under consideration can be written as:

$$y = x\beta + \varepsilon \tag{8.4}$$

where ε is an $n \times 1$ column vector of errors in the determination of y.

The p least-squares equations for the parameter estimates may then be expressed in matrix form as:

$$x'xb = x'y \tag{8.41}$$

where x' is the matrix transpose of x. The $p \times p$ symmetric matrix $x'x$, consisting of the corrected sums of squares and products S_{ij}, is often referred to as the Information Matrix.

Then in matrix notation, the solution to the least-squares equations is simply the $p \times 1$ column vector:

$$b = (x'x)^{-1}x'y \tag{8.42}$$

where $(x'x)^{-1}$ is the inverse matrix of $x'x$. Predicted values of the response are then given by:

$$Y = xb \tag{8.43}$$

Correlation matrix

8.321 Despite the centring of the original data, the elements S_{ij} of the information matrix can be of quite different sizes, which may lead to appreciable round-off errors in the computation when the variables cover widely differing ranges and particularly when there are a large number of variables in the model. This difficulty can be overcome by working with standardized deviates [§ **2.43**] as follows:

$$u_{ji} = \frac{x_{ji} - \bar{x}_{j.}}{\sqrt{\{S_{jj}/(n-1)\}}} \qquad \begin{array}{l} j = 1, 2, ..., p \\ i = 1, 2, ..., n \end{array}$$

$$z_i = \frac{y_i - \bar{y}}{\sqrt{\{S_{yy}/(n-1)\}}} \qquad i = 1, 2, ..., n$$

The information matrix then takes the form:

$$(n-1) \times \begin{bmatrix} 1 & r_{12} & \cdots\cdots & r_{1p} \\ r_{21} & 1 & \cdots\cdots & r_{2p} \\ \vdots & \vdots & & \vdots \\ r_{p1} & & \cdots\cdots & 1 \end{bmatrix}$$

where $r_{12} = S_{12}/\sqrt{(S_{11}S_{22})} = r_{21}$ is the correlation coefficient [§ **7.81**] between x_1 and x_2, and similarly for each pair of independent variables. This array, apart from the $(n-1)$ multiplying factor, contains the correlation coefficients for all possible pairs of variables and it is termed the Correlation Matrix. All its elements have values between -1 and $+1$. In practice the correlation matrix is obtained directly from the information matrix $x'x$, rather than by standardizing all the original individual observations.

The most accurate numerical procedure for estimating the parameter vector β is then based on the inversion of the correlation matrix rather than the information matrix. However, in addition to its importance as a step in the computation of multiple regression relationships the correlation matrix also has a valuable part to play in the interpretation of these relationships, as we shall see later.

An alternative linear transformation for improving the accuracy of the numerical matrix inversion is obtained by scaling each independent variable to lie in the range -1 to $+1$. This can occasionally lead to difficulties if one value x_{ji} of variable x_j is much higher or much lower than the remainder. Nevertheless, this procedure brings all the independent variables on to a common basis, with the result that their relative effects on the dependent variable are indicated directly by the sizes of the partial regression coefficients.

Variances and covariances of the coefficients

8.33 From the matrix Equation (8.42) we see that each estimated regression coefficient can be regarded as a weighted linear sum of the observed values of the dependent variable y; the weights are rather complicated functions of the values of the independent variables. This is similar to the form given in § **7.25** for a single regression coefficient, and it follows that the variance associated with each multiple regression coefficient can be derived by an extension of the method given there.

Furthermore, since each of the coefficients b_1, b_2, ..., b_p depends on the *same* set of observed y-values, they cannot normally be estimated independently of each other. (There is one exception, which is discussed in the next section.) The degree of inter-dependence of the estimates is measured by the covariance between each pair of coefficients.

The variances and covariances of the regression coefficients can be set out in a single $p \times p$ matrix array, with the variances forming the elements in the leading diagonal and the covariances the off-diagonal elements. This is termed the Variance-Covariance Matrix and it will be denoted by $V.C. (b)$. In fact, it can be shown from (8.42) that:

$$V.C. (b) = \{(x'x)^{-1}x'\}\{(x'x)^{-1}x'\}'\sigma^2$$

$$= (x'x)^{-1}x'x(x'x)^{-1}\sigma^2 \quad \text{(since } x'x \text{ is symmetric)}$$

$$= (x'x)^{-1}\sigma^2 \quad\quad\quad (8.44)$$

Thus the variances and covariances of the coefficients are simply the elements of the inverse of the information matrix multiplied by the error variance of the dependent variable.

Let us denote the element in the ith row and jth column of the inverse matrix $(x'x)^{-1}$ by C_{ij}. Since the matrix is symmetric, $C_{ji} = C_{ij}$. Then:

$$\left.\begin{aligned}
V(b_1) &= \sigma^2 C_{11} \qquad S.E.(b_1) = \sigma\sqrt{C_{11}} \\
V(b_2) &= \sigma^2 C_{22} \qquad S.E.(b_2) = \sigma\sqrt{C_{22}} \\
&\vdots \qquad\qquad\qquad\quad \vdots \\
V(b_p) &= \sigma^2 C_{pp} \qquad S.E.(b_p) = \sigma\sqrt{C_{pp}}
\end{aligned}\right\} \qquad (8.441)$$

and
$$C(b_1, b_2) = \sigma^2 C_{12} \text{ etc.} \qquad (8.442)$$

Usually, σ^2 will not be known in advance. If, however, the regression equation is correctly formulated, i.e. if all relevant terms have been included in the equation, then σ^2 may be estimated from the residual deviation of the observed y-values from the values predicted from the fitted relationship. Denoting the estimate of σ^2 by s^2, then the estimated standard error of b_1 is $s\sqrt{C_{11}}$, and so on. The estimation of σ^2 is discussed in more detail in § **8.36** below.

The standard errors give a measure of the uncertainties associated with the separate estimates, while the covariances show how chance errors in the estimation of any one coefficient will affect the estimates of the other coefficients. Both variances and covariances are required in assessing the accuracy of prediction using the regression equation. It is a fortunate fact that their enumeration by means of Equation (8.44) is a by-product of the matrix inversion which is required anyway to estimate the coefficients by means of Equation (8.42).

Orthogonality

8.34 Of the coefficients in the least-squares equations, the sums of squares are necessarily positive and, in practice, different from zero; for if a sum of squares, say S_{11}, were equal to zero, this would imply that all observed values of x_1 were the same, and hence that S_{12}, S_{13}, ..., S_{1p} were also zero. Thus all coefficients of b_1 in the least-squares equations would be zero and b_1 would be completely indeterminate; whatever value were assigned to it, the equations would still be satisfied. The indeterminacy accords with the common-sense view that, if x_1 does not vary in the set of observations, no information can be obtained from these observations about the effects of variations in x_1.

The vanishing of a sum of products does not, however, impose any such drastic limitations on the observed values, and the presence of the zero somewhat simplifies the arithmetic in the solution of the equations. The special case when *all* sums of products vanish is of particular interest. The least-squares equations reduce to:

$$b_j S_{jj} = S_{yj} \qquad (j = 1, 2, ..., p) \qquad (8.45)$$

and each b_j is obtained directly from the appropriate equation. The solution, it will be noted, is the same as that for the total regression coefficient of the response upon the variable concerned, i.e. it does not matter whether we consider the variables separately or together. The reason for this is that the vanishing of the sums of products means that the observed values of any independent variable are completely uncorrelated with those of any other, and hence that changes in any one of them are not associated, overall, with changes in any of the others. Each total regression coefficient thus reflects only the effect of the variable concerned.

A further simplification arises in the expressions for the variances and covariances of the estimates; the variances reduce to:

$$V(b_j) = \sigma^2/S_{jj} \qquad (j = 1, 2, ..., p) \qquad (8.46)$$

while all covariances are zero. The mutual independence of the estimates implies that each may be tested separately for significance [§§ **8.36**, **8.39**], while the effect of adding or suppressing a term in the regression equation may be ascertained at once.

In the matrix formulation of multiple regression analysis, this special case corresponds to an information matrix of diagonal form. Inverting the matrix to determine the regression coefficients and their variances therefore becomes a very simple operation.

The vanishing of a sum of products defines what is known as the Orthogonality of the two sets of values. Such orthogonality may arise by chance, but this is rather unlikely. When, however, the values of the independent variables may be assigned at will, as in a planned investigation rather than an examination of accumulated data, orthogonality may be deliberately imposed, and will lead to considerable saving of effort in the analysis of the results, as well as to some gain in the precision of estimation. This principle underlies the whole development of the statistical design of experiments, and is most clearly in evidence in the balanced factorial designs discussed in, for example, Chapters 7 and 8 of *Design and Analysis*. A more general discussion of methods of attaining orthogonality is given in [8.2], where it is shown how orthogonal designs for any number of factors and tests may be derived.

It should be noted that the vanishing of the product terms need not in any sense imply a functional independence of the variables considered; this will be evident when we come to consider curvilinear relationships [§ **8.6**].

Example 8.2. Flow of sulphate of ammonia

8.35 It is important that sulphate of ammonia should flow freely when being packed, in order that automatic sack filling and weighing machines should be able to function correctly. Occasionally, however, the crystals stick, and cause difficulties in the packing sheds. The sticking is caused in part by dampness, but may depend also on crystal shape or size. The adsorption of moisture by the material may also be affected by traces of impurity present in the product. To investigate causes of sticking, a test was devised, corresponding approximately to packing conditions, in which a quantity of sulphate was allowed to flow through a small funnel, and its rate of flow determined. Rates of flow and other data on the samples examined are given in Table 8.2.

TABLE 8.2.   RATES OF FLOW AND OTHER DATA ON SAMPLES OF SULPHATE OF AMMONIA

Sample Number	Flow Rate (g./sec.) (y)	Initial Moisture Content (in units of 0·01 %) (x_1)	Length/ Breadth Ratio for Crystals (x_2)	Percent Impurity (in units of 0·01 %) (x_3)
1	5·00	21	2·4	0
2	4·81	20	2·4	0
3	4·46	16	2·4	0
4	4·81	18	2·5	0
5	4·46	16	3·2	0
6	3·85	18	3·1	1
7	3·21	12	3·2	1
8	3·25	12	2·7	0
9	4·55	13	2·7	0
10	4·85	13	2·7	0
11	4·00	17	2·7	0
12	3·62	24	2·8	0
13	5·15	11	2·5	0
14	3·76	10	2·6	0
15	4·90	17	2·0	0
16	4·13	14	2·0	0
17	5·10	14	2·0	1
18	5·05	14	1·9	0
19	4·27	20	2·1	2
20	4·90	12	1·9	1
21	4·55	11	2·0	2
22	5·32	10	2·0	7
23	4·39	10	2·0	2
24	4·85	16	2·0	2
25	4·59	17	2·2	3
26	5·00	17	2·4	4
27	3·82	17	2·4	0
28	3·68	15	2·4	2
29	5·15	17	2·2	3
30	2·94	21	2·2	4
31	3·18	23	2·2	10
32	2·28	22	2·0	7
33	5·00	21	1·9	4

TABLE 8.2 (*continued*)

Sample Number	Flow Rate (g./sec.) (y)	Initial Moisture Content (in units of $0 \cdot 01\%$) (x_1)	Length/ Breadth Ratio for Crystals (x_2)	Percent Impurity (in units of $0 \cdot 01\%$) (x_3)
34	2·43	24	2·1	8
35	0	37	2·3	14
36	4·10	21	2·4	2
37	3·70	28	2·4	5
38	3·36	29	2·4	7
39	3·79	23	3·6	7
40	3·40	32	3·3	8
41	1·51	26	3·5	4
42	0	28	3·5	12
43	1·72	21	3·0	3
44	2·33	22	3·0	6
45	2·38	34	3·0	8
46	3·68	29	3·5	5
47	4·20	17	3·5	3
48	5·00	11	3·2	2

From these figures we obtain:

$$\bar{y} = 3\cdot843 \qquad \bar{x}_1 = 18\cdot98 \qquad \bar{x}_2 = 2\cdot550 \qquad \bar{x}_3 = 3\cdot125$$

$$S_{yy} = 74\cdot15 \qquad S_{11} = 2{,}052\cdot98 \qquad S_{22} = 12\cdot46 \qquad S_{33} = 577\cdot25$$

$$S_{12} = 49\cdot15 \qquad S_{13} = 782\cdot12 \qquad S_{23} = 13\cdot50$$

$$S_{y1} = -257\cdot59 \qquad S_{y2} = -11\cdot72 \qquad S_{y3} = -141\cdot37$$

The least-squares equations are thus:

$$2{,}052\cdot98b_1 + 49\cdot15b_2 + 782\cdot12b_3 = -257\cdot59$$
$$49\cdot15b_1 + 12\cdot46b_2 + 13\cdot50b_3 = -11\cdot72$$
$$782\cdot12b_1 + 13\cdot50b_2 + 577\cdot25b_3 = -141\cdot37$$

and the solutions are:

$$b_1 = -0\cdot04882$$
$$b_2 = -0\cdot56877$$
$$b_3 = -0\cdot16545$$

Hence the estimated dependence of flow rate (y) on percent moisture (x_1), crystal length/breadth ratio (x_2) and percent impurity (x_3) is:

$$(Y - 3\cdot843) = -0\cdot04882(x_1 - 18\cdot98) - 0\cdot56877(x_2 - 2\cdot550) - 0\cdot16545(x_3 - 3\cdot125)$$

i.e. $$Y = 6\cdot737 - 0\cdot0488x_1 - 0\cdot5688x_2 - 0\cdot1655x_3$$

The inverse matrix is:

$$10^{-3} \times \begin{bmatrix} 1\cdot0931 & -2\cdot7775 & -1\cdot4160 \\ -2\cdot7775 & 89\cdot4009 & 1\cdot6725 \\ -1\cdot4160 & 1\cdot6725 & 3\cdot6119 \end{bmatrix}$$

where the multiplication by 10^{-3} is understood to apply to each term of the matrix separately.

Standard errors and confidence limits

8.36 For one independent variable the sum of squares due to regression is given by $b\Sigma(x-\bar{x})(y-\bar{y})$, i.e. by bS_{xy} in the notation of Chapter 7. For more than one independent variable the sum of squares due to regression is given similarly by:

$$b_1 S_{y1} + b_2 S_{y2} + ... + b_p S_{yp} \tag{8.5}$$

It represents p degrees of freedom; the residual sum of squares:

$$S_{yy} - b_1 S_{y1} - b_2 S_{y2} - ... - b_p S_{yp} \tag{8.51}$$

thus gives an estimate of residual variance based on $n-p-1$ degrees of freedom, where n denotes the number of sets of observations used in deriving the regression equation, since in all $p+1$ parameters have been estimated. It should be noted, however, that the residual variance will only estimate the true "error" mean square if the regression equation is correctly formulated. If significant variables have been omitted, or if the true relationship is non-linear, the estimate will be biased.

The estimate of the residual variance is:

$$s^2 = (S_{yy} - b_1 S_{y1} - ... - b_p S_{yp})/(n-p-1) \tag{8.511}$$

As described earlier, the standard errors of the estimated regression co-efficients are obtained by way of the inverse matrix. Thus:

$$S.E.(b_1) = s\sqrt{C_{11}}$$
$$S.E.(b_2) = s\sqrt{C_{22}}$$
$$S.E.(b_3) = s\sqrt{C_{33}} \text{ etc.}$$

The non-diagonal terms of the inverse matrix give, similarly, the covariances of the estimates. We have:

$$C(b_1, b_2) = s^2 C_{12} = s^2 C_{21}$$

and similar expressions for the other covariances. The covariances of the estimates indicate the degree of their inter-correlation, i.e. they measure the extent to which chance errors in the estimate of any one coefficient will affect the estimates of other coefficients; they reflect any correlations between the observed or chosen values of the independent variables.

Confidence limits within which the true regression coefficients $\beta_1, ..., \beta_p$

probably lie are calculated from the standard errors by means of the appropriate t-multipliers (Table C). Thus the $(1-2\alpha)$ confidence limits for β_1 are:

$$b_1 \pm t_\alpha s \sqrt{C_{11}} \qquad (8.52)$$

where t has the same number of degrees of freedom $(n-p-1)$ as the estimate of s^2. The precision with which β_1 is estimated depends on the magnitude of s^2, i.e. on the residual scatter about the regression, on the number of observations available (which determines the appropriate t-multiplier) and also on the inverse matrix term C_{11}. This last depends on the spread of the observed x_1-values, and also on the extent to which the variations in x_1 are correlated with variations in the other "independent" variables. The greater the spread, and the less the correlation, the smaller will be the value of C_{11}. Its reciprocal, $1/C_{11}$, is in fact the residual sum of squares of the x_1-values about a regression of x_1 upon the other independent variables $x_2, ..., x_p$.

Considering the first regression coefficient in isolation, the significance of b_1 can be measured by its ratio to its standard error:

$$\cdot \ t = |b_1|/s\sqrt{C_{11}} \qquad (8.521)$$

This is a double-sided t-test, and the degrees of freedom are $n-p-1$.

For the example discussed in § **8.35**, the sum of squares due to regression is:

$$(0\cdot0488 \times 257\cdot59)+(0\cdot5688 \times 11\cdot72)+(0\cdot1655 \times 141\cdot37) = 42\cdot63$$

The residual sum of squares is thus:

$$74\cdot15-42\cdot63 = 31\cdot52$$

and corresponds to $(48-3-1) = 44$ degrees of freedom. The estimated residual variance is thus:

$$s^2 = 31\cdot52/44 = 0\cdot7164$$

giving:

$$s = 0\cdot8464$$

Hence:

$$S.E.(b_1) = 0\cdot8464\sqrt{(1\cdot0931 \times 10^{-3})} = 0\cdot0280$$
$$S.E.(b_2) = 0\cdot8464\sqrt{(89\cdot4009 \times 10^{-3})} = 0\cdot2531$$
$$S.E.(b_3) = 0\cdot8464\sqrt{(3\cdot6119 \times 10^{-3})} = 0\cdot0509$$

95% confidence limits for the true regression coefficients are:

For β_1: $-0\cdot0488 \pm 0\cdot0564$, i.e. $-0\cdot1052$ and $+0\cdot0076$

β_2: $-0\cdot5688 \pm 0\cdot5100$, i.e. $-1\cdot0788$ and $-0\cdot0588$

β_3: $-0\cdot1655 \pm 0\cdot1026$, i.e. $-0\cdot2681$ and $-0\cdot0629$

while the calculated t-values for testing the significance of the b's are $1\cdot74$, $2\cdot25$ and $3\cdot25$ respectively. We see from the confidence limits that none of the coefficients is very precisely determined, and that the estimate of β_1 is so imprecise that it is not unlikely that the true value is zero. The t-value for b_1 fails to reach the $0\cdot05$ significance level.

Confidence regions for simultaneous estimates

8.361 The standard errors of the regression coefficients, and the confidence limits calculated therefrom, measure the overall uncertainty of each estimated regression coefficient taken separately. The uncertainty as to the true value of (say) β_1 springs in part, however, from uncertainty about the true values of β_2 and β_3; the estimate b_1 of β_1 is obtained after making allowance for the *apparent* effects of the other two factors. Any errors of estimation in b_2 and b_3 will cause related errors in b_1, so that the separate estimates will be correlated one with another. In the example just discussed, the covariance of b_1 with b_3 is given by $s^2 C_{13}$, i.e. by $s^2(-1\cdot4160 \times 10^{-3})$. The correlation coefficient measuring the correlation between b_1 and b_3 is thus:

$$r = C_{13}/\sqrt{C_{11}C_{33}}$$

$$= -1\cdot4160/\sqrt{(1\cdot0931 \times 3\cdot6119)}$$

$$= -0\cdot71$$

This correlation coefficient is associated with $(48-4) = 44$ degrees of freedom and is highly significant.

The negative correlation between b_1 and b_3 implies that errors of estimation which cause the estimate b_1 to be high will tend also to cause the estimate b_3 to be low, and conversely. The correlation is not perfect, so that there will not be perfect association between estimates of β_1 and the corresponding estimates of β_3. It is high enough, however, for the scatter in the values of b_1 associated with any particular value of b_3 to be appreciably less than the overall scatter of possible values of b_1. Thus confidence limits for β_1 at any specified value of b_3 will be narrower than those for b_3 unspecified. Because of the association, also, the positions of the limits will change as the value of b_3 is changed. In general, therefore, we should not consider b_1 without also considering b_3.

It may be shown that, if the true values of the regression coefficients are β_1, β_3, then:

$$\frac{C_{33}(\beta_1 - b_1)^2 - 2C_{13}(\beta_1 - b_1)(\beta_3 - b_3) + C_{11}(\beta_3 - b_3)^2}{2s^2\{C_{11}C_{33} - (C_{13})^2\}} \qquad (8.53)$$

is distributed as an F-ratio with 2 & (in this instance) 44 degrees of freedom. The last figure is of course the number of degrees of freedom on which the estimate of s^2 is based. A $(1-\alpha)$ confidence region for the simultaneous uncertainty of β_1 and β_3 is obtained by setting the above ratio equal to the value of F, with the appropriate degrees of freedom, exceeded with probability α. With $\alpha = 0\cdot05$, and degrees of freedom as stated, this value is $3\cdot21$. Thus, the boundary of the region is given by:

$$C_{33}(\beta_1 - b_1)^2 - 2C_{13}(\beta_1 - b_1)(\beta_3 - b_3) + C_{11}(\beta_3 - b_3)^2 = 6\cdot42s^2\{C_{11}C_{33} - (C_{13})^2\}$$

or, substituting numerical values:

$$3 \cdot 61(\beta_1 - b_1)^2 + 2 \cdot 83(\beta_1 - b_1)(\beta_3 - b_3) + 1 \cdot 09(\beta_3 - b_3)^2$$

$$= 0 \cdot 8936 \times 10^{-2} \qquad (8.531)$$

The equation defines an ellipse, with axes inclined, centred on the point (b_3, b_1), as shown in Figure 8.5. The region bounded by the ellipse is one which may be expected, with 95% confidence, to include the point representing the two true regression coefficients (β_3, β_1).

Also shown in the figure is a rectangle bounded by (broken) lines parallel to the axes. These lines indicate the separate $(1 - \frac{1}{2}\alpha)$ confidence intervals for β_1, β_3 taken separately; the confidence coefficient is taken as $1 - \frac{1}{2}\alpha$, since

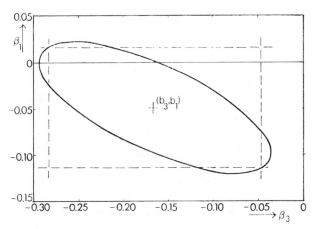

FIG. 8.5.   CONFIDENCE REGIONS FOR SIMULTANEOUS
ESTIMATION OF TWO REGRESSION COEFFICIENTS

the chance that both β_1 and β_3 are contained by the relevant limits is $(1 - \frac{1}{2}\alpha)^2 \simeq 1 - \alpha$. The rectangle is an alternative 95% confidence region for the pair of values, but the area contained is greater than that bounded by the ellipse, a result which may be shown to be generally true. We obtain the smallest possible confidence region, for a given confidence coefficient, when we plot the ellipse defined as above. It may be noted that the confidence coefficient $(1 - \alpha)$ applies to the joint statement that both β_1 and β_3 are represented by some point in the region. The statement is thus equivalent to two separate statements with confidence approximately equal to $1 - \frac{1}{2}\alpha$ concerning β_1 and β_3 separately, for if we make n statements with confidence $1 - \Delta$, the chance that all n will be simultaneously correct is only $(1 - \Delta)^n \simeq 1 - n\Delta$ for Δ small. This last is a point all too frequently overlooked in practice.

Confidence regions for a regression on many variables

8.362 The concept of simultaneous confidence regions may clearly be extended to the simultaneous estimation of several regression coefficients. The bounding surfaces are ellipsoids or hyperellipsoids in a space of many dimensions. A joint $100(1-\alpha)\%$ confidence region for all p parameters is given by the inequality:

$$\sum_{i=1}^{p}\sum_{j=1}^{p} S_{ij}(\beta_i-b_i)(\beta_j-b_j) \leqq ps^2 F_\alpha(p, n-p-1) \qquad (8.54)$$

where the S_{ij} are the elements of the information matrix. Such surfaces cannot easily be represented diagrammatically. It will be noted, however, that the concept is principally of value when the estimated regression coefficients are highly correlated; when the correlation is fairly small, say less than 0·5, the limits for one coefficient are very little affected by the value taken for another. Thus, even when many regression coefficients are estimated simultaneously, it will usually be sufficient to measure the precision of most of them by their standard errors or conventional confidence limits, and to consider confidence regions only for those pairs of coefficients which are closely correlated with one another.

Omission of a variable

8.37 The covariance $s^2 C_{13}$ of the estimates b_1 and b_3 implies, as we have seen, that in repeated sampling particular values of b_3 tend to be associated with particular values of b_1, and vice versa. It implies that there is a regression of b_3 upon b_1, or of b_1 upon b_3. The slope of the regression of b_3 upon b_1 is given in the usual way by (Covariance)/(Variance), i.e. by C_{13}/C_{11}. Thus a unit change in b_1 is accompanied on average by a change of C_{13}/C_{11} in b_3. If we decide to omit a non-significant independent variable x_1 from the regression equation, we are in effect "adjusting" the value of b_1 to zero. To allow for this adjustment we must also adjust b_3 by an amount $-b_1 C_{13}/C_{11}$, giving:

$$b_3' = b_3 - b_1 C_{13}/C_{11} \qquad (8.55)$$

This is the regression coefficient describing the estimated dependence of y on x_3 when x_1 is not included in the regression equation, and the equation shows how regression coefficients must be adjusted to allow for the omission of a variable. The variance of b_3 at a given value of b_1 is the variance about the regression of b_3 upon b_1; from § **7.86** this is:

$$s^2 C_{33}(1-r^2) \quad \text{or} \quad s^2\{C_{33}-(C_{13})^2/C_{11}\} \qquad (8.551)$$

This shows that the inverse matrix element C_{33} should be modified to:

$$(C_{33})' = C_{33}-(C_{13})^2/C_{11} \qquad (8.552)$$

to allow for the dropping of x_1 from the regression equation. Other diagonal elements are similarly adjusted, while the form of adjustment for non-diagonal terms, taking C_{23} as an example, may be shown to be:

$$(C_{23})' = C_{23}-C_{12}C_{13}/C_{11} \qquad (8.553)$$

It is thus possible to derive the whole of the new inverse matrix without the necessity of recalculation from the least-squares equations.

When observed values of two independent variables are highly correlated, it may well happen that neither regression coefficient, tested against its own standard error, is significantly different from zero, but that the regression on both variables accounts for a significant part of the total variation of the dependent variable. This is equivalent to saying that the appropriate confidence ellipse for joint estimation of both coefficients cuts both axes, but does not include the point $(0, 0)$, so that we may not reasonably conclude that both true regression coefficients are simultaneously zero. *Both* variables cannot therefore be ignored; indeed by omitting one of the variables, a significant regression relationship would be found between the response and the remaining variable. The point is discussed in more detail later [§ **8.39**].

Prediction using the regression equation

8.38 From the fitted regression Equation (*8.371*), the predicted value Y of y corresponding to an assigned set of values $(x_{10}, x_{20}, ..., x_{p0})$ of the "independent" variables $x_1, x_2, ..., x_p$ is given by:

$$Y = \bar{y} + b_1(x_{10} - \bar{x}_1) + b_2(x_{20} - \bar{x}_2) + ... + b_p(x_{p0} - \bar{x}_p) \qquad (8.6)$$

This value is subject to uncertainty, since it is derived by using coefficients which are themselves subject to uncertainty. It may be shown that the estimate $\bar{y}$ is independent of the estimates $b_1, ..., b_p$, but the estimates $b_1, ..., b_p$ are not, as we have seen, independent of one another. Hence [§ **3.71**]:

$$V(Y) = V(\bar{y}) + (x_{10} - \bar{x}_1)^2 V(b_1) + ... + (x_{p0} - \bar{x}_p)^2 V(b_p)$$
$$+ 2(x_{10} - \bar{x}_1)(x_{20} - \bar{x}_2) C(b_1, b_2) + ... \qquad (8.61)$$

We may write this as:

$$V(Y) = s^2 \{1/n + C_{11}(x_{10} - \bar{x}_1)^2 + ... + C_{pp}(x_{p0} - \bar{x}_p)^2$$
$$+ 2C_{12}(x_{10} - \bar{x}_1)(x_{20} - \bar{x}_2) + ...\} \qquad (8.611)$$

where s^2 is the residual variance about the regression. Since the estimates $b_1, ..., b_p$ are not, in general, independent of one another, we must allow for their $\frac{1}{2}p(p-1)$ covariances, as well as their p variances, in arriving at the variance of the regression estimate. We see here another advantage of orthogonality.

For Example 8.2, where there are three independent variables, we have:

$$y = 6.737 - 0.0488x_1 - 0.5688x_2 - 0.1655x_3$$

Thus the rate of flow estimated for an initial moisture content of 0.18% ($x_1 = 18$), length/breadth ratio (x_2) of 2.7 and an impurity of 0.03% ($x_3 = 3$) is:

$$6.737 - (0.0488 \times 18) - (0.5688 \times 2.7) - (0.1655 \times 3) = 3.83 \text{ g./sec.}$$

For the calculation of the variance of this regression estimate we have:

$$n = 48 \qquad s^2 = 0.7164$$

$$x_{10} - \bar{x}_1 = -0.98 \qquad x_{20} - \bar{x}_2 = 0.15 \qquad x_{30} - \bar{x}_3 = -0.125$$

The inverse matrix elements are as given earlier. Hence:

$$V(Y) = 0.7164[\tfrac{1}{48} + 10^{-3}\{1.09(0.98)^2 + 89.40(0.15)^2 + 3.61(0.125)^2$$
$$+ 5.55(0.98)(0.15) - 2.83(0.98)(0.125) - 3.34(0.15)(0.125)\}]$$
$$= 0.7164 \times 0.0244$$
$$= 0.0175$$

The standard error of Y is thus 0.132, and 95% confidence limits are found by using the appropriate value of t for 44 degrees of freedom. They are thus:

$$3.83 \pm (2.015 \times 0.132)$$

$$\text{i.e.} \quad 3.56 \text{ and } 4.10$$

These limits refer to the errors of the regression estimate only; the result of a future individual test will be subject additionally to the residual variance of 0.7164 (see § **7.281**).

Variance of the constant term

8.381 The constant term b_0 in the regression equation, when this is written in the form:

$$Y = b_0 + b_1 x_1 + \ldots + b_p x_p$$

is simply the estimate of y for $x_{10} = x_{20} = \ldots = x_{p0} = 0$. In some problems this constant term may be expected to be zero. It may be tested for significance against its standard error, which may also be used to derive appropriate confidence limits. The standard error is given by:

$$\sqrt{\{V(\bar{y}) + \bar{x}_1^2 V(b_1) + \ldots + \bar{x}_p^2 V(b_p) + 2\bar{x}_1 \bar{x}_2\, C(b_1, b_2) + \ldots\}} \qquad (8.62)$$

or $\qquad s\sqrt{\{1/n + C_{11}\bar{x}_1^2 + \ldots + C_{pp}\bar{x}_p^2 + 2C_{12}\bar{x}_1\bar{x}_2 + \ldots\}} \qquad (8.621)$

Analysis of Variance for multiple regression

8.39 The expressions for the sums of squares due to and about the regression given in § **8.36** may conveniently be arranged in the form of an Analysis of Variance Table, as shown in Table 8.3 overleaf.

One measure of the overall goodness of fit of the multiple regression equation is the Multiple Correlation Coefficient, usually denoted by R. It is defined by:

$$R^2 = \frac{\text{Sum of squares due to regression}}{\text{Total sum of squares of } y\text{-values about their mean}} \qquad (8.63)$$

so that the residual sum of squares is $(1 - R^2)$ times the original sum of squares. Often the value of R^2 is multiplied by 100 to give a Percentage Fit figure to the regression.

The F-ratio calculated from Table 8.3 may be used to test for the significance of the overall dependence of y on the variables $x_1, ..., x_p$. But such a test of the combined dependence is not usually sufficient, since it tells us nothing about the significance of particular terms in the regression equation.

TABLE 8.3. ANALYSIS OF VARIANCE FOR MULTIPLE REGRESSION

Source	Sum of Squares	Degrees of Freedom	Mean Square
Due to regression	$b_1 S_{y1} + b_2 S_{y2} + ... + b_p S_{yp}$	p	
About regression	$S_{yy} - b_1 S_{y1} - ... - b_p S_{yp}$	$n-p-1$	
Total	S_{yy}	$n-1$	

To test whether the introduction of a particular term, say that involving x_p, had led to any significant improvement in the fit of the representation, we could test b_p against its estimated standard error, using the t-test described in § 8.36. An alternative procedure, which is instructive because it shows more clearly what is being done in such a test, is as follows:

Using only $x_1, ..., x_{p-1}$ to predict y, we obtain estimates $b'_1, ..., b'_{p-1}$ of the coefficients in the regression equation, either by the method of least squares or by the adjustment procedure described in § 8.37. The sum of squares attributable to this regression is:

$$b'_1 S_{y1} + b'_2 S_{y2} + ... + b'_{p-1} S_{y,p-1}$$

with $p-1$ associated degrees of freedom. Consequently, the degrees of freedom for the residual sum of squares are $n-p$. If, however, we use $x_1, ..., x_p$ to predict y, we have estimates $b_1, ..., b_p$ of the regression coefficients, and the sum of squares due to regression is given by:

$$b_1 S_{y1} + b_2 S_{y2} + ... + b_p S_{yp}$$

In general, the coefficients will be different from those for the regression on $p-1$ independent variables. The "full regression" sum of squares represents p degrees of freedom and the corresponding residual sum of squares represents $n-p-1$ degrees of freedom.

The effect of introducing x_p is thus measured by the variance associated with the extra degree of freedom which is abstracted from the residual. This variance may be found by difference, and is tested for significance against the new residual mean square in the usual way. If it is not significant, we conclude that the introduction of x_p into the representation does not significantly improve the fit; the extra variation explained is not significantly greater than could reasonably be expected to be associated, purely by chance, with the variations in x_p, if x_p in fact bore no relationship to y. The Analysis of Variance table corresponding to the test is shown in Table 8.4. The test is exactly equivalent to the t-test of b_p against its standard error; the F-ratio with 1 & $n-p-1$ degrees of freedom is the square of the corresponding value of t.

TABLE 8.4. TESTING FOR SIGNIFICANCE OF AN ADDITIONAL INDEPENDENT VARIABLE: ANALYSIS OF VARIANCE TABLE

Source	Sum of Squares	Degrees of Freedom	Mean Square
Regression upon first $(p-1)$ variables	$b'_1 S_{y1} + b'_2 S_{y2} + \ldots \\ \qquad + b'_{p-1} S_{y,p-1}$	$p-1$	
Additional effect due to pth variable	$b_1 S_{y1} + b_2 S_{y2} + \ldots + b_p S_{yp} \\ \quad - b'_1 S_{y1} - \ldots - b'_{p-1} S_{y,p-1}$	1	
Residual	$S_{yy} - b_1 S_{y1} - \ldots - b_p S_{yp}$	$n-p-1$	
Total	S_{yy}	$n-1$	

The difference listed in Table 8.4 for the additional effect due to the pth variable is rather cumbersome. Fortunately this expression can be formulated much more simply as:

$$b_p^2 / C_{pp} \qquad (8.64)$$

which is easier for practical calculation.

Examples of Analysis of Variance

8.391 In Example 8.1—regression of abrasion loss upon hardness and tensile strength—the sum of squares due to the multiple regression on two variables is:

$$(6 \cdot 571 \times 22{,}946) + (1 \cdot 374 \times 27{,}857) = 189{,}054$$

The sum of squares due to regression on hardness alone [§ **7.241**] is 122,446. There are thirty sets of results. Thus the Analysis of Variance is as shown in Table 8.41 overleaf.

The F-ratio for the additional effect of tensile strength is $66{,}608/1{,}332 = 50 \cdot 0$, with degrees of freedom 1 & 27. For these degrees of freedom the $0 \cdot 01$ probability level is only $7 \cdot 68$, so that there is no doubt of the significance of the effect. In fact, the introduction of tensile strength reduces the residual variance from 3,663 to 1,332, i.e. by 64%.

The percentage fit for the simple regression upon hardness is 54%, but for the combined regression upon strength and hardness it rises to 84%.

TABLE 8.41.   TESTING   FOR   SIGNIFICANCE   OF   EFFECT   OF
TENSILE STRENGTH

Source	Sum of Squares	Degrees of Freedom	Mean Square
Regression upon hardness	122,446	1	122,446
Additional effect of tensile strength	66,608	1	66,608
Residual	35,957	27	1,332
Total	225,011	29	

In Example 8.2, for the flow of sulphate of ammonia, the sum of squares attributable to the regression upon three variables is 42.63 [§ **8.36**]. If, however, we fit a regression upon x_2 (length/breadth ratio) and x_3 (per cent impurity) only, the least-squares equations are:

$$12 \cdot 46 b_2 + \ 13 \cdot 50 b_3 = \ -11 \cdot 72$$
$$13 \cdot 50 b_2 + 577 \cdot 25 b_3 = \ -141 \cdot 37$$

which give $b_2 = -0 \cdot 6928$, $b_3 = -0 \cdot 2287$, and a sum of squares due to regression of:

$$(0 \cdot 6928 \times 11 \cdot 72) + (0 \cdot 2287 \times 141 \cdot 37) = 40 \cdot 45$$

Thus the Analysis of Variance for the significance of the additional effect of

TABLE 8.42.   TESTING FOR SIGNIFICANCE OF EFFECT OF INITIAL
MOISTURE CONTENT

Source	Sum of Squares	Degrees of Freedom	Mean Square
Regression upon L/B ratio and percent impurity	40·45	2	20·23
Additional effect of initial moisture content	2·18	1	2·18
Residual	31·52	44	0·72
Total	74·15	47	

initial moisture content is as given in Table 8.42. This addition could also have been calculated directly from Formula (*8.64*) as:

$$(-0.04882)^2/0.0010931 = 2.18$$

The *F*-ratio for the additional effect of initial moisture content is 3.03, with degrees of freedom 1 & 44. Since the 0.05 probability level for these degrees of freedom is 4.06, the additional effect cannot be adjudged significant. The conclusion is the same as that of § **8.36** above, the *F*-ratio of 3.03 being the square of the *t*-value of 1.74 then calculated. The percentage fit for the two variables is 55%, but for all three it only rises to 57%.

There is in fact a high correlation between initial moisture content and percent impurity ($r_{13} = 0.72$), and it is clear that, in practice, both the initial moisture content and the rate of flow reflect a hygroscopicity engendered by the traces of impurity. So far as the regression relationship is concerned, a knowledge of initial moisture content, in addition to length/breadth ratio and percent impurity, does not appreciably improve prediction of flow rates. So far as plant operation is concerned, if sticking is to be avoided, it is more important to limit the traces of impurity than to attempt a closer control of the drying of the salt.

Alternative analyses

8.392 The Analyses of Variance described above are by no means the only analyses that may be applied. For example, in considering the dependence of abrasion loss on hardness and tensile strength we might have first fitted a regression upon tensile strength and then tested for the significance of the additional effect of hardness, while in the example on flow of sulphate of ammonia the additional effect of any of the three variables after allowing for a regression upon the other two might have been tested. There are thus two possible Analyses of Variance like that of Table 8.41, and three like that of Table 8.42. Each analysis is equivalent to the *t*-test of one regression coefficient against its standard error.

It is instructive to carry out an Analysis of Variance, corresponding to that of Table 8.41, to test for the significance of the effect of hardness on abrasion loss. The sum of squares attributable to a regression upon tensile strength alone is:

$$(S_{y2})^2/S_{22} = (-27,857)^2/38,733 = 20,035$$

The residual sum of squares and the sum of squares due to regression upon both variables remain as before. Thus the Analysis of Variance is as shown in Table 8.43 overleaf.

The improvement consequent on the introduction of hardness after tensile strength is even more significant than the improvement consequent on the introduction of tensile strength after hardness; clearly, neither variable by itself is as satisfactory for prediction of abrasion loss as the two taken together.

It is noteworthy, however, that before any allowance is made for hardness the apparent residual variance is:

$$204,976/28 = 7,321$$

Thus the F-ratio for the effect of tensile strength, had this been taken alone, would have been only $20,035/7,321 = 2.74$, which does not reach even the 0.10 probability level. Had tensile strength and not hardness been considered first, it might well have been concluded that it bore no relation to abrasion loss, and the effect of tensile strength might not have been found.

TABLE 8.43.   TESTING FOR SIGNIFICANCE OF EFFECT OF HARDNESS

Source	Sum of Squares	Degrees of Freedom	Mean Square
Regression upon tensile strength	20,035	1	20,035
Additional effect of hardness	169,019	1	169,019
Residual	35,957	27	1,332
Total	225,011	29	

Selection of the best subset of variables

8.4 The last example illustrates a difficulty in testing for significance which becomes progressively more marked as the number of independent variables is increased, namely the order in which the variables should be considered. Moreover, any attempt to determine the "best" regression model for a given set of data requires a prior definition of what is meant by "best". One desirable feature in a final model is that it shall be as simple as possible; in general, variables should only be retained if they contribute significantly to the goodness of fit of the data. Various techniques for tackling this problem are described below. They each have some degree of theoretical validity, but no single method can claim to be the only correct one and unfortunately each can lead to a different equation.

It can be argued, however, that the statistical theory supporting a procedure is less important than a consideration of the purpose for which the equation is to be used. One procedure might be more appropriate when a good predictive equation is required, because it leads to small standard errors for predicted values. Another may be preferred as a method for deriving an empirical equation for use on a chemical plant, because it involves variables which are more fundamental or which can be conveniently changed or controlled.

The major techniques which have been advocated and used for selecting variables are as follows:

(a) All possible regression models

Ideally, all possible numbers and combinations of the independent variables should be examined, since otherwise there is always the possibility that a good combination may be overlooked. However, this approach becomes impracticable, even using a computer, when there are many independent variables, and particularly when a quadratic regression model is being fitted [§ 8.7]. For example, with six independent linear variables, $2^6 - 1 = 63$ separate regressions would have to be fitted, and solutions would be required, among others, of one set of six equations, six sets of five equations and fifteen sets of four equations. Various criteria have been suggested for selecting the "best" model from the $2^p - 1$ alternatives but they all have theoretical or practical difficulties. For instance, the model with the smallest residual mean square may be chosen, but there may be a number of quite different models with nearly the same residual mean squares. Alternatively, we might select the equations containing one, two, three, ..., p independent variables which have the smallest residual sums of squares of all the possible equations containing one, two, three, ..., p terms. Unfortunately, the p equations resulting from this analysis can contain very different groups of variables and this causes difficulties in any process of significance testing; however, one of them may appear to be more realistic than the remainder.

Of more widespread application are the methods which involve some form of sequential strategy for selecting variables and which "automatically" lead to a unique model. Such techniques are easily programmed for computers. Notable among these are the following:

(b) Forward selection

Variables are added into the model one by one. At each stage in the procedure the variable brought in is the one which contributes most (out of those not yet included) to the regression sum of squares. The selection stops when the contribution of such a variable ceases to be significant at a predetermined level, say 5%. The final regression equation then consists of the earlier terms only.

(c) Backwards elimination

After fitting the full model with all the variables included, variables are eliminated in turn if their additional contribution to the regression sum of squares is not significantly large. At each elimination stage, the variable rejected is the one whose contribution is smallest. An alternative (but equivalent) way of performing this technique is to reject the variable whose associated t-value is the lowest of the non-significant t-values. When a variable is eliminated the remaining regression coefficients and their standard errors can be recalculated by the adjustment technique described in § **8.37**.

Methods (*b*) and (*c*) suffer from the disadvantage that no "second thoughts" are permitted. The significance or non-significance of a particular variable in the presence of one set of other variables may be quite different in the presence of a larger or smaller set of the other variables; all depends on the pattern of the correlations between the independent variables, and this may be quite complex. The following two techniques largely surmount this difficulty.

(*d*) Stepwise regression

In this method, variables are added one by one as in Forward Selection, but, at each stage, any variable which is already included in the model but whose extra sum of squares contribution has declined to a non-significant level is eliminated. However, this variable may possibly be brought in again at a later stage. Selection stops when all unused variables are non-significant and all the included variables are significant.

(*e*) Stabilized elimination

This technique is the one currently used in the ICI standard computer program for multiple regression analysis. It starts as a Backwards Elimination procedure, but when no more variables can be rejected the program determines whether any of the rejected variables should now be reinstated. This is sometimes the case, because the variables comprising the current reduced model will differ from the comparison set when a variable was first rejected. As before, variables are considered one by one using the appropriate significance levels. When no further variables can be reinstated, the program attempts to reject variables again (not necessarily those that were rejected originally). This oscillatory process continues until it stabilizes on one particular model which can neither be added to, nor subtracted from, at the preselected significance level. The method is clearly similar to the Stepwise Regression method, but it operates in a reverse sense; these two techniques frequently lead to the same final model.

A radically different approach to the problem of variable selection has recently been suggested as follows [8.3]:

(*f*) Element analysis

This method is based on a set of "building bricks" termed Elements, which can be put together to yield a variety of regression models. The particular structure chosen depends on the intended end-use, but a study of the elements can give greater insight into the inter-relationships involved in the regression analysis.

According to this method, for each variable x_i a Primary Element I is calculated from the formula:

$$I = b_i^2/C_{ii} \qquad (i = 1, 2, ..., p) \qquad (8.64)$$

where b_i is the partial regression coefficient of y upon x_i when all variables are included in the model and C_{ii} is the corresponding term in the leading diagonal of the inverse matrix $(x'x)^{-1}$. Thus I is the additional sum of squares due to fitting x_i when all other variables have already been fitted. In this sense it may be described as the independent contribution of variable x_i to the regression. If x_j is then omitted from the regression, the sum of squares now due to x_i may be written as:

$$I+IJ = b_i'^2/C_{ii}' \qquad (8.641)$$

where b_i' and C_{ii}' are calculated for the regression equation excluding x_j. The quantity IJ (which is *not* an implied multiplication) is termed the Second-Order Element associated with variables x_i and x_j; it is the extra information that x_i now brings to the regression because of its association with x_j. It is easily shown that $IJ = JI$. The definition of second-order elements extends to any pair of variables available for the regression equation.

Further, if variable x_k is omitted from the equation as well as x_j, the contribution due to x_i can now be written as:

$$I+IJ+IK+JK+IJK = b_i''^2/C_{ii}'' \qquad (8.642)$$

and successively higher-order elements follow in a similar fashion. Clearly all the primary elements are non-negative, but secondary elements (a generic term for all higher-order elements) can be either positive or negative, depending on the correlation pattern. The $2^p - 1$ elements defined in this way are such that their sum is equal to the total sum of squares due to the regression when *all* variables have been included.

The technique of element analysis enables the analyst to consider the information contributed by each variable when all or most of the other variables are present. Elements can be manipulated by simple addition operations (much faster than matrix inversion), and they are useful in computing sums of squares due to a regression of *any* order and assessing the effect of the order of introduction or deletion of the variables. The "simple" regression sum of squares due to x_i when no other variables are present is in fact the most complicated of the expressions arising in element analysis, since it involves all the elements which include the letter I:

$$I+(IJ+IK+...)+(IJK+IJL+...)+(IJKL+...)+...+(IJKL...p) \qquad (8.643)$$

Given all the elements, construction of the regression equation becomes largely a matter of experience and the application of a set of rules. The following have been found to be useful guides:

  (i) The best operationally effective variables have large primary elements and negligible secondary elements.

(ii) Good predictive, but operationally ineffective, variables have negligible primary elements and large positive second-order elements.

(iii) Variables which are ineffective on their own, but which should be retained as operationally significant, have large primary elements and one or more large negative second-order elements.

The technique of element analysis is, however, new and relatively untried at present. For examples of applications the reader should study reference [8.4].

Notwithstanding the merits of the above six techniques ((a) to (f)) for the selection of variables, the advantages of orthogonality are clear; for in an orthogonal design, all coefficients can be calculated and tested separately, and they are unaffected by the omission or inclusion of other variables. In practice, however, difficulties inherent in multiple regression analysis with many independent variables can often be resolved by common sense and an appreciation of the technical background of the problem.

Readers who require further information on techniques for the selection of variables are referred to [8.5].

Use and misuse of significance tests

8.41 It must be appreciated that the non-significance of a particular regression coefficient does not in any way imply that the independent variable concerned does not affect, or is not related to, the dependent variable. It implies merely that, at the level of significance adopted, the confidence limits for the estimated effect, or slope, include zero as a possible value. Where, additionally, the confidence limits are so narrow that any true effect which may exist is small and of little importance over the range for which the relationship is required, we may be justified in ignoring the term. In other cases, however, we may only be justified in concluding that further work is required to define the relationship more precisely.

As stated earlier there are two classes of problems to which regression analysis may be applied; in one, as in the example above on abrasion loss of rubber samples, a regression equation is required only for purposes of prediction, and there is no assumption of an underlying functional relationship. In such problems, provided that the data on which the regression equation is based correspond to a random sample of individuals *from the Population for which prediction is required*, we may usually ignore non-significant terms, which do not add significantly to the precision of prediction and may even lessen it. The other class of problems, however, is that which makes use of regression analysis to derive, from experiment or from routine plant data, exact or approximate functional relationships. In such problems we cannot always ignore non-significant terms; it would be, for example, absurd to conclude from the non-significance of a regression coefficient that pressure did not affect the rate of a chemical reaction if kinetic theory could be used to show that it must. The important feature is not the significance of the effect, but how precisely the effect is determined. If the confidence limits are wide (whether or not they include zero), then further work is almost certainly necessary—possibly over a wider range of conditions—to estimate the effect more closely. The uncritical use of tests of significance, without thought to the implications of the corresponding Null Hypothesis, should be scrupulously avoided, particularly when using "automatic" procedures for variable selection. Because of the arbitrary nature of the significance level specified for variable rejection (or insertion) procedures, it may be worth while repeating the analysis at a second level of significance, e.g. at both 5% and 1% levels.

Further aspects of model building

8.5 For any given set of data relating a response and a number of independent variables, a regression equation can generally be obtained by the automatic application of the procedures described in § **8.4**. It is, however, of vital importance to determine whether the relationship adequately represents the real-life situation by considering questions such as:

  (i) Are the variables in the right form?
  (ii) Does the equation contain superfluous variables?
  (iii) Are important variables missing?
  (iv) Is the model grossly distorted by the inclusion of a few erroneous data points?
  (v) Are the regression coefficients, and indeed the whole model, completely fictitious because of exceptionally high correlations between some of the variables?

The following sections provide a detailed discussion of these problems.

Transformation of variables

8.51 A multiple linear regression equation is a convenient way of expressing the joint relationship between variables, but if an underlying cause-and-effect relationship really exists, it is unlikely that it is of such a simple form as the weighted sum of a number of linear terms in the observed variables. The laws of chemistry and physics often contain exponentials, logarithms, powers, product terms, etc. A more readily acceptable and understandable model, and indeed a better fitting model, may therefore be obtained if the data are suitably transformed at the beginning of the analysis. This comment applies to the response as well as to the independent variables. Only rarely, however, will the data themselves suggest which mathematical transformations are desirable; the statistician must rely on his past experience and on the experimenter for suitable leads.

Some common examples of transformations are as follows:

(i) Instead of including a linear term in temperature T, it may be more meaningful to use $\exp(-T)$.

(ii) The logarithm of a concentration level may be more appropriate than just concentration.

(iii) A square-root transformation may help if a circular cross-sectional area is actually measured but the diameter is more relevant.

(iv) In a batch plant, instead of using the cycle time of a particular process, it may be preferable to use its reciprocal, the production rate (e.g. in lb./hr.).

(v) If the relationship is thought to be of the form $y = Kx_1^a x_2^b$, then by a simple logarithmic transformation, $\log y = \beta_0 + \beta_1(\log x_1) + \beta_2(\log x_2)$ would be a suitable regression model.

(vi) Use "heat input" rather than temperature and time separately. This is an example of a new variable being derived from two or more of the original variables.

The inclusion of simple curvature in a multiple regression relationship is dealt with separately in § **8.7**.

Testing the validity of the model by a determination of the goodness of fit

8.52 The treatment of data which consist of more than one value of the dependent variable for each value of a single variable x was described in § **7.32**. It was explained that one must differentiate between replicates which are merely confirmations and those which provide genuinely new information. In the latter case the residual mean square can be partitioned into "pure error" and "lack-of-fit" mean squares, and it is possible to test whether a linear model is a valid representation of the data.

Equivalent situations arise in multiple regression analysis when there are replicate observations of the dependent variable at any combination of values of the independent variables. The validity of the multiple regression model can be tested in precisely the same way by partitioning the residual mean square, but in this multivariable case the nature of the inadequacy of the model is not immediately obvious. The difficulty can, however, usually be resolved by plotting the "residuals" against each variable in turn, as will be described in § **8.54**.

Testing the validity of the model by an examination of inverse diagonals

8.53 Mention has already been made in § **8.321** of the need to avoid the build-up of round-off errors in the solution of the least-squares equations and that this may be overcome by standardizing the variables.

The solution of the least-squares equations may also be dubious for several other reasons. Thus there may be strong simple correlations between some or all of the independent variables, or multiple correlations may exist which are not readily apparent from the original data. This can cause the information matrix (and the correlation matrix) to be nearly singular, and round-off errors become dominant. If the matrix is actually singular because one variable is exactly equal to a linear combination of some of the other variables, the inversion routine will fail entirely and no regression equation will be produced. This would occur, for example, if the weight of each constituent of a raw material were included as a variable as well as the total batch weight, or if separate time cycles for all the individual stages of a manufacturing process were included as well as the overall time cycle. Such *faux pas* are obvious with hindsight, but even the most experienced statistican is liable to fall into this kind of trap occasionally.

"Near singularity" can also occur, however, in obscure ways which are completely coincidental in the particular set of data. Special difficulties arise when second-order models are being fitted, as will be described in § **8.7**. In cases of "near singularity" not only are the estimated regression coefficients rather dubious but the application of an automatic variable selection procedure can produce complete nonsense.

Fortunately, the existence of a near singularity and of potentially large round-off errors can be determined from the numerical values of the terms in the leading diagonal of the inverse of the correlation matrix. Let C^{ii} denote the ith term in this diagonal, corresponding to the variable x_i; a superscript is used to distinguish this quantity from the ith term C_{ii} in the diagonal of the inverse of the information matrix. In fact, $C^{ii} = C_{ii}S_{ii}$; numerical values of C^{ii} are always ≥ 1. Then it can be shown that the residual sum of squares in a multiple linear regression of x_i upon all the remaining variables is simply:

$$S_{ii}/C^{ii} \text{ or } 1/C_{ii} \qquad (8.65)$$

If this residual is small compared with the total sum of squares S_{ii} for variable x_i, i.e. if $C^{ii} \gg 1$, then the dependence of variable x_i on the other so-called independent variables is high.

An alternative way of expressing this residual sum of squares relationship is [§ **8.39**]:

$$1/C^{ii} = 1 - R_i^2 \qquad (8.66)$$

where R_i is the multiple correlation coefficient of x_i on all the other independent variables. Thus at one extreme, if the inverse diagonal element $C^{ii} = 1$, then $R_i^2 = 0$ and variable x_i is orthogonal to all the other independent variables. But, on the other hand, if $C^{ii} = 1,000$, then $R_i^2 = 0.999$ and $R_i = 0.9995$; hence, in this case variable x_i is almost completely predictable from the other independent variables and is able to contribute very little additional information to the regression model. Even when $C^{ii} = 100$, $R_i^2 = 0.99$ and $R_i = 0.995$, which is still high. Such high values do not necessarily imply that x_i is dependent on *all* the other variables; it may be dependent on any subset of them or even on just one of them.

It is thus clear that any variables with large inverse diagonals should be treated with suspicion, and indeed in a computer program for general use it may be advisable to terminate the analysis if any inverse diagonal is greater than a predetermined level. Reasonable limits are 500 for a single diagonal element or 100 for any pair of elements.

Often, if there are any large inverse diagonals there will be more than one. If there are exactly two, then this is usually due to a high correlation between the two corresponding variables. When there are more than two large inverse diagonals, the reasons may be difficult to detect, but the results of the regression are no less open to suspicion. In general, the occurrence of large inverse diagonals should make the user doubt the validity of his initial model. He should study the data again, for example by looking at the correlation matrix, and amend the terms in his equation accordingly.

Examination of residuals

8.54 A residual is defined as the algebraic difference between an observation y_i and the predicted value Y_i obtained by substitution in the fitted regression equation. There is one residual for each of the original observed values, and the regression coefficients were actually estimated in § **8.3** by minimizing the sum of squares of these residuals. Residuals are therefore a form of error which cannot be explained in terms of the independent variables used in the model. The Analysis of Variance for the multiple regression [§ **8.39**] assumes that the errors in the response values are independent with zero mean and constant variance; furthermore, t- and F-tests of significance are based on an underlying assumption that the errors are Normally distributed.

After fitting a regression equation it is therefore worthwhile examining the residuals to see if there has been an obvious violation of the basic assumptions. This may lead to an improvement in the model by suggesting a transformation of the initial variables, or an addition of terms to the equation. Also, if outlying observations are omitted a much improved fit may be obtained.

The following methods should be considered:

(*a*) Plot the residuals as a histogram. Their mean is necessarily zero as a consequence of the method of least squares, but the investigator should check that their distribution "looks" approximately Normal. If any observation results in an exceptionally large residual according to the criterion described in § **3.5**, it may be omitted and the parameters in the model should be re-estimated. The percentage fit can be severely depressed by a freak observation and indeed it can seriously distort the fitted equation. A thorough enquiry should be made by the experimenter to try to discover the causes of the freak. Note, however, the possibility that the actual observation y_i may be genuine, but that the model is inadequate in the region of the doubtful point.

(b) Plot the residuals as a scatter diagram against the predicted values, Y_i. The points should all lie within a horizontal band. If, however, the band slopes steadily upwards or downwards, it suggests the presence of some systematic error in the analysis, such as wrongly omitting the constant term β_0. from the model. If the band follows a simple curved path, a transformation of the observations y_i is required. Finally, if the band gets progressively wider or narrower, this indicates that the error variance is not constant, i.e. the observations have been obtained with different degrees of precision; a transformation of the observations or some form of weighted regression analysis is therefore required.

(c) Plot the residuals against each of the independent variables in turn. The interpretation is similar to (b) above, with the additional possibility that a curved path suggests the need for a curvilinear relationship rather than a linear one.

(d) If the original data arose in a natural time sequence, plot the residuals as a time series. Again a horizontal band would be expected, but if it slopes steadily upwards or downwards there is some additional variable involved which changes with time. It is then feasible to re-estimate the regression equation, merely using time as an extra variable, but to enhance the future value of the model every effort should be made to identify the variable concerned. A simple curved band would imply that both linear and quadratic terms in time are needed in the model, while a tapering band would indicate the need for a weighted regression analysis. A cyclic pattern in the residuals suggests that a "seasonal" variation in the observations requires investigation. Step-changes in the time sequence can sometimes be revealed by a cumulative sum plot of the residuals [§ **10.6**].

(e) Other graphs may suggest themselves to the analyst in particular cases. For example, residuals could be plotted against a potentially relevant variable which has not been included in the present model, or they could be divided into sub-groups associated with, say, different machines or different operatives.

A complete examination of residuals on the above lines may be rather tedious, particularly when there are many variables or many observations. However, this technique is very revealing and it can lead to models which represent the true situation more accurately. Even when replicates suggest no significant lack-of-fit [§ **8.52**], it is still desirable to examine the residuals. For a detailed exposition of this topic the reader is referred to [8.5]. The uncritical acceptance of a final model, particularly one which has emerged from a sophisticated computer program with automatic selection procedures, can be unwise, and certainly it is wasteful of the full information contained in the data.

Blocking of data

8.55 Cases frequently arise where the data for a multiple regression analysis are an amalgamation of different sets of observations. These sets can result, for example, from the use of several different machines, different lots of raw material, different work-shifts or different laboratory analysts. Moreover, when data are collected over a long period of time, it is advisable to split them deliberately into weekly or monthly periods.

If it is thought possible that the level of the response could vary from one set to another, regardless of the values of the independent variables within each set, then this should be taken into account by separating this effect from the residual mean square. Such a procedure is called Blocking, and it is an integral part of many computer programs for multiple regression analysis. Assuming that the block differences are in no way associated with the values of the independent variables within the blocks, the numerical procedure is simply to calculate the corrected sums of squares and products for each block separately and then to accumulate these sums to form an overall information matrix. The final regression equation, therefore, has common regression coefficients for all blocks, but each block has its own constant term β_0. The block differences can be tested for significance against the reduced residual mean square which results from this analysis. It should also be noted that the significance tests on the variables contained in the model are more sensitive once real block effects have been removed.

An alternative procedure is to incorporate the different sets into the model as an additional *qualitative* variable or group of variables by the method given in the next section. This method has the advantage that the restriction on the nature of the block effect can be relaxed. When it is thought possible that the effects of the independent variables could be different for each set, e.g. the effect of variable x_i could vary from one machine to another, then either completely separate models can be constructed for each set or the sets can be included as a qualitative variable in one large model *together with appropriate interaction terms*, i.e. product terms of an independent quantitative variable x_i with the qualitative variable x_j.

Qualitative variables

8.56 It is sometimes necessary to include a qualitative or descriptive variable in a regression model. Such variables are not normally defined on a continuous scale of measurement, but consist of a group of discrete items; for example, different machines, vessels, men, suppliers, lots of raw material, etc.

If there are only two items, they can be accommodated by simply creating a variable x_i which takes the value -1 for one of the items and $+1$ for the other. When appropriate, interactions between x_i and any other variable may also be included in the model. The regression analysis is then carried out in the standard manner, but of course in the final equation the variable x_i may still only take the specific values of ± 1.

When there are three items in a group, they may be introduced into the model by means of a quadratic variable, i.e. the equation includes an x_i and an x_i^2 term; interactions with other variables may also be incorporated. The three items are conveniently given the values -1, 0, $+1$. In the interpretation of the final model, the x_i and x_i^2 terms must be considered jointly.

For the more general case of k (>3) descriptive items, the usual procedure is to create $k-1$ linear variables associated with the first $k-1$ items. Each observation involving a given item is then represented by a value of $+1$ for the corresponding variable and a zero value for the other $k-2$ variables. For the kth item, all the $k-1$ created variables are given the value -1. The regression analysis then proceeds normally. In effect, this device provides a comparison of each item with the last item in the set. Again, interactions with other variables can be included in the model, but their interpretation is rather complex.

Applicability of the regression method

8.57 There is one major assumption implicit in the above treatment of multiple regression which should be noted. This is the assumption that all the residual variation is, or may be considered to be, attributable to the dependent variable. The implications of this assumption for regression on a single variable were dealt with in § 7.3; they may be summarised as follows. Where the residual variation represents inherent variability of the individuals tested (as is largely true in Example 8.1 on the testing of rubber samples) and there is only a statistical relationship between the values of "dependent" and "independent" variables, the regression equation is a convenient and legitimate representation of this. Where a relationship is required for the prediction of values of the dependent variable from observed values of the independent variables and the data fitted are a random sample of possible sets of observations, the regression equation is again applicable. Where, however, there is an underlying functional relationship which is obscured only by errors of measurement, the regression equation only estimates this without bias if the errors are all in the values of the dependent variable. The problem of the estimation of a functional relationship in the multivariate case when there are errors in the independent as well as in the dependent variables is by no means fully solved, and a general discussion is outside the scope of the present book; a brief survey is, however, attempted in § 8.92.

CURVILINEAR REGRESSION

8.6 Where a linear representation of the relationship between observed values of two variables appears inadequate, an equation of higher order may be sought to improve the fit. This will be of the general form:

$$Y = \beta_0 + \beta_1 x + \beta_2 x^2 + \ldots + \beta_p x^p \qquad (8.7)$$

A similar generalisation of the multiple regression equation, involving powers and products of the several independent variables, is also possible. It arises, for example, in the representation of response surfaces in the neighbourhood of a maximum or minimum [*Design and Analysis*] or whenever it is thought that there may be curvature in the relationship between a response and a variable over the region of interest. A similar approach is generally desirable in the analysis of routine plant data. Initially, however, we will consider the simple case of dependence between two variables.

Fitting of curvilinear relationships

8.61 The fitting of a curvilinear relationship is exactly the same as the fitting of a multiple regression, with the powers $x, x^2, \ldots, x^p$ of the observed values of the independent variable x replacing the observed values of the independent variables $x_1, x_2, \ldots, x_p$. Thus we wish to choose values of the β's which will minimize:

$$S.S. = \Sigma(y - \beta_0 - \beta_1 x - \beta_2 x^2 - \ldots - \beta_p x^p)^2 \qquad (8.71)$$

where the summation is over all the observations. By partially differentiating $S.S.$ with respect to each parameter in turn, equating to zero and rearranging, we obtain the $p+1$ least-squares equations for the estimates $b_0, b_1, \ldots, b_p$:

$$\left.\begin{array}{l} b_1\Sigma(x-\bar{x})(x^k-\overline{x^k}) + b_2\Sigma(x^2-\overline{x^2})(x^k-\overline{x^k}) + \ldots \\[4pt] \quad + b_p\Sigma(x^p-\overline{x^p})(x^k-\overline{x^k}) = \Sigma(y-\bar{y})(x^k-\overline{x^k}) \quad (k = 1, 2, \ldots, p) \end{array}\right\} (8.72)$$

and $\quad b_0 = \bar{y} - b_1\bar{x} - b_2\overline{x^2} - \ldots - b_p\overline{x^p}$

$\overline{x^2}$ is used to denote the mean value of x^2, and so on.

Equations (8.72) are exactly equivalent to Equations (8.37) with the powers $x, x^2, \ldots, x^p$ of the independent variable x fulfilling the roles of the independent variables $x_1, x_2, \ldots, x_p$ in the multiple regression. Thus a curvilinear regression for the dependence of y on the successive powers $x, x^2, \ldots, x^p$ of x is fitted precisely as if it were a multiple regression on p independent variables $x_1, x_2, \ldots, x_p$. Each power of x is treated as if it were a separate independent variable. In solving the least-squares equations a vital first step to prevent rounding errors is to standardize the variable x [§ **8.321**]. Even so, the odd powers of x will be fairly highly correlated with each other and so will the even powers.

Tests of significance and measures of precision for curvilinear regressions

8.62 Standard errors and confidence limits are calculated, and tests of significance carried out, exactly as if a multiple linear regression were being fitted. There are, however, certain differences in interpretation and presentation which arise from the fact that the estimated coefficients are not the effects of separate variables, but are each part of the overall effect of a single variable.

The order of the equation required will not, in most cases, be known or predictable in advance. It is usual, therefore, to introduce successive powers of x into the representation and to test for the significance of the additional effect of each. The fitted equation is taken to that order beyond which the addition of extra terms produces no significant reduction in the residual variance. It is common to check that neither the next higher odd power of x nor the next higher even power adds significantly to the regression sum of squares, since odd and even polynomials have quite different shapes. Such tests of significance, to determine the order of the equation necessary for adequate representation of the data, are quite legitimate. When, however, the order of the equation has already been established, it is not in general either legitimate or necessary to test the significance of lower-order terms, and all terms up to the highest order required should be included in the representation. If we suppress a lower-order term we are imposing on the relationship a constraint which we have no reason to suppose corresponds with reality. If, for example, we suppress a non-significant linear term in a parabolic regression, we imply that the minimum or maximum of the parabola occurs precisely at the origin—possibly only a working origin—that we have chosen for our x-values. Thus once the order of the equation has been established, tests of significance of particular terms should be avoided.

The regression coefficients attached to the separate powers of x will generally be highly correlated with one another. Thus the standard errors of the coefficients are, in themselves, of little value as measures of the precision with which the overall effect of x has been estimated. The calculation of confidence regions for sets of coefficients is also somewhat unsatisfactory: it is difficult, for example, to assess mentally the practical effect of a decrease in the coefficient of x accompanied by an increase in the coefficient of x^2. The most convenient representation of the precision with which the effect of x has been determined is thus a plot of the fitted relationship together with confidence limits for the predicted values of y. These are calculated in the same way as for a multiple regression, except of course that specification of a value of x specifies also the associated values of x^2, x^3, etc. Such a plot is shown later (Fig. 8.7) in connection with Example 8.3.

Extrapolation

8.63 There is one point which applies to all use of fitted regression equations for prediction, but particularly to curvilinear regressions; namely that prediction outside the range covered by the data should be made only with considerable reserve. Consider, for example, a set of data in which y generally increases with x, but the relationship exhibits some negative curvature. A parabolic equation may adequately represent the data over the observed range of x values. However, the particular equation may have a maximum near the upper boundary, beyond which predicted values of y decline, whereas the true relationship may become asymptotic or continue upwards at a different rate. The only occasion on which extrapolation may be made with greater confidence is when it is known *a priori* that the fitted relationship is of the correct functional form; but this is seldom the case when a curvilinear regression is fitted.

Example 8.3. Efficiency of a water-gas plant

8.64 The data given in Table 8.5 are monthly "efficiencies" for a water-gas plant, expressed in working units. Air and steam are blown alternately through a bed of hot coke to produce a mixture of gases of which the principal

TABLE 8.5. MONTHLY COKE USAGE AND AIR/STEAM RATIOS FOR A WATER-GAS PLANT

Month	Coke Usage (working units)	Air/ Steam Ratio	Month	Coke Usage (working units)	Air/ Steam Ratio
1	120	2·11	15	51	1·76
2	122	2·29	16	53	1·33
3	128	2·32	17	50	1·23
4	124	2·31	18	34	1·40
5	118	2·25	19	68	1·38
6	114	2·22	20	70	1·96
7	119	2·20	21	49	1·47
8	149	2·41	22	50	1·42
9	141	2·19	23	66	1·33
10	86	2·06	24	46	1·65
11	78	1·99	25	40	1·26
12	31	1·62	26	51	1·61
13	51	1·59	27	51	1·74
14	72	1·70			

constituents are nitrogen, hydrogen and carbon monoxide. The measure of efficiency is coke used per 1,000 m.^{3} of ($H_2 + CO$) produced; since the nitrogen collected is already present in the air used, and is collected incidentally, it does not enter into the calculated efficiencies. Also given are the corresponding air/steam ratios (1,000 m.^{3} air/tonne steam). The data are plotted in Figure 8.6.

While there is considerable scatter, which is attributable to errors of sampling and analysis of coke and gas as well as to any real variations in efficiency, the plot shows clearly that coke usage is related to air/steam ratio. It also suggests that the relationship is not linear, but tends to level out at the lower air/steam ratios used. A simple parabolic regression was therefore fitted.

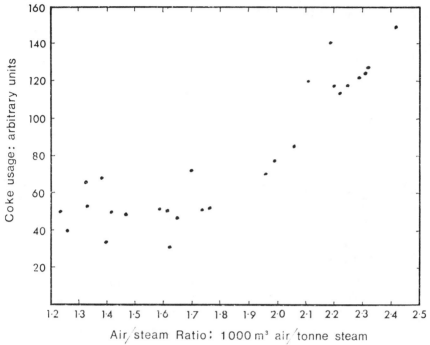

FIG. 8.6. SCATTER DIAGRAM. COKE USAGE v. AIR/STEAM RATIO FOR A WATER-GAS PLANT

If we denote coke efficiency by y, air/steam ratio by x, and (air/steam ratio)^2 for convenience by z, then:

$$\bar{y} = \quad 78\cdot96 \qquad \bar{x} = \quad 1\cdot807 \qquad \bar{z} = \quad 3\cdot409$$
$$S_{yy} = 34{,}849\cdot0 \qquad S_{xx} = \quad 3\cdot8559 \qquad S_{zz} = 51\cdot2192$$
$$S_{yx} = \quad 325\cdot29 \qquad S_{yz} = 1{,}215\cdot99 \qquad S_{xz} = 14\cdot0118$$

The x variable was not standardized in this instance, since the range of x values is large compared with the mean, and hence rounding errors will be negligible.

The least-squares equations for the regression of y upon x and z are:

$$3\cdot8559b_1 + 14\cdot0118b_2 = 325\cdot29$$
$$14\cdot0118b_1 + 51\cdot2192b_2 = 1{,}215\cdot99$$
$$b_0 = 78\cdot96 - 1\cdot807b_1 - 3\cdot409b_2$$

Hence, from the first two equations:

$$b_1 = -323 \cdot 54, \qquad b_2 = 112 \cdot 25$$

The fitted regression equation is thus:

$$(Y - 78 \cdot 96) = -323 \cdot 54(x - 1 \cdot 807) + 112 \cdot 25(z - 3 \cdot 409)$$

i.e.
$$Y = 280 \cdot 9 - 323 \cdot 54x + 112 \cdot 25z$$

$$= 280 \cdot 9 - 323 \cdot 54x + 112 \cdot 25x^2$$

This represents a parabola with a minimum at an air/steam ratio of:

$$x = \frac{323 \cdot 54}{2 \times 112 \cdot 25} = 1 \cdot 44$$

The sum of squares attributable to the curvilinear regression is:

$$b_1 S_{yx} + b_2 S_{yz} = (-323 \cdot 54 \times 325 \cdot 29) + (112 \cdot 25 \times 1{,}215 \cdot 99)$$

$$= 31{,}250 \cdot 6$$

If a linear regression only were fitted, the sum of squares attributable to the regression upon x would be:

$$(S_{yx})^2 / S_{xx} = (325 \cdot 29)^2 / 3 \cdot 8559 = 27{,}442 \cdot 0$$

Thus the Analysis of Variance for the significance of the parabolic term $z \, (= x^2)$ is as given in Table 8.6. The sums of squares for the additional

TABLE 8.6.   ANALYSIS OF VARIANCE FOR SIGNIFICANCE OF PARABOLIC
TERM IN REGRESSION OF COKE USAGE UPON AIR/STEAM RATIO

Source	Sum of Squares	Degrees of Freedom	Mean Square
Due to linear regression only	27,442·0	1	27,442·0
Addition due to x^2 term	3,808·6	1	3,808·6
Total due to parabolic regression	31,250·6	2	
Residual	3,598·4	24	149·9
Total	34,849·0	26	

effect of the x^2 term and for the residual are both obtained by differencing from the values already computed. The F-ratio for the additional effect of the quadratic term is $3808 \cdot 6 / 149 \cdot 9 = 25 \cdot 4$, which is highly significant for 1 & 24 degrees of freedom [Table D].

Standard errors and predicted values for Example 8.3

8.641 The variances and covariance of the regression coefficients b_1 and b_2 are obtained from the elements of the inverse of the matrix of corrected sums of squares and products. For the simple case of a regression on two variables the inverse matrix is:

$$\frac{1}{\Delta}\begin{bmatrix} S_{zz} & -S_{xz} \\ -S_{xz} & S_{xx} \end{bmatrix}, \qquad \text{where } \Delta = S_{xx}S_{zz}-(S_{xz})^2 = 1 \cdot 1656$$

Hence:

$$V(b_1) = 149 \cdot 9 \times 51 \cdot 2192/1 \cdot 1656 \quad = 6{,}591$$
$$V(b_2) = 149 \cdot 9 \times 3 \cdot 8559/1 \cdot 1656 \quad = \quad 496$$
$$C(b_1, b_2) = -149 \cdot 9 \times 14 \cdot 0118/1 \cdot 1656 = -1{,}803$$

The standard errors of the estimates are therefore:

$$S.E.(b_1) = 81 \cdot 2 \qquad S.E.(b_2) = 22 \cdot 3$$

but because of the high correlation between the estimates these give little obvious information on the precision of predicted values obtained from the regression equation.

Now the fitted regression in centred form is:

$$(Y-\bar{y}) = -323 \cdot 54(x-\bar{x}) + 112 \cdot 25(x^2 - \overline{x^2})$$

Hence if Y_0 denotes the regression estimate corresponding to a value x_0 of x, the uncertainty in Y_0 is measured by a variance:

$$V(Y_0) = V(\bar{y}) + 6{,}591(x_0 - \bar{x})^2 + 496(x_0{}^2 - \overline{x^2})^2 - 3{,}606(x_0 - \bar{x})(x_0{}^2 - \overline{x^2})$$
$$= 5 \cdot 56 + 6{,}591(x_0 - 1 \cdot 807)^2 + 496(x_0{}^2 - 3 \cdot 409)^2$$
$$- 3{,}606(x_0 - 1 \cdot 807)(x_0{}^2 - 3 \cdot 409)$$

The 95% confidence limits for Y_0 are given by $\pm S.E.(Y_0)$ multiplied by the appropriate value of t for 24 degrees of freedom, i.e. $2 \cdot 06$, and are shown graphically in Figure 8.7. Note that they describe only the uncertainties of the regression estimates; the uncertainty of a future individual result is obtained by adding the residual variance ($= 149 \cdot 9$) to $V(Y_0)$. For example, the observed coke usage at an air/steam ratio of $1 \cdot 7$ would be expected, with 95% confidence, to be within the limits:

$$55 \cdot 3 \pm 2 \cdot 06\sqrt{(149 \cdot 9 + 14 \cdot 4)}$$
$$\text{i.e. } 28 \cdot 9 \text{ and } 81 \cdot 7$$

Such limits would be used in applying the regression equation as a check on the future performance of the plant.

Applicability of the methods

8.65 It should be noted that the assumption throughout the above treatment of curvilinear regression is that the residual variation is, or may be considered to be, entirely attributable to the dependent variable. Difficulties arise in the estimation of nonlinear functional relationships when observed values of both variables are affected by errors of measurement. These are discussed briefly in § **8.92** below.

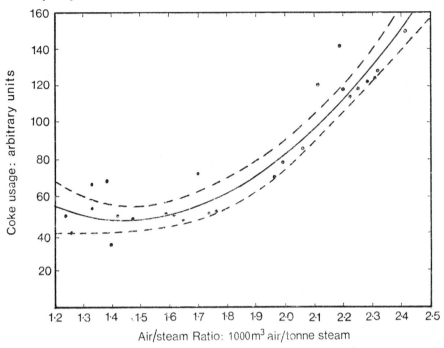

FIG. 8.7. PARABOLIC REGRESSION OF COKE USAGE UPON AIR/STEAM RATIO, WITH 95 % CONFIDENCE LIMITS FOR PREDICTED VALUES

MULTIPLE QUADRATIC REGRESSION

8.7 Multiple curvilinear regressions involving more than one independent variable are fitted in a precisely similar manner. Each power or product entering into the regression equation is treated as if it were a separate independent variable in setting up the least-squares equations for the regression coefficients. The only difficulties which arise spring from the number of equations which may have to be solved. Even for as few as three independent variables the fitting of a general quadratic relationship requires the estimation of nine regression coefficients and hence the solution of a set of nine simultaneous equations, while for four independent variables the number of possible coefficients rises to fourteen, and for five independent variables to twenty. There is an equally rapid increase in the number of possible coefficients as the order of the regression equation is increased, and even with electronic computers, fitting a multivariable model containing cubic or higher-order terms is rarely attempted. Fortunately, for practical purposes a multiple quadratic equation is usually adequate.

A multiple curvilinear regression analysis is an empirical attempt to represent an unknown underlying relationship as a simple polynomial. In this sense it is analogous to Maclaurin's Theorem, whereby almost any mathematical function can be expanded as an infinite series. The linear terms in the expansion form a first-order approximation to the true function, and if we also include all the second-degree terms (squares and products) we have a second-order approximation which is usually sufficiently accurate over the region of interest.

The general quadratic equation in p variables is of the form:

$$Y = \beta_0 + \beta_1 x_1 + \beta_2 x_2 + \ldots + \beta_p x_p + \beta_{11} x_1^2 + \beta_{22} x_2^2 + \ldots + \beta_{pp} x_p^2$$
$$+ \beta_{12} x_1 x_2 + \beta_{13} x_1 x_3 + \ldots + \beta_{p-1,p} x_{p-1} x_p \qquad (8.8)$$

It contains p linear terms, p squared terms and $\frac{1}{2}p(p-1)$ product terms, the latter corresponding to interaction effects of the variables on the response. Some computer programs automatically calculate all the necessary squares and products for the full quadratic model; in other programs the second-degree terms have to be treated as transformations of the linear terms and specified individually.

The amount of space devoted specifically to multiple quadratic regression in this chapter belies its importance as a major tool, but no new fundamental principles are involved beyond those given in earlier sections. Consideration should always be given to fitting a quadratic model to multivariable data arising from experimentation or from routine plant observations, and certainly whenever it is thought possible that:

(i) There may be curvature in the relationship, or
(ii) The process may be operating somewhere near to an optimum, or
(iii) The variables may interact with each other.

This recommendation must not be taken to imply that *every* variable must necessarily appear in its full quadratic form; indeed this may not always be possible as, for example, when a variable appears in the data at only two distinct levels. Similarly it is not possible to fit an interaction term between x_i and x_j when all the different values of x_j occur only in conjunction with one particular value of x_i. (The consequences of attempting to fit invalid models were discussed in § **8.53**.)

All the techniques and difficulties of model building described earlier in the chapter apply equally to quadratic regression. In particular, the scaling of the variables to avoid correlation between x_i and x_i^2, the examination of inverse diagonals and the scrutiny of residuals are of great importance. Since all the quadratic terms are effectively treated as if they were separate independent variables, the processes for automatic selection and rejection of variables [§ **8.4**] will now select or reject individual terms from the full quadratic model; this raises special difficulties which are discussed in the next section.

Balancing a final quadratic model

8.71 Regarding a quadratic model as a second-order approximation to the true relationship, it may be argued that despite certain individual terms having been discarded by an automatic testing procedure, the final model should nevertheless still have a balanced form. Thus if an interaction (product) term remains between two variables, then both the squared terms should also be present, and if squared terms remain for two variables, then their product term should be included. Linear terms are however allowed to stand alone. For example, if the remaining regression contained terms in

$$x_1, x_3, x_4, x_1^2, x_3^2, x_1 x_2 \text{ and } x_2 x_3$$

then the final equation would also include terms in

$$x_2, x_2^2 \text{ and } x_1 x_3$$

However, some analysts may prefer to maintain the simple model because of its easier subsequent manipulation and application. Furthermore, it is known that in certain circumstances the retention of such non-significant terms can markedly widen the confidence limits associated with predicted values of the response. So from these points of view balancing the equation is not really recommended.

However, it *is* considered desirable to replace any linear terms which have been discarded whenever the corresponding product or squared terms remain because:

(i) The original data will usually have been scaled, and the process of descaling the final equation will generally re-introduce linear terms. The simplification aspect will thereby be lost, and it is preferable to estimate the coefficients directly rather than re-introduce them by algebraic manipulation.

(ii) If we suppress a non-significant linear term in one of the variables of a quadratic relationship, then we imply that the vertex of the parabola in the hyperplane concerned occurs precisely at the origin for that variable, which is generally only a working origin. This is a constraint which we have no reason to suppose corresponds to reality.

(iii) If two variables appear solely in an interaction term, the model has a degree of symmetry which is probably spurious, and again the model is not independent of the arbitrarily chosen scaling.

Interpretation and presentation of a multiple regression analysis

8.8 Fitting a satisfactory multiple regression equation for a set of data is not an end in itself—the equation must be interpreted and applied. This final stage of analysis will generally include one or more of the following steps:

(i) Examine which variables are included in the final model, i.e. which are statistically significant and which are not.

(ii) Obtain some idea of the relative effects of the individual variables from the size of the regression coefficients, bearing in mind the ranges covered by the data.

(iii) Obtain an appreciation of the shape of the response surface, i.e. the extent to which the variables have curvature and interaction effects.

(iv) Determine whether there is a maximum or minimum of the response surface within the experimental region or whether the highest (or lowest) value of the response occurs at boundary values of one or more variables.

(v) Determine whether the response surface is relatively flat over any particular region; it may, for instance, contain a saddle point.

(vi) Determine whether there are *any* combinations of the variables which can produce specified values of the response.

(vii) See how reliably the model can be used to make predictions.

These seven steps will generally lead to more executive actions of the type:

(viii) Decide which variables need to be carefully controlled in order to lower the variability in the response. Near an optimum this may also have the effect of improving the achieved mean response.

(ix) Decide which combinations of the levels of the variables to use in the future in order to obtain some desirable pre-specified value of the response. This may be a combination which yields the maximum (or minimum) value of the response, or at least a step away from the current operating conditions in the direction of substantial improvement.

(x) Decide which variables should be investigated more carefully in a (further) planned experiment. Specify the ranges of the variables to be explored and whether provision should be made for further exploration of curvature and interactions in such an experiment.

Except in the simplest cases, the mere visual examination of a multiple regression equation is not very rewarding. The relationship is too complex to be fully comprehended by just looking at the regression coefficients. The following sections describe techniques which aid the interpretation; they can all be performed on a computer, and some programs include them as "optional extras" to be carried out once a satisfactory model has been obtained.

A grid of predicted values

8.81 It is often helpful to calculate predicted values at a systematic set of points in the p-dimensional space. If the original data were scaled so as to occupy the range ± 1 in each variable, then predictions can be made at all possible combinations of ± 1, or even at points slightly in excess of this range, say at $\pm 1 \cdot 25$ units. The central point $(0, 0, ..., 0)$ may also be included, or even all combinations of the three values -1, 0, $+1$, but the number of points increases rapidly with the number of variables concerned. It is also desirable to calculate a measure of the reliability of the predictions at each point, e.g. the 95% confidence limits; these limits are often surprisingly wide.

Such a grid of extreme points is of particular value for a purely linear model, since there is no mathematical optimum or turning value; the highest or lowest response value must lie on whatever rectangular boundary is chosen for the grid and can be read off directly from the list of predicted values. However, if the model is quadratic, a simple grid may obscure the true shape of the response surface and hide the existence of optimum settings in some of the variables.

Contour diagrams

8.82 It is very instructive to see a pictorial representation of the relationship between a response and the independent variables. Just as the relationship between y and a single variable x may be represented by a two-dimensional curve drawn on a sheet of paper, so the relationship between y and two factors x_1 and x_2 may be represented by a three-dimensional surface called the Response Surface. A solid model of this surface may be constructed, but it is much simpler to draw *lines of equal response* on a two-dimensional graph whose co-ordinates denote the values of the independent variables. These lines are termed the Response Contours. This type of representation is commonly used in maps to indicate the height of land, and in weather charts to show the pattern of atmospheric pressure. Contour lines which are close together in any region imply a rapid change in the response for relatively small changes in the variable or variables.

A multiple linear regression equation yields a set of straight-line contours; distances between these lines are proportional to differences in the response levels. For a quadratic relationship, contour lines generally consist of either concentric ellipses or concentric hyperbolas, although these figures may not be complete within the experimental region. When ellipses occur, the response has a definite maximum or minimum, which may or may not be within the region plotted. The occurrence of hyperbolas implies that there is a saddle point somewhere on the response surface [*Design and Analysis*, § **11.52**].

Figure 8.8 shows the contours for variables x_1 and x_3, each over the range ± 1.5, when x_2 and x_4 are set equal to zero in the complete four-variable quadratic equation:

$$Y = 7{\cdot}184 + 1{\cdot}207x_1 - 0{\cdot}754x_2 + 1{\cdot}398x_3 - 0{\cdot}157x_4 - 2{\cdot}113x_1{}^2 - 0{\cdot}314x_2{}^2$$

$$- 0{\cdot}944x_3{}^2 + 0{\cdot}863x_4{}^2 - 0{\cdot}521x_1x_2 + 0{\cdot}492x_1x_3 - 0{\cdot}033x_1x_4$$

$$+ 0{\cdot}013x_2x_3 - 0{\cdot}076x_2x_4 + 0{\cdot}044x_3x_4 \qquad (8.81)$$

These contours have been printed out as individual characters by a computer at response levels pre-specified by the analyst; normally the user would join the asterisks up with smooth curves. (A computer equipped with graphical output facilities could be programmed to yield a virtually smooth curve directly.) This family of ellipses represents a simple "hill" with a maximum at $x_1 = 0.383$, $x_3 = 0.840$, where the response has a value of 8.00 units.

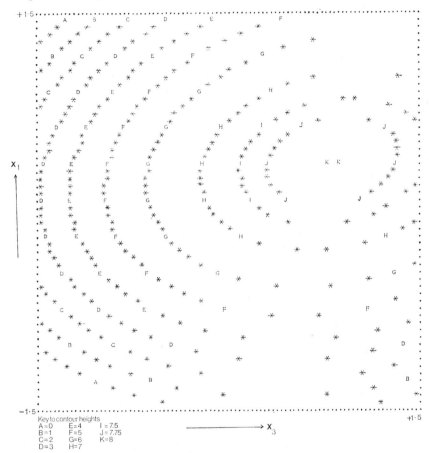

FIG. 8.8.   COMPUTER-PRINTED CONTOUR DIAGRAM

A quadratic regression relationship involving three independent variables is more difficult to appreciate; the direct equivalents to contour lines are three-dimensional quadric surfaces of equal response. The simplest approach in this case is to obtain and compare separate two-variable contour diagrams at a number of fixed values of the third variable. The selected third variable should preferably be the one which has the smallest and simplest effect on the response. It is helpful if the contour diagrams are drawn on separate sheets of a rigid transparent material and held one above the other with suitable spacing; the full quadric contour surfaces are then easily envisaged. Different colours can help to distinguish between the various response levels. If the third variable appears solely as a linear term, then a single contour diagram at, say, $x_3 = 0$ is sufficient, since for any other value of x_3, all the contour heights are merely increased or decreased by a fixed amount.

The effects of four independent variables may be displayed as follows. Suppose, for simplicity, that the variables have all been scaled so that they each have a mean of approximately zero and a range of approximately ± 1 unit. First obtain a contour diagram for, say, x_1 and x_3 when x_2 and x_4 are both set equal to zero, and place it at the centre of a much larger chart. (The larger chart could, for example, cover a wall of a room.) Let the horizontal direction of this large chart represent variable x_4, let the vertical direction represent x_2 and let the centre of the chart be the origin for x_2 and x_4. Obtain another contour diagram for x_1 and x_3, but this time with x_2 and x_4 both set equal to $+1$, say, and place it in the upper right-hand quadrant of the chart. Compute three further contour diagrams for the other symmetric combinations of x_4 and x_2, i.e. $(-1, +1)$, $(-1, -1)$, $(+1, -1)$, and place them in their appropriate positions. Then by viewing the set of five diagrams simultaneously a picture emerges of the effects and interplay of all four variables. This murographic method is illustrated in Figure 8.9, which is based on the four-variable quadratic relationship given in Equation (8.81). Additional diagrams for intermediate values of x_2 and x_4 may of course be incorporated for any region of special interest or of rapid change in response values.

This approach could be extended to five variables by using, say, three large charts to represent three different levels of a fifth variable. Further extensions are conceivable, if somewhat fanciful.

Before ending the discussion of this useful interpretive tool, it is appropriate to stress again the potential dangers of extrapolating a quadratic response surface beyond the region of the actual data. Extrapolation is a quick route to misleading conclusions. Further, even though two variables both cover the range ± 1 the data may not actually include points at extreme *combinations* of the variables, and in these circumstances it is more appropriate to bound a contour diagram by a unit circle rather than a square.

Canonical analysis of a fitted quadratic equation

8.83 If Y_s denotes the response value at the Centre of Symmetry of a quadratic equation, e.g. at the centre of elliptic contours, then the equation can be expressed more simply in the Canonical Form:

$$Y - Y_s = B_{11}X_1^2 + B_{22}X_2^2 + \dots + B_{pp}X_p^2 \qquad (8.82)$$

where the derived variables X_1, X_2, ..., X_p are linear functions of the original

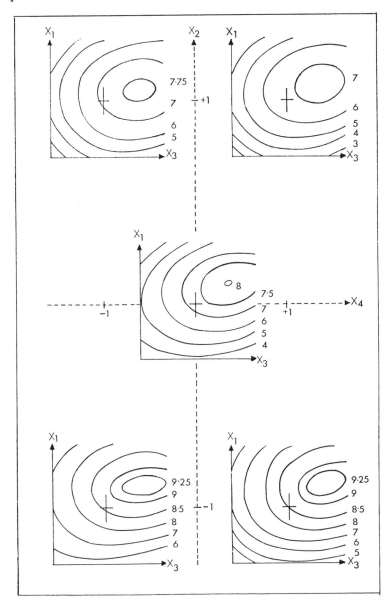

FIG. 8.9. REPRESENTATION OF EFFECTS OF FOUR QUADRATIC
VARIABLES BY SPACIALLY ARRANGED CONTOUR DIAGRAMS

variables $x_1, x_2, ..., x_p$. This transformation is effected by moving the origin
to the centre of symmetry and rotating the axes until they lie along the
Principal Axes, i.e. the axes of symmetry of the system.

The signs and the relative sizes of the canonical coefficients provide a key to the nature of the response surface. Consider first the two-variable case. If B_{11} and B_{22} are both negative, the surface forms a hill with a distinct maximum and the contour diagram comprises a family of ellipses; if $|B_{22}| < |B_{11}|$ the ellipses are elongated along the X_2 axis and vice versa. If the coefficients have opposite signs, the contours consist of hyperbolas with a minimax (saddle) at the centre. When B_{11}, say, is negative and B_{22} is zero (or essentially zero) the contours represent a stationary ridge; there is no unique maximum point, but a line of maxima given by the equation $X_1 = 0$. Similarly a rising ridge would result from one small negative coefficient when the centre of the system is well outside the experimental region.

These inferences may be extended to three or more variables. For example, if in a three-variable equation, two canonical coefficients are relatively small, say, B_{22} and B_{33}, then for most practical purposes there is a plane of extrema given by the equation $X_1 = 0$. It follows that despite changes in any one or two of the original variables, suitable changes in the other variables can maintain the response at its optimum level. Sometimes quadratic regression analysis predicts a maximum remote from the experimental region with the surface elongated along an axis which passes close to the data. This finding implies that the previous runs have brought the experimenter not to a maximum but close to a rising ridge. No conclusion about the position or even the existence of the remote maximum should be drawn, but attention should be focused on the local ridge and further experiments should be carried out along it.

When the principal axes are inclined to the original co-ordinate axes and the surface is elongated along one of its axes, interaction between the variables is indicated. In other words, when x_1 is varied, the value at which a response optimum is obtained is not independent of the level of x_2. However, when the interaction coefficient b_{12} is zero, the X_1 and X_2 axes coincide with the x_1 and x_2 directions and there is no interdependence arising from these two variables.

For a more detailed exposition of canonical analysis the reader should refer to Chapter 11 of *Design and Analysis*.

Path of steepest ascent to optimum response

8.84 The location of any maximum or minimum value of a response can be determined from a quadratic regression equation by setting the partial derivative with respect to each variable equal to zero and solving the resulting set of simultaneous linear equations. When four or more variables are involved this is best done on a computer.

However, the position of an extremum is frequently remote from the original data as regards some or all of the variables and, as stated earlier, it is then of little direct interest or use. More often we require the maximum or minimum value obtainable either within the region of the data or on its boundaries; in any specific case this optimum may be within the range of some variables and at the extremes of others. With each variable scaled to the range -1 to $+1$, the region may be taken as a rectangular hypercube centred on the origin, or preferably as the unit hypersphere [§ **8.82**]. The location of the optimum response value is then best obtained by an iterative numerical method which starts at some pre-specified point—usually the centre—and takes small steps up the Path of Steepest Ascent in the p dimensions [8.6]. A computer program for this calculation may be written so as to reduce the step size progressively as the improvement in response becomes small. When the path reaches a boundary the local region may be explored more closely.

If the regression equation is a model of an industrial process with the data centred around the current target operating conditions, it may be considered imprudent to make a large change to operate immediately at the optimum location calculated as above. It would then be reasonable to move provisionally to a point, say, half-way along the path of steepest ascent, bearing in mind that there will be many other combinations of the variables which would give the same modest improvement.

In addition to imposing boundaries appropriate to the data, it is sometimes necessary to impose a further condition that another response, dependent on some or all of the same variables, should not exceed or fall below some specified value (or should actually have some specified value). Examples of this are cost limitations in a search for improved product quality and impurity levels in a quest for increased yield. Such additional restrictions can be built into a computer program which uses either the method of steepest ascent or some other suitable numerical technique.

MISCELLANEOUS TOPICS IN MULTIPLE REGRESSION ANALYSIS

Extracting useful information from plant records

8.91 It is common in industrial processes to record, by various means, a large volume of data relating to the operation of these processes and to their products. At first sight, multiple regression analysis appears to be a powerful technique for ascertaining the effects of operating conditions on, for example, yield, conversion efficiency or product properties. However, it is frequently claimed that such statistical analyses are fruitless exercises and the phrase "garbage in—garbage out" is even used to describe the nature of the calculation. Whilst not denying that there is a certain amount of truth in this criticism, it would perhaps be more appropriate to describe the raw data as "garbled" rather than "garbage".

It is quite true that the study of plant records is much inferior to carrying out properly planned, and realistic, experiments on either the full-scale plant or even on a pilot plant. Despite this reservation, there is frequently no immediate alternative to such an analysis; the data are already there and could contain useful information. Even if the analysis does not always directly result in improvements to the process, it can at least give a few hints on how to obtain a more stable product, or indicate which variables are likely to be worth looking at in a subsequent experiment. Certainly many successful applications are known to the authors.

Foremost among the difficulties is the inherent unreliability of the data, which are often recorded with little care by staff who have scant appreciation of their potential worth or with little knowledge of what they are really all about. The listed data are frequently only a gross approximation to the true state of affairs and may, on investigation, relate to something quite different from that intended by the management. It is not unknown for them to be wholly or partly fictitious, and missing values are numerous. In many instances, however, one has advance warning of an investigation and it is sometimes possible, and very worth while, to organize special supervision and immediate checking of the records for a sufficiently long period.

A second difficulty arises because the ranges of values of the variables are often quite small relative to the errors of recording and measurement—if only because the usual aim in plant operation is to maintain conditions as steady as possible. Deliberate changes in operating conditions may be determined more by the ease with which the factors concerned can be controlled than by their importance. Some outstanding successes for regression analysis have been achieved in newly installed and highly complex plants where multiple quadratic models have given excellent fits to the observed data, and predictions of more advantageous processing conditions have proved to be correct. However, as the processes were gradually brought under closer control over a period of months, the ranges of values of the variables were reduced and later regression equations were less satisfactory or useful.

The third difficulty arises from the fact that all the measurements or observations on any one variable arise in a definite sequence, that is in the form of a Time Series, and adjacent results tend to be correlated with each other by the very nature of the process and the supply of its raw materials. The data are generally full of trends of different amplitudes and durations, together with what appear to be sudden jumps in mean level. If two different time series show the same trend pattern, say, for example, a simple linear trend over the whole period considered, then even if these trends are in fact merely coincidental and completely unconnected with each other, regression analysis will give a statistically significant relationship between the two variables concerned. This relationship would, of course, be completely spurious in the sense that it does not imply a "cause and effect" relationship. One never really knows whether any specific relationship is spurious and for this reason it is desirable to confirm any such apparent effect by subsequent direct experimentation. The more complicated the general pattern of the series, the less is the likelihood that any apparent relationship is spurious. Apart from the purely coincidental case, even if one variable is not a true "cause" and the other not the resultant "effect" but both are really the effects of some unknown third variable, then the knowledge of even this indirect relationship may be of great use in understanding and controlling the process, provided that the indirect relationship is a permanent one.

Various devices have been suggested for minimizing these spurious relationships. If the trend is a linear one, or even a simple curve, "time" can be included as an additional independent variable in the regression; thus the analysis is based on the *deviations* about the best fitting straight line rather than on the observations themselves. If a relationship is still found after the major time trend has been eliminated in this way, it is desirable to determine whether the slope of the "spurious" relationship is the same as that of the "genuine" one, in which case they are both probably the same genuine overall relationship. Alternatively it may be desirable to divide the data into two or three distinct periods of time, or preferably collect the data from quite separate periods, and then compare the results obtained from a separate statistical analysis for each period. Spurious correlations have been known to reverse their directions under this treatment. A further technique which eliminates virtually all trends (but which may leave you with nothing much more useful than recording and measurement errors) is to carry out the regression analysis using the algebraic differences between adjacent values of responses and variables as the basic data.

The fourth major problem of multiple regression analysis of plant data is concerned more with the interpretation of the analysis than with its validity. It arises because in complex processes some of the measurements recorded on the plant cannot always be treated as independent variables in the sense of being basic entities which each define some separate aspect of the conditions under which the product is made. They may themselves be wholly or partially dependent on the amounts and characteristics of the raw materials or on the external conditions applied in earlier stages of the process, and if these other variables are also included in the regression one must tread very warily. This topic is discussed again in the next section.

Whenever a multiple regression analysis of plant data is attempted, it is recommended that the equation fitted should be of multiple quadratic form rather than multiple linear, since in many industrial processes the relationships between the variables are markedly curved, particularly when the process is close to an optimum. Moreover, interactions between the variables are very common and these are not investigated in a purely linear model. Models which explain 60–80% of the observed variability in the responses (measured as a proportion of the total sum of squares) have been obtained quite often, and occasionally more than 90% of the plant variability has been explained in this way. When the percentage fit is only, say, 50, it is not possible to place very much reliability on derived estimates, such as the location of the optimum operating conditions or the value of the response to be expected at this point; however, these statistics do give some useful pointers to the plant and development staff.

Initial editing of data and choice of variables

8.911 It is essential to edit the data and choose the variables carefully before starting the formal multiple regression analysis. Otherwise the analysis will probably fail, as evidenced by a number of unacceptably high values in the diagonal elements of the inverse correlation matrix [§ **8.53**]. The following procedures are mainly self-evident, but are vital to avoid "nonsense" conclusions, or even to produce any conclusions at all:

(i) All variables and responses should be plotted out as time series and examined for simple trend patterns as discussed earlier.

(ii) Values of all variables and responses should be displayed as histograms. The greater the variation the more likely will real effects on the responses be detected, but it is advisable to discard as "outliers" any results which are further away from the mean than approximately 3σ units (or some other arbitrarily higher limit), since a few such points, if erroneous, can cause gross distortion of the model.

(iii) The different measurements recorded on the plant need not all be included in the analysis; the plant staff may be able to state with certainty that some variables could not possibly affect the responses to be investigated. Any variable should also be excluded whose total variation is little greater than its known or expected sampling and testing errors.

(iv) Two supposedly different plant variables may really be different measures of essentially the same quantity. Discuss this possibility with the plant staff, and if necessary carry out simple correlation analyses between the likely variables. If a correlation coefficient greater than, say, 0·95 exists between two variables, then only one of the variables should be included in the multiple regression analysis. Any effect of this variable on the response could, of course, really be due to either or both of these variables.

(v) As stated in § **8.53**, it is only too easy to construct an information matrix that turns out to be singular, because one variable can be calculated identically from some linear combination of other variables. In particular, avoid the traps of including:

(a) All the components of a time cycle of operations as well as the total cycle time.

(b) The weights of all the component raw materials when the total weight is constant. The same difficulties arise when the amounts of raw materials are expressed as ratios, since the sum of the ratios is unity.

(c) Two factors, when their product (which will be present as an interaction term in the complete quadratic regression model) is also included as a variable in its own right. This mistake occurs most often when the factors are ratios of primary measurements, or a ratio and a direct observation.

(vi) Hard and fast rules cannot be laid down for the number of complete sets of observations which should be included in the analysis. In general, the longer the period of time over which the data are collected the more likely is the analysis to be meaningful and fruitful, provided that the process or equipment has not been materially changed during this time. The period should be long enough to embrace all the normal types of variation, including several trends, oscillations and jumps. As a working rule, it is advisable to have at least as many residual degrees of freedom as there are coefficients in the complete quadratic regression on all the variables.

(vii) Mention has been made earlier of the fact that some plant observations are often more in the nature of "intermediate stage responses" than truly independent variables. If these intermediate responses are of technical interest in themselves, clearly they should be separately analysed as responses dependent on all earlier variables. In any case they should be also included as separate variables in the larger analysis unless at least 95%, say, of their variability is explained by other variables which are also to be included in the main analysis. An actual example will clarify this point.

Consider a chemical process where one solid and several liquids are to be reacted together at high temperature and pressure. The basic chemical reaction is believed to be completed as soon as the solid is fully dissolved, but this is by no means the end of the whole process. It is desired to reduce the variability of the time taken to go into solution; this period can be regarded both as an intermediate response, probably dependent on earlier variables including raw materials, and as an "independent" variable, possibly having its own effect on some subsequent major response by which the ultimate product is characterised.

The "time-to-solution" was found to depend significantly on two of the early variables—the ratio of two different raw materials and the reaction temperature at that stage; the relationship accounted for 81 % of the variability in "time-to-solution". It was, nevertheless, included as a variable in a larger analysis with "chemical yield" as the response. As well as being dependent on the two early variables (and some others) the yield was also dependent on the time-to-solution itself to such an extent that its inclusion increased the yield sum of squares accounted for by the regression from 62% to 85%. This finding was interpreted as meaning that part of the unexplained 19% of time-to-solution variability was due to some factor not currently measured (in this case it could well be associated with the particle-size distribution of the solid raw material), and that this unknown factor itself had a large effect on the subsequent yield. Until this factor can be isolated and separately measured the time-to-solution must be used as an indirect measure of it. The analysis is further interpreted as meaning that some or all of the early variables also have effects on yield which take place *after* the solid has gone into solution as well as *before* it.

Use and interpretation of models from plant data

8.912 Despite the initial editing of the data it is still highly desirable to examine the residuals after fitting the model, as described in § **8.54**. It is also unwise to ignore completely variables which are rejected as non-significant. All regression coefficients should be examined with their associated confidence limits and covariances. These limits provide an indication of the likely magnitudes of suspect effects, and thus suggest where further work would be advantageous.

Even greater dangers than the apparent non-significance of real effects may sometimes arise in the analysis of plant records; in some circumstances the plant may be run in such a way as to introduce a completely spurious apparent dependence. Consider, for example, a study of conversion efficiency for particular exothermic reaction. Since the reaction was exothermic and the converters were uncooled, the conversion was very closely proportional to the temperature rise through the converter. The aim in plant operation had usually been to keep the exit temperature steady, since at unduly high temperatures undesirable side reactions occurred: regulation was effected by varying the temperature of the inlet gas. A study of accumulated data was proposed to ascertain the effects on the conversion efficiency of, for example, the composition of the inlet gas. One of the factors put forward for study was temperature; since the temperature varied through the converter, the most representative figure appeared to be the average temperature, which was adequately approximated by the average of inlet and exit temperatures. What was not immediately obvious, but soon became so, was that the use of this average temperature as an independent variable in the regression analysis would lead to completely nonsensical conclusions, such as, for example, that the conversion showed no dependence on the composition of the inlet gas. The reason lies in the mode of operation of the converter. Because exit temperatures are held steady, the average temperature is almost perfectly correlated with temperature rise, which in turn closely reflects conversion. Thus a regression of conversion efficiency upon average temperature is effectively a regression of conversion upon itself; all the variation except that attributable to errors of measurement is explained by the regression, and there is no chance of any other effects being found. The slope of the regression of conversion efficiency upon average temperature does not, moreover, give the real effect of temperature; it is merely, at second or third remove, a statement that $x = x$. Such pitfalls are not uncommon, and are an added reason why regression analysis of process records should be undertaken with some caution and why full attention should be given to all details of plant running.

Regressions and functional relationships

8.92 A regression equation is a statistical relationship between observed values of two or more variables, and *no more than this*. This point should not be lost sight of in any discussion on regressions and functional relationships. Though a regression equation may be, and frequently is, used as an estimate of a functional relationship, the circumstances for such use must be closely defined, otherwise misleading results may be obtained. There is, for example, a class of problems in which regressions exist but there is no such thing as a functional relationship; in these problems the scatter about a fitted regression arises from real variation in the individuals studied, and is not the result of errors of observation or measurement. This is commonly the case in biometric and biological work, but it is also sometimes true in industrial work, as illustrated by the example discussed earlier on abrasion loss and hardness of rubber samples. The key word for this class of problems is "individuals"; where relationships are sought between observations on a series of separate individuals or items, it is usually inappropriate to envisage other than regression relationships. Where such relationships are the only relationships which can exist, regression equations may legitimately be used for purposes of description and prediction. In some instances more than one regression equation may be fitted to each set of data, since each variable may equally be considered as the "dependent" variable, but all regressions are equally valid and may be used as appropriate.

In most industrial and scientific work, however, a series of observations will represent a series of "states of nature" rather than a series of individuals or separate items. It is common then to postulate or expect a functional relationship between the quantities studied; examples of simple relationships are Ohm's Law and the perfect gas laws, while more complex examples arise in, for instance, reaction kinetics.

The conditions under which a regression equation may be used to estimate a functional relationship, when one exists, are specified from the manner in which the equation is derived. The least-squares equations giving the coefficients are those which minimize the sum of squares of deviations of observed values of the "dependent" variable from the values predicted by the regression, the dependent variable always being denoted above by y. The assumption is thus that only errors in y need be considered; that it is the observed value of y which deviates from the corresponding true value, and not the value of any other variable. If this is in fact so, then the regression of y upon the "independent" variables is simultaneously the best estimate of the underlying functional relationship. Only if all values of the independent variables are determined without error, and the only variations which arise are errors of measurement of values of the dependent variable, is the regression of the dependent variable upon the independent variables the best estimate of the functional relationship. An exception to this is if the variables have been preset according to Berkson's definition as discussed in § 7.64. In other circumstances the regression gives a biased estimate of the true relationship, the bias depending on the errors of measurement for the independent variables and the ranges covered by these. The regression may still be used for prediction of "observed" values from observed values (*cf.* § 7.3) if the observations on which it is based may be considered as a random sample of possible observations, but is otherwise of little value.

In practice it is not necessary to insist too rigorously on the absence of errors in the measured values of the independent variables; provided that these are small compared with the ranges covered and with the errors in the dependent variable, the regression equation gives a reasonably satisfactory estimate of the functional relationship. These conditions are most likely to be met in controlled experiments, where wide ranges can usually be covered and the values of the independent variables can be obtained with a high degree of accuracy. Thus, wherever possible, functional relationships should be estimated from the results of controlled experiments rather than from accumulated uncontrolled observations. Also, it is often possible to arrange in a controlled experiment for the orthogonality of the separate effects, which is an added advantage.

Whenever the error variance of the dependent variable is known it is worth while comparing it with the residual mean square for the multiple regression. If it is a substantial fraction of the residual, the regression equation can be used with greater confidence as though it were a functional relationship. Any variability which remains is not necessarily associated with errors in the independent variables; it can equally well arise from (unknown) variables which have not been included in the model, or from existing variables which do not appear in their best algebraic form.

Where controlled experiments cannot be carried out, or where for other reasons the conditions for use of a regression equation cannot be satisfied, other methods must be used to estimate a functional relationship, and these are, in general, outside the scope of this book. The simplest case is that of a multiple linear relationship, when the error variances for the independent variables are known. Let these be denoted by $\sigma_1^2, ..., \sigma_p^2$, there being n sets of observations. Then the usual least-squares equations for the regression coefficients should be modified to:

$$
\left.
\begin{aligned}
b_1\{S_{11}-(n-1)\sigma_1^2\}+b_2S_{12}+\ldots+b_pS_{1p} &= S_{y1} \\
b_1S_{12}+b_2\{S_{22}-(n-1)\sigma_2^2\}+\ldots+b_pS_{2p} &= S_{y2} \\
\vdots \qquad \vdots \qquad \qquad \vdots \qquad \qquad \vdots \\
b_1S_{1p}+b_2S_{2p}+\ldots+b_p\{S_{pp}-(n-1)\sigma_p^2\} &= S_{yp}
\end{aligned}
\right\}
\qquad (8.9)
$$

i.e. each sum of squares should be "corrected" for the extra variation attributable to error. The result is a generalisation of that given in § **7.62**. Where only the *ratios* of the error variances are known, no generalisation of the results [§ **7.621**] for the two-variable case is available, but the relationship may be estimated by an iterative method given in [8.7]. No generalisation to the multivariate case is available of Bartlett's method [§ **7.63**] for estimating a linear functional relationship when the error variances are unknown.

The estimation of curvilinear functional relationships when all variables are measured with error is more difficult, and there is a further difficulty in that, taking the bivariate case as an example, even if y is measured without error, the regression of x upon y is not readily obtainable for use as an estimate of the functional relationship. In the bivariate case, a functional relationship of known form may be estimated, given one of the error variances or the ratio of the two error variances; the appropriate least-squares equations can always be set up. These will not, however, be linear, and an iterative method of solution will be necessary. Multivariate curvilinear relationships introduce still further complexity.

In view of the computing facilities now generally available, considerable developments on this topic can be expected in the next few years.

Nonlinear estimation

8.93 Expressions to be fitted by regression analysis may be regarded either as functions of the independent variables, x_i, or of the unknown parameters, β_i. The methods described so far can be applied to any expression which is a linear function of the unknown parameters, regardless of its form as a function of the independent variables. The problem of Nonlinear Estimation arises when the expression is a nonlinear function of the unknown parameters; the alternative term "nonlinear regression" should be avoided, as it can mean either nonlinear estimation or the fitting of a function which is linear in the unknown parameters, but other than a straight line or plane. In nonlinear estimation the least-squares equations are no longer linear in terms of the unknown parameters, and alternative techniques for their solution are required. The available methods will be discussed under the following headings:

(i) Transformations to linear forms
(ii) Approximate methods

(iii) Tabulations for standard forms

(iv) Iterative methods for exact solutions

The technique selected in any particular case will largely depend on whether or not an approximate solution is acceptable.

Transformations to linear forms

8.931 The idea of transforming the regression equation to give a linear form has been introduced in § **8.51**. This method is appropriate if the inherent assumption about the distribution of errors can be justified.

Consider, for example, the estimation of the rate constant for the first-order chemical reaction:

$$B \xrightarrow{\beta_1} A$$

Suppose a number of experiments are performed which yield a set of independent observations, y_i, of the concentration of B for various reaction times, x_i, using a constant initial concentration β_0. The equation to be fitted is:

$$Y = \beta_0 e^{-\beta_1 x} \qquad (8.91)$$

Whether or not the initial concentration β_0 is known makes little difference to the calculation. A logarithmic transformation gives:

$$\log_e Y = \log_e \beta_0 - \beta_1 x \qquad (8.92)$$

If it is assumed that errors in $\log_e y$ rather than y are distributed with zero mean and constant variance, then $\log_e \beta_0$ and β_1 may be fitted by the calculations of § **7.22**. This is equivalent to the assumption that the error standard deviation of the concentration measurement is proportional to the level of concentration, i.e. the coefficient of variation of y is independent of x. This may be a perfectly valid assumption and is often more justifiable than that of supposing a constant standard deviation independent of concentration level.

A transformation can be combined with a technique for successive approximation to provide estimates in equations such as:

$$Y = \beta_0 + \beta_1 e^{-\beta_2 x} \qquad (8.93)$$

An initial guess, $\beta_0{}^0$, is used for the parameter β_0 to enable

$$\log_e (Y - \beta_0{}^0) = \log_e \beta_1 - \beta_2 x \qquad (8.94)$$

to be fitted. The values of the three parameters can then be modified iteratively so as to minimize:

$$S.S. = \sum_{i=1}^{n} (y_i - \beta_0 - \beta_1 e^{-\beta_2 x_i})^2$$

Approximate methods

8.932 If the number of sets of observations is equal to the number of unknown parameters the regression curve will pass exactly through all the data points. The parameters can thus be estimated by solving a set of simultaneous non-linear equations, the values at the data points being substituted to produce one equation per point. Sometimes these equations can be solved easily. Hence a technique for handling the general case, in which there are more observations than parameters, is to average neighbouring data points to reduce the number of observation-equivalents to the number of parameters. If the curve computed from the reduced set of data appears to be a close approximation to the curve drawn through the full data, then the approximation can be expected to be adequate for many purposes.

A useful curve in the field of long-term forecasting is the simple modified exponential function:

$$Y = \beta_0 - \beta_1 \beta_2{}^x \tag{8.95}$$

Clearly three simultaneous equations of this form can easily be solved for β_0, β_1 and β_2 by successive elimination. When the observations are sales data for successive periods the division into sets for averaging is also easy and the approximate estimates of the parameters are quickly derived. An equation of the form (8.95) is also used in fertiliser trials, where it is known as the Mitscherlich Equation.

When the number of data points is only marginally greater than the number of parameters, it is possible to obtain simple functions of the observations which provide estimates only slightly inferior to those calculated exactly. For the modified exponential curve, a good approximation for β_2 from four observations at $x = 0, 1, 2$ and 3 is the rational function [8.8]:

$$\beta_2 = (4y_3 + y_2 - 5y_1)/(4y_2 + y_1 - 5y_0) \tag{8.96}$$

Estimates of β_0 and β_1 can subsequently be found by linear regression analysis.

Tabulations

8.933 The Mitscherlich equation is of such frequent application in agricultural trials that the solution of the estimation problem for equally-spaced data has been tabulated in a form where interpolation can be used to derive estimates very rapidly. The problem is reduced to solving for β_2 the equation:

$$y_1 J_{n,1} + y_2 J_{n,2} + \dots + y_n J_{n,n} = 0 \tag{8.97}$$

where n is the number of data points and the functions J, known as Gomes J functions, are tabulated for n and β_2 [8.9].

Tabulation of solutions for data which are not equally spaced or for other models is only worthwhile if the equation and the experimental design are in frequent use.

Iterative methods for exact solutions

8.934 The general technique for deriving least-squares estimates for non-linear functions is to minimize the sum of squared residuals by successive improvements to initial guesses. This technique is required for exact solutions for curves which can be fitted approximately by the methods described above and for equations for which approximate solutions are not readily obtained. For example, two typical equations arising in chemical kinetic studies are:

$$Y = \beta_0\beta_1 x_1/(1 + \beta_1 x_1 + \beta_2 x_2) \qquad (8.98)$$

and

$$Y = \beta_0 x_1 x_2/(1 + \beta_1 x_2 + \beta_2 x_3)(1 + \beta_3 x_3) \qquad (8.99)$$

The estimation of the parameters in these equations is a particular case of the numerical optimization or "hill climbing" problem and as such is in the field of numerical analysis rather than statistics. The use of an electronic computer is required for all but the most trivial problems; a program using an algorithm specifically for the least-squares case will have advantages over general hill-climbing programs. For further information the reader is referred to [8.6].

Experimental design in nonlinear situations

8.935 As with other estimation problems, nonlinear estimation methods benefit considerably from well-designed experiments. Unfortunately, the designs which are recommended for linear situations are not the most informative in nonlinear situations. Furthermore, the choice of the most informative experiments in the nonlinear case depends not only on the form of the regression curve, but also on the values of certain of the unknown parameters, which, at the time of design, can only be estimated roughly [8.10]. Again, for all but relatively simple problems, the use of a computer and an optimization program are required at the design stage.

References

[8.1] WOOLF, B. "Computation and Interpretation of Multiple Regressions", *J. R. Statist. Soc.* B, **13**, 100-119 (1951).

[8.2] BOX, G. E. P. "Multi-factor Designs of First Order", *Biometrika*, **39**, 49-57 (1952).

[8.3] NEWTON, R. G., and SPURRELL, D. J. "A Development of Multiple Regression for the Analysis of Routine Data", *Applied Statistics*, **16**, 51–64 (1967).

[8.4] NEWTON, R. G., and SPURRELL, D. J. "Examples of the Use of Elements for Clarifying Regression Analysis", *Applied Statistics*, **16**, 165–172 (1967).

[8.5] DRAPER, N. R., and SMITH, H. *Applied Regression Analysis* (second edition). John Wiley (New York, 1981).

[8.6] BOX, M. J., DAVIES, D., and SWANN, W. H. *Non-Linear Optimization Techniques*. I.C.I. Monograph No. 5. Institute of Manpower Studies (Brighton, 1981).

[8.7] QUENOUILLE, M. H. *Associated Measurements*. Butterworth (London, 1952).

[8.8] PATTERSON, H. D. "A Simple Method for Fitting an Asymptotic Regression Curve", *Biometrics*, **12**, 323–329 (1956).

[8.9] GOMES, F. P. "The Use of Mitscherlich's Regression Law in the Analysis of Experiments with Fertilisers", *Biometrics*, **9**, 498–516 (1953).

[8.10] BOX, G. E. P., and LUCAS, H. L. "Design of Experiments in Nonlinear Situations", *Biometrika*, **46**, 77–90 (1959).

Chapter 9

Frequency Data and Contingency Tables

In the majority of examples discussed in this book the data have consisted of observations that, apart from the rounding off, were measured on a continuous scale. The present chapter is concerned with data that occur as frequencies—the numbers of times certain events have happened, or counts of objects—and are therefore bound to consist of integers 0, 1, 2, 3, etc.

DISCRETE DISTRIBUTIONS

Binomial distribution

9.1 Since the Binomial Distribution was discovered through the study of games of chance, it will be appropriate to introduce it by considering the following problem: "Five 'poker' dice are thrown: what is the probability that the number of aces will be five, four, three, two, one, or zero?"

The probability that the number of aces is five is the product of the probabilities that each die shows an ace, i.e.

$$\tfrac{1}{6} \times \tfrac{1}{6} \times \tfrac{1}{6} \times \tfrac{1}{6} \times \tfrac{1}{6} = (\tfrac{1}{6})^5$$

provided that the dice are unbiased and that the events are *independent*, i.e. that the way one die falls does not influence the way another die falls.

The probability that four *specified* dice show aces and the remaining one "not-ace" is:

$$\tfrac{1}{6} \times \tfrac{1}{6} \times \tfrac{1}{6} \times \tfrac{1}{6} \times \tfrac{5}{6} = (\tfrac{1}{6})^4(\tfrac{5}{6})$$

The die which does not show the ace may, however, be any one of the five dice, so that the probability that the number of aces is four is $5(\tfrac{1}{6})^4(\tfrac{5}{6})$.

The probability that three specified dice show aces and the remaining two not-ace is $(\tfrac{1}{6})^3(\tfrac{5}{6})^2$. But the three showing aces may be any three out of the five, and the number of ways of choosing three out of five is:

$$\frac{5!}{2! \ 3!} = \frac{5.4.3.2.1}{2.1.3.2.1} = 10$$

Hence the probability of three aces is $10(\tfrac{1}{6})^3(\tfrac{5}{6})^2$.

The remaining probabilities may be calculated in the same way (Table 9.1).

TABLE 9.1.   THROWS OF FIVE POKER DICE

Number of Aces	Probability	Numerical Value of Probability
5	$(\frac{1}{6})^5$	0·00013
4	$5(\frac{1}{6})^4(\frac{5}{6})$	0·00321
3	$10(\frac{1}{6})^3(\frac{5}{6})^2$	0·03215
2	$10(\frac{1}{6})^2(\frac{5}{6})^3$	0·16075
1	$5(\frac{1}{6})(\frac{5}{6})^4$	0·40188
0	$(\frac{5}{6})^5$	0·40188
Total		1·00000

The probabilities are in fact the successive terms of the binomial expansion of $(\frac{1}{6}+\frac{5}{6})^5$.

In the general case, if

$$p = \text{probability that an event will happen}$$

and $\qquad q = 1-p = \text{probability that it will not happen}$

then the probability that it will happen a times, and fail to happen b times in n trials, where $n = a+b$, is:

$$P = \frac{n!}{a!\,b!}\,p^a q^b \qquad\qquad (9.1)$$

which is the general term of the binomial expansion of

$$(p+q)^n$$

The distribution defined by such a sequence of probabilities is accordingly known as the Binomial Distribution. Numerical values of P are listed in Table 37 of [9.1] for $n = 5(5)30$ and $p = 0.01, 0.02(0.02)0.1(0.1)0.5$.

Mean and variance of the binomial distribution

9.11 If the distribution of a is given by the terms of the binomial expansion of $(p+q)^n$, the mean or expected value of a is given by:

$$\mu = np \qquad\qquad (9.21)$$

This can readily be seen from the example of the dice. Since in a very large number of throws the number of aces must average one-sixth of the number of dice thrown, the average number of aces *per set of five dice* must be $5 \times \frac{1}{6}$, i.e. np, where $n = 5$ and $p = \frac{1}{6}$.

It will be noticed that although a is always a whole number, its mean value in a very long series of trials, i.e. its expected value, is not necessarily a whole number.

The variance associated with a can be determined by applying Formula (3.94) to the sum of the outcomes of the n individual trials. In a single trial the event variate assumes the values 1 and 0 with probabilities p and q respectively, and the expected value is p. The deviations from the mean value are thus:

$$(1-p) \text{ with probability } p$$

$$\text{and } p \text{ with probability } q$$

Since variance is by definition the average of the squared deviations from the mean, it follows that the variance associated with the result of a single trial is given by:

$$p(1-p)^2 + qp^2 = pq(q+p) = pq$$

The total number, a, of occurrences of the event is n trials in the sum of the numbers in the individual trials, and when these trials are made independently the variance associated with a is the sum of the variances for the separate trials, i.e.

$$\sigma^2 = npq \qquad (9.22)$$

This is also the variance associated with b.

The binomial distribution differs from the Normal distribution, not only because it is discontinuous, but also because, unless n is large, it can be very markedly asymmetrical when either p or q is small. Nevertheless, provided n is large and neither p nor q is small, that is provided both a and b are large, sufficiently accurate tests of significance can be made by treating a as though it were a Normal variate with mean and variance given by the above formulae. These tests, based on "large sample" theory, are illustrated in the succeeding section, and are followed by the exact tests suitable for "small" samples.

Example 9.1. Factory accident rates

9.12 The data in Table 9.2 opposite show the number of lost-time accidents to women and men in an I.C.I. factory over a period of many years. To tell whether one sex is more prone to accidents than the other, it would be natural to express these figures as rates per hundred thousand "man-hours" worked. However, if the hours of work a day and number of days worked a year are taken to be the same for the two sexes, and if there have been no violent fluctuations in the numbers employed, virtually the same conclusions will be drawn if we calculate the accident rate per hundred persons. The number employed will conveniently be taken as an average over the period of observation, say the average of the numbers on the payroll at the beginning of each year of the period.

The rate for men is apparently higher, but it is necessary to apply a statistical test to decide whether the difference between the two rates really is greater than can reasonably be attributed to chance.

The proportion of women employed is:

$$p = 246/824 = 0.2985$$

and of men:

$$q = 578/824 = 0.7015$$

$$\overline{1.0000}$$

If men and women were equally prone to accident, the *expected number* of accidents to women would be:

$$np = 175 \times 0.2985 = 52.24$$

and to men:

$$nq = 175 \times 0.7015 = 122.76$$

$$\overline{175.00}$$

TABLE 9.2. ACCIDENTS IN AN ICI FACTORY

	Number of Accidents	Average Number Employed	Rate per Hundred
Women	$a = 48$	246	19·5
Men	$b = 127$	578	22·0
Total	$n = 175$	824	21·2

The expected number is, of course, a mathematical concept, since there cannot be 0·24 of an accident. With a total of 175 it is clear that the numbers of accidents to women and men respectively could not be more strictly proportional to the average numbers employed than 52:123. A small variation either way can be attributed to chance, and it only remains to decide whether the actual numbers, 48 and 127, deviate from the expected numbers more than would often happen by chance.

The variance of the binomial distribution is:

$$npq = 175 \times 0.2985 \times 0.7015 = 36.65$$

The standard deviation is the square root of the variance, viz. 6·05. Now the observed deviation from expectation is:

$$127 - 122.76 = 4.24$$

(The deviations in the two classes are, of course, equal in magnitude but opposite in sign). Hence the ratio of observed deviation to standard deviation is:

$$u = 4.24/6.05 = 0.70$$

Provided the numbers in either class are not too small, the binomial distribution approximates to the Normal distribution, and the above ratio may therefore be referred to a table of the Normal distribution [Table A] to judge its significance. More briefly, it may be considered significant if it exceeds 2, this being the two-sided Normal deviate corresponding to a level of significance of approximately 0·05. It is obvious that, in the present example, the deviation is not significant: the probability of encountering a deviation from expectation (in either direction) larger than 0·70 times the standard deviation is about 0·48, a probability so large that there is no reason, *on the evidence of these figures alone,* for claiming that the rates are different for the two sexes.

An alternative approach

9.121 The following method, although mathematically equivalent and necessarily leading to the same conclusion, provides an alternative way of looking at the same problem.

The proportion of women employed is, as we have already seen:

$$\mathrm{p} = 0·2985$$

The proportion of accidents which happen to women is:

$$p = 48/175 = 0·2743$$

If men and women are equally prone to accident, these two proportions should be the same, and the question resolves itself into: "Is p significantly different from p?"

As already stated, the variance of a or b in a binomial distribution is $n\mathrm{p}\mathrm{q}$, from which it follows that the variance of p ($= a/n$) or of q ($= b/n$) is:

$$\mathrm{p}\mathrm{q}/n = (0·2985)(0·7015)/175 = 0·0011966$$

Hence the standard deviation is 0·0346.

But
$$|p - \mathrm{p}| = |0·2743 - 0·2985| = 0·0242$$

Thus the ratio of the deviation of p to its standard deviation is:

$$u = 0·0242/0·0346 = 0·70$$

This is the same ratio as was obtained before and leads to the same assessment of the significance.

Precision of the estimate of a proportion

9.13 In some problems we are concerned, not with testing whether an observed proportion differs significantly from a hypothetical value p, but merely in estimating p and determining the precision of our estimate. The maximum likelihood estimate [Appendix 4B] of p is the value which maximizes the Expression (*9.1*) given that the event has occurred a times and failed to happen b times. This value of p will also maximize the logarithm of P, i.e.

$$\log P = \text{const.} + a \log p + b \log (1 - p)$$

and can be determined from the solution of the equation:

$$\frac{d \log P}{dp} = a/p - b/(1 - p) = 0$$

Hence the observed proportion in the sample, $p = a/(a+b)$, is the maximum likelihood estimate of p, the true proportion in the Population.

We do not know the values of p and q to insert in the formula for the variance, pq/n, of the proportion. However, we shall not be seriously in error if we use the *estimates* p and q for insertion in the formula, i.e. if we take the variance to be pq/n.

Let us suppose (in order to work with the same figures as before) that a Public Opinion Survey has shown that 48 persons in a random sample of 175 persons answered "Yes" to a certain question. Then the estimate of the proportion in the Population who would answer "Yes" is:

$$p = 48/175 = 0.2743$$

The estimated variance of p is accordingly:

$$pq/n = (0.2743)(0.7257)/175 = 0.0011375$$

Estimated standard deviation $= 0.0337$

The estimated standard deviation can be used for setting limits to p. Thus the 95% confidence limits are:

$$0.2743 \pm 1.96 \times 0.0337 = 0.208 \text{ and } 0.340$$

Hence it can be asserted, with 95% confidence, that the proportion in the Population lies between 21% and 34%.

Continuity correction

9.14 The methods described above depend on substituting for the discontinuous binomial distribution a continuous Normal distribution with the same mean and variance. This approximation can be improved by the application of a Continuity Correction.

For the given number of employees in the accident data, the only possible deviations from expectation for the number of accidents to men are:

$$\ldots -0.76, \ 0.24, \ 1.24, \ 2.24, \ 3.24, \ 4.24, \ 5.24, \ 6.24 \ldots$$

What we are doing is to replace a series of discrete values by a continuous distribution. On this convention a number like 4.24 really stands for a continuum of numbers from $(4.24 - 0.5)$ to $(4.24 + 0.5)$.

Since the probability, P, which measures significance, is supposed to tell us the probability of getting a deviation *as large as* or larger than the one observed, we expect to get a closer approximation if, in calculating the corresponding standard Normal deviate, we use a deviation just half a unit smaller than the one observed. Thus:

$$u = (4·24 - 0·5)/6·05 = 3·74/6·05 = 0·62$$

The probability of encountering a larger deviation from expectation (in either direction) is now obtained from Table A as 0·54.

Exact method

9.15 If the number of occurrences, a or b, is less than 15, the Normal approximation to the binomial distribution will be unsatisfactory. In Quality Control work the number of defective articles in a sample is often as small as two or three, and in such cases the mechanical application of the above methods will lead to grossly misleading conclusions.

An exact test of significance is always provided by the sum of the appropriate number of terms of the binomial distribution, as the following example will illustrate.

9.151 A person claiming "clairvoyant" powers offers to predict the fall of a die. In five successive attempts he is right three times. How strong is the evidence in favour of his claim? The Null Hypothesis to be tested is that his "predictions" are pure guesses. On this hypothesis, the probability of guessing right is exactly $\frac{1}{6}$. The probabilities of any number of correct guesses from 0 to 5 in five attempts are given by the terms of the expansion of $(\frac{1}{6} + \frac{5}{6})^5$, and in fact have already been worked out in Table 9.1. If any result is to be judged significant, then so must any greater number of correct guesses. The significance of his result is therefore judged by the probability of guessing right three, four, or five times in the five attempts, i.e.

$$P = 0·03215 + 0·00321 + 0·00013 = 0·03549$$

The probability of getting by chance a result as good as or better than he actually achieved is approximately 0·035 and according to the usual convention is sufficiently small for the result to be judged significant. However, in view of the extravagance of the claim, perhaps a much *higher* level of significance (i.e. a much smaller value of P) should be demanded.

Example 9.2

9.152 Let us now consider a more practical problem. A sample of 25 articles is drawn at random from a bulk or from a day's production, etc., and three of them are found to be defective (or underweight, or possessing some other relevant characteristic). We wish to draw conclusions about the proportion of defective articles in the bulk. In this and the following example we assume that the sample is only a small fraction of the bulk (lot or consignment) being sampled, certainly not more than a fifth. In other words, we must reasonably be able to treat the bulk as an infinite Population in comparison with the sample. The maximum likelihood estimate of proportion defective is:

$$p = 3/25 = 0.12$$

If the true proportion defective is p, the probability of encountering 3 *or fewer* defectives in a sample of 25 is the sum of the last four terms of the binomial expansion of $(p+q)^{25}$, i.e.

$$P = \frac{25!}{3!\,22!}\,p^3q^{22} + \frac{25!}{2!\,23!}\,p^2q^{23} + 25pq^{24} + q^{25}$$

For any chosen level of significance, say $P = 0.025$, it is possible to solve this equation and find a value of p such that the 3 defectives in 25 are just on the borderline of being significantly *too few*. For any larger value of p the result would be significant, and for any smaller value not significant. In other words, the value of p given by this equation sets an upper limit beyond which one can state, with reasonable confidence, that the true value of p does not lie. Similarly the probability of encountering 3 *or more* defectives in a sample of 25 is given by the sum of the first twenty-three terms of the same binomial expansion:

$$P = p^{25} + 25p^{24}q + \frac{25!}{23!\,2!}\,p^{23}q^2 + \ldots + \frac{25!}{3!\,22!}\,p^3q^{22}$$

Putting $P = 0.025$ we can solve this equation to find a value of p such that 3 defectives in 25 are just on the borderline of being significantly *too many*. This value of p sets a lower limit below which one can state, with reasonable confidence, that the true value of p does not lie.

The numerical solution of these equations is clearly tedious and Table F (Confidence Limits for mean μ of the Binomial and Poisson Distributions) has therefore been prepared. The table is used when either a or b is less than 15, and the smaller of the two numbers is always denoted by a. It is entered for a and for p, the observed proportion, interpolating if necessary with respect to p. Thus to find the 95% confidence limits we enter with $a = 3$ and $P = 0.025$ and find:

p	Lower	Upper
0.2	0.650	7.21
0.1	0.634	7.96

Simple linear interpolation between $p = 0.1$ and $p = 0.2$ gives the limits corresponding to $p = 0.12$, the observed proportion defective:

p	Lower	Upper
0·12	0·637	7·81

These are the limits of the *expected* number defective (i.e. of np) in a sample of twenty-five, and the limits of p are accordingly:

$$0.637/25 = 2.55\%$$
$$7.81/25 = 31.2\%$$

Hence it can be stated that the percentage defective in the bulk from which the sample was drawn is not less than 2·55% nor greater than 31·2%. The extraordinarily low precision of determination of p will be noted. Evidently a much larger sample than twenty-five would be necessary to obtain useful information about p. This point is often overlooked by those responsible for drafting sampling clauses in specifications.

Example 9.3

9.153 A manufacturer has to meet a specification which lays down that 200 articles shall be drawn at random from the lot, or consignment, and that the lot shall be rejected if there are more than 3 defective articles in the sample. He wants to be sure that not more than 10% of the lots he submits shall be rejected. Assuming that he can manufacture under controlled conditions, to what level must he reduce his Process Average (= average percentage defective)?

A lot will be rejected if $a = 4$ or more. The observed proportion defective when $a = 4$ and $n = 200$ is $p = 4/200 = 0.02$. The lower confidence limit for $\mu = np$ is found from the table against $a = 4$, under $P = 0.1$. It is hardly necessary to interpolate for p: when the observed proportion is very small it will be sufficient to read from the bottom line corresponding to $p = 0$, giving the value 1·74. Hence the process average must not be permitted to exceed $p = 1.74/200 = 0.87\%$ defective.

The protection accorded to the consumer by the specification is indicated by the upper $P = 0.1$ limit for $a = 3, p = 0$; this gives a value of 6·68. Hence even if the percentage defective in a lot were as high as 3·34%, the consumer would still run a one-in-ten risk of accepting it.

Poisson distribution

9.2 When the n of the binomial distribution is very large and the p very small, the terms of the binomial, from the last backwards, tend to the values:

$$e^{-\mu}, \ e^{-\mu}\mu, \ e^{-\mu}\mu^2/2!, \ e^{-\mu}\mu^3/3!, \ e^{-\mu}\mu^4/4!, \ldots$$

the general term being

$$e^{-\mu}\mu^a/a! \qquad\qquad (9.3)$$

where $\mu = np,$

e = base of natural logarithms.

These are the probabilities that the event will happen respectively 0, 1, 2, 3, 4, ... times, and the general term is the probability that it will happen a times. This distribution is known as the Poisson Distribution.

The mean and variance of the Poisson distribution can be found from those of the binomial distribution by putting $np = \mu$ and letting p tend to zero and q to unity:

$$\text{Mean} \quad = np \qquad = \mu$$

$$\text{Variance} = npq \to np = \mu$$

The parameter μ therefore represents both the mean and the variance of the distribution.

The Poisson distribution may be considered as a convenient approximation to the binomial distribution when n is very large compared with a. We were in effect using the Poisson distribution in Example 9.3, when we read the limits of expectation appropriate to $p = 0$. It is also the distribution which will be followed if the events are independent and there is no absolute upper limit to a, or at least if the upper limit is obviously very large though not exactly known. The following example will illustrate this.

Example 9.4. Density of particles in a dusty gas

9.21 In using an ultramicroscope to count the number of suspended particles in a dusty gas, it is obviously impossible to count more than a few at a time, because they are in motion. The device is adopted of illuminating the field of the microscope by a flash of light and adjusting the volume of the field so that only a very few particles are seen simultaneously. By making a sufficiently large number of counts, in successive flashes of light, it is possible to determine the average number in the field, and hence the number per cubic centimetre, to any desired degree of accuracy.

The data in Table 9.3 show the number of occasions on which the number of visible particles equalled 0, 1, 2, 3, ... out of a total of 143 occasions.*

It can be understood why the conditions of the experiment are such as to make it likely that the distribution will follow the Poisson law. The total number n of particles in the gas is many millions. The probability p that any particular particle will be in the field at a given instant is exceedingly small, but with so many particles available we shall expect to see a few. The total number of particles seen is 205, and the mean number per occasion is therefore $205/143 = 1\cdot4336$.

* Data kindly supplied by Mr. E. H. M. Badger, Gas Light and Coke Company.

Using this as our estimate of μ, we find that the probabilities that the field will contain 0, 1, 2, 3, ... particles are given by:

$$e^{-1\cdot4336} = 0\cdot23845$$

$$e^{-1\cdot4336}(1\cdot4336) = 0\cdot34184$$

$$e^{-1\cdot4336}(1\cdot4336)^2/2 = 0\cdot24503, \text{ etc.}$$

The expected number of occasions (out of 143) on which the field should contain 0, 1, 2, 3, ... particles is found by multiplying these probabilities by 143, and is given in the last column of Table 9.3. It is seen that the observed frequencies follow the expectations given by the Poisson law rather closely.

TABLE 9.3.   PARTICLE COUNTS IN A DUSTY GAS

Number of Particles Seen in Field	Number of Occasions	Total Number of Particles Seen	Expected Number of Occasions
0	34	0	34·1
1	46	46	48·9
2	38	76	35·0
3	19	57	16·7
4	4	16	6·0
5	2	10	1·7
$\geqslant 6$	0	0	0·5
Total	143	205	142·9

The variance of the number of particles seen on one occasion is μ. The variance of the mean number on 143 occasions is therefore $\mu/143$. To obtain an estimate of the variance we insert the estimate of μ, giving:

$$1\cdot4336/143 = 0\cdot010025$$

The standard deviation of the estimate of μ is therefore $\sqrt{0\cdot010025} = 0\cdot1001$.

To find the number of particles per cubic centimetre, we must divide by the volume of the illuminated field. The standard deviation of the number of particles per cubic centimetre will of course be found by dividing the above standard deviation by the same factor. Alternatively we may note that the coefficient of variation is $100 \times 0\cdot1001/1\cdot4336 = 7\cdot0\%$. Using 95% confidence limits, we may say that it is reasonably certain that the result will be correct to within $\pm14\%$.

9.211 The variance of the distribution could also be estimated from the sum of the squares of the observations according to the method of § **3.34**. This estimate of variance is found to be $1\cdot3318$. This is a little smaller than the mean ($1\cdot4336$), but the difference is not significant [9.2]. An estimated variance significantly too small might be due to a tendency to miss some particles when the number seen is large: an excessive estimated variance would usually indicate lack of uniformity in the dust-gas mixture. In either case the experimental technique is at fault. When the technique is satisfactory, it is better to use the mean as the estimate of variance than to estimate the latter from the sum of squares.

Negative binomial distribution

9.3 In a binomial distribution the variance, npq, of the number of occurrences of an event is less than the mean, np. For events which follow a Poisson law, these parameters are both equal to $\mu = np$. However, in some practical situations the Population variance is larger than the mean.

As an example, consider the frequency of faults observed in pieces of textile fabric of uniform size. Such fabrics may be manufactured from many hundreds of bobbins of yarn, which (for reasons associated with the method of production of the yarn) may have different fault rates. Counts of the numbers of faults per piece of fabric will then reflect not only the variations due to the random occurrence of faults in each bobbin but also the fluctuation in fault rate from one bobbin to another. In this situation the variance of the numbers of faults per piece exceeds the mean, and the difference $(\sigma^2 - \mu)$ may be regarded as arising from fluctuations in the average fault rate between bobbins.

Several theoretical distributions have been derived with this property, each relating to a particular type of mechanism producing the events. Probably the most useful of these is the Negative Binomial Distribution, which is characterised by two positive parameters, c and k. (In the example, if the average bobbin fault rates are distributed in the Gamma form, then the distribution of faults per piece is negative binomial [9.3].)

The mean and variance of the negative binomial distribution are given in terms of c and k by:

$$\mu = k/c \qquad\qquad (9.41)$$
$$\sigma^2 = k/c + k/c^2 \qquad\qquad (9.42)$$

Noting that $\sigma^2 = \mu(1 + 1/c)$, the parameters of the distribution can be estimated from the observed mean, $\bar{x}$, and variance, s^2, of a sample by the formulae:

$$c = \frac{\bar{x}}{s^2 - \bar{x}}; \quad k = c\bar{x} \qquad\qquad (9.5)$$

(Maximum likelihood estimation does not lead to simple formulae in this case.)

It is also convenient to introduce the related parameters:

$$p = \frac{c}{c+1}; \quad q = 1 - p = \frac{1}{c+1} \qquad\qquad (9.6)$$

Then the probabilities of 0, 1, 2, etc. events are given by the terms of the expansion of $p^k(1-q)^{-k}$, namely:

$$p^k \left\{ 1, \quad kq, \quad \frac{k(k+1)}{2!} q^2, \quad \frac{k(k+1)(k+2)}{3!} q^3, \text{etc.} \right\} \qquad (9.7)$$

Thus the distribution is generated by a binomial with a *negative* index. It tends to the Poisson law as c and $k \to \infty$ whilst maintaining a finite k/c ratio.

The negative binomial distribution can be derived from several theoretical bases, and it is thus applicable to a variety of problems, in addition to the fabric faults example described above [9.4]. A classic case is concerned with the application of some "treatment" to a finite set of items. If there is a probability p that any individual will be "affected" at any one application and that after being affected k times the individual is removed from the set, then the Formulae (9.7) describe the proportions of items withdrawn at the kth, $(k+1)$th, etc. applications of the treatment. This model is obviously applicable to the recurring attacks of a disease, and to bactericide and insecticide situations. It can also describe the pattern of failure of components where there is a risk p that a component will receive some damage, wear, etc., from a single cycle of use, and that such damage or wear can be repaired $k-1$ times. The exposure of consumers to advertising is also amenable to this kind of model.

Example 9.5. Fabric fault rates

9.31 To illustrate the fitting of a negative binomial distribution, Table 9.4

TABLE 9.4.   FREQUENCIES OF FAULTS IN TEXTILE FABRIC

Number of Faults	Number of Occasions	Total Number of Faults	Expected Number of Occasions	
			Negative Binomial	Poisson Law
0	51	0	49·2	36·5
1	68	68	71·7	72·9
2	61	122	61·9	72·8
3	44	132	41·2	48·4
4	24	96	23·4	24·2
5	15	75	11·9	9·7
6	3	18	5·6	3·2
7	1	7	2·4	0·9
8	0	0	1·0	0·2
9	1	9	0·4	0·1
$\geqslant 10$	1	10	0·3	0·0
Total	269	537	269·0	268·9

lists the number of occasions on which the number of faults observed in 100 sq. yd. pieces of textile fabric equalled 0, 1, 2, 3, ... out of a total of 269 occasions.

The mean and variance of the observed numbers of faults are 1·996 and 2·735 respectively. Hence from the Equations (9.5) and (9.6), the estimated parameters of an underlying negative binomial distribution are:

$$c = 1·996/(2·735 - 1·996) = 2·702$$

$$k = 2·702 \times 1·996 = 5·394$$

$$p = 2·702/3·702 = 0·730$$

$$q = 1 - 0·730 = 0·270$$

The expected number of occasions each number of faults will occur can then be calculated from (9.7). It is seen from the table that they agree well with the observed numbers. In contrast, a single-parameter Poisson distribution with $\mu = 1·996$ has too few occurrences of zero faults and too many of 2 faults. Thus a Poisson distribution does not describe this fabric fault data adequately.

CHI-SQUARED STATISTIC

The χ^2 distribution

9.4 In the succeeding examples use will be made of a criterion which has not yet been mentioned specifically. It will now be defined.

Let a sample of ϕ quantities, $u_1, u_2, ..., u_\phi$, be drawn at random from a Normal Population with zero mean and unit standard deviation. Then χ^2 (chi-squared) is their sum of squares:

$$\chi^2 = u_1^2 + u_2^2 + ... + u_\phi^2$$

If we continue drawing samples, each of ϕ quantities, from the same Population, we can build up the distribution of the resulting values of χ^2 in the form of a histogram. As the number of samples tends to infinity, the histogram tends to a smooth distribution, which is called the χ^2 distribution. Clearly the form of the distribution depends on ϕ, which is referred to as the Number of Degrees of Freedom.

Numerical values of χ^2 are listed in Table B corresponding to various probabilities for values of ϕ from 1 to 30. An approximate formula is provided for use when ϕ exceeds 30. Thus for $\phi = 8$ and level of significance $P = 0·01$ the table gives $\chi^2 = 20·1$. This means that the probability that a χ^2 variate with 8 degrees of freedom will exceed 20·1 is just one in a hundred.

When there is only one degree of freedom, χ^2 is simply the square of a Normal deviate with mean zero and standard deviation unity. Thus the values of χ^2 in the top line of the table are simply the squares of the Normal deviates for the same levels of significance:

$$\chi^2 = 1{\cdot}96^2 = 3{\cdot}84 \text{ for } P = 0{\cdot}05, \text{ etc.}$$

(where P is double-sided for the Normal deviate).

Two useful properties of the χ^2 distribution follow directly from its definition:

(i) The sum of a χ^2 variate with ϕ_1 degrees of freedom and a χ^2 variate with ϕ_2 degrees of freedom is a χ^2 variate with $\phi_1 + \phi_2$ degrees of freedom.

(ii) The average value of a χ^2 variate with ϕ degrees of freedom is simply ϕ.

Now let us see how the chi-squared statistic can be used to draw conclusions from sets of discrete data.

Single classification: More than two classes

Example 9.6. Piston-ring failures

9.51 The data below show the number of times piston-rings failed on each

TABLE 9.5.   FAILURE OF PISTON-RINGS ON STEAM-DRIVEN COMPRESSORS

Compressor No.	1	2	3	4	Total
Failures	46	33	38	49	166

of four steam-driven compressors during a period of some years. Assuming, as was approximately the case, that the compressors were used equally, we shall want to know if we can conclude that there are real differences between them in respect of piston-ring failure—whether, for example, the fact that compressor No. 2 had the smallest number of failures really means that there is something about this machine which reduces the probability of failure, or whether, on the contrary, the differences between these numbers can reasonably be ascribed to chance.

If all four compressors are identical and subject to identical treatment, the probability that the next failure will be on No. 1 (or on any one of the four compressors) is $p = 0{\cdot}25$. The expected number of failures on No. 1 compressor out of a total of $n = 166$ is therefore $np = 166 \times 0{\cdot}25 = 41{\cdot}50$, and of course this is also the expected number of failures on any other compressor.

Here, as before, the "expected number" is a mathematical concept—there cannot be $0{\cdot}5$ of a failure. The total of 166 cannot be distributed more equally between the four machines than 42:42:41:41, in some order. The question to be answered is: "Do the data deviate from equal numbers sufficiently to cause us to reject the hypothesis that the machines are behaving identically, and to admit the alternative hypothesis that there are real differences?"

The statistic used to measure and to test the significance of the deviations from expectation is:

$$\chi'^2 = \sum \frac{(\text{Deviation})^2}{\text{Expectation}} \tag{9.8}$$

The computation of χ'^2 may be set out as in Table 9.51.

TABLE 9.51.   CALCULATION OF χ'^2: EQUAL EXPECTATIONS

Compressor No.	1	2	3	4	Total
Failures observed	46	33	38	49	166
Failures expected	41·50	41·50	41·50	41·50	166·00
Deviation	+4·50	−8·50	−3·50	+7·50	0·00
$\dfrac{(\text{Deviation})^2}{\text{Expected value}}$	0·488	1·741	0·295	1·355	$\chi'^2 = 3·88$

Notice that if the deviations are recorded with sign, they should sum to zero.

It is apparent from Formula (9.8) that χ'^2 is a general measure of deviation from expected values, being large when the deviations are large. Moreover it can be shown that if the expectations are not too small in number (say not less than five) this quantity is distributed, to a very close approximation, like the χ^2 variate defined in § 9.4. The prime will therefore be dropped from χ'^2 in the subsequent exposition. To judge the significance, it is only necessary to refer to Table B and see if the calculated χ^2 is larger than the value listed for the 0·05 or 0·01 level of significance.

To enter the table one needs to know ϕ, the number of degrees of freedom. This is always the number of *independent* deviations. Although there are four deviations, the method of calculation implies that their total is identically zero. Hence there are only three independent deviations, since when any three are known the fourth can be determined. The number of degrees of freedom is accordingly three.

Entering the table with $\phi = 3$ and making a rough interpolation, we find that the significance of the calculated χ^2 is $P = 0·28$.

It appears that even if the probabilities of failure are equal in all four compressors, the chances of getting a value of $\chi^2 = 3·88$ or larger are 28 in 100. There is, then, no reason for believing that the compressors differ in respect of probability of failure, and there is no point in looking for explanations of the apparent differences.

Alternative method of calculating χ^2

9.511 A shorter method of calculating χ^2 when the expectations are equal is provided by the formula:

$$\chi^2 = \frac{\Sigma(\text{No. failures})^2}{\text{Expectation}} - n \qquad (9.81)$$

$$= \frac{46^2 + 33^2 + 38^2 + 49^2}{41 \cdot 50} - 166 = 3 \cdot 88$$

However, the other method has the distinct advantage of showing the contributions to χ^2 made by each class. The exceptional classes are then indicated, as will be shown more clearly in the next example.

Unequal expectations

Example 9.7. Factory accident rates in sections

9.52 In the more general case, the expectations in the different classes are unequal. In Table 9.52 is shown another classification of the 175 accidents in an ICI factory [Example 9·1]. The sections have been arranged in order of decreasing rate per hundred, although this step is not essential to the subsequent computation.

There appear to be large differences between the rates in different sections, but as the numbers in several sections are quite small it is necessary to apply a test of significance. The χ^2 statistic (9.8) provides the criterion.

TABLE 9.52. ACCIDENTS CLASSIFIED BY SECTIONS

Section	Average Number Employed	Number of Accidents	Rate per Hundred	Expected Number	Deviation	Contribution to χ^2
Samplers	21	11	52·38	4·46	+6·54	9·59**
By-products	22	10	45·45	4·67	+5·33	6·08*
Laggers	18	6	33·33	3·82	+2·18	1·24
Fitters	103	29	28·16	21·88	+7·12	2·32
Fitters (Labs.)	106	26	24·53	22·51	+3·49	0·54
Riggers	21	5	23·81	4·46	+0·54	0·07
Refinery	145	33	22·76	30·80	+2·20	0·16
Laboratories	66	14	21·21	14·02	−0·02	0·00
Conversion	153	29	18·95	32·49	−3·49	0·37
Elect. and Inst.	89	8	8·99	18·90	−10·90	6·29*
Day Gang	45	3	6·67	9·56	−6·56	4·50*
U.S. Plant	35	1	2·86	7·43	−6·43	5·56*
Total	824	175		175·00	0·00	36·72
General Rate			21·238			

The general rate for the whole works is $175/824 = 21\cdot238$ per hundred average employed. The expected number of accidents in any section (fifth column) is simply the average number employed multiplied by the general rate ($\div100$), e.g. for "Samplers" it is:

$$21 \times 0\cdot21238 = 4\cdot46, \text{ etc.}$$

In the sixth column are shown the deviations from expectation, e.g. for Samplers:

$$11 - 4\cdot46 = +6\cdot54$$

As a check, the sum of this column must be zero.

In the last column are shown the contributions to χ^2:

$$(\text{Deviation})^2/\text{Expectation} = 6\cdot54^2/4\cdot46 = 9\cdot59, \text{ etc.}$$

The total is $\chi^2 = 36\cdot72$.

The number of degrees of freedom, as in the previous example, is one less than the number of classes:

$$\phi = 12 - 1 = 11$$

Reference to Table B shows that the probability of encountering as large a value of χ^2 or larger due to chance is less than $0\cdot001$. The differences between rates are therefore almost certainly real.

In order to pick out the sections with rates which differ significantly from the general rate, we may regard each contribution as a χ^2 of one degree of freedom, to a first approximation. Clearly there is an error in this assumption, because if each contribution were a χ^2 of one degree of freedom, their sum would be a χ^2 of 12 degrees of freedom, although, as we have already noted, the sum has only 11 degrees of freedom. The effect of this assumption is therefore slightly to overestimate the significance; but this is not a serious matter, since the borderline cases can be checked by the exact method of § **9.15**.

The $0\cdot05$ and $0\cdot01$ levels of χ^2 with one degree of freedom are respectively $3\cdot84$ and $6\cdot63$. Individual contributions passing these levels are marked in the table with one and two asterisks respectively. Five sections show significant deviations, of which the Day Gang is the least significant.

If we exclude the five significant sections and calculate χ^2 for the remaining seven, by exactly the same process as was used for the whole twelve, we get:

$$\chi^2 = 3\cdot29, \quad \phi = 6$$

If we include Day Gang with these seven, we get, for testing the differences between eight sections:

$$\chi^2 = 8\cdot65, \quad \phi = 7$$

It was noted in § **9.4** that the expected value of χ^2 with ϕ degrees of freedom is ϕ. If Day Gang is excluded, the variation as measured by χ^2 is a little less than expected, whereas if Day Gang is included, the variation is a little more than expected, although the χ^2 still does not approach anywhere near significance. It is therefore doubtful whether Day Gang should be regarded as significantly different from the other seven non-significant sections. Of the remaining sections, it is reasonably safe to conclude that Samplers and By-products show significantly high rates and Electrical and Instrument and U.S. Plant show significantly low rates. Further investigation would therefore be directed to explaining the peculiarities of these four sections, and no time would be lost in looking for "causes" of the apparent differences between the other sections.

It will be observed that the most significant sections are not necessarily those which show greatest deviation from the general mean rate. A section rate may show a large deviation from the general rate and yet, being based on small numbers, may fail to reach significance. In the present example, Electrical and Instrument shows stronger evidence of departing from the general rate than Day Gang, although the latter has apparently the lower rate. This illustrates how the statistical investigation modifies and corrects general impressions.

It may also be noticed that χ^2 has been calculated in spite of the general rule which says that the approximation is not reliable when any expected value is less than five [§ **9.51**]. This rule need not be applied too rigidly: when there are many degrees of freedom it does not matter if a few of the expectations are a little below five; but if any doubt is felt, it is always possible to test individual comparisons by the exact method described in § **9.15**.

Alternative method of calculating χ^2

9.521 When the expectations in the classes are unequal, the value of χ^2 may also be found by a short method which is a generalisation of (*9.81*). Multiply the number of accidents by the corresponding rate in each class, and accumulate:

$$11 \times 52{\cdot}38 + \ldots + 1 \times 2{\cdot}86 = 4{,}496{\cdot}41$$

Divide by the general rate:

$$4{,}496{\cdot}41 \div 21{\cdot}238 = 211{\cdot}72$$

Subtract total accidents:　　　　　　　　175

Then　　　　　　　　　　　　　　$\chi^2 = 36{\cdot}72$, as before.

The disadvantage of this method is that it does not show the individual contributions to χ^2 and therefore fails to direct attention to the more significant classes.

Testing goodness of fit

9.53 The single classification chi-squared statistic can also be used to test the agreement between a histogram of observations and a fitted theoretical distribution. The aim is to decide whether the discrepancies between observed frequencies and those derived from the model are greater than could reasonably be attributed to chance fluctuations.

In Example 9.5 it was concluded that a Poisson distribution did not fit the textile fabric fault data very well. This statement can be quantified by calculating χ^2 from Formula (9.8) as shown in Table 9.53. The classes for 6 or more faults have been combined so that the expected value is close to five.

TABLE 9.53.   CALCULATION OF GOODNESS OF FIT χ^2

Number of Faults	Number of Occasions	Poisson Law Expected No.	Deviation $(O-E)$	$\dfrac{(\text{Deviation})^2}{\text{Expectation}}$
0	51	36·5	+14·5	5·72
1	68	72·9	− 4·9	0·34
2	61	72·8	−11·8	1·92
3	44	48·4	−4·4	0·41
4	24	24·2	−0·2	0·00
5	15	9·7	+5·3	2·96
≥6	6	4·4	+1·6	0·56
Total	269	268·9	0·1	$\chi^2 = 11 \cdot 91$

With seven classes six degrees of freedom are available, but one of these is utilised to estimate the parameter μ of the Poisson distribution. Hence $\phi = 5$, and reference to Table B shows that the probability of obtaining a value of $\chi^2 > 11 \cdot 91$ is less than 0·05. Noting also the ordered sequence of signs in the deviations, we conclude that the Poisson distribution is not at all a good fit to these fault data.

On the other hand, when a negative binomial distribution is fitted the value of χ^2 from 8 classes is only 2·78. This distribution is defined by two parameters, so ϕ is again 5. The calculated value of χ^2 is smaller than the expected value for five degrees of freedom, so we conclude that the negative binomial distribution provides a good description of the fault data.

CONTINGENCY TABLES

Example 9.8.   Piston-ring failures in a 4×3 classification

9.61 The data in Table 9.61 show the number of times piston-rings have failed in each leg (North, Centre, and South) of the four compressors discussed in Example 9.6.

The four compressors are apparently identical and are orientated the same way in the Compressor House. Each leg consists of two cylinders arranged vertically: the lower cylinder deals with the first stage of compression, and the upper cylinder with the second stage. The South leg is, in every case, adjacent to the drive. Since the machines are apparently identical, it is legitimate to subtotal the data vertically and to use these subtotals to decide whether one leg is generally more likely to fail than another. The horizontal subtotals were used for comparing the compressors in Example 9.6.

TABLE 9.61. PISTON-RING FAILURES IN FOUR COMPRESSORS

Comp. No.	Leg			Total
	North	Centre	South	
1	17	17	12	46
2	11	9	13	33
3	11	8	19	38
4	14	7	28	49
Total	53	41	72	166

A table of frequency data which can be subtotalled in two directions is known as a Contingency Table. In particular, the example given is known as a 4×3 contingency table to denote that there are 4 rows and 3 columns of data.

The data should provide answers to the following questions:

(i) How does the probability of failure differ in the different compressors?
(ii) How does the probability of failure differ in the different legs?
(iii) Is the answer to (ii) generally true of all the compressors, or do the different compressors behave differently?

It is evident on reflection that it is best to answer question (iii) first because, if the compressors do not behave in a similar way, no general statement about compressors of this type can be valid.

Examination of the data suggests that compressors 2, 3, and 4 are similar (each has most failures in the South leg and fewest in the Centre leg), while compressor No. 1 appears to be anomalous in having fewest failures in the South leg. The statistical treatment will be directed to confirming or rejecting this impression.

In the first instance we are not concerned with whether the differences between the compressor subtotals or between the leg subtotals mean anything, but taking these subtotals as given, we ask whether there is any evidence that the proportion of failures in any leg differs significantly in the different compressors.

The proportion of failures in the North leg over all compressors is:

$$53/166$$

If the proportion of North leg failures were the same for all compressors, the expected number of failures in the North leg of No. 1 compressor would be:

$$46 \times 53/166 = 14 \cdot 687$$

The argument may be put in a slightly different way. The probability of a failure in some North leg is estimated at:

$$53/166$$

The probability of failure in some leg of No. 1 compressor is estimated at:

$$46/166$$

If these events are independent, i.e. if the compressors behave in the same

TABLE 9.62. EXPECTED NUMBERS OF PISTON-RING FAILURES

Comp. No.	North	Centre	South	Total
1	14·687	11·361	19·952	46·000
2	10·536	8·151	14·313	33·000
3	12·133	9·386	16·482	38·001
4	15·645	12·102	21·253	49·000
Total	53·001	41·000	72·000	166·001

way, the probability of a failure in the North leg of No. 1 is the product of these two probabilities:

$$(53/166) \times (46/166)$$

The expected number of failures in the North leg of No. 1 out of a total of 166 failures is therefore:

$$(53/166) \times (46/166) \times 166 = 14 \cdot 687, \text{ as before.}$$

In Table 9.62 are shown the expected numbers of failures in each leg of each compressor.

Here again the expected number is a mathematical fiction. Obviously the number of failures in the North leg of No. 1 cannot be closer to expectation than 15, and of course, because of chance fluctuations, the numbers will generally not even be as close as this. The statistical test must decide whether, on the whole, the data deviate from expectation by more than can reasonably be ascribed to chance.

9.611 The χ^2 statistic is calculated from Formula (9.8) as in the earlier examples. For the first "cell" of the table we have:

Number observed   =   17

Number expected   =   14·687

Deviation from expectation   =   +2·313

Contribution to χ^2 = (deviation)^2/expectation =   0·364

The full calculation is set out in Table 9.63. It will be noticed that the vertical and horizontal subtotals of the deviations are all zero, apart from rounding errors; this provides a check.

TABLE 9.63.   PISTON-RING FAILURES: COMPUTATION OF χ^2

Comp. No.	North	Centre	South	Total	
1	17	17	12	46	
	14·687	11·361	19·952	46·000	
	+2·313	+5·639	−7·952	0·000	
	0·364	2·799	3·169		6·332
2	11	9	13	33	
	10·536	8·151	14·313	33·000	
	+0·464	+0·849	−1·313	0·000	
	0·020	0·088	0·120		0·228
3	11	8	19	38	
	12·133	9·386	16·482	38·001	
	−1·133	−1·386	+2·518	−0·001	
	0·106	0·205	0·385		0·696
4	14	7	28	49	
	15·645	12·102	21·253	49·000	
	−1·645	−5·102	+6·747	0·000	
	0·173	2·151	2·142		4·466
Total	53	41	72	166	
	53·001	41·000	72·000		
	−0·001	0·000	0·000		
	0·663	5·243	5·816		$\chi^2 = 11\cdot722$

The number of degrees of freedom is equal to the number of *independent* deviations. It is evident, from the method of computing the deviations, that not more than six of them could be assigned arbitrarily. If we know any two in the first row, we also know the third, since their sum is zero. Similarly any two in the second row will determine the third, and any two in the third row will determine the third. When the deviations in three rows are determined, so are the deviations in the fourth row, because vertical subtotals are also zero. Hence the number of degrees of freedom $\phi = 6$. In general, the number of degrees of freedom for a $j \times k$ table is $(j-1)(k-1)$.

It will be found, by reference to the table of the χ^2 distribution, that the 0·10 and 0·05 levels for 6 degrees of freedom are 10·6 and 12·6 respectively. (Interpolation with the aid of the Nomogram at the end of the book gives $P = 0·07$ as the probability that χ^2 will exceed 11·722). According to the usual convention, the χ^2 for the present example does not reach significance, although since it passes the 10% level one can admit that there is some rather weak evidence that the compressors are not all behaving alike. The biggest contribution to χ^2, amounting to more than half the total, comes from No. 1 compressor. If one were asking specifically whether this compressor differed from the remaining three, one would compare the data for No. 1 with the data obtained by subtotalling Nos. 2, 3, and 4, as shown in Table 9.64.

TABLE 9.64.   PISTON-RING FAILURES: NO. 1 COMPARED WITH THE REST

Comp. No.	North	Centre	South	Total
1	17	17	12	46
2 + 3 + 4	36	24	60	120
Total	53	41	72	166

The value of χ^2 for this table can be calculated by the same method as was used for the whole table. Since it is a 2×3 table, there will be $(2-1)(3-1) = 2$ degrees of freedom:

$$\chi^2 = 8·760, \quad \phi = 2$$

This value of χ^2 passes the 0·025 level, showing fairly conclusively that No. 1 is different. On the other hand, the differences between the remaining compressors are tested for significance by applying the χ^2 test to Table 9.65.

TABLE 9.65.   PISTON-RING FAILURES: COMPRESSORS NOS. 2, 3 AND 4

Comp. No.	North	Centre	South	Total
2	11	9	13	33
3	11	8	19	38
4	14	7	28	49
Total	36	24	60	120

For this table we find:

$$\chi^2 = 3\cdot106, \quad \phi = 4$$

so there is evidently no reason for believing that these three compressors are dissimilar.

The conclusion that No. 1 is exceptional would have been on a firm logical foundation if the decision to compare it with the other three had been made *before examining the data*, i.e. if the comparison had been made because, for example, Nos. 2, 3, and 4 were of the same model but No. 1 of a different model. In the present case, however, the comparison was made only because it was suggested by the data themselves. As has often been pointed out, the danger of this procedure is that even when the data do not, as a whole, show significant departure from the hypothesis we are testing, we can usually discover some particular comparisons which appear significant. In fact, one would expect to find that one-twentieth of all possible comparisons pass the one-in-twenty ($0\cdot05$) level of significance. The present example should be contrasted with Example 9.7. There the total χ^2 was indubitably significant, and one had no hesitation in picking out the classes showing significant departure from the general mean. Here the total χ^2 is not significant, although it approaches significance, and one must therefore be very cautious before claiming significance for any special comparisons suggested by the data themselves.

An interesting situation arises if a plausible explanation for the anomalous behaviour of No. 1 is advanced *after the data have been examined*. A strict logician would probably not admit that this could affect the issue, but it must be conceded to common sense that the evidence would be strengthened if it could be shown, *a posteriori*, that No. 1 differed from the other three in a way in which the other three did not differ among themselves.

Summarising the conclusions, we may say that there is no evidence that the proportions of failures in the three legs differ in compressors Nos 2 3, and 4, but that there is some evidence that the proportions in No. 1 are anomalous, and that this evidence would be correspondingly strengthened if a plausible explanation of the anomaly were found. These conclusions accord with the general impressions noted at the beginning of the discussion, but the statistical test shows that less weight should be given to the apparent anomaly of No. 1 than most investigators would probably give it if they relied solely on an inspection of the figures.

We may now turn to a consideration of the marginal subtotals. The differences between compressors have already been tested in § 9.51. The procedure for testing the differences between legs will depend on whether we do or do not choose to accept the conclusion that No. 1 is anomalous. If we do not, we shall simply apply the method of § 9.51 to the lower marginal subtotals of Table 9.61, and find $\chi^2 = 8\cdot831$ with $\phi = 2$.

In the former case we shall use the subtotals 36:24:60, obtained by excluding No. 1, and find $\chi^2 = 16\cdot800$ with $\phi = 2$.

In both cases the value of χ^2 is significant, so we may conclude that the probability of failure is not the same in the three legs. The second value of χ^2 is the more significant because the exclusion of the possibly anomalous compressor permits the differences between legs to show up more clearly on the others.

Since the three legs are clearly proved to be different, one may complete the analysis by testing comparisons between any two, by the method of § **9.1**. It will be found that the evidence for a difference between North and Centre is very weak, but that the failure rate for South is certainly higher than for the other two.

Significance test for small expected frequencies

9.612 The distribution of χ^2 calculated from contingency data is fairly close to the χ^2 defined in § **9.4** and tabulated in Table B, provided that the expected values are not too small. The traditional working rule for the test of significance to be reasonably accurate is that no expectation in a cell should be less than 5. However an examination of the exact distribution of Formula (*9.8*) suggests that this rule is too stringent [9.2] and that it may be relaxed to the following: "If relatively few expectations are less than 5 (say fewer than one in five) a minimum expectation of 1 is allowable when using the χ^2 table".

When even these relaxed restrictions are not satisfied, the significance of a value of χ^2 calculated from Formula (*9.8*) can be judged against a set of individual tables [9.5]. These tables have been constructed by simulation [Chapter 12] for up to 5×5 size classifications. The sample contingencies were generated from a Population with equal expected frequencies per cell and there may be some error when the probabilities are applied to contingency tables with very unbalanced marginal subtotals.

More detailed analyses of contingency tables, including partitioning the degrees of freedom, testing for trends and combining the results from different investigations, are described in [9.6].

$2 \times k$ contingency table

9.62 When a contingency table contains only two rows or two columns, there are a number of short methods for calculating χ^2. One such method is illustrated in [9.7], Example 11. These special methods are not exemplified here, because one may always use the general method given in the example of the 4×3 table above.

An array of this kind is obtained if samples are drawn at random from k binomial Populations. The value of χ^2 is then known as the Binomial Index of Dispersion. If the calculated χ^2 is significant at some chosen level, the conclusion is that the k Populations are not homogeneous, i.e. the data cannot be described by a single binomial parameter.

Example 9.9.   Article counts in a 2 × 2 contingency table

9.63 A rather special case arises when a contingency table contains only two rows and two columns. In a 2 × 2 table a mechanical application of the method described in § **9.61** to test the association between two attributes can sometimes produce seriously misleading results.

Table 9.66 shows the number of defective and effective articles in two

TABLE 9.66.   DEFECTIVE AND EFFECTIVE ARTICLES
IN TWO SAMPLES

	Defective	Effective	Total
Before	18	162	180
After	4	96	100
Total	22	258	280

samples, one taken before and one after the introduction of a modification intended to improve the process of manufacture. The proportion of defective articles has clearly fallen, and it is desired to test whether this apparent decrease is significant.

Calculating the deviations from expectations in the usual way, we shall find that they are all equal in magnitude, but two are positive and two negative.   For the top left cell we have:

$$\begin{aligned}
\text{Number observed} &= 18 \\
\text{Number expected} = 22 \times 180/280 &= 14 \cdot 143 \\
\hline
\text{Deviation from expectation} &= 3 \cdot 857
\end{aligned}$$

Now the χ^2 for a 2 × 2 table has only one degree of freedom and is therefore simply the square of a Normal deviate. This suggests that instead of finding χ^2, an exactly equivalent test would be provided by comparing the deviation from expectation with its standard deviation. The variance of the number in any cell is given approximately by the formula:

$$\text{Variance} = \frac{\text{Product of marginal subtotals}}{\text{Grand total cubed}}$$

$$= \frac{180 \times 100 \times 22 \times 258}{280^3}$$

$$= 4 \cdot 654$$

The standard deviation, being the square root of the variance, is therefore 2·157.

The ratio:

$$u = \frac{\text{Observed deviation}}{\text{Standard deviation}} = \frac{3·857}{2·157} = 1·788$$

may therefore be used to judge the significance of the process modification; it is the square root of the χ^2 which would have been found by the usual method. From the table of the Normal integral (Table A) we find that this corresponds to a probability of $P = 0·037$. This is the probability of finding a deviation as large as the one actually observed, or larger, in the direction *favourable to the second sample*. If we are sure that the modification cannot make matters worse, and are merely asking whether it effects an improvement, this single-sided value is appropriate. If, however, we are asking whether the proportions defective were different in the two samples, we should have to take into account the possibility of a deviation in the opposite direction, and accordingly assess the significance by a double-sided test as:

$$P = 2 \times 0·037 = 0·074$$

This is the level of significance obtained if we had calculated χ^2 from Formula (9.8); χ^2, being a square, cannot distinguish between deviations in the two directions. The conclusion is that there is some evidence that the modification has effected an improvement, but the question is not settled beyond reasonable doubt.

Significance testing of the 2×2 table

9.631 A 2×2 contingency table can arise in three ways:

(a) Both sets of marginal subtotals are random variables and only the total frequency is given (the *double dichotomy* case). For example, failures of a random set of specimens in industrial boiler tests may be classified as cracked or uncracked, and also according to whether or not an additive was included in the feedwater.

(b) Either row or column subtotals are given, but the second classification is a random variable. The analysis then tests for the *homogeneity* of two proportions. The data in Table 9.66 are of this kind, assuming that the numbers of articles inspected before and after the modification were fixed in advance.

(c) Both sets of marginal subtotals are given. This case is rarer than the other two. The classic example involves an assessor repeatedly distinguishing between two objects (e.g. taste tests of butter and margarine), when he is told the total frequency of each object.

The correct significance test for the association between the two attributes depends on which of these conditions constrains the data. The u-ratio and the calculated value of χ^2 are approximate criteria which hold when the expected frequencies in each cell are sufficiently large.

In case (c), significance levels may be determined exactly by the treatment given in [9.7], § 21.02. The exact probabilities are given in [9.1], Table 38, for marginal numbers up to 15; the approximate method of [9.8], Table VIII is also quite accurate. However, the simplest improvement in this case is to subtract 0·5 from the deviation before calculating u or χ^2—the so-called Correction for Continuity, which was introduced in § **9.14**.

It has recently been established that in cases (a) and (b) subtraction of 0·5 is too severe a correction for continuity. Other corrections are allowed for in [9.9] and applying these to the inspection data of Table 9.66 results in a more accurate one-sided probability level:

$$P = 0.0335$$

The exact probabilities in case (b) are given in [9.10] for marginal numbers up to 20, and extended to 50 in [9.11]. The exact probabilities in case (a) are given in [9.12] for total frequencies up to 20.

Estimation of proportions from a 2×2 table

9.632 The data in Example 9.9 have been analysed from the point of view of establishing the significance of the apparent decrease in the proportion of defective articles. We can also regard such data as a problem of estimating proportions and of setting up confidence limits for the improvement effected by the modification.

The maximum likelihood estimate of the initial proportion of defective articles is $18/180 = 0.10$; after the process modification has been introduced the proportion drops to $4/100 = 0.04$. Hence the maximum likelihood estimate of the change is 0·06. However, because of the sampling variation associated with each of the recorded proportions, it was concluded from the analysis of § **9.63** that the effectiveness of the alteration to the process had not been established beyond reasonable doubt. The best estimate of the proportion of defective articles over the whole period would therefore be $22/280 = 0.079$.

Nevertheless, since the data suggest that the modification may prove to be beneficial over a longer run, the process manager may decide to retain it in the hope that further inspections will confirm its merit. This would be a sensible decision if the cost of the modification was small or if there were no alternative modifications to be tried. Suppose he decides to review the position after a further 200 articles have been sampled and inspected, and that this leads to the aggregate data shown in Table 9.67.

Now if the proportion of defective articles is not affected by the modification, we would have a deviation from expectation of 7·875 in each cell, and the variance associated with the number in each cell is 5·972. The test criterion has now increased to:

$$u = 7.875/2.444 = 3.222$$

From Table A we find that this corresponds to a probability:

$$P = \begin{cases} 0.0006 \text{ (single-sided test)} \\ 0.0013 \text{ (double-sided test)} \end{cases}$$

Thus with the evidence of the additional 200 articles it can be concluded that there is a highly significant improvement following the process modification.

The maximum likelihood estimate of the later proportion of defective

TABLE 9.67.   DEFECTIVE AND EFFECTIVE ARTICLES
IN EXTENDED INSPECTION

	Defective	Effective	Total
Before	18	162	180
After	9	291	300
Total	27	453	480

articles is now $9/300 = 0.03$.   The variances associated with the "before" and "after" proportions are:

$$0.10 \times 0.90/180 = 0.000500 \text{ and } 0.03 \times 0.97/300 = 0.000097$$

respectively.   So the standard error of the change in the proportion of defectives is approximately:

$$\sqrt{(0.000500 + 0.000097)} = 0.0244$$

Hence the 95% confidence limits on the change are approximately:

$$0.07 \pm 1.96 \times 0.0244 = 0.022 \text{ and } 0.118$$

The manager may now decide to adopt the process modification permanently.   He must balance the cost of installation against the value of an expected 7% fewer articles being defective, but remember that on present evidence the reduction may only be 2%.

Three-dimensional contingency tables

9.64 Tables of frequency data may have a triple or multiple classification. Thus the full data on piston-ring failure [§ **9.61**] would show a classification under four machines, three legs and two stages (high-pressure and low-pressure); it would therefore be a $4 \times 3 \times 2$ contingency table.   Such examples may be treated by an extension of the methods already explained [9.13].

An electronic computer may be used to carry out the calculations when several large two- or multi-dimensional contingency tables are to be analysed. Although this saves the labour of hand calculation, the print-out from the computer should be studied critically. The piston-ring failure example illustrates the need for great care in the interpretation of the analysis.

General considerations on frequency data

9.7 In conclusion, it is worth while to emphasise that although valid conclusions can always be drawn from frequency data by applying the correct tests of significance, yet the results of these tests are generally of a low order of accuracy. It is usually better to avoid having to present the data as a contingency table if there is any reasonable alternative. Two examples will make this clear.

9.71 If in measuring a machine-made article we use gauges set to upper and lower tolerances, we can classify the members of a sample into three classes—undersize, within tolerance, and oversize. Comparing samples from, say, six different machines, we may therefore present the results in a 6×3 contingency table. However, more precise comparisons can be made if we have the actual measurements and use the methods described in Chapter 6. Admittedly, in practice this higher precision may be offset by the fact that direct measurement takes more time than applying "go" and "no-go" gauges, but if this consideration is advanced, a statistician should be consulted so that the relative costs per unit of information can be determined.

9.72 In some cases, one classification of the contingency table may be on a qualitative scale. Thus we may be subjecting test-pieces to rough treatment and classifying them at the end of the test (preferably by comparison with a scale of standards) into undamaged, slightly damaged, damaged, very badly damaged, and completely disintegrated. Instead of using contingency table methods, it will often be found more profitable simply to score these five grades as 0, 1, 2, 3, and 4, and to treat the data by Analysis of Variance as though the score x were a continuous measurement. It is perhaps rather surprising that the Analysis of Variance can be employed on data which depart so violently from the Normal distribution, but it is a fact that the methods remain very nearly correct provided we have suitably chosen our qualitative scale, i.e. provided there is not a large excess of test-pieces at either end of the scale.

References

[9.1] PEARSON, E. S., and HARTLEY, H. O. *Biometrika Tables for Statisticians*, Vol. 1 (third edition). Charles Griffin (High Wycombe, 1976).
[9.2] COCHRAN, W. G. "Some Methods for Strengthening the Common χ^2 Tests", *Biometrics*, **10**, 417–451 (1954).

[9.3] KENDALL, M. G., and STUART, A. *The Advanced Theory of Statistics*, Vol. I (fourth edition). Griffin (High Wycombe, 1971).

[9.4] BISSELL, A. F. "Analysis of Data Based on Incident Counts", *The Statistician*, **19**, 215–247 (1970).

[9.5] CRADDOCK, J. M., and FLOOD, C. R. "The Distribution of the χ^2 Statistic in Small Contingency Tables", *Applied Statistics*, **19**, 173–181 (1970).

[9.6] MAXWELL, A. E. *Analysing Qualitative Data*. Methuen (London, 1961).

[9.7] FISHER, R. A. *Statistical Methods for Research Workers* (fifteenth edition). Hafner (New York, 1973).

[9.8] FISHER, R. A., and YATES, F. *Statistical Tables for Biological, Agricultural and Medical Research* (sixth edition). Longman (London and New York, 1974).

[9.9] GARSIDE, G. R., and MACK, C. "A Quantitative Analysis of All Sources of Correction in the Homogeneity Case of the 2×2 Contingency Table", *New J. of Stats. and O.R.*, **6**, Pt. 1, 16-25 (1970).

[9.10] GARSIDE, G. R., and MACK, C. "Correct Confidence Limits for the 2×2 Homogeneity Contingency Table with Small Frequencies", *New J. of Stats. and O.R.*, **3**, Pt. 2, 1-25 (1967), **4**, Pt. 1, 2-18 and Pt. 2, 9-34 (1968).

[9.11] GARSIDE, G. R. "An Accurate Correction for the χ^2 Test in the Homogeneity Case of 2×2 Contingency Tables", *New J. of Stats. and O.R.*, **7**, Pt. 1, 1–26 (1971).

[9.12] GARSIDE, G. R. "Correct Confidence Limits for the Double Dichotomy Case of the 2×2 Contingency Table", *New J. of Stats. and O.R.*, **5**, Pt. 3, 1-22 (1969).

[9.13] LEWIS, B. N. "On the Analysis of Interaction in Multi-dimensional Contingency Tables", *J. R. Statist. Soc.*, A, **125**, 88-117 (1962).

Chapter 10

Control Charts

Quality control charts are graphs on which the quality
of the product is plotted as manufacture is actually
proceeding. By enabling corrective action to be taken
at the earliest possible moment and avoiding unnecessary corrections, the charts help to ensure the
manufacture of a uniform product which complies
with specification.

Introduction

10.1 Control charts are a very useful statistical tool for analysing data
obtained during production or research investigations on a plant where a
large number of individual readings (say 50 or more) may be obtained. A
feature of the activity known as Quality Control is the use of control charts
as part of a continuous inspection system where readings are plotted on
charts as manufacture proceeds. It must be emphasised, however, that
control charts may be used for the statistical analysis of data in many types of
problem and they are not confined to the use normally associated with
Quality Control.

A control chart is a chart on which the values of the quality characteristic
being analysed are plotted in sequence. The type of chart initiated by W. A.
Shewhart [10.1] consists of a central line and two pairs of limit lines spaced
above and below the central line. These are usually termed the Inner and
Outer Control Limits. The distribution of the plotted values in relation to
the control limits provides valuable statistical information on the quality
characteristic being studied. More recently a type of chart has been developed
in which the cumulative differences of the quality characteristic from a target
level are plotted in sequence. This is known as a Cumulative Sum Chart
and its use can lead to tighter control of a manufacturing process.

When data are recorded on a number of loose sheets or in separate books,
it is not easy to appreciate the significance of the values of the various product
properties and process variables. But when the values are plotted on the
appropriate charts, a clear picture of the performance of the plant emerges

from the mass of information. A control chart is not of course an end in itself: it is an aid to prediction and a basis for action which supplements technical knowledge of the process.

The control chart may take a variety of forms: what is suitable for one process may not be suitable for another, and a detailed knowledge of the process is required before successful application of control charts can be made. This chapter is concerned with the basic principles; detailed explanations and examples are confined to the control of the average quality and the variability of a process where the criterion of quality is a continuous measure. This is usually referred to as "the control of variables", and it represents the most common application in the chemical industry. Other applications are outlined in § **10.9**, and detailed accounts of these will be found in the references cited.

General purpose of control charts

10.2 A set of data obtained under plant research or production conditions usually consists of measurements of a quality characteristic such as weight, a dimension, moisture content, concentration, etc. Control charts enable one to test whether a set of data is statistically uniform or statistically controlled, i.e. whether the data are consistent with the hypothesis that they are random values from the same Population or whether changes in level and variability have taken place.

Variation in the values of the quality characteristic under examination is bound to occur. If this variation arises solely from a constant system of chance causes, then the data are said to be Statistically Uniform or Homogeneous and the variation is said to be Statistically Controlled: in other words, the data belong to the same Population. Chance causes are permissible variations which cannot be identified, either because of lack of knowledge or because such identification would be uneconomic. There may be a large number of these, each exercising a small effect on the total variation. They are inherent in the production system and cannot be reduced or eliminated without modification in the system itself; examples are small variations in reaction conditions, small variations in quality of raw material, etc. In batch production, such random variations may arise both between and within batches. More generally they may arise from within or between samples which are taken from the production stream. Each case must be considered on its merits, and it may be necessary to carry out an experiment to determine the components of random variability.

If the variations are of this homogeneous type and remain statistically controlled, valid predictions may be made about further data from the same source. Such assurance is invaluable in deciding whether the values of the quality characteristic will comply with a specification which may have been laid down, or in helping to determine the extent of the variation which may be expected from the process. In turn this knowledge permits an economic level to be fixed for the frequency of sampling and testing.

On the other hand, there may be assignable causes of variation which lead to marked changes in the average level or the variability of the process. It is usually economically worth while to track these down and to eliminate them. They may arise, for example, from sudden or abnormal variation in the properties of raw materials or reaction conditions, or from mechanical faults. In a manufacturing process they cause difficulties during production and interfere with smooth running. Assignable causes of variation may operate during a laboratory experiment or test as well as on a plant, and may vitiate the results accordingly.

The purpose of control charts is to test when variation ceases to arise solely from a constant system of chance causes and when assignable causes intervene. The occurrence of such changes can be spotted particularly quickly if cumulative sum charts are used; moreover the sizes of the changes can be estimated very simply. In some processes it may be known that a deterioration in the quality level can be corrected by adjusting a plant variable, for example a persistent drop in yield may be recovered by lengthening a process cycle time. An estimate of the size of property change can then be combined with the knowledge of the plant mechanism to forecast how large an adjustment is needed to the plant variable to bring the property back on to target.

SHEWHART CONTROL CHARTS

10.3 The procedure to be followed in analysing data by means of Shewhart control charts is set out more fully in § **10.4** when dealing with control charts for production Quality Control. It is also given in several publications on the subject [10.1]–[10.6]. The main steps to be taken are briefly as follows.

Set out the individual readings in the order in which they were obtained. Sometimes they can be divided into a number of Rational Sub-groups or samples, *within* which there is reason to believe that only chance causes of variation have operated but *between* which assignable causes may have operated. Thus a sub-group may consist of the results emanating from one batch of raw material. Data from different plants, machines, or experiments could clearly not be combined to form a rational sub-group. Estimate the standard deviation due to chance causes from the within-sample ranges as described in § **3.35**; estimate the process average from the mean of all the data.

Construct one control chart for sample averages and another for sample ranges. Plot these values for the samples to hand. If no points fall outside the outer control limits it is concluded that the data are statistically uniform, i.e. no major assignable causes of variation have intervened during the period. Predictions may then be safely made about the data, and other statistical techniques such as tests of significance may be employed and inferences drawn.

Sometimes, however, chance causes of variation may arise not only *within* but also *between* rational sub-groups or samples, and any assignable cause of variation must be judged against the combined between and within sample variation of chance causes. This situation also arises when there is no clear indication from the process as to how the data should be divided into rational sub-groups. In these cases the sample means are calculated in the order in which they were obtained and the successive differences between them are used to estimate the Population standard deviation of chance variations. A control chart is then constructed for the process averages, in the same way as described above. If necessary a check could also be kept on the variation within samples, but it must be borne in mind that this is not the sole source of chance variation in these cases.

A particular example of the procedure described in the last paragraph occurs when there is only one measurement per sample, so that the chart of process averages is the chart of all the individual measurements.

Control charts for process control

10.4 This use of control charts is now widely known under the title of Quality Control. Control charts, when used as part of a process-testing system, provide a continuous graphical record of the quality characteristic being charted. The results of measurements are recorded in the order in which they are obtained, and as soon as they are obtained. Samples consisting of groups of, say, 5 consecutive readings are taken from the system in the order of production. Control charts then test whether the samples meet the criteria associated with samples drawn at random from the same Population. If they do, then clearly the order of taking the samples did not affect the sampling results and the product did not appear to change with time. Therefore there is reason for believing that the production system is in a state of statistical control. The charts simply show whether the data are statistically uniform or not. They may be uniform and at the same time show very wide variation which may make them of little value. It must not therefore be assumed that because statistical uniformity has been established the data are satisfactory in all other respects.

Control charts are most easily applied where measurements may be made on discrete units taken from the producing system. This is the reason why they have been widely used in light engineering work for inspecting the output of machine-tool processes. In the chemical industry the procedure is not usually so straightforward. Sampling problems are more complex, and the method and frequency of sampling may have to be carefully investigated before control charts can be introduced. Batch processes are frequently employed for the manufacture of chemicals and often every batch is sampled

and tested. A batch may take only a few hours or several days to manufacture, and frequently analytical tests on a batch to assess its quality take a
long time. It follows that charts for batch processes are more suitable for
long-term control and less suitable for day-to-day or hour-to-hour control.

A chart consisting of the individual batch results is usually required, and
in many cases this is sufficient. Charts of averages and ranges of consecutive
groups of 4 or 5 batches are useful additions for the more established
manufactures.

The main purposes of control charts for batch processes are:

(a) To give a clear picture of the performance of the process.

(b) To indicate whether the process is under control and, if not under
control, to indicate the extent of the departure from control.

(c) To indicate what the process is capable of doing if operated under
conditions of statistical control.

A valuable application is in the testing of raw materials, where the charts
can be used to provide evidence of the variation in quality of the materials
being supplied. From these charts compliance with specification may be
judged, and it may be ascertained whether the supplier can produce to limits
close enough to ensure that practically all his raw material conforms to
specification. Furthermore, control charts enable raw materials from various
suppliers to be compared and the most uniform source selected for future
purchases. The use of charts to check raw materials may enable a customer
to draw a distinction between variation which arises from his production
process and that introduced by defective raw materials, and "trouble-
shooting" may be reduced in consequence. Control charts are primarily
producers' tools, however, and suppliers of raw materials should be encouraged to use control charts themselves, if these can be applied to their
processes.

Most chemical processes at some stage involve filling, packing and weighing
operations. Control charts are ideally suited to these activities to test
whether the operation is in a state of statistical control.

Taking of samples

10.41 The principles of sampling are discussed in detail in Chapter 11. For
processes where individuals may be selected, e.g. bags or containers for control
of weight, it is sufficient to take, say, 5 individuals in sequence from a conveyor.
Samples should not be taken at absolutely regular intervals in case there is
some assignable cause of variation operating with a periodicity which coincides with the interval between the taking of successive samples. The proportion to be sampled will depend on the degree of control shown and on the
factors dealt with in Chapter 11.

Construction of Shewhart control charts

10.42 The procedure outlined in § **10.3** is followed when constructing charts for process control, and the method will now be described in more detail. It is imperative that the results be analysed and plotted in the order in which they are obtained. It is also important to take measurements at those points in the process at which any action consequent on points falling outside the control limits would rectify the trouble in the shortest possible time.

Unless the Population average and standard deviation have been defined in a specification or are known from previous experience, it is necessary as a first step to obtain sufficient readings to enable good estimates of these parameters to be made. We require an estimate of the standard deviation which measures the inherent variation of the producing system, and 50–100 readings should be suitable for this purpose. The symbols μ and σ will be used for these preparatory estimates, to distinguish them from the sub-group values which will be denoted by $\bar{x}$ and s.

The Population average μ is simply estimated by the arithmetic mean of the 50–100 individual readings. The observations are then divided into rational sub-groups or samples and the mean value of the property in each of them is calculated. To estimate the standard deviation associated with the chance causes of variation we must decide whether these occur only *within* samples. If we are satisfied that this is the case and that only assignable causes operate *between* samples, then σ is estimated from the within sample variation by either of the following methods:

(*a*) Combined variance within samples [§ **3.342**]
(*b*) The average sample range, $\bar{w}$ [§ **3.35**]

Method (*a*) provides the more accurate estimate, but involves more calculation. If the sample size n is less than 12, σ can be estimated sufficiently accurately from $\bar{w}$ using the formula $\sigma = \bar{w}/d_n$; values of the d_n factor are listed in Table G.1. Moreover, the position of the control lines can be calculated direct from $\bar{w}$ without computing σ first. Hence method (*b*) is used more frequently and all calculations are thereby made very simple.

However, if we believe that part of the source of the chance variations is *between* samples, then the variations within samples would by themselves give an underestimate of the standard deviation, σ, against which to judge assignable movements. In such cases we must use the variation in the sequence of k sample means $\bar{x}_1, \bar{x}_2, \ldots, \bar{x}_k$ to estimate the short-term variability of the series and apply the following method:

(*c*) Calculate $\sigma = \left\{ n \sum_{i=1}^{k-1} (\bar{x}_{i+1} - \bar{x}_i)^2 / 2(k-1) \right\}^{\frac{1}{2}}$

This estimate will give a high figure if there are appreciable assignable shifts of mean in the initial set of samples; so if there are any obvious step movements in the sequence $\bar{x}_1, \bar{x}_2, \ldots, \bar{x}_k$, the associated differences should be omitted in applying method (*c*). In view of the smaller number of degrees of freedom, the precision of this estimate is considerably lower than that obtained by either method (*a*) or (*b*).

Drawing limit lines

10.421 When the chance causes of variation occur only within samples, two charts should be constructed. One provides a picture of the sample averages $\bar{x}$, and indicates any lack of statistical control in the process mean level. The second chart is a plot of the sample ranges *w*, which indicates any lack of statistical control within samples, e.g. an increase in process variability compared with the figure computed initially. When chance causes may also arise between samples, it is usual only to construct the chart for sample averages, since the range chart would give an incomplete picture of the chance variations.

(i) *Chart for sample averages*

The drawing of the limit lines for sample averages is based on the following principles. It has been shown in Chapter 3 that if individual observations from the same Population are distributed about a mean level μ with a standard deviation of σ, then the averages of samples, each containing *n* individuals drawn at random from that Population, are distributed round μ with a standard error of $\sigma/\sqrt{n}$. Whatever the distribution of the individuals, the distribution of sample averages approaches Normality as *n* increases. Even when *n* is as small as 4 or 5, the distribution of averages will be close to Normal provided the distribution of the individuals is not extremely asymmetric [§ **2.42**]. Thus by taking averages of samples of size *n* instead of the individuals themselves, the probabilities associated with the Normal distribution curve may be used [Table A].

Provided the data are statistically uniform or controlled, only 1 sample mean in 40 will on average lie above the limit $\mu + 1.96\sigma/\sqrt{n}$ and 1 in 40 below the limit $\mu - 1.96\sigma/\sqrt{n}$. These limits are referred to respectively as the upper and lower 1 in 40 or 0·025 limits. Further, only 1 sample mean in 1,000 will lie above the limit $\mu + 3.09\sigma/\sqrt{n}$ and 1 in 1,000 below $\mu - 3.09\sigma/\sqrt{n}$, i.e. 1 in 500 outside $\mu \pm 3.09\sigma/\sqrt{n}$. These limits are known as the 1 in 1,000 or 0·001 limits. Lines drawn at a distance of $1.96\sigma/\sqrt{n}$ above and below μ provide the Inner Control Limits. Lines drawn at a distance of $3.09\sigma/\sqrt{n}$ above and below μ constitute the Outer Control Limits. Table G at the end of the book facilitates the drawing of limit lines for various values of *n*; if the sample size is less than 12 the control limits can be calculated direct from $\bar{w}$.

Since with statistical control only 1 sample mean in 20, on average, lies outside the inner control limits, the occurrence of a point outside these limits may be regarded as a warning of possible lack of statistical control, i.e. that an assignable cause of variation has intervened. A point outside the outer control limits is very strong evidence of lack of statistical control and calls for action to discover the assignable cause of variation and to eliminate it. Moreover, while points lie inside the limits no deliberate changes should be made to the plant, thereby avoiding unnecessary corrections. It is a good dictum that unless a process has clearly indicated (by a point or points outside the control lines) that something is amiss, it is best to leave it alone.

To simplify the calculation and the verbal description, practitioners sometimes round the multiplying factor 1·96 up to 2 and the 3·09 down to 3. They then refer to the inner and outer limits as the Two-Sigma and Three-Sigma limits respectively. The corresponding tail probabilities are slightly changed, but this is of negligible practical importance.

B.S. 600R:1942 [10.3] recommends the use of both sets of limit lines, whereas B.S. 1008 [10.4] (which deals with American practice) refers to outer control limits only, spaced symmetrically above and below μ at a distance of $3\sigma/\sqrt{n}$. Another common practice is to base action on a single point falling beyond the outer limits or two successive points outside the same inner limit.

The justification for these choices of control limits to indicate a variation which should not be allowed to pass is that practical experience has shown that the occurrence of points beyond them is a very useful signal for action. When points have fallen outside, trouble has been experienced and detected on investigation, and its elimination has proved economically worth while. The fact that control limits correspond to explicit probabilities derived from the Normal distribution curve may be of value when interpreting the charts, but too much attention should not be paid to the precise probability values associated with the particular multiple of the standard error which is adopted. This remark is particularly apt when only a single measurement is made on each sample ($n = 1$) and the distribution may not approximate very closely to the Normal form.

(ii) *Chart for sample ranges*

Tables G.1 and G.2 for drawing limit lines for sample ranges have been compiled on the assumption that the distribution of the individual observations is approximately Normal. Table G.2 permits the position of the lines to be calculated direct from $\bar{w}$, which will usually be the most convenient route. The same probability levels have been used as in the chart for sample averages, viz. 0·001 in each tail for the outer limits and 0·025 in each tail for the inner limits. Note that the lower and upper limits are not symmetric about the average sample range.

Practical experience has shown that using these limit lines provides an effective method of studying and controlling the dispersion within a sample. If the distribution of individuals is not Normal, the probabilities associated with the limit lines will be different to a greater or lesser extent, but this will not invalidate the use of the outer limits as action limits.

Plotting the points

10.422 After the limit lines have been drawn, the points for the samples which have been used to derive μ and σ may be plotted. It is advisable to plot both charts on one piece of paper, with one chart under the other, and to cover as long a period of time as possible to show up trends. The points for sample averages may be distinguished from sample ranges by using dots for the former and circles for the latter.

It may well happen that the points show lack of statistical control for either or both the charts, with one or more points outside the outer control limits. This shows that assignable causes of variation have intervened during the acquisition of the initial data. Any sample which shows lack of control in relation to sample range should be discarded if it is desired to know the inherent variation of the process, and the value of σ should be recalculated. It is, of course, advisable to try to locate and eliminate the cause of the trouble which led to the sample value being out of control. New control lines should be drawn and the procedure repeated until all the remaining points come within the control limits on the ranges chart.

Whether a sample showing an average value out of control with its range under control is discarded or not will depend on the purpose for which the chart is being used. If no specification exists for the process and it is desired to know the best that can be expected, samples showing lack of control on sample averages would also be discarded. If an average has been specified for the process, the calculated limit lines could be drawn round the specified average without reference to the average μ achieved during the trial period. A detailed numerical example of the construction is given in Appendix 10A for the setting and hardening properties of a plaster.

If the points derived in the trial period or from past production data show a state of statistical control, the charts can be put into use and test results for the process recorded as they come along. Provided the points lie within the inner control limits, production is statistically controlled and the product is as uniform as the process is capable of making it. Points outside the inner control limits are regarded as a warning that assignable causes of variation may be present, and a further sample should be taken to see whether the warning is confirmed. Points outside the outer control limits should be regarded as justifying action to find the assignable cause and to eliminate it.

Interpretation of control charts

10.43 Lack of control as exhibited by the two control charts may be interpreted by reference to Figures 10.1–10.3.

Figure 10.1 shows two distribution curves with different Population means μ_1 and μ_2, but the same standard deviation, $\sigma_1 = \sigma_2$. The distribution curves are of the same shape, but No. 2 is displaced to the right of No. 1, i.e. the mean has shifted. This state of affairs would be reflected in the control charts by the sample averages being out of control while the sample ranges were under control. This is a common occurrence in practice, and means that the system is inherently capable of operating under control but that some

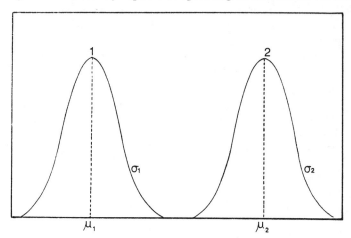

FIG. 10.1.   LACK OF CONTROL OF AVERAGES

factor or factors have caused the average to shift. An example would be a control chart for a series of sample weights where the sample average shifted owing to material sticking in the pan of the weighing apparatus.

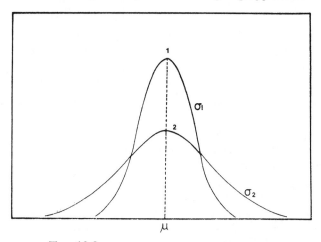

FIG. 10.2.   LACK OF CONTROL OF RANGES

A movement of the average property level may either occur gradually, due perhaps to the drifting of a process variable, or there may be a fairly sudden alteration caused by some radical change in the system. A slow downward drift is exemplified in Appendix 10A by the average values of the initial setting time of the plaster [Figure 10A.2]. The average hardening rate of the plaster also exhibits drifting, but the most marked feature in Figure 10A.3 is the dramatic rise of 99 units from Sample No. 13 to 14. The discovery that either of these types of movement exists is a signal to search for the assignable cause. In some instances it may be possible to rectify the fault once and for all; in others it may be necessary to adjust a process variable from time to time to ensure that the average property value does not move too far from a target level.

Figure 10.2 corresponds to the condition of lack of control on the chart for ranges while sample averages are under control. This is a more fundamental lack of control and suggests something inherently wrong with the

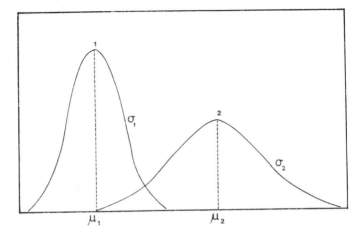

FIG. 10.3.   LACK OF CONTROL OF AVERAGES AND RANGES

system. A mechanical example would be that no sticking was occurring in the weighing-machine pan but bearings were loose or knife-edge movement was taking place. In practice the conditions suggested by Figure 10.2 would probably result in the state of affairs shown in Figure 10.3. Figure 10.3 corresponds to the conditions of lack of control in both sample averages and sample ranges, exemplified in the above case by both sticking of material and worn machine parts. These illustrations represent special cases, and in general lack of control may imply the presence of many different Populations.

When examining the range chart for lack of control, too much attention should not be paid to points below the lower limits, unless an investigation is in hand to determine the conditions under which some samples are more uniform than others. Zero values of the range may occasionally be recorded owing to limitations in the precision of measurement or to rounding. For routine work the lower control limits may be omitted entirely.

A cyclic movement of the sample averages may sometimes be observed.

The sample range appears in control, but a clear periodic pattern may be visible in the averages, with points falling successively beyond the lower and upper control lines. In such cases the length of the period should be noted and a search should be made for an underlying feature of the process which changes in the same way, e.g. the usage of successive batches of raw material. A cyclic movement may also be induced by over-correcting a process when really no adjustment is needed. This phenomenon has sometimes been observed in the control of the weight of powders and tablets of pharmaceutical products. Note that imposing "corrections" for chance variations does nothing more than impose a man-made variation on the inherent variations in the process. Hence the practical rule is: "Do not make adjustments unless they are shown to be necessary".

In general, the chart for averages and the chart for ranges should be examined together to form an opinion of the kind of trouble to look for when lack of statistical control is evident. Interpretation of the charts will clearly depend in a large measure on knowledge of the process factors involved.

Level of control

10.44 It must not be thought that because a process is under control it is necessarily operating satisfactorily. A plant may be manufacturing a product most of which does not comply with specification, and yet the process may be under control. If such circumstances obtained, control would exist at too low a level. The level of control is often as important as the control itself. A process may be under control at the wrong level for one or both of the following reasons:

(i) The achieved average value of the quality characteristic does not coincide with the specified value;

(ii) The inherent variation of the producing system is too great, so that too much of the product falls outside the specified limits, even when they are plotted round the specified average.

The level of control in relation to the specified average may often be suitably adjusted. But if the level of control in relation to the specified limits about the average is too low, the specification cannot be met without alteration to the process, which may entail redesign of equipment, modification to the process recipe or the blending of batches.

Compliance with specifications

10.45 The procedure given in the previous sections has been directed chiefly to discovering the inherent variation of the process and the fluctuations in the quality characteristic about its achieved average value. This method quantifies what the process is doing, and is of value in fixing standards where none already exists. For many processes, however, specifications exist which define the average to be aimed at and also define the limits above or below the average which may only occasionally be exceeded, e.g. 1 in 1,000 samples. With this information it is possible to construct a chart for sample averages in which the central line is the specification average and on which, in addition to the control chart limit lines calculated from σ, specification limit lines are also included. If the specification allows a tolerance of $\pm b$ on the specified average for single tests, the specification limit lines are spaced at a distance of $b/\sqrt{n}$ above and below the target average for samples of n tests.

It is clear that if the product is to comply with the specification, $3 \cdot 09\sigma$ must not exceed b. If $3 \cdot 09\sigma > b$, the process is incapable of manufacturing a product practically all of which will meet the specification. Possible remedies include increasing b by widening the specification limits or reducing σ by modifying the process. Procedures for selecting acceptable batches or blending the product may also be considered.

TABLE 10.1.   SCHEME FOR FREQUENCY OF TESTING

Tolerance, b	Test Frequency
Less than $3 \cdot 09\sigma$	Every sub-group
Between $3 \cdot 09\sigma$ and $3 \cdot 5\sigma$	Alternate sub-groups
Between $3 \cdot 5\sigma$ and $4 \cdot 0\sigma$	1 sub-group in 5
Between $4 \cdot 0\sigma$ and $4 \cdot 5\sigma$	1 sub-group in 10
Between $4 \cdot 5\sigma$ and $5 \cdot 0\sigma$	1 sub-group in 25
Greater than 5σ	1 sub-group in 100

On the other hand, if $3 \cdot 09\sigma < b$, practically all the product is satisfactory as long as the average remains on target. It may be safe to reduce the frequency of sampling and thus lower operating costs. One chemical company has adopted the guideline shown in Table 10.1 for testing sub-groups of raw materials, process intermediates and finished products [10.7]. Under this plan, if a sample mean from a sub-group is found to be outside specification, the immediately preceding sub-groups are tested (if this is possible) followed by every sub-group up to the next scheduled one. If all these values of $\bar{x}$ fall within the specification, the test frequency reverts to the plan, but otherwise a process change would be suspected and a new schedule could be devised.

If the producing system is under statistical control at the correct level, valid predictions may be made from the results of tests on small samples on the extent to which the whole product is meeting the specification and the proportion of the product which is failing to meet the specification. This aspect of predicting the limits within which various proportions of the total product must lie is one of the most valuable practical results arising from statistical control; but it is important to stress that unless statistical control is achieved, valid predictions relating to the whole future product cannot be made from the results of the tests on small samples. Predictions made in the absence of a state of statistical control are usually unreliable unless 100% inspection is carried out, and even 100% inspection is not foolproof.

Rate of detection of changes in average level

10.5 One of the chief merits of the Shewhart chart is that a plant manager can see at a glance whether the process variables and product properties are in a state of statistical control. Let us now consider how quickly he will discover that something is amiss.

Suppose the average value suddenly shifts away from μ by an amount equivalent to $3.09\sigma/\sqrt{n}$. Then there is a 1 in 2 chance that the very next sample average will fall beyond the outer control limit and initiate action. There is a probability of $0.50 + 0.25 = 0.75$ that the movement will have been detected after two samples, and so on. The average number of samples which will be tested before the change is detected is termed the Average Run Length (*A.R.L.*). Hence for a $3.09\sigma/\sqrt{n}$ shift of mean we have:

$$A.R.L. = 0.5 \times 1 + 0.25 \times 2 + 0.125 \times 3 + \ldots = 2$$

This is reasonably short.

In general, for a given level of the long-term process average, the *A.R.L.* associated with the upper control line is simply $1/A$, where A is the tail area of the Normal distribution beyond the control line. Thus if the process average moves to a level $\mu + u\sigma/\sqrt{n}$, A is the area beyond the $(3.09 - u)$ ordinate in Table A. In particular, when $u = 1$ the *A.R.L.* is 55, which implies that a large number of samples are likely to be produced before an out-of-control signal is sounded. In some plants it may be known that this shift of property mean can be corrected by adjusting a process variable and we would not want to wait an average of 55 samples before taking action.

The reason for the tardy response to moderate size changes is that action depends on an individual point falling outside the control limit. This situation can be improved by adopting the additional rule that two consecutive values of $\bar{x}$ beyond one of the inner control lines also require investigation. Other improved schemes have been based on counting the number of sample averages all on the same side of μ and on a moving average of the last few values of $\bar{x}$. However, we should really base our decision on the whole of the recent sequence of process measurements. This can be done most effectively by constructing a different type of chart which will now be described.

CUMULATIVE SUM CHARTS

10.6 This method of presenting process data requires a small amount of prior calculation. Instead of plotting the sample averages $\bar{x}_1$, $\bar{x}_2$, ..., a reference value k is chosen and the differences $(\bar{x}_1 - k), (\bar{x}_2 - k), \ldots$ are formed. These are accumulated in sequence to form the Cumulative Sums:

$$\left.\begin{aligned}
S_1 &= (\bar{x}_1 - k) \\
S_2 &= (\bar{x}_1 - k) + (\bar{x}_2 - k) = S_1 + (\bar{x}_2 - k) \\
S_3 &= S_2 + (\bar{x}_3 - k) \\
& \cdot \quad \cdot \quad \cdot \quad \cdot \quad \cdot \quad \cdot \\
S_r &= S_{r-1} + (\bar{x}_r - k) = \sum_{i=1}^{r} (\bar{x}_i - k) \\
& \cdot \quad \cdot \quad \cdot \quad \cdot \quad \cdot \quad \cdot
\end{aligned}\right\} \qquad (10.1)$$

Thus each sum is obtained from its predecessor by the simple operation of adding on the new difference $(\bar{x}_r - k)$.

The calculation is exemplified in Table 10.2 for a set of cycle times measured on a batch chemical plant. Here the sample size n is 1, and a reference value of 2 hours has been chosen.

TABLE 10.2.   BATCH CYCLE TIMES

Observed Cycle Time		Difference from k	Cumulative Sums
hr.	min.	min.	min.
2	15	+15	+15
2	10	+10	+25
1	55	−5	+20
2	05	+5	+25
2	30	+30	+55
2	15	+15	+70

The cumulative sums are generally plotted as each sample average is incorporated, the resulting graph being known as a Cumulative Sum Chart. The words "cumulative sum" are frequently abbreviated to "cusum". The reference value k will sometimes be set equal to the estimated Population average μ, but calculation of the cusums is simplified if a rounded value is chosen, e.g. exactly 2 hours in Table 10.2; other choices for k will be discussed in §§ **11.4** and **11.6**.

Interpretation of cumulative sum charts

10.61 If the mean value of the sample averages remains close to the reference value, some of the $(\bar{x}_r - k)$ differences will be positive and some negative, so that the cumulative sum chart will be essentially horizontal and the process is in a state of statistical control. However, if the average value of the process rises to a new constant level, more of the differences will become positive and the mean chart path will be a straight line sloping upwards. Similarly, if the average value of the process falls to a constant level below the reference value the general slope of the chart will be downwards.

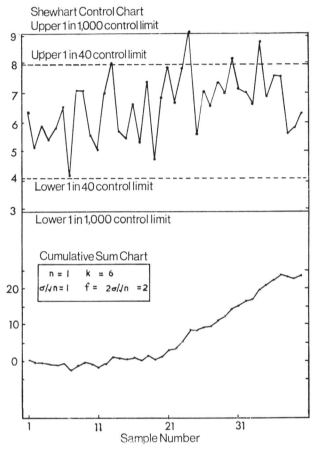

Fig. 10.4. COMPARISON OF CUMULATIVE SUM AND SHEWHART CHARTS. MEAN OF FIRST TWENTY RESULTS = 6; MEAN OF SECOND TWENTY RESULTS = 7

This is illustrated in Figure 10.4. Points on the upper chart have been sampled randomly from a Normal Population with a mean of 6 units and a standard deviation of 1; half-way along the series the mean level has been deliberately increased to 7. (The data were actually generated by taking the sums of 12 successive values from Table K using the method described in § **12.321**.) One point in the second phase falls above the outer control line;

three points fall above the inner control line, one of them in the first group of 20 values and two in the second group. The lower part of the Figure charts the cumulative sums of the differences of the data from a reference level of 6 units. During the first phase the cusum path is essentially horizontal, but then it rises steadily, corresponding to the higher level of the mean in the second half of the data. The cusum plot clearly reveals the two phases of the original series.

Cumulative sum charts are thus interpreted by the average slopes of the line which is graphed. A horizontal sequence of points indicates that the mean value of the process at that period is the same as the reference value, irrespective of the level of the points on the chart. The further the current mean process level is from the reference value, then the steeper will be the slope of the cumulative sum chart. In fact, the slope of the line joining the ath and bth $(b > a)$ points on the chart is given by:

$$\frac{S_b - S_a}{b - a} = \frac{\sum_{i=1}^{b} (\bar{x}_i - k) - \sum_{i=1}^{a} (\bar{x}_i - k)}{b - a} = \frac{\sum_{i=a+1}^{b} (\bar{x}_i - k)}{b - a} \qquad (10.2)$$

which is the average difference from the reference value of all the results from $\bar{x}_{a+1}$ to $\bar{x}_b$ inclusive. Thus the chart quickly indicates averages over *any* number of sequential results.

10.62 One of the main virtues of cumulative sum charts is that relatively small changes in the process mean value appear as quite clearly different slopes. However, the visual picture depends to some extent on the scales chosen for the axes of the chart. Regarding the horizontal distance between the plotted points as 1 unit, it is recommended that the same distance on the vertical scale should represent approximately $2\sigma/\sqrt{n}$ property units, where σ is the Population standard deviation and n is the sample size. With this system of scaling, the mean path of the chart will make an angle of 45° with the horizontal when the sample averages are separated by $2\sigma/\sqrt{n}$ from the reference value, and the chance variations will appear quite small. It is generally desirable that no slope should exceed 60°, since angles greater than this are relatively insensitive to changes in the mean value of the series. Further, the adoption of a standard system of scaling makes it easier to recognise the relative importance of movements on a set of charts covering a variety of process properties.

The unit distance on the vertical scale does not have to be exactly equivalent to $2\sigma/\sqrt{n}$ property units and in fact it eases the task of plotting the cusums if a suitably rounded lower value is chosen. For example, if $\sigma = 14$ and $n = 5$, then $2\sigma/\sqrt{n} = 12 \cdot 52$ and we would probably choose the unit distance to be equivalent to 10 property units. The scale factor f actually used should be noted on the cumulative sum chart, as indicated in Figure 10.4.

As well as providing easy visual detection of changes in the process mean level, cumulative sum charts also pin-point the time at which the change occurred. This is identified by the position of the change in cusum slope, and in practice a knowledge of this location is remarkably useful in helping to discover its cause. The precision of the estimated time is much higher than can be achieved with a Shewhart chart, and there have been many examples where some cause came to light at the very sample indicated by the change of cusum slope.

Decision making with a V-mask

10.7 Just as action decisions from a Shewhart Chart are signalled by points falling outside the control limits, a rule is needed to decide when a major change of slope has occurred on the cusum chart, resulting from some assignable cause in the process. We do not want to be misled by small apparent changes which are only due to the chance variations of the process.

One method for taking these decisions involves superimposing a V-shaped mask over the chart. The axis of the V is set in a horizontal position with the vertex of the V at a distance d ahead of the most recent point on the chart. After each new sample average has been plotted, the mask is moved diagonally across the chart so as to preserve the lead distance d, whilst maintaining the horizontal attitude of the V. This is shown in Figures 10.51–3. The mask can be constructed as a solid truncated V in a material such as hardboard, or preferably the V can be engraved on a sheet of transparent perspex.

If all the previously plotted points lie within the limbs of the V it is assumed that the process is in a state of statistical control [Figure 10.51]. But if the cusum path (including the origin) cuts across one of the limbs of the V (or its extrapolation) then the decision is reached that the process mean has moved significantly away from the reference level, k. When the lower limb is crossed, an increase in the process mean is indicated [Figure 10.52] and the slope of the cusum path must exceed $\tan\theta$, where θ is the angle between each of the limbs of the V and the horizontal. But if the sequence of points straddles the upper limb [Figure 10.53], then a reduction in the process mean level has occurred. Either event is a signal to make a corrective adjustment to the process if its mechanism is sufficiently well understood, or to initiate a search into the cause of the movement.

Thus the two limbs of the V-mask superimposed on a cusum chart provide double-sided control of the process mean about the level k, just as the parallel control lines do on a Shewhart chart.

Performance of V-mask schemes

10.71 The properties of a V-mask control scheme clearly depend on the values selected for the lead distance, d, and the semi-vertex angle, θ. The larger each of these parameters is, the fewer will be the interruptions to the

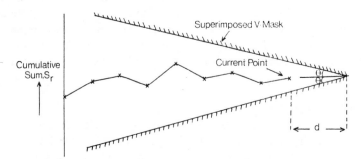

FIG. 10.51.   PROCESS MEAN CLOSE TO REFERENCE VALUE

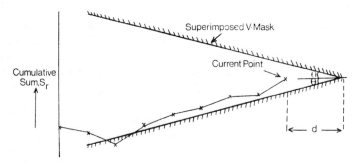

FIG. 10.52.   PROCESS MEAN HIGHER THAN REFERENCE VALUE

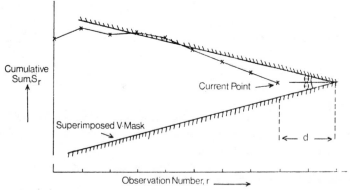

FIG. 10.53.   PROCESS MEAN LOWER THAN REFERENCE VALUE

FIGS. 10.5.   V-MASK SUPERIMPOSED ON A CUMULATIVE SUM CHART

process. This is advantageous when the plant is working satisfactorily, but when a genuine change takes place we want to detect it as quickly as possible.

Accordingly, d and θ are usually chosen according to the average run lengths [§ **10.5**] which will result from a scheme based on the selected shape. We require a high *A.R.L.*, L_a, when the process mean is at the reference (or target) level k and a low *A.R.L.*, L_r, when the sample averages move off to a rejectable level. L_a is usually fixed at several hundreds, while L_r is generally between 3 and 10. If sufficient cost data are also available it may be possible to make monetary comparisons between alternative schemes.

Curves of average run lengths resulting from 16 combinations of d and θ are given in Figure 5 of an I.C.I. Monograph [10.8]. They are based on the assumption that the sample averages are Normally distributed. A typical mask with a lead distance equal to twice the horizontal plotting interval and a semi-vertex angle of 30° has an L_a of 500 samples. Thus, as regards false signals when the process is truly running at the reference level, this mask is equivalent to the two outer control lines of a Shewhart chart. Note that in making this statement and in the curves of [10.8], the value of θ is based on the standard scaling of cusum charts [§ **10.62**]. More generally, if f property units on the cusum scale occupy the same length of chart as one horizontal plotting interval, the values of $\tan \theta$ in [10.8] should be multiplied by $2\sigma/f\sqrt{n}$, where the Population standard deviation of chance variations, σ, is estimated as in § **10.42**. The value of d is measured in units of the horizontal plotting interval and is not dependent on the scaling factor f.

When the mean level of the process sample averages moves away from the reference level, a cusum control scheme will pick this up relatively quickly. This is shown in Table 10.3, which lists *A.R.L.*s at various displacements of the current mean for:

(i) A cusum scheme based on a V-mask with $d = 2$ and $\theta = 30°$ (standard scaling);

(ii) A Shewhart chart using the combined rule: "Take action if one point falls beyond the outer control lines or two successive points fall outside the same inner control lines";

(iii) A Shewhart chart using outer control lines at $\pm 3{\cdot}09\sigma/\sqrt{n}$ only.

The displacements of the process mean from the target (reference) level are given in multiples of the standard error of the sample averages.

This cusum scheme and the Shewhart chart with the outer lines only have the same chance of false action when the process is running at the reference level, but for all except the largest shifts of process mean the cusum scheme reveals movements more quickly. Thus when the mean shifts by an amount equal to the standard error of a sample average, an out-of-control signal will be sounded three times more quickly by a past cusum point appearing beyond a limb of the V-mask. Alternatively, a process manager could safely reduce his sample size and know that he would still achieve efficient control based on the cusum scheme.

The Shewhart approach is improved by the use of the combined rule. It has a higher chance of false action signals when the process is running at the reference level, but it reveals shifts in the mean level more quickly than just waiting for a point beyond the outer control lines. Nevertheless, the cusum scheme discloses moderate sized movements more quickly than the Shewhart combined rule; this is because the cusums reflect the average of all the recent sample values rather than just the last one or two points.

More details of comparisons between schemes are given in [10.9]. Note that whilst the average run length has been used in the above comparisons, the run length itself has a statistical distribution with a wide dispersion.

TABLE 10.3.   COMPARISON OF AVERAGE RUN LENGTHS FOR CUSUM AND SHEWHART CONTROL SCHEMES

Deviation of Process Average from Target	Cusum Scheme	Shewhart "Combined" Rule	Shewhart Outer Lines
0	500	320	500
$\pm 0.50\sigma/\sqrt{n}$	98	108	201
$\pm 1.00\sigma/\sqrt{n}$	18	26.3	54.6
$\pm 1.50\sigma/\sqrt{n}$	6.3	8.92	17.9
$\pm 2.00\sigma/\sqrt{n}$	3.5	4.14	7.25
$\pm 2.50\sigma/\sqrt{n}$	2.4	2.46	3.60
$\pm 3.00\sigma/\sqrt{n}$	1.8	1.75	2.15
$\pm 3.09\sigma/\sqrt{n}$	1.8	1.67	2.00
$\pm 4.00\sigma/\sqrt{n}$	1.3	1.19	1.22

Sampling size and sampling interval

10.72 In many applications of control charts, the sample size and the sampling interval are predetermined. If the measurements are non-sampled production data, the recording interval is generally in work-shifts or days, e.g. the amount of waste material or the machine efficiency during that interval. However, where there is a choice of sample size and sampling interval, this should be determined from a consideration of the size of the change in mean level, say δ, which is regarded as important and which the control scheme should be capable of detecting, and the average length of time, τ, that may elapse at this quality level before it is detected [10.9].

Since cusum chart schemes are particularly effective for detecting changes in the region $\sigma/\sqrt{n}$ to $2\sigma/\sqrt{n}$, the sample size, n, should be chosen so that $\sigma/\sqrt{n} < \delta < 2\sigma/\sqrt{n}$, and the sampling interval should be between $\tau/10$ and $\tau/4$. The exact choice will depend on the precision with which one attempts to define δ and τ. In practice these are often defined rather hazily, so a reasonable choice for the sample size is such that $\delta \simeq \frac{3}{2}\sigma/\sqrt{n}$ and the sampling interval $\leqslant \tau/6$. For example, if a change equal to σ is considered important if it can persist for more than 2 or 3 days, then a sample of 2 items should be taken once per 8-hour shift.

Corrective action

10.73 When a decision is reached that the process mean value has moved away from the target, the kind of action to be taken will depend on the process concerned and on the state of knowledge of how the process really works. In some cases the action may merely be to investigate what really has gone wrong; in others it may be to replace a faulty component in the equipment. However, in many modern processes the corrective action consists of turning a knob to alter a plant variable, e.g. temperature, speed or rate of flow.

In the latter case we have to estimate how far to turn the knob. This is fairly straightforward if cumulative sum charts are used. When a previous sample point first falls outside the limbs of the V, draw a straight line between the most recent points, back as far as the point where the slope of the graph appears to change significantly. Measure the slope of this line. The correction to be applied is then proportional to this slope, the constant of proportionality being a known feature of the cause-and-effect relationship in the system. However, it is often advisable slightly to under-correct the plant variable to avoid process "hunting". Further, it is possible to calibrate the edges of the V in such a way that the necessary correction is read off at the point where the cusum path crosses the boundary of the V.

More details of this procedure and a number of practical considerations are discussed in Chapter 4 of [10.8].

Decision limit schemes

10.74 Instead of superimposing a V-mask on a cusum chart, there is an alternative (but equivalent) method of deciding when a significant movement away from the reference level has taken place in the mean level of the process. To control increases in the level we may apply the rule: "Take action if the current point on the chart rises more than a certain amount, h, above the lowest point on the cusum path since the last such decision". The quantity h is termed the Decision Limit. However, discussion of this method will be deferred until § **11.6**, since it can be applied numerically to the sample data without constructing a control chart at all.

Choice of charting method

10.8 We have seen that a control chart based on a cusum graph and a V-mask is straightforward to operate and that it has several advantages over conventional Shewhart procedures:

(a) The ease with which changes in mean level can be detected visually by a change in the slope of the chart.

(b) If an out-of-control signal is sounded, it can be seen from the graph when the change started.

(c) For the same number of false alarms, cusum tests can be made to react more quickly to medium-sized changes in the process mean level.

(d) The size of the change, and hence the size of corrective action, can be estimated from the slope of the graph between the point at which the slope changed and the last recorded point.

With these various advantages, cusum charts for process averages have superseded the classical type of plot in many chemical plants. However, cumulative sums do involve a small amount of arithmetic and the continued movement of the V-mask. Also the "slopes" interpretation of a cusum chart is not quite so immediate as the direct plot provided by a Shewhart chart. The user must therefore weigh these advantages and disadvantages in deciding which type of chart to construct. For a new process the Shewhart procedure may be selected initially; later, when the system has settled down, cusum charts may be adopted for processes where it is known how to correct any drifts which may occur.

The above discussion of cusums has concentrated on the control of the process averages. A Shewhart chart for sample *ranges* [§ **10.421**] can be run in parallel with a cusum chart for the averages. If this is not regarded as providing sufficient discrimination, then a cusum procedure for ranges is available [10.10].

FURTHER APPLICATIONS OF CONTROL CHARTS

10.9 In this section other types of control chart are briefly described and references are given where detailed treatments will be found.

Defect charts

10.91 So far we have only dealt with the control of quality when this may be represented by a continuous variable. Another important measure of quality which applies to a different type of manufacture is the proportion or the number of defective items in a batch consisting of a number (usually large) of similar articles, for example a day's run in the manufacture of cartridge cases or of tubes of antiseptic. For these products each article should pass a specification in order to be suitable for sale, or for use in a subsequent process. When the article fails to satisfy the specification it is said to be "defective", and a convenient measure of the quality of a batch of articles is then the "proportion defective". The higher the proportion of defective articles, the greater will be the trouble experienced in the later stages of the process.

Closely allied to "proportion defective" is another measure of quality referred to as the "number of defects"; this applies to products such as yarn, cloth, metal or plastic sheets, etc., where it is important that the number of defects per unit length or area of the material should be kept low.

Shewhart charts for "proportion defective" and "number of defects per unit" are based respectively on the binomial and Poisson distributions, which were discussed in Chapter 9. Both types of chart are dealt with in a similar way; a sample from each batch is taken and examined in detail for the number of defectives or the number of defects as the case may be, and the results are plotted in chronological order. The samples from each batch are of the same size. The central line on the chart is the mean of previous manufacture, and the control limits corresponding to any required probability may be determined from Table F. For instance, suppose samples of 100 are taken from each batch and the Population mean number of defectives is 5. Now the number of defectives in a sample of 100 giving a lower single-sided $97\frac{1}{2}\%$ confidence limit of 5 is between 10 and 11. Hence the upper inner control limit for this Population is 10. Tables exist which will enable these limits to be obtained directly (see [10.3] and [10.5], where a full discussion of these charts is also given).

Cumulative sum charts for numbers of defectives can be constructed to give earlier warning of moderate shifts in the average level of defectives or to reduce the sample size for given risks of wrong decisions. For each sample the difference between the number of defectives and a reference number k is calculated. These differences are accumulated; if the sum reaches or exceeds a decision limit h, action is taken to discover the cause and if possible to eliminate it. This method of process control is discussed in § **11.4**.

Simultaneous control of mean and standard deviation

10.92 The following method is sometimes useful for controlling the proportion of defective articles. It applies in those cases where an article is defective if it fails to reach a certain specification based on a quantitative measurement. For example, if strength is the criterion, it may not matter how high the strength is, provided the specified figure is exceeded. In a batch of articles the strengths will vary from one individual to another according to a certain statistical distribution, which can usually be assumed to be Normal or may readily be transformed into one that can be assumed Normal. Given the mean μ and the standard deviation σ, we can calculate the expected proportion of articles below any specified value x_0. This is obtained by first calculating $u = (\mu - x_0)/\sigma$ and then referring to Table A. Conversely, given x_0 and σ we can determine the value of μ for which the proportion less than x_0 is equal to a given value.

In practice μ and σ are often unknown and have to be estimated from an initial set of samples. Thereafter each sample mean and standard deviation are obtained and the ratio $u = (\bar{x} - x_0)/s$ is calculated. This statistic is distributed as a Non-central t, which is fully tabulated; in fact tables exist which give directly the Shewhart control limits for u [10.3]. An efficient cusum chart scheme for u can also be constructed; details are given in [10.10]. The proportion defective derived from the value of u is a more efficient estimate of the proportion defective in the batch than is the actual proportion defective in the sample.

Double cusum charts

10.93 In certain circumstances it may be advantageous to vary the horizontal plotting interval in cumulative sum charts. Suppose, for example, there is a budgetted production level in each of the next twelve months and we keep a cusum chart of the deviations between the achieved and the planned levels; this highlights the current excess or shortfall from target. If the planned production is approximately constant from month to month, straight lines on the chart indicate that the performance is running at a fixed percentage above or below the target level. We might, for example, draw lines on the chart which the points would have to follow to be 5% or 10% above plan. However, such an interpretation is incorrect when the planned performance is not substantially constant, as happens with seasonally dependent products. In these cases, instead of plotting the months at equal distances along the horizontal axis, the intervals should be made proportional to the budget figure for the months. Straight lines drawn on the chart do then describe fixed percentage deviations from the target level.

This procedure is illustrated in Figure 10.6 for the data of Table 10.4. An inch of the horizontal scale represents 300 production units.

Thus the plan calls for higher production levels in the spring and summer months. The chart of cumulative differences shows that for the first five months of the year the plant ran about 5% above the target level and at the end of May it looked as if the summer peak would be reached comfortably. However, for the second five months of the year production fell to about 9% below target, and at the end of October there was a net shortfall of 22 units.

The values plotted against each other in Figure 10.6 are both cumulative sums and this presentation of data has therefore been termed a Double Cusum Chart. Another application of this type is to the conversion efficiency of a chemical plant. It may be known that 1 lb. of raw material is normally transformed into g lb. of product in a particular process, and that the volume of production varies from week to week. Then a weekly plot of the cumulative deviations:

$$\sum\{(\text{End-product weight}) - g \times (\text{Raw material usage})\}$$

against the cumulative raw material usage will highlight changes in the conversion efficiency in the process. The statistical properties of double cusum charts are discussed in [10.11].

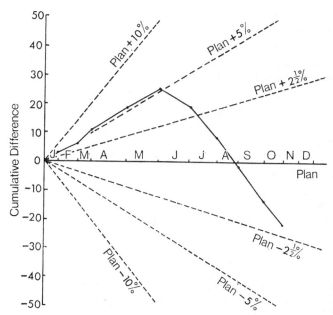

FIG. 10.6.   DOUBLE CUSUM CHART FOR A SEASONAL PRODUCT

TABLE 10.4.   COMPARISON OF ACHIEVED AND PLANNED PRODUCTION LEVELS

Month	Planned Production		Actual Production		Cumulative Difference
	Monthly	Cumulative	Monthly	Cumulative	
January	60	60	63	63	3
February	80	140	83	146	6
March	90	230	95	241	11
April	120	350	128	369	19
May	140	490	146	515	25
June	130	620	124	639	19
July	120	740	109	748	8
August	90	830	80	828	−2
September	110	940	98	926	−14
October	80	1020	72	998	−22
November	70	1090			
December	60	1150			

Concluding remarks

10.94 The above discussion does not exhaust the possible types of control charts that arise in practice. We emphasize that each process has to be examined in detail to find the best means of controlling the quality or performance, and that each process must be dealt with on its own merits. It will rarely be found that a type of control used in one process can be used in another without at least some modification.

Careful consideration must be given to deciding at what stage or stages in the process the control should be applied and what particular measurements should be used. Ideally, the control should be applied at the earliest stage where defects or changes can be detected or prevented, or where corrective measures can be applied. The criterion for control need not necessarily be a measurement on the finished product; it can be a measurement taken at an earlier stage that is known to be correlated with the measurement on the finished product. To find such a measure will require detailed examination of the process variables and may require specially planned experiments to establish the relationship. A control chart should not only be devised to suit the process but must also be acceptable to the process chemist or engineer. It may well be worth using a criterion that is not quite 100% efficient if it is easier to operate, more acceptable to the process men, or more in line with the previous methods of control. A control chart must therefore not be regarded as an end in itself; it is an aid to judgment additional to, but not in place of, technical knowledge of the process.

References

[10.1] SHEWHART, W. A. *Economic Control of the Quality of Manufactured Product*. Van Nostrand (New York, 1931).

[10.2] PEARSON, E. S. *The Application of Statistical Methods to Industrial Standardisation and Quality Control* (B.S. 600: 1935). British Standards Institution (London, 1960).

[10.3] DUDDING, B. P., and JENNETT, W. J. *Quality Control Charts*. B.S. 600R: 1942. British Standards Institution (London, 1942). (Now withdrawn as a British Standard).

[10.4] AMERICAN DEFENCE EMERGENCY STANDARDS. *Guide for Quality Control and Control Chart Method of Analysing Data* (Reproduced by courtesy of the American Standards Association). B.S. 1008: 1942. British Standards Institution (London, 1942).

[10.5] DUDDING, B. P., and JENNETT, W. J. *Control Chart Technique when Manufacturing to a Specification*. B.S. 2564: 1955. British Standards Institution (London, 1955).

[10.6] DUNCAN, A. J. *Quality Control and Industrial Statistics* (third edition). R. D. Irwin Inc. (Homewood, Illinois, 1965).

[10.7] LOWE, C. W. *Industrial Statistics*, Vol. 1. Business Books (London, 1968).

[10.8] WOODWARD, R. H., and GOLDSMITH, P. L. *Cumulative Sum Techniques*. I.C.I. Monograph No. 3. Institute of Manpower Studies (Brighton, 1981).

[10.9] EWAN, W. D. "When and How to Use Cu-Sum Charts", *Technometrics*, **5**, 1–22 (1963).

[10.10] BISSELL, A. F. "Cusum Techniques for Quality Control", *Applied Statistics*, **18**, 1–30 (1969).

[10.11] BISSEL, A. F. "Process Monitoring with Variable Element Sizes", *Applied Statistics*, **22**, 226–238 (1973).

Appendix 10A

Use of Shewhart control charts

10A.1 This appendix illustrates the use of Shewhart control charts in a research investigation on a plant manufacturing a plaster. The investigation was concerned with both the setting and hardening properties of the plaster.

The results are set out in the following tables, which show how the Population average and standard deviation are estimated:

Table 10A.1. Measurements made on initial setting time v in minutes, and hardening rate u in arbitrary units.
Table 10A.2. Division of measurements into rational sub-groups or samples.
Table 10A.3. Control chart calculations for v.
Table 10A.4. Control chart calculations for u.

TABLE 10A.1. MANUFACTURE OF A PLASTER

(*Measurements made on initial setting times v in minutes and hardening rate u in arbitrary units*)

Run*	Batch	v	u	Run*	Batch	v	u
48	1	$18\frac{3}{4}$	225	49	1	$14\frac{3}{4}$	211
	2	$22\frac{1}{4}$	222		2	17	211
	3	$23\frac{1}{2}$	263		3	$18\frac{1}{2}$	263
	4	$23\frac{3}{4}$	278		4	20	253
	5	$17\frac{3}{4}$	244		5	$19\frac{1}{2}$	238
	6	$17\frac{1}{2}$	270		6	$21\frac{1}{2}$	240
	7	$17\frac{1}{2}$	250		7	$20\frac{1}{2}$	255
	8	$33\frac{1}{4}$	250		8	19	232
	9	29	204		9	$18\frac{1}{4}$	238
	10	18	233		10	19	234
	11	$17\frac{1}{4}$	288		11	$14\frac{3}{4}$	208
	12	$20\frac{1}{4}$	267		12	$18\frac{1}{2}$	218
	13	$16\frac{1}{2}$	260		13	$19\frac{3}{4}$	227
	14	18	263		14	$18\frac{3}{4}$	235
	15	$21\frac{1}{4}$	244		15	$20\frac{1}{4}$	244
	16	$22\frac{1}{2}$	213		16	$18\frac{1}{2}$	226
	17	$21\frac{1}{2}$	232		17	18	200
	18	$21\frac{1}{2}$	230		18	17	207
	19	$21\frac{1}{2}$	213		19	$18\frac{1}{2}$	200
	20	19	210		20	18	190
	21	$11\frac{1}{2}$	212		21	$18\frac{1}{2}$	187
	22	$11\frac{1}{2}$	229		22	$18\frac{1}{4}$	176
	23	$12\frac{1}{2}$	240		23	19	183
	24	$15\frac{1}{4}$	232		24	21	180
	25	12	225		25	17	192
	26	$15\frac{1}{2}$	220		26	$15\frac{1}{2}$	194
	27	$13\frac{3}{4}$	246				
	28	16	202				

* A run represents a convenient number of batches for record and costing purposes.

TABLE 10A.1. MANUFACTURE OF A PLASTER (*continued*)

(*Measurements made on initial setting times v in minutes and hardening rate u in arbitrary units*)

Run	Batch	v	u	Run	Batch	v	u
50	1	24	282	51	1	$17\frac{1}{4}$	260
	2	25	293		2	$20\frac{1}{2}$	287
	3	$25\frac{3}{4}$	276		3	$21\frac{1}{2}$	292
	4	$22\frac{3}{4}$	270		4	15	262
	5	$21\frac{1}{2}$	275		5	$15\frac{3}{4}$	254
	6	15	266		6	15	259
	7	18	296		7	$14\frac{3}{4}$	263
	8	19	290		8	14	262
	9	$15\frac{1}{2}$	256		9	$13\frac{1}{2}$	254
	10	16	236		10	21	256
	11	21	296		11	$20\frac{3}{4}$	268
	12	$16\frac{1}{2}$	307		12	$15\frac{1}{2}$	247
	13	$15\frac{1}{2}$	300				
	14	$15\frac{1}{2}$	306				
	15	18	317				
	16	$17\frac{1}{4}$	296				
	17	16	306				
	18	$15\frac{1}{2}$	281				
	19	$14\frac{1}{2}$	266				
	20	$15\frac{3}{4}$	276				
	21	16	271				
	22	15	277				
	23	15	224				
	24	$15\frac{1}{2}$	289				
	25	16	253				

TABLE 10A.2. DIVISION OF MEASUREMENTS INTO RATIONAL SUB-GROUPS OR SAMPLES (BASED ON TABLE 10A.1)

No. of Sample	No. of Batch	No. of Sample	No. of Batch
	v		u
1	48/1, 2, 3, 4	1	48/1, 2, 3, 4
2	48/5, 6, 7, 10	2	48/5, 6, 7, 8
Omitted	48/8, 9	3	48/9, 10, 11, 12
3	48/11, 12, 13, 14	4	48/13, 14, 15, 16
4	48/15, 16, 17, 18	5	48/17, 18, 19, 20
5	48/19, 20, 21, 22		
		6	48/21, 22, 23, 24
6	48/23, 24, 25, 26	7	48/25, 26, 27, 28
7	48/27, 28; 49/1, 2	8	49/1, 2, 3, 4
8	49/3, 4, 5, 6	9	49/5, 6, 7, 8
9	49/7, 8, 9, 10	10	49/9, 10, 11, 12
Omitted	49/11		
10	49/12, 13, 14, 15	11	49/13, 14, 15, 16
		12	49/17, 18, 19, 20
11	49/16, 17, 18, 19	13	49/21, 22, 23, 24
12	49/20, 21, 22, 23	Omitted	49/25, 26
13	49/24, 25, 26; 50/1	14	50/1, 2, 3, 4
14	50/2, 3, 4, 5	15	50/5, 6, 7, 8
15	50/6, 7, 8, 9		
		16	50/9, 10, 11, 12
16	50/10, 11, 12, 13	17	50/13, 14, 15, 16
17	50/14, 15, 16, 17	18	50/17, 18, 19, 20
18	50/18, 19, 20, 21	19	50/21, 22, 23, 24
19	50/22, 23, 24, 25	Omitted	50/25
20	51/1, 2, 3, 4	20	51/1, 2, 3, 4
21	51/5, 6, 7, 8	21	51/5, 6, 7, 8
22	51/9, 10, 11, 12	22	51/9, 10, 11, 12

Batches were omitted where knowledge of the process suggested that assignable causes of variation had operated.

The calculations for the samples are summarised in Tables 10A.3 and 10A.4.

Note the reference to the units used in Table 10A.3.

TABLE 10A.3.   ESTIMATES OF μ_v AND σ_v (BASED ON TABLES 10A.1 AND 10A.2)

(*Units: Minutes except Col.* (3), *where unit is* $\frac{1}{4}$ *minute*)

(1)	(2)	(3)		(4)	(5)
No. of Sample	No. in Sample n	Sums of Differences from 18 min. $+$	$-$	Average $\bar{v}$	Range w
1	4	65		22·1	5·00
2	4		5	17·7	0·50
3	4	0		18·0	3·75
4	4	59		21·7	1·25
5	4		34	15·9	10·00
6	4		67	13·8	3·50
7	4		42	15·4	3·25
8	4	30		19·9	3·00
9	4	19		19·2	2·25
10	4	21		19·3	1·75
11	4	0		18·0	1·50
12	4	7		18·4	1·00
13	4	22		19·4	8·50
14	4	92		23·8	4·25
15	4		18	16·9	4·00
16	4		12	17·3	5·50
17	4		21	16·7	2·50
18	4		41	15·4	1·50
19	4		42	15·4	1·00
20	4	9		18·6	6·50
21	4		50	14·9	1·75
22	4		5	17·7	7·50
Totals	$N = 88$	$+324$ -337 $= -13$		395·5	79·75
Average Values	$n = 4$	$-13/88 = -0·1$		18·0	3·63

From Col. (3), $\mu_v = 18·0 - 0·1 \times \frac{1}{4}$ (min.) $= 18·0$ min.

From Col. (5), $\bar{w}_v = 3·63$, $d_4 = 2·059$ [Table G.1], $\sigma_v = \bar{w}_v/d_4 = 1·76$ min.

TABLE 10A.4.   ESTIMATES OF μ_u AND σ_u (BASED ON TABLES 10A.1 AND 10A.2)

(*Units: Arbitrary scale of hardness*)

(1)	(2)	(3)		(4)	(5)
No. of Sample	No. in Sample n	Sums of Differences from 250 $+$	$-$	Average $\bar{u}$	Range w
1	4		12	247	56
2	4	14		254	26
3	4		8	248	84
4	4		20	245	50
5	4		115	221	22
6	4		87	228	28
7	4		107	223	44
8	4		62	234	52
9	4		35	241	23
10	4		102	224	30
11	4		68	233	18
12	4		203	199	17
13	4		274	181	11
14	4	121		280	23
15	4	127		282	30
16	4	95		274	71
17	4	219		305	21
18	4	129		282	40
19	4	61		265	65
20	4	101		275	32
21	4	38		260	9
22	4	25		256	21
Totals	$N = 88$	$+930$ $-1{,}093$ $= -163$		5,457	773
Average Values	$n = 4$	$-163/88 = -1.9$		248.0	35.1

From Col. (3), $\mu_u = 250 - 1.9 = 248.1$ units

From Col. (5), $\bar{w}_u = 35.1$, $d_4 = 2.059$ [Table G.1], $\sigma_u = \bar{w}_u/d_4 = 17.0$ units

Initial setting time v

10A.2 From Table 10A.3 it will be seen that estimates of the Population average and standard deviation, etc. are:

$$\mu_v = 18\cdot0 \text{ min.}, \quad \sigma_v = 1\cdot76 \text{ min.}, \quad \overline{w}_v = 3\cdot63 \text{ min.}, \text{ and } n = 4$$

1. *Control Limits for Sample Averages* (*see Table G at end of volume*)

    (*i*) *Inner Control Limits*

    These are $18\cdot0 \pm 1\cdot76 \times 0\cdot980$, or $18\cdot0 \pm 3\cdot63 \times 0\cdot476$
    $= 19\cdot7$ and $16\cdot3$ min.

    (*ii*) *Outer Control Limits*

    These are $18\cdot0 \pm 1\cdot76 \times 1\cdot545$, or $18\cdot0 \pm 3\cdot63 \times 0\cdot750$
    $= 20\cdot7$ and $15\cdot3$ min.

2. *Control Limits for Sample Ranges* (*see Table G.1 or G.2*)

    These are:

    (*i*) *Inner Control Limits*—Lower: $1\cdot76 \times 0\cdot59 = 1\cdot0$ min.
    Upper: $1\cdot76 \times 3\cdot98 = 7\cdot0$ min.

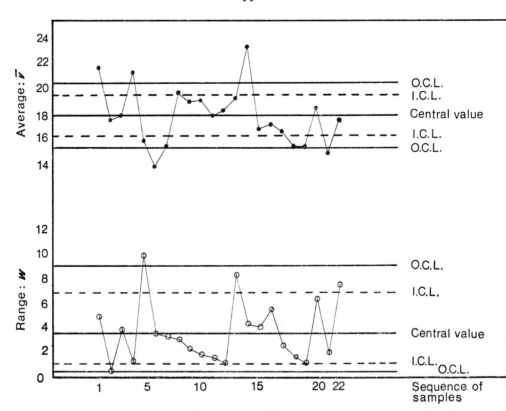

FIG. 10A.1.   CONTROL CHARTS.   INITIAL SET

(*ii*) *Outer Control Limits*—Lower: $1\cdot76 \times 0\cdot20 = 0\cdot4$ min.

Upper: $1\cdot76 \times 5\cdot31 = 9\cdot3$ min.

Central Value: $3\cdot63$ min.

Control charts for sample averages and ranges are plotted on Figure 10A.1. It will be seen that the range for Sample 5 is outside the upper outer control limit and should therefore be omitted, as the purpose of the investigation was to determine the inherent variation of the process freed from the effect of assignable causes of variation. The assignable causes were identified in the investigation. It was found that Sample 13 was also out of control on ranges when the revised limits were calculated. When Samples 5 and 13 are omitted and μ_v and σ_v recalculated, we have:

$$\mu_v = 18\cdot0 \text{ min.,} \quad \bar{w}_v = 3\cdot06 \text{ min.,} \quad \sigma_v = 1\cdot49 \text{ min.}$$

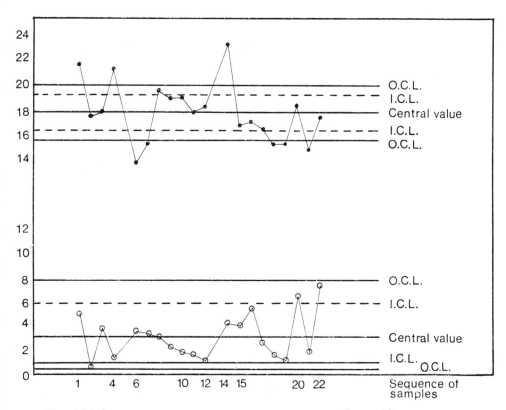

Fig. 10A.2. CONTROL CHARTS. INITIAL SET, SAMPLES 5 AND 13 OMITTED

The revised limit lines are as follows:

| Statistic | Control Chart Limits | | | | |
| | Lower | | Central Value | Upper | |
	O.C.L.	I.C.L.		I.C.L.	O.C.L.
$\bar{v}$	15·7	16·5	18·0	19·5	20·3
w_v	0·3	0·9	3·1	5·9	7·9

In practice a small amount of preliminary calculation would show which samples should be discarded, and only the final limits and charts need be drawn. It will be seen from Figure 10A.2 that sample ranges show good control, but that sample averages show pronounced lack of control; this is discussed below [§ **10A.4**].

Hardening rate u

10A.3 The control limits for u are as follows:

$$\mu_u = 248\cdot1 \text{ units}, \quad \bar{w}_u = 35\cdot1 \text{ units}, \quad \sigma_u = 17\cdot0 \text{ units}$$

| Statistic | Control Chart Limits | | | | |
| | Lower | | Central Value | Upper | |
	O.C.L.	I.C.L.		I.C.L.	O.C.L.
$\bar{u}$	222	231	248	265	274
w_u	3	10	35	68	90

The control charts are given in Figure 10A.3. Sample ranges show good control, whereas the sample averages show pronounced lack of control.

Interpretation of charts

10A.4 The charts show that in respect of initial setting time and hardening rate the process is inherently capable of being controlled with standard

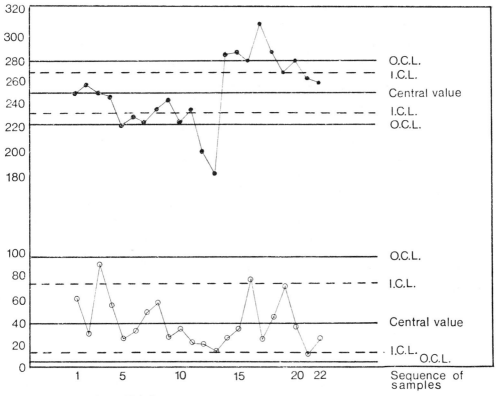

FIG. 10A.3. CONTROL CHARTS. HARDENING RATE

deviations of 1·5 min. and 17 units respectively. The sample averages, however, were not under control during the trials. This is the state of affairs illustrated in Figure 10.1.

Lack of control was due to the fact that no attempt was made during the trials to control the sample averages except within the specification limits. It will usually be found in investigations of this kind, where data are being obtained for a process where no control charts exist, that sample averages are out of control. If the quality characteristics had been plotted on charts during the plant trials they could have been brought under control. The research investigation discussed here has provided estimates of the variation to be expected from the process, and the next step would be to introduce the charts as part of the routine testing system to enable assignable causes of variation to be eliminated and to improve the uniformity of the product. The frequency of sampling would depend mainly on the degree of control achieved.

From practical experience it has been found that if the plaster is to be satisfactory it should comply with the following requirements:

Initial setting time: Between 13 and 30 min., i.e. 21·5 min. ± 8·5 min.

Hardening rate: Between 168 and 308 units, i.e. 238 units ± 70 units.

If the plant is to be capable of making a product practically all of which complies with the above specification, we must have:

Initial setting time: $3 \cdot 09\sigma_v$ less than $8 \cdot 5$ min.

But, in fact, $3 \cdot 09\sigma_v = 4 \cdot 6$ min., which is less than $8 \cdot 5$ min.

Hardening rate: $3 \cdot 09\sigma_u$ less than 70 units

But, in fact, $3 \cdot 09\sigma_u = 53$ units, which is less than 70 units.

Thus the required conditions are met and the specification limits lie outside the outer control limits, thus ensuring that if the process is operated under control practically all the product will meet the specification. It will be noted that the achieved averages ($\mu_v = 18 \cdot 0$ min. and $\mu_u = 248 \cdot 1$ units) differ from the specified averages. There would be no difficulty in adjusting the achieved averages to comply with the specification averages and thus in obtaining statistical control at the correct level. If it is desired to know the proportion of the product which will meet the specification, this can easily be calculated from Table A, as shown in § **2.44**.

Chapter 11

Sampling and Specifications

To maintain a product at a consistently high level of quality a sampling inspection scheme may be employed. In such a scheme a sample is taken from each manufactured lot and tested; on the results of these tests the lot is either accepted or rejected. In this chapter the various considerations that are relevant to the selection of an efficient scheme are discussed.

Introduction

11.1 The term "sampling" is customarily taken as indicating the segregation of a small fraction from a large bulk of material in such a way that the characteristics of the bulk can be estimated by studying those of the sample. The object of this chapter is to discuss relationships between size of fraction and precision of estimation, taking into account the characteristics being studied and the nature of the material.

Sampling may be carried out, for instance, on a chemical product, such as a fertiliser, the characteristic to be determined being its nitrogen or potash content. If we require the average value for a lot of 100 bags, sampling might be carried out by extracting a suitable quantity from representative bags, mixing and analysing. If, however, the nitrogen or potash contents of individual bags are required, analyses must be carried out separately on samples drawn from different bags. Statistical treatment of the results is clearly different in the two cases. In other problems material may be defective in a way which it is impossible or inconvenient to measure, for example because of chips, scratches, breaks, or other qualitative imperfections. A proportion of the lot would then be inspected and the number of imperfections counted.

11.11 The results obtained on a sample are often compared with standard figures, such as those quoted in a relevant specification. In fact, much sampling is carried out to ensure that materials comply with specification requirements, although there are instances where tolerances are not imposed but where the characteristics must be known so that the material may be priced or used to the best advantage, e.g. coal and other minerals.

The fact that the results obtained on a sample are compared with a standard value implies that sampling has been carried out previously on similar material. In practice it is generally a continuing process, carried out on successive lots or batches, so that prior information on the mean and standard deviation, or on the average per cent defective, is available. Such information is required in drawing up sampling plans. At the same time further information is continually accumulating, and may well be used in setting up standards, such as specification limits, for future use. Sampling and specifications cannot be considered independently, each having its influence on the other, and they are therefore discussed together in this chapter. Furthermore, no lot or batch should be regarded as an entirely separate entity, but should preferably be related to preceding or ensuing batches. This leads to procedures based on the cumulative sums which were introduced in the last chapter.

Random sampling and systematic sampling

11.2 A random sample is defined as a set of observations drawn from a Population in such a way that every possible observation has an equal chance of being drawn at every trial. In this chapter the Population is a lot or batch, each part or element of which should be equally likely to appear in the sample; this randomness is necessary in order to avoid possible bias.

The technique of obtaining a random sample obviously depends on the material being sampled and also on the particular properties that are being studied. If we are interested, for instance, in the weights of drums of material leaving a packing point, it is probably satisfactory to allow the check-weigher to go to the band conveyor at random intervals and each time take the first drum which comes along. If, however, we are interested in the condition of the paint film on the outside of the drum, there is a considerable risk that the man's choice may be influenced by the appearance of the drum, so that well-painted drums have a better (or possibly worse) chance of being chosen than badly painted drums. Such a sample would be said to be biased, and could not be regarded as a random sample. In spite of the bias for appearance, the sample might well be random for weight, as long as there was no connection between weight and appearance.

Practical considerations frequently make it difficult to obtain a truly random sample, but care should be taken to ensure that material from portions which do not come readily to hand stands a fair chance of being chosen. The judgment of the inspector or sampler may introduce a bias, and it may be considered desirable, in critical cases, to assign numbers to different portions of a lot and to select from these by a random method. This can be done by means of the random numbers given in Table K. In some instances practical considerations make it impossible to avoid bias, as for example in taking samples from sheets or rolls. They must then be taken from the ends or sides, but the amount of bias introduced should be determined initially, and at intervals, by thorough overall sampling of typical rolls or sheets, and correction applied in the case of subsequent samples.

It is common practice in sampling to work according to a systematic plan instead of in a truly random manner. Thus individual items or fractions may be taken at approximately equal intervals in time, in space, or by number. Such samples may frequently be perfectly satisfactory, but in their departure from the rules of random sampling they introduce a possibility of bias, and individual sampling schemes of this type should be carefully examined with this in mind. Three cases may be distinguished:

(i) Either the variations in the material are completely random, i.e. the material is "controlled" in the sense used in Quality Control [Chapter 10], or they have a period which is short compared with the sampling interval. In this case the systematic sample will be to all intents and purposes a random sample.

(ii) The variations in the material have a period which is long compared with the sampling interval. In this case systematic sampling is generally satisfactory.

(iii) The variations have a period comparable with the sampling interval. In this case there is a danger of the sampling and the variations getting into step and thus introducing a bias.

Where sampling has to be carried out at equal intervals according to a systematic plan, the position or time of the initial sample of the series should if possible be chosen in some random manner.

11.21 Statistical aspects of sampling are frequently concerned with relationships between sample size, lot size, and the accuracy with which the characteristics of the bulk should be estimated. The common assumption that a defined level of accuracy can only be attained by relating sample size to lot size is wrong, since truly random samples of the same size provide equally reliable estimates of bulk characteristics whatever the size of the lot when, as is generally the case, the sample is a small fraction of the lot.

However, when larger lots are being considered the amount at risk is greater, and a greater precision is therefore required, leading to an increased sample size. The indication is that the best sampling schemes are based on economic considerations, that is on minimum overall costs, following the principles already discussed in Chapter 5. As the information necessary to derive such schemes is not always available, only brief consideration will be given to them in this chapter. More attention will be devoted to schemes derived from consideration of defined risks of reaching wrong conclusions regarding the acceptance or rejection of a lot on the characteristics determined from a sample. The permitted risk of a wrong conclusion for a large lot will naturally be less than for a smaller lot. Manufactured lots of one product are generally of the same size, and so one set of defined risks can be adopted for this product.

If it is known or suspected that a lot is not uniform, it is advisable to divide it into approximately homogeneous sections, taking a random sample from each, i.e. to employ stratified sampling. The size of a lot is generally determined by the capacity of receptacles, storage facilities or transport arrangements. In theory, the optimum size is the largest which is likely to be homogeneous, depending on constancy of manufacturing conditions. The producer, knowing the order of manufacture, has the opportunity of building up homogeneous lots and of planning his sampling accordingly. The consumer, on the other hand, having no assurance that a lot can be regarded as uniform, may be faced with a heavy programme of stratified sampling, engendered by his lack of knowledge of order of manufacture. He may be able to establish that the production process is stable by applying Control Chart techniques to results obtained on a run of lots; but an alternative solution is that the producer should make his records available to the consumer and that they should be used as a basis for acceptance. In these circumstances the functional type of specification becomes redundant, but the methods of sampling, testing and recording should be carried out on an agreed basis and according to sound statistical principles.

In many instances a non-uniform lot can be made uniform by mixing and blending, and such operations are frequently included as part of normal production. Similarly, a number of samples drawn from different parts of the lot may be combined to form one large sample when the average quality only is being investigated. This presupposes, however, that variations in quality within the lot are of no consequence in the subsequent operation.

SAMPLING OF ATTRIBUTES

11.3 In Chapter 9 consideration was devoted to the binomial and Poisson distributions for treatment of data classified by frequency of occurrence or non-occurrence of certain events or attributes. Many sampling problems are concerned with data of this type, and the statistical theory developed previously is therefore applicable. As already indicated, the attribute may be a defect, such as a scratch or imperfection on a discrete article, or a break in a thread, a crack in a piece of metal, a weaving fault in cloth, a bubble in a plastic sheet, an improperly soldered joint, etc. By examining a portion of a lot, one can assess the percentage or number of such attributes and compare it with a standard figure. Circumstances may also arise in which it is inconvenient or uneconomic to measure a dimension on each unit of a sample of similar articles, and they may then be classified as being greater or less than a certain fixed size by means of a gauge. Such a procedure is common in engineering industries, and the results may again be recorded as a percentage or count of defective articles.

In the last example the gauge defines a defect by fixing a measured and repeatable limit, but it is also necessary to set up similar inspection standards for many of the less definite examples quoted previously. While a break is either present or absent, a scratch, fault or crack may vary in size, and a decision has to be made as to the maximum size which does not render the articles or material unacceptable. Furthermore, "standard defects" may be required which can be constantly referred to, so that successive lots are subjected to inspection of the same rigour; this, however, is a problem of inspection rather than of sampling.

As a simple example, consider the examination of a lot of discrete articles for the incidence of chips and breakages. A random sample ensures that repeat tests will provide the same results, within the limit of sampling error expressed by the binomial distribution [Chapter 9]. The examination is expected to reveal a small number of defective articles in the sample, and the result will be expressed as a percentage of defective articles. In the majority of instances the percentage will be small, so that the distribution of sampling errors will approximate to the Poisson type, for which there is a small chance of an occurrence in each of a large number of trials. The Poisson distribution is a sufficiently good approximation, in fact, when the percentage defective is not more than 10%.

With defects such as breaks in thread, the result of an examination is expressed not as a percentage but as the number of faults in a certain length or area. These numbers are usually subject to the Poisson or the negative binomial distribution; there is a large, although indefinite, number of occasions on which the fault may occur, but the probability of occurrence associated with each occasion is very small.

It should be noted that an implicit assumption has been made that a small number of defects is admissible. If not, then all the material must be examined and the question of sampling procedure does not arise.

Single sampling. Consumer's and producer's risks

11.31 The principles underlying the various methods of sampling for attributes may be expounded from a consideration of simple examples, although a complete theoretical treatment is outside the scope of this book; more details are given in [11.1] and [11.2]. Suppose in the first place that a sample of 100 pieces is being inspected and that it will be accepted if 2 or fewer defective articles are found and rejected if there are 3 or more. The Poisson distribution [§ **9.2**] defines the probabilities with which 0, 1, 2, 3, ... defective articles will be found in the sample when the percentage in the lot is known; conversely, a specified number of defectives may arise, with different probabilities, from lots having different percentages of defective articles. Thus three or more defectives will be found with a probability of 0·1 when the proportion in the lot is such that the expected number in the sample is 1·10, a value obtained by entering Table F with $a = 3$, $p = 0$ (for the Poisson distribution) and the lower confidence limit for probability P of 0·1. In other words, if samples of 100 are taken from a number of lots all containing 1·10% defective articles, then on average one sample in ten will contain 3 or more defectives and will be rejected. There is, however, no difference between the lots, and the producer, in adopting a sampling plan, undertakes a risk that it may reject some lots unjustifiably. This is known as the Producer's Risk. Table F also indicates that 2 or fewer defectives would be obtained with a probability of 0·1 when the expected number is 5·32. The consumer, therefore, also undertakes a risk of 0·1 that a lot containing 5·32% defectives will be accepted. This is known as the Consumer's Risk. Producer's and consumer's risks are seen to be the same as errors of the first and second kinds respectively [§ **5.43**]. Thus the producer's risk is the probability of asserting that there is a difference when none exists, while the consumer's risk is the probability of asserting no difference when a difference does in fact exist.

If a sample of 500 had been taken instead of 100, 9 or more defectives would have been found with a probability of 0·1 when the expected number was 5·43. The percentage of defective articles in the lot would then be 1·09%, and it will be seen, therefore, that in this particular illustration rejection with 3 defectives out of 100 or with 9 defectives out of 500 provides nearly the same producer's risk, 0·1, when the lot contains 1·1% defectives. The consumer's risks, however, are very different. Thus, the latter sampling plan would accept a sample containing 8 or fewer defects with a probability of 0·1 when the expected number is 13·0, equivalent to a percentage of 2·60% when the sample size is 500. The two plans will thus accept, with 0·1 probability, lots containing respectively 5·32% and 2·60% defective articles. Increasing the sample size has therefore made the inspection more efficient in the sense that an inferior lot is less likely to be accepted.

Attention need not be restricted to producer's and consumer's risks of 0·1, as the probabilities of accepting or of rejecting a lot containing any specified percentage of defective articles can be calculated, from the Poisson distribution, for a sampling plan defined by the sample size and the acceptable number of defective articles therein. Such calculations have, in fact, been made for the two sampling plans considered above, and the results are illustrated in Figure 11.1, which shows the relationship between the percentage of defective articles in the lot and the probability of acceptance.

Operating characteristic curve

11.32 The curves illustrated in Figure 11.1 are particular examples of Operating Characteristic Curves or Power Curves already considered in

Chapter 5, but it is worth while recalling some of the points made there. The operating characteristic curve gives the probabilities of accepting or of rejecting lots containing specified percentages of defective articles when a defined sampling plan is employed. Any sampling plan has an Operating Characteristic associated with it, defined uniquely by either:

(i) Specifying the sample size and the acceptable number of defective articles therein, or

(ii) Choosing two points $(p_a, 1-\alpha)$ and (p_r, β), where α is the probability of rejecting a batch in which the proportion of defective articles is p_a, and β is the probability of accepting a batch in which the proportion of defective articles is p_r, the "rejectable" level p_r being greater than the "acceptable" level p_a.

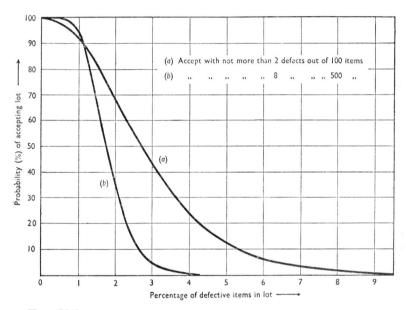

FIG. 11.1. OPERATING CHARACTERISTIC CURVES OF TWO SINGLE
SAMPLING PLANS

The argument in § 11.31 derives the second of these alternatives from the first. If p is less than p_a the probability of rejection is less than α; similarly, if p is greater than p_r the probability of acceptance is less than β. It is worth stressing that the curve for any given sampling plan is completely defined by two points only, and that when these are fixed the probabilities of accepting lots of differing qualities are also determined. The Operating Characteristic therefore contains complete information on the probable sentencing of a lot of any quality that may be presented. Furthermore, any two points may be chosen initially; thus curve (b) of Figure 11.1 is defined equally well by $(p_a = 1.1\%,\ 1 - \alpha = 0.9)$ and $(p_r = 2.0\%,\ \beta = 0.35)$ as by $(p_a = 1.1\%,\ 1 - \alpha = 0.9)$ and $(p_r = 2.6\%,\ \beta = 0.1)$. The value of p for which there is a probability of 0·5 that the lot will be accepted is termed the Indifference Quality Level, since there is also a probability of 0·5 that a lot with this percentage of defectives will be rejected.

The Operating Characteristic is of value in indicating the discriminatory power of a sampling plan. Thus, of the two plans illustrated in Figure 11.1, that with the larger sample size discriminates more effectively between good and inferior lots, as discussed in § **11.31**. However, the Operating Characteristic by itself provides no information on the quality of accepted material, but only indicates the probability of accepting a lot containing a specified proportion of defectives when such material is presented for inspection. If, in fact, the consumer's risk is 0·1 of accepting a lot containing 2·6% defectives, and 1% of the lots are of this quality, then 0·1% of the accepted material will be at this inferior level. The distribution of quality in production lots must be known, therefore, before the qualities of accepted and of rejected material can be determined. The mean quality of the accepted lots is termed the Average Outgoing Quality (A.O.Q.). For some sampling plans [11.1], the quality controller specifies an upper limit to the A.O.Q. instead of the two points $(p_a,\ 1 - \alpha)$ and $(p_r,\ \beta)$.

Example 11.1. Inspection of metal cups

11.33 The dimensions of a small metal cup were important from the point of view of its insertion into a cavity during an assembly operation. In particular, it should be of the correct diameter, which could be checked most readily by means of maximum and minimum gauges. Producer and consumer in this case were part of the same organisation, and the cups were inspected in lots of 10,000 at the end of their manufacture before passing to the next stage. The consumer was of the opinion that lots containing 2·5% defectives or more might well entail too frequent stoppages of his assembly machines: the consumer's risk was chosen, therefore, with his consent, as a 0·1 probability of accepting lots containing 2·5% defectives. As rejected lots would be scrapped, 100% inspection being uneconomic, the probability of rejecting a lot containing 0·5% defectives was fixed at 0·025. With this information regarding producer's and consumer's risks, acceptance and rejection numbers may be found from Table F. The result is that a sample of 320 is inspected, accepted if there are 4 or fewer defectives, and rejected if there are 5 or more. This result is readily obtained by trial from Table F. Thus the tabulated value for $a = 4$, $p = 0$, P (upper limit) = 0·1, is 7·99, i.e. with an average or expected number of defectives of 7·99 the probability of obtaining 4 or fewer in one sample is 0·1. This, however, is postulated as the probability with which a lot containing 2·5% defectives should be accepted, and the size of sample is therefore:

$$7.99 \times 100/2.5 = 320$$

Similarly, the tabulated value for $a = 5$, $p = 0$, P (lower limit) $= 0.025$, is 1·62, and as a lot containing 0.5% defectives should be rejected with a probability of 0·025, the sample size is $1.62 \times 100/0.5 = 324$. The procedure, therefore, is to find from Table F acceptance and rejection numbers (the latter being necessarily one greater than the former), that lead to approximately the same sample size for the defined probabilities. There is no need to determine sample sizes with any great accuracy, and the size could well be 300 instead of 320 without any appreciable loss of discrimination. Thus, a lot containing 2.66% defectives would then be accepted with a probability of 0.1.

Inspection had previously been carried out on large samples from 366 different lots during a period of six months. No sample contained more than 1.2% defectives, so that for this series of lots there was virtually no risk of passing a lot containing 2.5% defectives, which would be unacceptable to the consumer. Provided that the conditions of manufacture remain unchanged and the overall quality of the product consequently unaltered, no sampling is required for the purpose of deciding whether a lot is satisfactory from the consumer's point of view. Sampling is, in fact, needed only to ensure that production remains controlled at the same level, and to detect any assignable causes of variation that might arise. For such purposes, which could be dealt with by a Control Chart system, the sample size could safely be reduced below that of 300 deduced above, and this matter is considered further in § **11.35**. The sample size of 300 was deduced by considering one lot as a separate entity, unrelated to previous or ensuing batches. Considerable advantage may be gained, however, by regarding the manufacture as a whole, particularly taking into account such objective information as is available.

Decision-Theory Schemes

11.34 In the example just considered, the distribution of percentage defectives between lots could be estimated from the information provided by the lots produced over a period of six months. This is the Prior Distribution, used in conjunction with the Operating Characteristic Curve, to assess completely a given sampling scheme, as described in § **5.2**. The prior distribution defines the proportion of lots of a given quality, and the operating characteristic the probability of acceptance, from which the proportions of satisfactory batches which will be rejected and of faulty batches which will be passed may be calculated. Both are wrong decisions which will cost money either directly or indirectly. If these costs can be evaluated, then, together with the cost of sampling, the cost of the whole scheme can be found, which will form a basis for the derivation of the optimum size of sample.

A reasonable method of attack is by trial and error, starting with a few different sampling schemes and choosing that which provides minimum costs. Information on the costs of wrong decisions is frequently indefinite, so that calculation need not be pursued to the limit of precision. On the other hand, mathematical theory is available for many practical situations; useful papers have been published by Barnard [11.3] and Hald [11.4]. It may be worth while programming a computer to evaluate the economics of alternative schemes.

Double and multiple sampling

11.35 A discouraging feature of sampling by attributes is the large sample size that is necessary to discriminate between lots containing even widely different percentages of defectives, such as 0·5% and 2·5%. The amount of inspection may be reduced by using Control Charts to establish statistical control and to ensure its maintenance by examining comparatively small samples. Other methods for reducing sample size may be illustrated by considering the plan for the inspection of cups, in which the sample size is 300.

If a lot of inferior quality is presented for inspection, the number of defectives for rejection may be found when only a proportion of the sample has been examined. Inspection might then be stopped and the lot rejected, although the size of the sample examined is smaller than 300. It can also be shown that a lot of very good quality may be accepted legitimately when a relatively small sample has been examined, and the full sample of 300 will be needed only with lots of intermediate quality, more particularly those near the range 0·5% to 2·5% defectives associated with producer's and consumer's risks. It is reasonable, then, to adopt a Double Sampling Plan, examining a first sample in relation to assigned acceptance and rejection numbers, and a second only when a firm decision is not indicated by the first results. The advantage is that the total amount of inspection is less than with a Single Sampling Plan, since some lots should be accepted or rejected after inspecting the first sample.

The relative sizes of the first and second samples may be chosen in any convenient way; a ratio of 1:2 is generally satisfactory. In the example under discussion an acceptance number of 0 defectives in a first sample of 100 items corresponds to a 0·1 probability of accepting a lot containing 2·30% defectives. Similarly, a rejection number of 3 corresponds to a 0·025 probability of rejecting a lot containing 0·62% defectives. These probabilities, however, are only an approximation to the producer's and consumer's risks associated with the full Double Sampling Plan, in which a second sample of 200 is examined if there are 1 or 2 defectives in the first, and the lot is accepted if there are not more than 4 defectives in both samples combined. It may be shown, in fact, that this plan rejects lots with 0·5% defectives with a probability of 0·026, and accepts those with 2·9% defective with a probability of

0·1. The differences in risk levels introduced by subdividing the sample may well be greater in other instances, and those seeking further guidance are advised to study [11.2].

Inspection effort may be still further reduced by the use of three or more successive samples, reaching a decision to accept, reject or continue inspection when each sample has been examined. These schemes are termed Multiple Sampling Plans and consist of a list of the sample sizes and the acceptance and rejection numbers at each stage. Computers can be programmed to calculate the operating characteristic of any plan, and if cost information and the incoming quality curve are known a complete economic evaluation of the scheme is feasible.

Multiple sampling plans call for careful recording of results, and to inspectors who are not fully aware of the statistical background they may appear to be indecisive in doubtful cases. If the inspection procedure is complicated or destructive, making it costly in relation to the selection of the sample, it would be preferable to proceed to the logical limit and to examine the accumulated results after the inspection of each individual piece, i.e. to employ Sequential Sampling. Sequential methods are outlined in the succeeding sections; a fuller discussion will be found in Chapter 3 of *Design and Analysis*.

Sequential sampling

11.36 In sequential sampling the size of sample is not fixed in advance; a test is applied to the accumulated data after each article has been examined, and sampling is terminated as soon as this shows that a decision either to accept the lot or to reject it can be made with the required degree of certainty. The characteristics of a Sequential Plan, like those of plans involving single or double sampling, are dependent on the producer's and consumer's risks, α and β, and on the fractions of defective articles, p_a and p_r, associated with these risks. In general the values of p_a and p_r are sufficiently small to allow the Poisson approximation to the binomial distribution. With imperfections in wire, thread, sheet, cloth or similar products, the number of faults in a certain length or area is recorded and the distribution is often then also of the Poisson type.

We use the letter m to denote either the average number of defects or the average number of defective items per unit inspected. In the latter sense it is the proportion of defectives, which has been denoted above by p. Then:

α is the risk of rejecting a lot in which m_a is either the proportion of defectives or the number of defects per unit;

β is the risk of accepting a lot in which m_r is either the proportion of defectives or the number of defects per unit $(m_r > m_a)$.

These four quantities must be chosen on economic considerations, and they define two points on the operating characteristic curve.

The advantage of sequential sampling is that the total number of items inspected over a long run of lots is generally less than that in any other system for the same values of m_a, m_r, α and β. Nevertheless, attention should first be given to the size of the sample likely to be encountered. This will depend on the quality of material presented for inspection, very good or very poor lots needing relatively small samples to reach a decision. Maximum sample sizes normally occur with lots having a fraction of defectives intermediate between m_a and m_r, but the amount of inspection will differ from lot to lot, even when they are of the same quality. *Average* sample sizes encountered with lots of quality m_a and m_r are respectively:

$$\frac{(1 - \alpha)h_1 - \alpha h_2}{s - m_a} \quad \text{and} \quad \frac{(1 - \beta)h_2 - \beta h_1}{m_r - s} \tag{11.1}$$

where

$$h_1 = \frac{\ln\{(1 - \alpha)/\beta\}}{\ln(m_r/m_a)}$$

$$h_2 = \frac{\ln\{(1 - \beta)/\alpha\}}{\ln(m_r/m_a)} \tag{11.11}$$

$$s = \frac{m_r - m_a}{\ln(m_r/m_a)}$$

These expressions are derived from a consideration of likelihood ratios in Chapter 3 of *Design and Analysis*. Table K in *Design and Analysis* includes values of $\ln\{(1-\alpha)/\beta\}$ and of $\ln\{(1-\beta)/\alpha\}$ for various levels of α and β.

The greatest average sample size occurs with lots having a quality close to $m = s$, and approximates to:

$$h_1 h_2/s \tag{11.12}$$

For practical purposes it may be assumed that the amount of inspection required to reach a decision for any particular lot will not exceed three times the largest of the average sample sizes calculated above. If the result is regarded as excessive, a compromise must be worked out between degree of protection and amount of inspection by modifying the values of m_a, m_r, α or β. Average sample size will be reduced by increasing α or β or the interval between m_a and m_r.

Having fixed reasonable values for m_a, m_r, α and β, a suitable test can be worked out which may be applied to the accumulated inspection results to indicate whether to accept the lot, to reject it, or to continue sampling. When n items have been inspected the greatest allowable cumulative number of defects for the lot to be accepted is d_1 and the smallest cumulative number for it to be rejected is d_2, where d_1 and d_2 are given by:

$$\left.\begin{array}{l} d_1 = -h_1 + sn \\ d_2 = h_2 + sn \end{array}\right\} \tag{11.2}$$

and h_1, h_2 and s have the values defined above (*11.11*). Before inspection commences, values of d_1 and d_2 can be calculated for a series of values of n and a table drawn up of acceptance and rejection numbers corresponding to number of items inspected. As inspection of an actual sample proceeds, observed numbers of defects are compared with tabulated values, and a decision is taken to accept, to reject, or to continue sampling, depending on whether the observed number is not greater than d_1, not less than d_2, or intermediate between them. Alternatively, the straight lines represented by Equations (*11.2*) may be drawn on a graph on which points showing cumulative number of defects against number of items inspected are then plotted. Sampling is continued as long as the trace falls between the two lines but is terminated as soon as it crosses a line, the batch being accepted if it crosses the lower line and rejected if it crosses the upper line. It may sometimes be convenient to inspect items in groups instead of individually; this is a legitimate procedure, resulting only in slightly increased sample sizes.

Occasionally in the intermediate region very large samples may have to be taken before the trace crosses one of the lines. This possibility may be regarded as an unacceptable feature of the sequential sampling scheme. If so, an upper limit can be set to the number of items inspected, curtailing further inspection with a definite decision to accept or reject the lot according to the cumulative number of defects recorded at that point. The curtailment will slightly alter the risks α and β characterising the scheme.

Example 11.2. Sequential inspection of drums

11.361 Drums for containing liquids, received in batches of approximately 500, are inspected visually for external defects, such as indentations, score marks, assembly faults, etc. As the inspection of each drum occupies a few minutes, available manpower limited the total amount of inspection, restricting the average sample size to about 30 or 40. A sequential scheme was clearly desirable, not only because the physical process of inspection would in any case select drums one by one in sequence, but also because such a scheme would provide the best protection for a defined total amount of inspection. Statistical investigations were directed, therefore, to finding a reasonable balance between average sample size and the risks involved, presenting alternative schemes to producer, consumer and inspector. The plan eventually chosen gave a 0·05 probability of rejecting a lot containing 1% defectives and a 0·1 probability of accepting a lot containing 8% defectives, so that

$m_a = 0.01$, $m_r = 0.08$, $\alpha = 0.05$, $\beta = 0.1$. Therefore from (11.11):

$$h_1 = \frac{\ln (0.95/0.1)}{\ln (0.08/0.01)} = 1.0826$$

$$h_2 = \frac{\ln (0.9/0.05)}{\ln (0.08/0.01)} = 1.3900$$

$$s = \frac{0.07}{\ln (0.08/0.01)} = 0.0337$$

The average sample size expected with lots containing a fraction m_a of defectives is therefore, from (11.1):

$$\frac{0.95 \times 1.0826 - 0.05 \times 1.3900}{0.0337 - 0.01} = 40$$

Similarly, the average sample size expected is 25 with lots of quality m_r, and is 45 when $m = s$.

Adopting the graphical method for assessing inspection results, the equations of the two straight lines delimiting the zones of acceptance and rejection are, from (11.2):

$$d_1 = -1.0826 + 0.0337n$$

and
$$d_2 = 1.3900 + 0.0337n$$

These are included in Figure 11.2, from which it will be noticed that the lower line meets the n-axis at a value just exceeding 32, so that at least 33 drums must be inspected before a lot can be accepted. Similarly, rejection can only occur when at least two drums have been examined.

A limit of 140 was also fixed for the maximum number of drums which would ever be inspected from one batch. This bound closes the zone of continued sampling in Figure 11.2. If this number were reached, it was decided to accept the batch when just four defective drums had been noted and to reject it otherwise.

The consumer would have preferred a scheme in which there was a 0.1 probability of accepting a lot containing 5% defectives (i.e. $m_r = 0.05$), but the minimum sample size for acceptance would then have been 56, and the average sample size for lots of m_a quality, 83. These figures implied a greater inspection effort than could reasonably be undertaken, and the degree of protection was therefore relaxed, with the consumer's consent. Consideration of the costs consequent on this sampling scheme would have allowed a more precise definition of reasonable inspection effort, but the necessary information was not available.

11.362 Although the advantage of sequential sampling is that the total number of items inspected over a long run of lots is generally less than that in any other system with the same values of m_a, m_r, α and β, this economy may be obtained at the expense of a greater administrative effort in planning inspection procedures and in recording the result after inspecting each item. If the cost of obtaining the sample is large relative to that of carrying out the inspection or test, there may be little economic advantage in employing sequential methods. It is also essential that the sample should be random, i.e. each item in the lot should have the same chance of appearing in the sample, and care must be taken to maintain this requisite when items are being selected and inspected one by one.

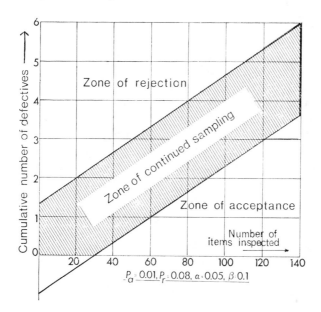

FIG. 11.2.   GRAPHICAL ASSESSMENT OF SEQUENTIAL
SAMPLING. (BINOMIAL DISTRIBUTION)

Preparing a plan for sampling for attributes

11.37 When choosing a plan for sampling for attributes, decisions based on technical or economic considerations must first be taken on the quality m_a of material which the producer does not wish to be rejected and the quality m_r which the consumer is not prepared to accept, and on the associated risks α and β. If the consequences of accepting a lot of poor quality or of rejecting a lot of good quality are serious or costly, the values of m_a and m_r should be close together, thus giving a fairly steep operating characteristic curve, and the requisite protection obtained at the expense of large samples. If the scheme becomes too expensive because of the sample size, the interval between m_a and m_r should be increased, but the scheme will then be less discriminatory.

The economic balance between acceptance and rejection levels and amount of sampling should be found by consultation between producer and consumer, and if possible by calculating costs consequent on the inspection scheme. The choice of a fixed sample size or of a sequential scheme will be indicated by the nature of the inspection problem, as discussed above. A scheme involving a fixed sample size can be worked out in detail from Table F for values of α and β of 0·005, 0·025 and 0·1; these should suffice for many applications. Appropriate formulae for sequential plans are given above.

Sampling clauses in specifications

11.38 When sampling inspection is required to ensure compliance with specification requirements, a clause to the effect that the proportion of defective articles must not exceed a certain figure, e.g. 2%, is not operationally practical. Thus it would require a very large sample to distinguish between 1·8% and 2%, and supplier and purchaser might well be involved in unresolvable arguments with batches of borderline quality. On the other hand, a specification sampling clause defining the size of the sample and the allowable number of defective articles, particularly if it includes a statement of the sampling risks involved, is unambiguous and readily put into operation. Nevertheless, a requirement such as "no defective articles should be found in a sample of 10" provides very limited protection. Such a sampling scheme, in fact, accepts 10% of lots containing 21% of defectives, if they are presented for inspection, and rejects 10% of those containing 1%. While it may be satisfactory in restricted circumstances, such as checking that an established level of quality is being maintained, the consequent risks should be fully appreciated, and in general a larger sample would be preferred. Sequential schemes can be included in specifications by reference to a table of acceptance and rejection numbers, preferably including a statement of the consequent risks.

Sampling clauses in specifications generally legislate for all eventualities, including the possibility of inspecting very inferior lots. However, the producer may be able to demonstrate, by means of control charts, for instance, that his process is stable and that unacceptable lots are presented only very occasionally. A reduction in sample size is then justified, by mutual agreement, to such a level as to ensure only that stability is being maintained, as discussed in § **11.33**. If it were established that production was no longer stable, the larger sample size would be reinstituted to reject lots of unacceptable quality. The specification for such a scheme would include a clause to the effect that the smaller sample size would be permissible, provided that no rejections had taken place in a run of lots (e.g. the last twenty), and that the average percent defective was less than a specified figure.

Process control of attributes

11.4 The producer should be able to reduce the number of unacceptable lots by operating a quality control scheme. Samples should be taken at each stage in the process to monitor the number of defective articles and to take appropriate action. If an increase in the level of defectives occurs, it may be possible to correct it by adjusting a process variable. Moreover, if inspection is carried out at an early stage in the production sequence, it may be possible to avoid the losses associated with a poor lot passing right through the process.

Shewhart control charts for plotting the number of defectives in each individual sample were outlined in § **10.9**, but more efficient schemes are based on the cumulative number of defectives counted in successive samples. These schemes are similar to the sequential sampling procedures described in § **11.36**, except that there is no equivalent to the zone of acceptance. They can be applied without plotting a graph and involve very little labour.

We continue to use the letter m to denote the average number of defective items or the average number of faults per unit inspected. A certain low level m_a will be regarded as the average number which is accepted during normal running, because they constitute inevitable imperfections in the process which would be far too costly to eliminate. But if the average number rises to a rejectable quality level m_r we want to know this quickly and to take corrective action.

As each sample is taken the number of defectives is compared with a Reference Level k, which is a rounded number approximately midway between m_a and m_r. As long as the observed number does not exceed k nothing further is done. But when a higher value occurs the difference from k is calculated and an accumulation of the algebraic deviations is started. This continues until either:

(a) The cumulative deviations fall below zero; in this case a new accumulation is not begun until the number of defectives next exceeds k.

or (b) The cumulative sum reaches a Decision Limit h (the term Decision Interval is also used); this is a signal to search for the cause of the higher number of defectives and if possible to eliminate it. It may also be necessary to downgrade recent material.

This type of control scheme has the merit that needless calculations are avoided when the level of defectives is satisfactory, but that significant increases in the level will be spotted quickly. The value of the limit h is fixed by reference to the Average Run Lengths [§ **10.5**] L_a and L_r of the scheme when the average numbers of faults are m_a and m_r respectively. We require L_a to be high and L_r to be low. Table N at the end of the book lists combinations of m_a, $R(= m_r/m_a)$, k and h corresponding to $L_a = 500$, 250 and 125 and $L_r = 5$, 7·5 and 10; other combinations are given in [11.5]. They are based on a Poisson distribution of numbers of faults.

For schemes to control the fraction of defective items p, m represents the average number of defective items per sample. To design a scheme when p_a

and p_r have specified values, calculate the ratio $R = p_r/p_a$, look up Table N to find the values of m_a, k and h which correspond to the desired levels for L_a and L_r and hence deduce the necessary sample size $n = m_a/p_a$. If this proves to be too large, the requirements on L_a and L_r will have to be relaxed. (This approach is based on the fact that the Poisson distribution is a convenient approximation to the binomial distribution when $p < 0.1$.)

When the number of defective articles in a sample or the number of defects per unit inspected follows a negative binomial distribution, the value of L_a is reduced for given m_a, k and h. Control schemes for these types of data are discussed in reference [11.6].

Example 11.3. Control of fabric faults

11.41 As an example of a scheme for a Poisson variate, consider the control of the level of faults which can be seen in lengths of fabric by an expert checker. Suppose that quality is regarded as acceptable if the faults do not exceed a mean of 5 per 100 yard length, but rejectable if the mean number of faults per 100 yards exceeds 15. We suppose also that we require $L_a = 500$ and $L_r = 7.5$. Then entering Table N at $R = 15/5 = 3$, we find:

$$m_a = 0.46, \quad k = 0.9, \quad h = 3.5$$

For this scheme we need to examine fabric lengths of

$$(0.46/5) \times 100 = 9.2 \text{ yards,}$$

count the number of faults in each length and accumulate their deviations from 0.9. Action is taken if the cumulative sum of deviations exceeds 3.5. This is illustrated in Table 11.1, which lists a sequence of fault counts leading to action after the 18th fabric length is inspected.

SAMPLING OF VARIABLES

11.5 When the criterion on which the sample is judged is a measurement as distinct from an attribute, the data are treated by the methods described in Chapters 3–5. The measurement may be any mechanical, physical or chemical property; the material may either consist of discrete articles or be a mass of liquid, gas or solid in lump, powder or sheet form. The object of sampling is to assess the mean value of the measurement and possibly its standard deviation.

The physical process of taking the samples will clearly depend on the type of material being dealt with, but the same statistical principles will apply whether the measurements comprise, for instance, the weights of pellets, the viscosity of oil, the ash content of coal, or the tensile strength of metal specimens. Further consideration need not be devoted, therefore, to the form of the material, except in so far as it may influence the statistical treatment.

TABLE 11.1. CUSUM CONTROL OF FABRIC FAULTS

Sample Number	Number of Faults	Deviation from 0·9	Cumulative Sum
1	0		
2	1	0·1	0·1
3	0	−0·9	−0·8
4	0		
5	2	1·1	1·1
6	0	−0·9	0·2
7	1	0·1	0·3
8	0	−0·9	−0·6
9	0		
10	1	0·1	0·1
11	0	−0·9	−0·8
12	2	1·1	1·1
13	0	−0·9	0·2
14	2	1·1	1·3
15	1	0·1	1·4
16	1	0·1	1·5
17	2	1·1	2·6
18	2	1·1	3·7

Random sampling

11.51 If random samples of n articles or individual units are taken from the batch, then it was shown in § **3.71** that if μ is the average value of any property in the batch and σ is the standard deviation from article to article, the average value $\bar{x}$ of the property in the sample will vary round μ with standard error $\sigma/\sqrt{n}$. It was also stated that, even when the original distribution is not Normal, the distribution of $\bar{x}$ rapidly approaches the Normal form as n is increased. Provided the original distribution is unimodal and not excessively skew, the distribution of $\bar{x}$ can be assumed to be Normal, even for sample sizes as small as 3. Thus, the limits for the value of μ at various probability levels can usually be determined with sufficient accuracy by establishing Confidence Limits [§ **4.21**] $\bar{x} \pm u\sigma/\sqrt{n}$, the value of u for a given probability level being obtained from Table A. When σ is not known it has to be estimated from the sample, and the confidence limits are then $\bar{x} \pm ts/\sqrt{n}$, where the appropriate value of t is obtained from Table C.

Example 11.4. Sampling of fertiliser

11.511 A batch of fertiliser in bags had to be sampled for chemical estimation of certain constituents. A representative portion was withdrawn from each of eight bags selected at random and subjected to analysis; the results for the estimation of soluble phosphorus pentoxide were 6·34, 6·23, 6·47, 6·27, 6·43, 6·52, 6·25, 6·54. The mean of these results is 6·381 and the standard deviation 0·125; the standard error of the mean is therefore 0·044. As the standard deviation has been determined from the sample itself, confidence limits must be derived by employing a value of t from Table C, which for a probability of 0·025 with 7 degrees of freedom is 2·36. The 95% confidence limits for expressing the precision of the mean are therefore:

$$6·381 \pm 2·36 \times 0·044$$

or
$$6·485 \quad \text{and} \quad 6·277$$

Precision of the average

11.512 If, in addition to the variation of the average value of the property in the sample, we consider also the errors in the determination of this property, two cases arise:

(a) If the property is measured on each article in the sample separately and then averaged arithmetically, the variance of the resulting average will be $(\sigma_1^2 + \sigma_2^2)/n$, where σ_1 is the standard deviation among the articles in the batch and σ_2 is the standard deviation of the determination on the individual articles.

(b) If, on the other hand, the average is determined on the sample as a whole (e.g. finding the average weight of a number of pellets by weighing the whole sample and dividing by the number of pellets), then if σ_3 is the standard deviation of the determination on the whole sample (e.g. total weight of pellets), the standard deviation of the determination of the average will be σ_3/n, and therefore the total variance of the resulting average will be $(\sigma_1^2/n + \sigma_3^2/n^2)$.

When the error in the determination is independent of the magnitude being determined, $\sigma_3 = \sigma_2$, and method (b) will give the more precise result. If, on the other hand, the error is proportional to the magnitude being determined (i.e. the coefficient of variation is constant), then $\sigma_3 = n\sigma_2$ and the variance for method (b) reduces to $(\sigma_1^2/n + \sigma_2^2)$. Hence in this instance the arithmetic average of determinations on individual articles will give the more precise result, though it will of course involve more work.

A case of particular importance arises when the property measured is composition as determined by a chemical analysis. The sample as a whole may be ground (in the case of solids), thoroughly mixed, and divided before analysis. The process of mixing averages the composition over the n individual items and the analysis gives the average composition direct. Since the precision of the analysis is independent of the size of the original sample, the total variance will be $(\sigma_1^2/n + \sigma_2^2)$, where σ_2 is the standard deviation of the analytical method.

In practice s will be used as an estimate of σ in the various cases discussed above. A more complete account of investigations into sampling and testing errors is contained in Chapter 4 of *Design and Analysis*.

Stratified sampling

11.52 When it is known or suspected that the average values of the properties of the material may not be the same in all parts of the bulk, Stratified (or Representative) Sampling should be employed. The bulk to be sampled is divided into a number of real or imaginary sections, and the sample is made up by drawing an amount from each section proportional to its size. The portions should be drawn from the individual sections in a random manner and are known as Increments. If the variation between the sections is con- siderable compared with the variation within the sections, it will be easily seen that, since each section is represented in the sample by its own increment, the composition of the sample will correspond more closely to that of the bulk than would be the case if each increment had been drawn at random from the bulk as a whole, which would probably result in some sections supplying a disproportionately large portion of the sample. Such a random sample would of course be unbiased in the sense that the average of a large number of such samples would tend to the average of the bulk; the sample mean would, however, have a higher standard deviation than that of the corresponding stratified sample.

Consider a batch of N articles divided up into a number of sub-groups containing respectively N_1, N_2, N_3, etc., articles. Consider some property whose average value μ for the whole batch we desire to measure, and let the average values of this property in the sub-groups be μ_1, μ_2, μ_3, etc., and the standard deviations of individual articles in the sub-groups be σ_1, σ_2, σ_3, etc. Then:

$$\mu = (N_1\mu_1 + N_2\mu_2 + N_3\mu_3 + ...)/N$$

From each sub-group relatively small numbers n_1, n_2, etc., of articles are extracted and measured, the proportion in each group being the same, that is, $an_1 = N_1$, $an_2 = N_2$, etc., where a is a constant. Let the total number of articles in the sample be n, where $n = n_1 + n_2 + n_3$, etc. The average value of the property in the sample is given by:

$$\bar{x} = (n_1\bar{x}_1 + n_2\bar{x}_2 + n_3\bar{x}_3 + ...)/n$$

and it will vary round μ with variance:

$$V(\bar{x}) = (n_1\sigma_1^2 + n_2\sigma_2^2 + n_3\sigma_3^2 + ...)/n^2 \qquad (11.3)$$

If the sub-groups have the same standard deviation, $\sigma_1 = \sigma_2 = \sigma_3 = ... = \sigma$, this formula reduces to:

$$V(\bar{x}) = \sigma^2/n \qquad (11.31)$$

It should be noted that the precision depends on the standard deviation *within* the sub-groups and not, as in the case of random sampling, on the standard deviation of the batch as a whole. The above reasoning may be applied when the material is a continuous medium, such as a powder or liquid, instead of being composed of discrete articles. If appreciable variations in the bulk are suspected, the bulk is divided into portions, the number of increments from each being proportional to its size. Since few materials can with certainty be regarded as homogeneous, stratified sampling should normally be used whenever a sample is taken in a number of increments that are to be mixed together before examination.

Example 11.5. Stratified sampling of pellets

11.521 A representative sample is to be taken from a consignment of 10 tins of pellets consisting of a mixture of sodium nitrite and ammonium chloride. This sample is to be used to determine the average chlorine content of the consignment. The pellets normally contain 29% of chlorine, and the variation from pellet to pellet in the same tin is known from previous work to correspond to a standard deviation of about 1·2% chlorine, the analytical error being negligible. How many pellets should be taken from each tin in order that the sampling error of the average for the consignment should have a standard deviation of less than 0·1% chlorine?

In this case we have $\sigma = 1·2\%$, and $V(\bar{x})$ should not exceed $(0·1)^2$.

From Formula (*11.31*) we have:

$$n = \sigma^2/V(\bar{x}) = (1·2/0·1)^2 = 144$$

The requisite accuracy would therefore be obtained by taking 15 pellets from each tin, making 150 in all. These would then be ground, mixed, subdivided, and analysed for chlorine.

If the standard error of analysis, σ_T, is appreciable, more than one analysis may have to be carried out. For m analyses:

$$V(\bar{x}) = \frac{\sigma^2}{n} + \frac{\sigma_T{}^2}{m} = 0·1^2$$

and n and m may be readily chosen by trial. Several solutions may be possible, and the choice between them can be made on a cost basis.

It should be noted that in using Formula (*11.31*) we have made the tacit assumption that all the pellets are of the same weight and that therefore the arithmetic average of the compositions of the individual pellets is identical with the average composition by weight. Only when the distribution is very skew would this assumption materially affect the result, in which case a weighted average would have to be used.

Producer's and consumer's risks

11.53 Most sampling will be carried out to ensure compliance with specification requirements in which maximum or minimum values, or both, are specified. Sampling errors considered in relationship to these limits necessitate consideration of producer's and consumer's risks, already discussed in § **11.31** for sampling by attributes.

Consider an example in which the 95% confidence interval for the mean value of the property is $(\bar{x}-3, \bar{x}+3)$, i.e. there is a probability of 0·05 that the true value will lie outside this range. If a sample mean happens to fall just on a specification limit, there is a 0·025 probability that the consumer will accept material in which the true value is 3 or more units on the wrong side of the specified limit, and conversely a similar probability that the consumer will reject material the true value of which is 3 or more units on the right side of the specified limit. As in the former discussion, these probabilities are the consumer's and producer's risks. The practical result is that the sentencing of a lot is indeterminate when the true value is close to a specified limit, but that the sentencing, either acceptance or rejection, becomes more definite when the true value is remote from this limit. The operating characteristic curve, showing the relationship between true value and probability of acceptance, is, in fact, a cumulative Normal curve for samples of reasonable size. It is defined either by the mid-point, standard deviation and sample size, or by producer's and consumer's risks and the levels of quality associated with them.

The range of uncertainty for defined producer's and consumer's risks can be reduced by increasing the size of the sample, but this may be uneconomic when the majority of lots have mean values which are well removed from the specified limits. In these circumstances it would be profitable to adopt a double sampling plan, taking a second sample when the result of the first falls in the range of uncertainty. Economy of sampling would also be achieved by the use of control charts. Whenever possible, the economics of a sampling scheme should be considered by reference to the distribution of quality of lots presented for inspection, and costs of inspection, rejection, and acceptance of inferior material.

Sequential sampling of variables

11.54 When every lot is tested, instead of adopting a multi-sampling technique it may be more logical to employ sequential sampling, as discussed in § **11.36**. The results are then assessed after each observation, and a decision is reached to accept the lot, reject it, or continue sampling.

The sequential plan is defined in terms of the five quantities m_a, m_r, α, β, and σ. When lots of worse quality than the "rejectable" level m_r are presented the probability of acceptance will be less than β, and similarly lots of better quality than the "acceptable" level m_a will be rejected with a probability less than α. The value of m_a may be greater or less than m_r, depending on whether the specified limit is a minimum or maximum figure, and the difference between them, as well as the values of α and β, should preferably be decided from technical and economic considerations.

The operating characteristic curve is fixed by two points, (m_a, α) and (m_r, β). Figure 11.3 illustrates the curves for both minimum and maximum specification limits.

The size of sample required to reach a decision will depend chiefly on the quality of lots presented for inspection. Thus, small samples will suffice for lots of very good or of very bad quality, which will be quickly accepted or rejected; whereas lots of intermediate quality, particularly those with average values between m_a and m_r, will require larger samples. Furthermore, samples of the same size will not, in general, be taken from lots of the same quality owing to sampling fluctuation; average sample sizes can, however, be calculated, the values for lots with mean values of m_a, m_r and $(m_a+m_r)/2$ being respectively:

$$\frac{(1 - \alpha)b - \alpha a}{(m_r - m_a)^2/2\sigma^2}, \quad \frac{(1 - \beta)a - \beta b}{(m_r - m_a)^2/2\sigma^2} \quad \text{and} \quad \frac{ab}{(m_r - m_a)^2/\sigma^2} \qquad (11.4)$$

where $a = \ln\{(1-\beta)/\alpha\}$ and $b = \ln\{(1-\alpha)/\beta\}$.

Sample sizes greater than three times the largest of these values are unlikely to be encountered. If the amount of inspection appears to be excessive, necessitating a reduction in sample size, an increase should be made in α, β or the difference between m_a and m_r, but there is then a greater risk of reaching wrong decisions. In the long run, sequential methods provide a valid basis for decisions with smaller samples than other procedures.

As the results become available one by one during sequential sampling, they may be assessed by reference to a table of acceptance and rejection values prepared beforehand from the formulae:

$$\left. \begin{array}{l} x_1 = (m_a+m_r)n/2 + h_1 \\ x_2 = (m_a+m_r)n/2 - h_2 \end{array} \right\} \qquad (11.5)$$

where
$$\left. \begin{array}{l} h_1 = b\sigma^2/(m_a-m_r) \\ h_2 = a\sigma^2/(m_a-m_r) \end{array} \right\} \qquad (11.51)$$

and n is the number of observations made.

When the cumulative total of the observations is between x_1 and x_2 for the corresponding value of n, a further sample is inspected; but the lot is rejected —assuming that the specified limit is a minimum value—when the cumulative total is less than x_2, or accepted when it is greater than x_1. Alternatively, a graphical procedure may be used in which the cumulative total is assessed against the two parallel lines formed by plotting x_1 and x_2 versus n [*Design and Analysis*, Chapter 3].

Example 11.6. Strength of plastic

11.541 The breaking load of a plastic, employing a test piece of defined shape and size, is of the order of 1,350 lb. with a standard deviation of 150 lb.

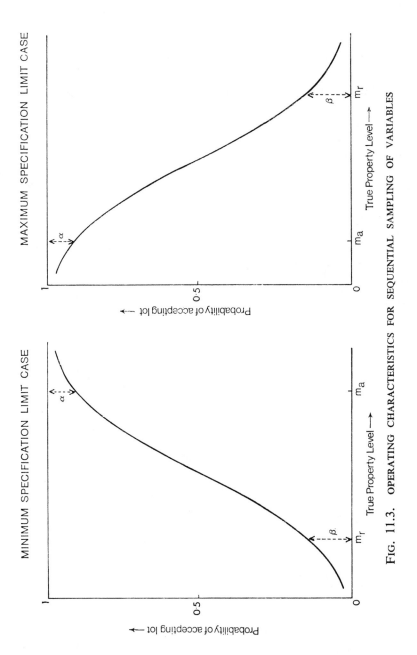

FIG. 11.3. OPERATING CHARACTERISTICS FOR SEQUENTIAL SAMPLING OF VARIABLES

Batches having a mean value as low as 1,150 lb. should not be accepted, and similarly values of 1,350 lb. should not be rejected. The associated risks were fixed at 0·05 and 0·1 respectively. Therefore $m_a = 1,350$, $m_r = 1,150$, $\alpha = 0·1$, $\beta = 0·05$ and $\sigma = 150$.

From Expressions (11.4), the average sample sizes for lots of quality m_a, m_r and $\frac{1}{2}(m_a+m_r)$ are 2·7, 2·2 and 3·7 respectively, sizes which are acceptable. From (11.5):
$$x_1 = 1,250n + 325 \qquad \text{and} \qquad x_2 = 1,250n - 253$$
and from these a table of acceptance and rejection values can be calculated for integral values of n.

In this example interest centres not so much on average sample size as on maximum sample size, since it is convenient to prepare a defined number of test pieces at one and the same time. Although most batches may be sentenced on the results of two or three tests, some may require several more, and a reasonable maximum should be assigned. From the above equations it will be seen that, if the cumulative value of breaking load after five tests lies between 5,997 and 6,575 lb., i.e. average values of 1,199 and 1,315 lb., no decision can be reached to accept or reject the lot. Economic considerations then determine whether (a) the lot should be arbitrarily accepted or rejected, implying a relaxation of the risks defined above, or (b) five more test pieces should be prepared and tested sequentially.

11.542 The case considered hitherto is that of a specification embodying either a maximum or a minimum limit. Producer's and consumer's risks are then assigned in relationship to this one limit. Many specifications, however, contain both maximum and minimum limits, and values of α, β, m_a and m_r will then have to be established with respect to both limits. In practice these will lead to two pairs of parallel lines on the graph relating cumulative value to number of samples. The area will then be divided into five sections, the two outermost indicating rejection, and the central one acceptance; sampling will be continued if the plotted point falls between either pair of parallel lines. Further information on this application of sequential methods is contained in *Design and Analysis*, § **3.25**.

Sequential methods are likely to be of most value when a clear-cut decision is needed and when the physical process of taking samples is easier and cheaper than the measuring operation. In cases where the mean value must be determined with a stipulated precision, a sample of fixed size is likely to be more satisfactory, particularly if the material is not homogeneous. The necessity for taking a truly random sample is as great with sequential as with any other sampling procedure.

Sampling clauses in specifications

11.55 Sampling clauses in specifications should, of course, indicate the quality characteristics which the consumer desires in the product. At the same time, however, they should be related to the level of quality which the producer is able to maintain in his manufacturing process and to the accuracy of testing methods. The inherent variability of batches should also be considered; this is best done by assessing confidence limits, a knowledge of which is very useful in interpreting test results in relation to specification limits. In fact, information obtained over a period in sampling batches of a specified material may well be used in drawing up sampling clauses for specification purposes.

These principles are exemplified in reference [11.7]. This specification deals with a number of physical properties, such as tensile strength, volume resistivity and water absorption, for which either maximum or minimum limits are indicated. These limits, however, are not specified, but are defined by means of basic values for average and standard deviation; values for these have been agreed at levels which are obtainable with the majority of well-established, acceptable commercial materials. Furthermore, sample sizes are not specified, but suggestions are brought forward with respect to recommended numbers of test specimens. For each size of sample, if a decrease in the value of a property denotes a worsening of quality, a lower limit is set for the mean such that there is a 0·01 probability that material of basic quality will be rejected. The lower limit is set, in fact, at a value:

$$\mu - 2{\cdot}33\sigma/\sqrt{n}$$

where μ and σ are the basic values for mean and standard deviation respectively, n is the sample size, and 2·33 is the value of the Normal deviate, u, for a 0·01 probability. Similarly, upper limits for sample standard deviation are set by means of a multiplying factor, depending on sample size, and applied to the basic standard deviation.

An important feature of this specification is that it recognises that evidence of conformity can be established by a study of test results obtained on a sequence of batches of material produced during a period not exceeding six months. One batch, assessed purely on its own test results, apparently may be of quite low quality and yet be accepted; but, if a consistently doubtful product is being presented for sampling, the results obtained on a sequence of lots may justify rejection when those on a single lot do not. The producer's risk is taken care of in setting the limit at 0·01 probability level below the basic value, whereas the consumer is protected in the initial setting of the basic value and in the assessment from a sequence of samples, which, by increasing the number of test results, brings the lower limit closer to the basic value. Furthermore, the clauses relating to standard deviation exercise some degree of control on the precision of the testing methods.

The specification [11.7] makes the inherent assumption that batches are homogeneous, and provides an instructive example of the use of statistical techniques in specifications. When lots are not homogeneous, the physical process of taking a sample should be clearly specified to avoid arguments between producer and consumer. Such materials, however, are likely to be sampled not so much to determine conformity with specification as to ascertain quality for the purpose of fixing price or the process for which the material is fitted.

The need for a realistic approach in drafting specifications must be emphasised. If limits are fixed without adequate consultation, in the hope that the producer will be able to meet them, arguments may well arise on the interpretation of the specification and the testing methods to be used, the only result of which will be to bring the specification into disrepute. A more logical approach is that in which test data are analysed to ascertain levels within which suitable values can be reasonably maintained and to determine the inherent variability of the process and of testing methods. In many instances a designed experiment of simple form including a number of lots, and replicate sampling and testing of variables, should be carried out. If agreement between producer and consumer cannot be reached on the results of such tests, then production methods may have to be altered. The designed experiment should also provide adequate information to enable sample sizes to be specified and to indicate the risks involved in the agreed sampling scheme. Further information may be obtained from *Design and Analysis*, Chapter 3.

Assumptions regarding standard deviation and Normality

11.56 In discussing the sampling of variables several assumptions have been made either implicitly or explicitly. Thus it is assumed that the standard deviation is known and that it maintains a constant value, subject to the usual sampling errors, over a considerable number of batches or lots. Similarly an approximately Normal distribution is assumed. The only way of knowing whether a single batch is likely to be homogeneous is by having a fairly complete knowledge of its previous history, which is not always available. In order to draw up a reasonable sampling plan, therefore, a considerable amount of prior information is required, which can be obtained from an examination of accumulated results. A simple and illuminating method of assessing a series of figures is to plot them on control charts for both means and either standard deviations or ranges [Chapter 10]. Thus, if the standard deviation chart indicates lack of stability, the most likely cause is that the lots are not homogeneous; this could be confirmed by a special investigation employing stratified sampling on a few questionable lots. Similarly the relationship of the mean values to the specified limits provides an indication whether the amount of sampling should be increased or decreased. For instance, reduced sampling could be instituted if it has been established that mean values are statistically stable and the amount of variation is well within that allowed by the specification. Mean values associated with producer's and consumer's risks can be included on the chart, and the plotted points judged with respect to them. It should always be remembered, however, that this procedure is only a graphical method for applying statistical tests of significance.

If the value of the standard deviation is required over a number of lots, it can be obtained directly from the control chart. Sampling is rarely carried out with the object of assessing variability rather than mean value, but sequential methods for detecting defined differences in standard deviation are available if required. Cases in which constant variability is assumed are much more common, and these are best dealt with by means of control charts.

The validity of the assumption that mean values are distributed Normally depends on the shape of the parent Population and on the number of measurements in the sample. This was discussed in § **2.42**, where it was concluded that mean values of samples of as few as three articles from a non-Normal Population are distributed in nearly the Normal form. The practical outcome is that probabilities associated with acceptance or rejection of lots may be slightly different from those indicated by theory. For instance, a sampling scheme which theoretically rejects material at a defined level with a probability of 0·05 may in reality operate with a probability of 0.07, a difference which is unlikely to be technically significant.

Process control of variables

11.6 Sampling clauses in specifications generally legislate for all eventualities, including the possibility of inspecting very inferior lots. However, the producer may be able to demonstrate, by means of control charts for instance, that his process is stable and that unacceptable lots are presented only very occasionally.

Process control schemes based on cumulative sums are the most efficient for the control of variables; they are similar in many respects to the sequential sampling procedures described in § **11.54**. Double-sided control by superimposing a V-shaped mark on a cumulative sum chart was described in § **10.7**. But we now consider a method which is purely numerical and which involves only a small amount of calculation.

Suppose we have a target level μ which the process property should achieve and that there is a higher level μ_1, which is defined as just unsatisfactory. A reference level k is then set *mid-way* between μ and μ_1. As each sample mean is measured it is compared with the reference level. When the observed figure is less than k nothing further is done. But when a higher value occurs, the difference from k is calculated and an accumulation of the algebraic deviations is started. This continues until either:

(a) The cumulative deviations fall below zero; in this case a new accumulation is not begun until a later sample mean exceeds k.

or (b) The cumulative sum exceeds a decision limit h; this is a signal to search

for the cause of the higher level in the process mean and to correct the
level if this is possible.

This type of control scheme has the advantage that during periods of
satisfactory production it will frequently be unnecessary to calculate cumulative sums at all, but that significant increases in the process mean level will
be discovered rapidly. This is quite an important practical consideration,
because it reduces the labour needed to operate the control schemes on large-
scale manufacturing plants. An example of the tabular presentation of the
results is given in Appendix 11A (for double-sided control). Further, with
the aid of a proprietary "Shannovue" display panel, process record sheets
can be used to compile the control sequence directly, thus avoiding the need
to transcribe the data [11.5].

The value of μ_1 is termed the Rejectable Quality Level (R.Q.L.), while the
target level μ is often described as the Acceptable Quality Level (A.Q.L.).
However, when manufacturing to a predetermined specification, it is sometimes more convenient to define the A.Q.L. at a level, μ_0, between μ and μ_1.
This is particularly the case when the process can easily comply with the
specification all the time the sample averages remain centred on the target
level. The reference level for cusums is then set at $\frac{1}{2}(\mu_0 + \mu_1)$ rather than
$\frac{1}{2}(\mu + \mu_1)$, and fewer deviations need to be calculated.

Design of decision limit schemes

11.61 To complete the design of the cusum control scheme we must determine
a suitable value for the decision limit h. If h is raised, it will clearly take
longer for an action signal to be given at R.Q.L.; on the other hand there will
be fewer inadvertent signals when the process average is truly at A.Q.L. It
may also be possible to fix the sample size n so that the scheme has desirable
characteristics.

The performance of a cusum control scheme is usually measured by the
average run lengths that will result from its application. There is a different
A.R.L. for each level of the current process average, but at most two of these
can be selected independently. Attention is therefore concentrated on the
A.R.L.s at A.Q.L. and R.Q.L. Indeed these two quality levels do not really
acquire meaning until A.R.L.s are associated with them. When the process
average deteriorates to the R.Q.L. quick action is needed, i.e. a low A.R.L.
is wanted; but when the process average is at A.Q.L., interruptions are to be
avoided, i.e. a high A.R.L. The parameters of the control scheme are selected
with these objectives in mind.

Table 11.2 lists the average run lengths for six typical schemes when the
successive sample averages are Normally distributed and independent of each
other.

Other values can be deduced from the Intersection Chart at the end of the book or from Tables 5 of [11.8]. When the sample size n is fixed beforehand, only one of the *A.R.L.*s can be defined arbitrarily. But in general the *A.R.L.* at both *A.Q.L.* and *R.Q.L.* can be chosen freely and the necessary values of n and h to achieve the desired degree of control can be calculated from the Chart.

TABLE 11.2. AVERAGE RUN LENGTHS OF CUSUM CONTROL SCHEMES

$\lvert\mu_1-\mu_0\rvert\sqrt{n}/\sigma$	$\lvert h\rvert\sqrt{n}/\sigma$	*A.R.L.* at *A.Q.L.*	*A.R.L.* at *R.Q.L.*
2·18	2·45	1,000	3
1·29	4·08	1,000	7
2·03	2·29	500	3
1·19	3·80	500	7
1·88	2·12	250	3
1·07	3·52	250	7

Example 11.7

11.611 Suppose we wish to maintain the mean value of a measured property of a certain manufactured article at 5·0 units or less, and that when the process is producing items at this level it should not be interfered with in error more than once in 500 hours on average. If, however, the mean rises to 5·2 units, then on average we wish to detect the change in $6\frac{1}{2}$ hours. It is only convenient to take samples once per hour, so that an average delay of $6\frac{1}{2}$ hours is equivalent to an *A.R.L.* of 7 samples. The property is known to be approximately Normally distributed with $\sigma = 0·48$.

The specified tightness of control corresponds to the fourth scheme in Table 11.2. Hence we require:

$$0·2\sqrt{n}/0·48 = 1·19, \quad \text{i.e. } n = 8·2$$

Rounding up to achieve greater discrimination, we would therefore take samples of 9 articles and accumulate differences of $\bar{x}$ from a reference value of 5·1 units. Then $(\mu_1-\mu_0)\sqrt{n}/\sigma = 1·25$ and from the Intersection Chart we find that any scheme between

$h = 0·582$ units with *A.R.L.* = 500 at 5·0 units and 6·5 at 5·2 units

and $h = 0·636$ units with *A.R.L.* = 770 at 5·0 units and 7·0 at 5·2 units

is suitable. For simplicity we would probably decide to initiate action when the cusums exceed the rounded decision interval of 0·6 units; the *A.R.L.*s are then 580 samples at 5·0 units and 6·7 samples at 5·2 units.

Double-sided control

11.62 The decision limit schemes discussed above are of the single-sided type. When however it is desired to maintain the process on a target level and to control both increases and decreases from this, then two such decision interval schemes can be run concurrently; the upper scheme monitors an increasing mean level, while in the lower scheme cumulative sums are plotted only when the lower reference value is passed.   Often the two schemes are symmetric about the target level, but sometimes the tightness of control required on the two sides is different.   A detailed illustration is given in Appendix 11A, relating to the manufacture of a synthetic fibre; in this example the two schemes are symmetric and both $A.Q.L.$s coincide with the target level.

If L_u and L_l denote the average run lengths at a given displacement of the current mean of the upper and lower schemes respectively, then the average run length L of the double-sided scheme is given by the reciprocal rule:

$$\frac{1}{L} = \frac{1}{L_u} + \frac{1}{L_l} \qquad (11.6)$$

at that value of the process mean.   (Formula (11.6) corresponds to the addition of tail-area probabilities A_u and A_l [§ 10.5].   Strictly it is only valid when $(k_u - k_l) \geqslant |h_u - |h_l||$, but this condition is always fulfilled by a pair of symmetric cusum tests.)   Hence if the upper and lower schemes have the same $A.Q.L.$, then the joint $A.R.L.$ at $A.Q.L.$ is half that of the single-sided schemes separately.   At either of the $R.Q.L.$s, the $A.R.L.$ of the double-sided scheme will be very nearly equal to that of the corresponding single-sided scheme.

When the upper and lower schemes are symmetric and have the same $A.Q.L.$, it can be shown that this numerical technique for double-sided control is exactly equivalent to the chart and superimposed V-mask scheme described in § 10.7.   The relationship between the two sets of parameters is:

$$k_u = \mu_0 + f \tan\theta \qquad h_u = fd \tan\theta \qquad (11.71)$$

for the upper decision limit scheme and:

$$k_l = \mu_0 - f \tan\theta \qquad h_l = -fd \tan\theta \qquad (11.72)$$

for the lower decision limit scheme, where f is the number of property units on the cusum scale occupying the same length of chart as one horizontal plotting interval, and the lead distance d is in units of the horizontal plotting interval.   This equivalence permits the $A.R.L.$s of V-mask schemes at $A.Q.L.$ and $R.Q.L.$ to be deduced from the Intersection Chart at the end of the book.

There is something to be said in favour of both methods of presenting the data when both increases and decreases from the target level are important. If a recorded history of the deviations is useful for technical investigation as well as for control purposes, the superposition of a V-mask on a complete cumulative sum chart is a powerful technique.   On the other hand, if we only want to know whether the process is running at an acceptable level, a pair of decision limit schemes involves less labour and the cusums do not have to be plotted.

References

[11.1] WETHERILL, G. B. *Sampling Inspection and Quality Control* (second edition). Chapman and Hall (London and New York, 1977).

[11.2] *Sampling Procedures and Tables for Inspection by Attributes* (B.S. 6000 and B.S. 6001:1972). British Standards Institution (London, 1972).

[11.3] BARNARD, G. A. "Sampling Inspection and Statistical Decisions", *J. R. Statist. Soc.* B, **16**, 151–174 (1954).

[11.4] HALD, A. "Bayesian Single Sampling Attribute Plans for Discrete Prior Distributions", *Mat. Fys. Skr. Dan. Vid. Selsk.*, **3**, Part 2 (1965).

[11.5] WOODWARD, R. H., and GOLDSMITH, P. L. *Cumulative Sum Techniques.* I.C.I. Monograph No. 3. Institute of Manpower Studies (Brighton, 1981).

[11.6] BISSELL, A. F. "Cusum Techniques for Quality Control", *Applied Statistics*, **18**, 1–30 (1969).

[11.7] *Synthetic Resin Moulding Materials* (B.S. 771: 1948). British Standards Institution (London, 1948).

[11.8] VAN DOBBEN DE BRUYN, C. S. *Cumulative Sum Tests: Theory and Practice.* Charles Griffin (London, 1968).

[11.9] DUNCAN, A. J. *Quality Control and Industrial Statistics* (third edition). R. D. Irwin Inc. (Homewood, Illinois, 1965).

Appendix 11A

Process control by decision limit schemes

11A.1 This appendix illustrates the use of a pair of symmetric cusum control charts in the manufacture of a synthetic staple fibre. A surface coating is sprayed on to the fibres to improve the processing characteristics and it is important to keep the level of this coating, expressed as a weight percentage of the fibre, within close bounds. Samples of the fibre are taken at regular intervals, termed Doffs, and a single measurement is made of the coating percentage ($n = 1$). When a significant movement occurs in the average level of coating, it can be corrected by a proportional change in the rate at which the spray is applied.

The control system is based on the following premises:

(i) The target coating level is 0·405%. If the process is *really* running at this mean level, then approximately once in every 600 results the sampling and testing errors will be such that the actual test results will falsely lead to a decision to take corrective action. This false action will nevertheless be rectified a few hours later.

(ii) If the process mean *really* moves to 0·380% or 0·430%, then the test results will lead to a decision to take corrective action in 3 results on average.

(iii) The short-term variability comprising test errors, true variations through a fibre bale and random variations which are inherent in the process. can be expressed as a standard deviation of 0·012%.

Thus for the upper control scheme:

$$\frac{(\mu_1 - \mu_0)\sqrt{n}}{\sigma} = \frac{0.025 \times 1}{0.012} = 2.08$$

From the Intersection Chart at the end of the book, the upper decision interval for the required average run lengths is given by:

$$\frac{h \times 1}{0.012} = 2.34, \quad \text{i.e.} \quad h = 0.028\%$$

The upper reference value is set at 0.418%, a rounded value approximately midway between 0.405 and 0.430. Similarly, the lower reference value is set at 0.392%.

Operating procedure

11A.2 The following procedure is applied separately for each production line.

As each result becomes available enter it on the appropriate control record sheet together with the identification particulars. Initially if a result is between the reference values, 0.392 and 0.418, no further action is required and the coating level can be assumed to be satisfactory. Continue entering the results until one exceeds 0.418 or falls below 0.392. Then proceed as follows:

1. *To Detect an Increase in the Average Coating Level*

Having entered a result greater than 0.418, calculate the Upper Score by subtracting 0.418 from the result. This score is entered in the column headed *U.S.* and also in the place marked Cumulative Upper Score (*C.U.S.*). On receipt of each subsequent result, the *U.S.* is calculated (maintaining its correct algebraic sign) and the *C.U.S.* is obtained by accumulating all the upper scores from the start of the sequential run. Continue accumulating until either:

(i) The *C.U.S.* becomes negative; this run of results will be then regarded as at an end, and in effect a definite decision will have been taken that all recent and current fibre is satisfactory.

or (ii) The *C.U.S.* exceeds 0.028, in which case the conclusion is reached that the coating level is running significantly high. A Warning Note is issued stating that corrective action is needed and this sequential run is terminated. The high coating level will normally be considered to have been in existence since the beginning of the present run.

In either case (i) or (ii) a new sequential run does not commence until the first result above 0.418 is received.

TABLE 11A.1.   DOUBLE-SIDED CONTROL FOR FIBRE COATING LEVEL

Date	Time	Doff Number	Result %	Paper Corr.	Corr. Result	Lower Control		Upper Control		Action	Control Knob Setting
						L.S.	C.L.S.	U.S.	C.U.S.		
16/10	0805	185	0·407								2.15
16/10	0830	186	0·414								
16/10	0930	187	0·413								
16/10	0945	188	0·386			−0·006	−0·006				
16/10	1015	189	0·393			+0·001	−0·005				
16/10	1120	190	0·386			−0·006	−0·011	+0·003	+0·003		
16/10	1250	191	0·394			+0·002	−0·009	+0·015	+0·018		
16/10	1345	192	0·421			+0·029	+0·020	+0·009	+0·027		
16/10	1515	193	0·433					−0·028	−0·001		
16/10	1615	194	0·427					+0·012	+0·012		
16/10	1715	195	0·390					+0·010	+0·022		
16/10	1830	196	0·430					+0·017	+0·039		
16/10	1920	197	0·428								
16/10	2010	198	0·435	−0·020	0·416					W.N.+0·3	
16/10	2115	199	0·436								
16/10	2240	200	0·431	−0·020	0·411						
17/10	0020	201	0·435	−0·020	0·415						
17/10	0100	202	0·427	−0·020	0·407						
17/10	0200	203	0·431	−0·020	0·411						
17/10	0245	204	0·418	Nil							
17/10	0315	205	0·413					+0·001	+0·001		2.45
17/10	0345	206	0·405					−0·001	0		
17/10	0415	207	0·411								
17/10	0510	208	0·419								
17/10	0620	209	0·417								

2. *To Detect a Decrease in the Average Coating Level*

Independently of the upper control chart, as soon as a result is received below 0·392%, a Lower Score (*L.S.*) is calculated by subtracting 0·392 from it; this *L.S.* will be negative. A Cumulative Lower Score (*C.L.S.*) is calculated on receipt of subsequent results, again maintaining the correct algebraic sign. The accumulation continues until either:

   (iii) The *C.L.S.* becomes positive, in which case the current run terminates and the fibre is considered satisfactory.

or (iv) The *C.L.S.* falls below −0·028%. A Warning Note is issued in this case and the present run is terminated.

The procedure is illustrated in Table 11A.1. Sometimes both *C.U.S.* and *C.L.S.* are being calculated simultaneously, but it is impossible for a Warning Note decision to arise simultaneously on both. When action is signalled, the current coating mean level is normally calculated by dividing the Cumulative Score by the number of results included in it and adding this figure to the corresponding reference value. However if the number of doffs involved exceeds 12, calculation of the current mean is based on the last 12 results only. To apply a correction of 0·010%, a change of 0·15 units is necessary in the spray control setting.

Owing to the time taken to carry out the measurements and to delays in the transmission of samples and results, there are always several doffs which have already been made when corrected action is decided upon, but whose results are not available at the time of the decision. When these results are received subsequently, they are utilized by applying a "paper correction" equivalent to the expected change in coating level which the change in control setting would have made if it had been applied in time. The relationship, 0·15 setting units ≡ 0·010% coating, is used. This requires two extra columns on the record sheet, for "paper correction" and "corrected per cent coating", as illustrated in Table 11A.1.

Chapter 12

Simulation

There are often difficulties in carrying out experiments
on large systems without spending considerable
amounts of time and money. One way of overcoming
these is to build a simulation model which provides a
numerical imitation of the real system. The model
consists of a set of rules describing the inter-
relationship of the pertinent parts of the system, for
example the successive stages of a batch chemical
plant. Performance is studied by processing real or
sampled data through the model. The effect of
changing parts of the system, for example the introduction of an extra vessel into the plant, can then be
assessed fairly easily. Simulations of large systems are
invariably carried out on an electronic computer.

Introduction

Example 12.1. A blending problem

12.1 The previous chapter was concerned with the construction of sampling
schemes which allow us to classify batches of material as satisfactory or
otherwise. In the examples, material was accepted or rejected depending
upon the sequence of results from the sampling scheme. However there are
many situations in the chemical industry where a compromise is possible by
blending some good material with some poor material so as to produce a
final product within specification. In fact, in some areas where plant control
is difficult this is the only way of ensuring that a large proportion of the
product is not discarded. The question then arises as to what form of blending scheme should be used.

Consider the case of a certain chemical polymer which is being manufactured as a series of batches. Over the years it has been recorded that one
measure of the quality of the polymer batches has followed a Normal distribution with a mean value of 45 and a standard deviation of 3 units. The
sales specification for the material states that the polymer quality level should
lie between 42 and 48. From Table A at the end of the book we find that
this implies that almost 32% of the batches are outside the specification
limit, 16% having too high a value and 16% being too low. It is known that

the quality property is additive so that, for example, the mixture of batches of equal size with values 40 and 46 gives a "blend" with the (mean) value of 43. Hence it is possible to devise a blending scheme which will enable all the material manufactured to be brought within specification.

If the quality of batches were distributed randomly with respect to time, we could calculate how many successive batches to mix together so that the great majority of the blends would fall inside the specification limits. From the Equation (3.8) for the standard error of a mean we know that for a two-batch blend the standard error will be $3/\sqrt{2} = 2\cdot12$, for a three-batch blend it will be $1\cdot73$ for a four-batch blend $1\cdot50$, and so on. From Table A we find that a series of four-batch blends would yield 95% of polymer blends inside specification.

However, it is seldom that batches are produced with a random distribution of quality. More often in practice there is a run of good quality material followed by perhaps a short run where the quality drops. It is obviously no use blending four of the low quality batches together and expecting these to produce satisfactory sales material. In these circumstances a different blending procedure is required. One possibility is to adopt a Selective Blending Scheme. In this the successive batches are not blended together in the order of manufacture but a selection procedure is used to choose which batches should go into particular blends. To operate such a scheme it is necessary to have storage vessels available to hold the material waiting to be blended. The use of different batch selection rules and specification ranges will call for different numbers of containers. Information is needed about the amount of storage required for any particular scheme. The chance that all the available storage will be occupied simultaneously must also be evaluated.

It is not possible to calculate these quantities from an analytical formula or by solving a set of equations which describe the blending scheme. Nor is it feasible to carry out expensive trials on the plant itself; this would take too long and could easily upset normal production. Instead an answer can be obtained by processing a sequence of batch values in a manner which numerically imitates the behaviour of the real blending system. This calculation can be based on actual plant data if there are sufficient available; otherwise it is necessary to generate representative sample values from the property values which are involved in the blending process.

This type of calculation is called Simulation. It will be discussed in some detail in this chapter. We will return to the particular blending problem in § **12.61** below.

Types of simulation

12.2 In its widest sense the term simulation refers to the use of a model to represent the behaviour of a real-life object or system. This could be a small-scale physical imitation of the real object, such as the model aircraft used in wind-tunnel tests, but it could also be a mathematical model where equations and logical rules represent the system under investigation. For example, many chemical reactions can be described by a set of differential equations and their solution is sometimes referred to as Continuous Simulation; this may involve the use of Analogue Computers.

In other cases the system involves discrete changes, for example, the movement of individual batches of material through a plant. This chapter is concerned with the simulation of such discrete systems by direct numerical imitation. A sequence of data is processed according to the logical rules which govern the operation of the system, and enables its behaviour to be studied when analytical solutions are not possible.

Sometimes long sequences of historical data are available which can be processed as they stand. On other occasions the data are limited and it may be better to make assumptions about the distribution of values, e.g. of polymer quality or of batch processing times, and to sample randomly from these distributions. In the latter case the method is known as Monte Carlo Simulation. When sampling is involved, repeat simulation runs will not give exactly the same result, and replication is necessary to determine its precision.

Initially the simulation model is constructed to represent the way the system operates at present. This task is a useful exercise in itself, because the investigator is forced to think carefully about the system logic and may gain unexpected insight into its working. It is always necessary to check that the numerical imitation faithfully reproduces the essential details of the observed behaviour of the system.

However, the chief virtue of simulation is that it permits experimentation to be made on the model rather than on the real system. Full-scale experimentation can be inconvenient, expensive, slow and sometimes even dangerous. A model is much more flexible and can be used to study changes to the real system more quickly and more cheaply than modifications to the system itself. For example, the effect of introducing an additional vessel in a multi-stage chemical plant can be conveniently examined by simulation. The model can be constructed at whatever level of detail is required for this purpose. Some applications will be outlined in § **12.6**.

On a different level, statisticians can use simulation to obtain solutions to sampling problems which cannot be resolved analytically. Much of the early theoretical work on the t-distribution (Table C) was preceded by W. S. Gosset ("Student") doing large quantities of sampling experiments; only later did he develop the underlying theory. An illustration of this use of simulation will be given in § **12.322**.

Building a simulation model

12.3 Two aspects of building a simulation model can be distinguished and will be examined separately at this stage. They are:

(a) Formulation of the model—the set of rules necessary to describe the system being studied.

(b) Generation of the data—the numbers used in the model in order to explore the system.

Formulation of the model

12.31 If the model is logically simple, for example the estimation of statistical functions by sampling from distributions, then a set of rules to describe its operation can easily be formulated. However the system may be complex, such as one of the queueing situations which are common in industry and which are the subject of a large proportion of simulation studies.

Systems for batch-chemical production, vehicle unloading, aircraft landing and telephone traffic have a number of features in common. Basically they can be described as a number of "machines" through which "materials" flow. The machines carry out a number of "activities", which in the simplest systems are either "processing" or "idling". Associated with each activity is a processing time, which is often variable. When an activity is completed a change of state takes place and material may flow to another machine if the latter's state is appropriate. The interaction of one machine with another is set down in a logical way, and by providing for all possible outcomes a comprehensive model can be built up.

Book-keeping activities are required to collect information as the simulation progresses. The system must be examined from time to time to see how it is operating. Two methods are available to do this. The most obvious, known as the Constant Time-Step Technique, monitors the system at regular intervals and collects information regarding the activity of each machine and each unit of material. If the system is to be recorded accurately, a large number of such steps are usually required and the method becomes inefficient. However it is a convenient method when simulating continuous systems. A more satisfactory approach for discrete simulation is known as the Next Event Technique, in which the system is only examined when a machine completes an activity. The particular machines involved in a transfer of material are noted and other relevant information is recorded. The simulated time in the model is then advanced to the next "event", which is found by scanning the current finishing times of all machines and selecting the earliest.

This brief explanation of the techniques of building simulation models will be made clearer by the following illustration.

Example 12.2. A simple production system

12.311 In a single-stage production system, batches of material arrive to be processed by one of a number of machines. If all machines are occupied on arrival a queue is formed. When a machine becomes available one of the waiting batches can then be processed and will subsequently leave the system.

Both the time between arrivals and the machine processing time are variable. The distribution of processing times may also vary from machine to machine. We require to study this system to estimate its throughput, the proportion of time the machines are idle and the average queue length.

The activities of this system can be defined by studying the behaviour of the batches of material; they arrive, queue, are processed and then depart from the system. Events which trigger off changes in activity can be seen to

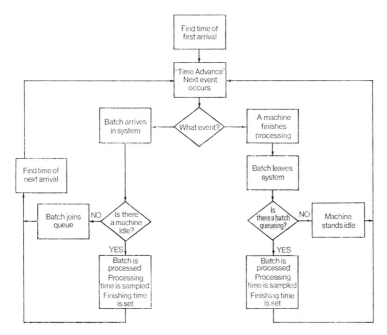

FIG. 12.1. THE OPERATION OF A SINGLE-STAGE PRODUCTION SYSTEM

be the arrival of a batch and the completion of its processing; further changes are dependent on these two. Thus the start of processing a batch depends either on its arrival (if a machine is idle) or on the completion of processing a batch (if the next batch is already queueing). The flow diagram in Figure 12.1 describes the operation of this system.

The process begins with the arrival of the first batch. Then a cycle is entered which is repeated for the length of the simulation. The starting point of this cycle can be termed the "time advance". During the course of the simulation, the model is examined only when an event occurs which triggers off a change in activity. The flow diagram shows the change in activities dependent on each event, and when all relevant activities have been changed control is returned to the "time advance".

The diagram does not include the necessary book-keeping operations. A count must be kept of the number of batches passing through the system, the length of the queue, the machine idle time, etc. More details of the actual operation may also be needed; for example, the selection procedure to be used if more than one machine is idle. Such complexities can be added to the basic model.

Generation of the data

12.32 The data required for a simulation study are usually of a historical nature, and if sufficient information is available real sequences of past data can be employed. In the example just described series of actual inter-arrival and processing times could be used for the simulation.

Very often, however, data in this form are insufficient for a long simulation run and in general are not as flexible as a sampled sequence. Monte Carlo simulation involves the representation of the actual data by a probability distribution from which a sequence can be sampled randomly.

It is possible to generate random numbers physically, for example by tossing dice, but in practice two other methods are commonly used in simulation studies. One involves using lists of tabulated random numbers; Table K will be adequate for small investigations but [12.1] offers a much larger list. These numbers can be used immediately as random samples from rectangular distributions. For example, the first line in Table K can be read as 83, 28, 78, 05, ... if numbers between 00 and 99 are required, or as 0·8328, 0·7805, ... if numbers with four decimal digits are needed.

An alternative method is to generate a sequence of apparently random numbers from an algebraic formula. Such a series is reproducible, which seems to contradict the requirement of a random sequence. In fact reproducibility is a valuable aid in carrying out statistically designed simulation experiments [§ **12.4**]. Series generated in this way also repeat themselves but the cycle is usually very long. For these reasons the sequences are not truly random and are known as Pseudo-Random Numbers.

A number of procedures for generating pseudo-random numbers are available. A commonly used one is known as the Multiplicative Congruential Method [12.2], in which the sequence is defined by:

$$r_{(i+1)} = \left| Kr_i \right|_{\mod M} \qquad i = 0, 1, 2, \ldots$$

where r_0 is a starting value and K and M are constants. Thus the next number in the series is obtained by multiplying the current number by K, dividing by M and taking the remainder. This yields a sequence of numbers which fall evenly in the range 0 to M. To generate rectangularly distributed variates between 0 and 1, r_i is divided by M. The constants K and M determine the periodicity of the sequence and values have been found which give cycles of up to 10^{12} numbers. Reference [12.3] gives suitable values of these parameters and also describes other generating procedures.

Before any pseudo-random number generating routine is used as the basis of Monte Carlo simulation it is clearly necessary to check that the series appears to behave randomly. Unfortunately it is not possible to test a sequence of numbers for randomness positively. The best we can do is to apply a finite group of tests to confirm that the numbers do not appear to behave *non*-randomly in certain specific ways. Thus some worthwhile simple tests include checking that the single digits 0 to 9 and that the pairs of digits 00 to 99 appear with equal frequency (apart from sampling variation), and that various run lengths conform to their theoretical probability distribution according to the null hypothesis. A selection of tests for pseudo-random numbers is given in [12.4].

Non-rectangular distributions

12.321 In many cases samples are required from distributions other than the rectangular. These can be derived by transforming rectangularly distributed variates into variates of the specified distribution. For example, we can obtain closer approximations to a Normal distribution by taking the mean of an increasing number of rectangular variates [Fig. 2.6]. The sum of N numbers rectangularly distributed in the range $(0, 1)$ has a mean of $N/2$ and a variance of $N/12$. Hence by taking the sum of 12 rectangular variates and subtracting 6 from the total, a very good approximation is obtained to a variate drawn from a Normal distribution with a zero mean and unit variance. This can be scaled to become a sample from a Normal distribution with any specified mean and standard deviation.

One method of generating a sample from a Poisson distribution with mean μ [§ **9.2**] is to multiply together successive rectangular variates in the range $(0, 1)$ until the product first drops below $\exp(-\mu)$. The required Poisson variate is then the number of terms in the product minus one. This procedure is very quick for low values of μ. Other examples of this type of transformation are given in [12.5].

Sometimes an analytical transformation is not possible. A method of obtaining a random variate x_i from a general distribution $p(x)$ is therefore required. If the cumulative distribution $P(x)$ is drawn and a random number r_i is taken from the rectangular distribution between 0 and 1, then the value x_i for which $P(x_i) = r_i$ can be found by graphical means. This technique is illustrated in Figure 12.2. If $P(x)$ is available in tabular rather than graphical form, interpolation will normally be required to obtain an accurate value of x_i. Some notes on suitable methods of interpolation are given in [12.5].

Example 12.3. Range factors

12.322 In § **3.35** the ranges of 17 sets of five measurements were used to estimate the within-sample standard deviation. The mean range was calculated and divided by the factor d_5, given in Table G.1. Now the values of

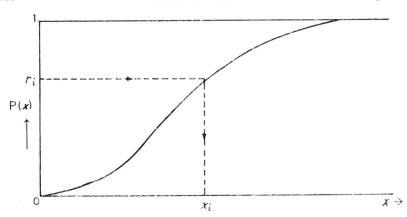

FIG. 12.2.   METHOD OF SELECTING A RANDOM VARIATE FROM A
CUMULATIVE DISTRIBUTION

the conversion factors have been determined from theoretical considerations based on a Normal distribution of sample values. If this theoretical approach had not been possible, the d_n values could have been estimated by simulation as follows.

Random Normal variates are generated in samples of five and the range is calculated for each sample. The mean range of the samples is an estimate of the conversion factor, d_5. Table 12.1 shows a series of 10 such samples with a mean range of 2·171, compared to the theoretical value of 2·326. By carrying out further sampling and taking the mean range of a total of 200

TABLE 12.1.   TEN SAMPLES OF FIVE STANDARD NORMAL VARIATES

Sample No.	Normal Variates					Range
1	−1·639	0·872	−0·205	0·626	−0·414	2·511
2	1·305	1·262	0·419	−0·414	−1·003	2·308
3	−2·388	0·635	−0·247	−1·421	−0·158	3·023
4	−1·233	−0·805	−0·094	−0·328	−0·474	1·139
5	−0·258	0·046	−0·459	−0·159	−1·060	1·106
6	0·291	−0·395	−0·138	1·056	−0·850	1·906
7	3·738	1·956	1·076	0·346	0·538	3·392
8	−1·288	1·702	0·829	−0·684	−0·068	2·990
9	−1·642	−1·238	0·102	−1·359	0·182	1·824
10	−1·472	−0·351	−0·917	0·036	−0·271	1·508
					Mean	2·171

samples an estimate of 2·297 was obtained. For higher numbers of samples the range factor estimated by simulation becomes increasingly close to the theoretical figure.

Validation of the model

12.33 Having built the simulation model, it must be thoroughly validated. The logic of the model and the data distributions must be representative of the real system. If, for example, processing times are distributed log-normally [§ **2.45**], but are assumed in the model to be Normal, incorrect results will be produced. Simulation models are often built to study an extension to an existing system and a satisfactory way of testing them is to compare the results when the model is representing the present situation with the results found in practice. Only if the results are very similar is it sensible to extend the model to study the proposed system.

In some systems it may be practicable only to compare simulation results with actual system performance under a standard set of operating conditions. Wherever possible, however, it is better to carry out simulation runs to imitate a variety of states of the real system—for example, the peak loading and the slack periods of a service facility. In this way the model can be checked more thoroughly, and the user acquires greater confidence for making predictions from the model concerning situations for which there is no direct observational evidence.

Simulation experiments

12.4 Consideration must be given to the initial conditions in the model and the length of time for which the simulation is to run; both of these depend on the type of system being studied. At the beginning of a run, time must be allowed for the system to settle in from the initial state, and to avoid bias in the final result it is often desirable to disregard the early part of the simulation.

Example 12.3 illustrates how the precision of a result increases with the number of simulation runs. The standard error of the mean of N results declines in proportion to $1/\sqrt{N}$. Thus in order to halve the standard error, and hence to double the precision of a simulation experiment, we would need to carry out four times as many runs. This implies that when a result is rather variable, e.g. when we want to investigate the occurrence of an in-frequent event, a large volume of computation is necessary.

When pseudo-random variates are used, a paired comparison experiment can be made using the same sequence of sample values. This gives somewhat greater precision in determining the effect of some change in the system, for example, the addition of an extra vessel in a multi-stage plant. Procedures used to increase the precision of estimates for a fixed sample size or to decrease the sample size without losing precision are known as Variance Reducing Techniques. A number of these are described in [12.6].

For some systems, the values of the parameters used in a simulation model are uncertain, e.g. a forecast of sales of a product, an estimate of future costs, the distribution of the random variations, etc. The estimated values may be largely subjective in some cases. To assess the effects of errors in these estimates, it is worth while carrying out a number of simulation runs, changing the value of each parameter in turn within its range of likely error. This is known as Sensitivity Analysis. Such experimentation enables the analyst to determine which parameters have a critical effect on the responses in the system, e.g. on the cash flow and the rate of return of a capital investment project.

More generally in simulation experiments, we will want to make changes in both the system parameters and in the configuration itself, e.g. the insertion of an extra vessel in a multi-stage multi-unit chemical plant with variable cycle times. There are likely to be interactions between these variables [§ **6.65**], and it is important that the experiments should be statistically designed. Factorial arrangements often prove efficient [*Design and Analysis*].

As with any type of forecasting, care must be taken to ensure that simulation runs do not represent too great an extrapolation of system conditions from those which have been validated earlier. When several, possibly large, changes in parameters or in the system configuration are made simultaneously, there is a distinct possibility that new constraints will arise which have not been foreseen in the original model and that certain operating equations are no longer valid.

The use of computers for simulation

12.5 The calculations for the simulation of a simple system can be made by hand or with the help of a desk calculator. But when the system is complex or where long runs are necessary in order to obtain sufficiently precise results, an electronic computer becomes an indispensable tool in simulation studies. Alternative system configurations or parameter values can be quickly investigated simply by changing a few items of data in the computer input.

Any general purpose digital computer can be used to simulate a discrete system. A flow diagram of the type illustrated in Figure 12.1 forms an ideal base from which to develop the computer program to describe the operation of the model in detail. The instructions are usually written in a "high-level" computer language; the user must check which languages are available at his particular installation. Sometimes a general language such as Fortran is used, but a number of languages designed specifically for simulation work have also been developed.

Such specialised languages are feasible because, although every simulation model is unique, they do have some common characteristics. Thus many systems can be described as a number of "machines" through which flow "materials". Hence a simulation language on a computer has facilities for directing the flow of materials through machines, thus updating the current state of the system. The discussion in § **12.31** emphasised the use of "events" in determining which activities should take place; the timing of these events is constantly scanned within the computer and the Next Event Technique is used to advance the simulation.

Sequences of pseudo-random numbers can be generated very quickly inside the computer and then transformed to provide samples from any specified probability distribution. Since many thousands of these numbers may be used in a simulation run it is especially important that the generation procedure be fast and efficient. These procedures will be an integral part of a specialised computer language. Such a language should also include tabulation facilities to record the results of the run, and diagnostic notes to enable the user to unravel any faults in the logic of his model. The ideal language should be both powerful, giving the maximum of help to the user, and flexible, allowing him to describe the idiosyncracies of his particular system. There is a difference of emphasis on these features in the languages currently available [12.7].

Applications

12.6 There are many areas in research and production where simulation offers a useful approach to an investigation, and in many cases the only approach. The four cases outlined here are typical of the problems which are tackled in practice.

Example 12.1 (continued). A blending problem

12.61 This example relates to the polymer blending problem introduced in § **12.1** and the reader should return to that Section to remind himself of the details.

Over a period of time quality measurements had been collected for a series of 1,000 batches. It was assumed that this sample could be considered representative of future manufacture. A set of blending rules had been proposed for the selective blending of the polymer into four-batch blends. Two possible schemes were being considered, the first designed so that all blends should lie within 1 unit of the specification of 45 and the second designed so that all blends should lie within 0·5 of the specification. Simulation was the only possible method to discover the polymer storage requirements in the two cases.

When the polymerisation of each batch has been completed, it is run off into a container, pending the result of an analytical test to determine its quality level. When this is known a search is made of all the waiting batches to see whether an acceptable blend can be formed. In the simulation study, the 1,000 readings of batch quality were taken in sequence and the two schemes were applied to these numbers as though the corresponding batches were being so blended on the real plant. After each acceptable blend had been found, the number of batches still waiting to be blended was counted. This indicated how many extra storage containers would be needed in the application of each scheme (over and above four initial holding vessels).

The results of such a calculation are shown in Table 12.2 which lists the percentage of times that a given number of extra containers were occupied. This shows that with the tighter specification, 9 or more extra containers are needed 10% of the time, whereas they are needed only 3% of the time with the wider margin. Depending on the losses associated with polymer which has to be down-graded because it is outside specification, these results can be used to decide how many containers should be installed in the two cases.

TABLE 12.2   OCCUPANCY OF ADDITIONAL CONTAINERS

No. of Extra Containers	Specification 45 ± 1	Specification 45 ± 0.5
0	11·5	4·6
1	27·9	17·2
2	23·6	19·6
3	13·0	13·8
4	8·5	11·6
5	5·4	8·4
6	3·5	6·3
7	1·8	4·8
8	1·6	3·8
9	1·3	2·9
10	0·8	1·8
> 10	1·1	5·2
Total	100·0	100·0

Example 12.4.   Total service requirements

12.62 A chemical plant often consists of several units each requiring supplies of services such as steam, water, ice, etc. Suppose it is known that there is a peak demand of 2,000 gals./hr. of water on a particular unit and that this peak will last for 2 hours. Other units have different peak usages for different durations. It would be wasteful to build a water main which would cope with the aggregate of all the peak demands, since the probability of all the peaks occurring together is usually small.

A simulation model can be constructed where the "activity" of each unit at any instant in time is given by its rate of water consumption at that time. By summing over all the units a measure of the total water usage is obtained. The activity of a unit is changed when the demand for water on that unit alters, and this may be at fixed intervals or at some randomly generated times based on actual observed data. By following the activities of the units over a long length of simulated time a distribution of total demand can be compiled. This is then used to estimate the optimum size of water main.

Example 12.5. Stockholding policy

12.63 An important problem for any manufacturer is to determine how much finished product should be held in stock ready to meet sales demands. This may involve the use of warehouses for some materials or storage tanks for bulk liquids. Increasing the stockholding improves the chance of selling more material. The value of the extra sales has to be balanced against the extra costs incurred in providing the storage facilities, insuring the material in stock, having capital tied up in stock and so on.

Simulation can be used to determine the optimum size and strategy to be used for stockholding and is especially valuable in the more complex cases. Here it is necessary to generate sequences representing the demands from customers and the supply of material into storage. The model can be run with different sizes of storage and account can be taken of any special features such as limitations on the storage life of the product. Some detailed examples are given in an I.C.I. Monograph [12.8].

Example 12.6. Plant extensions

12.64 In multi-stage chemical plants where demand exceeds capacity it is usually fairly easy to spot the bottleneck in the system. This may be a particular unit that is constantly overloaded. It would be useful if we could measure the benefit of replacing or duplicating this unit without doing any actual modifications to the plant.

The solution is to build a simulation model of the plant which allows investigations to be carried out on the effects of possible extensions to the plant. The model describes the flow of materials through the plants and takes account of charging, processing and discharging times for each unit. Rules regarding the operating procedure of the plant can be built into the model and the incidence and duration of plant breakdowns can also be included [12.9].

References

[12.1] THE RAND CORPORATION. *One Million Random Digits*. The Free Press (Glencoe, Illinois, 1955).

[12.2] DOWNHAM, D. Y., and ROBERTS, F. D. K. "Multiplicative Congruential Pseudo-Random Number Generators", *Computer J.*, **10**, 74–77 (1967).

[12.3] TAUSSKY, O., and TODD, J. "Generation and Testing of Pseudo-Random Numbers", *Symposium on Monte Carlo Methods*. John Wiley (New York, 1956).

[12.4] MILLER, J. C. P., and PRENTICE, M. J. "Additive Congruential Pseudo-Random Number Generators", *Computer J.*, **11**, 341–346 (1968).

[12.5] TOCHER, K. D. *The Art of Simulation.* English Universities Press (London, 1963).

[12.6] ACKOFF, R. L., and SASIENI, M. W. *Fundamentals of Operations Research.* John Wiley (New York, 1968).

[12.7] TOCHER, K. D. "Review of Simulation Languages", *Op. Res. Q.*, **16**, 189–217 (1965).

[12.8] MORRELL, A. J. H. (Editor). *Problems of Stocks and Storage.* I.C.I. Monograph No. 4. Institute of Manpower Studies (Brighton, 1981).

[12.9] DYSON, J., GOLDSMITH, P. L., and ROBERTSON, J. S. M. "A General Simulation Programme for Material Flow in Batch Chemical Plants", *Technometrics*, **3**, 497–507 (1961).

Statistical and Mathematical Symbols

Most of the symbols used in this book are those which are in general, if not universal, use. In some cases a symbol is adopted even though it is not the one most commonly used for the purpose; for example, u is used for the standard Normal variate rather than t, which has a special use.

A clear distinction should always be drawn between parameters and estimates, i.e. between quantities which characterize the Population and estimates of these quantities calculated from the observations. To emphasize this distinction a Greek letter is often employed for the parameter and the corresponding Latin letter for its estimate. Thus σ is the standard deviation in the Population and s is the estimate of σ. In the binomial distribution, where the observed proportions are usually denoted by p and q, it may be better to avoid the corresponding Greek letters π and χ in order not to clash with the usual meanings of these symbols. In this book Gothic $\mathfrak{p}$ and $\mathfrak{q}$ have therefore been adopted for the parameters of the binomial distribution, although in writing, or where no possibility of confusion arises, π and χ may be, and in fact quite commonly are, employed.

In some other books, and particularly in published papers, a circumflex is used to indicate a sample estimate of the corresponding Population parameter. Thus we might have used $\hat{\sigma}$ in place of s for the estimated standard deviation.

The general order in the following list is that Greek letters precede the corresponding Latin letters, and small letters precede the corresponding capital letters. Where the same symbol is used for more than one purpose, this is indicated by sub-headings (i), (ii), etc.

α (i) Probability when used for statistical significance statements. Error of the "first kind".

       (ii) Population intercept in regression equation.

a (i) Estimate of intercept in a regression equation.

a, b (ii) The numbers of individuals in two groups into which a sample has been classified, e.g. the numbers defective and effective in a sample drawn for Quality Control. (See also b.)

| A, A' | Multiples of the standard deviation and average range respectively in the construction of Shewhart chart Quality Control limits for the process average [Table G]. |

$A.Q.L.$ Acceptable quality level.

$A.R.L.$ Average run length.

β (i) Probability level used for error of the "second kind".

        (ii) Population slope coefficient for linear regression upon one independent variable.

β_1 (i) $\sqrt{\beta_1}$ denotes the skewness coefficient of a distribution.

β_2 (i) Kurtosis measure.

$\beta_1, \beta_2, \ldots$ (ii) Population partial coefficients of multiple linear regression upon two or more independent variables.

b (ii) Estimate of linear regression coefficient, the estimated regression equation being:

$$Y - \bar{y} = b(x - \bar{x})$$

$b_1, b_2, \ldots$ Estimates of multiple linear regression coefficients. The estimated regression of y upon x_1 and x_2 is, for example:

$$Y - \bar{y} = b_1(x_1 - \bar{x}_1) + b_2(x_2 - \bar{x}_2)$$

$b_{11}, b_{22}, \ldots$
$b_{12}, \ldots$ Estimates of quadratic regression equation coefficients for the square terms x_1^2, x_2^2, $\ldots$ and the interaction terms $x_1 x_2, \ldots$

c (i) One of two parameters describing the negative binomial distribution.

        (ii) Number of columns in a two-way cross-classification.

C_j Sum of values in jth column of a two-way cross-classification.

C_{ij} In multiple regression analysis, the element in the ith row and jth column of the inverse of the information matrix.

C^{ij} In multiple regression analysis, the element in the ith row and jth column of the inverse of the correlation matrix.

$C(x_1, x_2)$ Covariance of two variates x_1 and x_2.

$C.V.$ Coefficient of variation (%) or an estimate of it:

$$C.V. = 100\sigma/\mu \text{ or } 100s/\bar{x}$$

$C.V.(x)$ Coefficient of variation of x.

δ A difference; especially a pre-specified difference of interest, between two means, say.

d	(i) An observed deviation or difference, either with or without sign, according to context.
	(ii) Lead distance of vertex of a superimposed V-mask ahead of current point in a cusum control scheme.
d_n	The ratio of the expected range of a random sample of n values from a Normal distribution to the (known) Population standard deviation.
D, D'	Multiples of the standard deviation and average range respectively in the construction of Shewhart chart Quality Control limits for the process range [Tables G.1 and G.2].
e	The base of natural logarithms, $e = 2{\cdot}71828 \ldots$
$E(x)$	The expected value (arithmetic mean) of x.
$\exp(\ldots)$	The exponential function of the quantity enclosed in brackets: $$\exp(x) \equiv e^x$$
f	(i) A mathematical function, e.g. $y = f(x)$.
	(ii) Factor relating lengths on horizontal and vertical scales of a graph, especially of a cumulative sum chart.
$f_1, f_2, \ldots$	The frequencies of observations in a grouped distribution.
F	The variance ratio, defined as the ratio of two independent estimates of variance, the larger estimate always being the numerator [Table D]: $$F = s_1^{\,2}/s_2^{\,2}$$
g	Gain.
$g_1, g_2, \ldots$	Weight fractions.
G	Geometric mean.
h	Decision limit in cumulative sum control schemes [Table N and Intersection Chart].
H	Harmonic mean.
i	(i) Subscript for an individual item in a sample.
	(ii) Subscript for an individual row in a two-way cross-classification.
I	Primary element in multiple regression analysis.
IJ	Second-order element in multiple regression analysis.
j	Subscript for an individual column in a two-way cross-classification.

k	(i) The number of groups in a batch.
	(ii) One of two parameters describing the negative binomial distribution.
	(iii) Reference value for cumulative sum control schemes [Table N and Intersection Chart].
K	Estimate of missing value.
L	(i) Likelihood function.
	(ii) Ratio of two independent estimates of standard deviation:

$$L = s_1/s_2$$

L_a, L_r	Average run length at acceptable and rejectable quality levels respectively.
L_1, L_2	Multiples of sample standard deviation for lower and upper confidence limits of the Population standard deviation [Table H].
μ	The mean level of a Population probability distribution, especially the mean of a Normal distribution and the parameter of a Poisson distribution.
μ_0, μ_1	Standard and alternative levels of Population means, especially the acceptable and rejectable quality levels in Quality Control.
m	(i) Sometimes used for the mean of a sample. This is correct by the Greek-Latin convention, though the use of the "bar" is more common.
	(ii) Used by many writers as the parameter of the Poisson distribution, although μ is to be preferred.
m_a, m_r	Acceptable and rejectable quality levels, respectively, for counted data.
$M_0, M_1, \dots$	Mean squares in the Analysis of Variance. M_0 is usually the error mean square.
M_n, M_w	Number average and weight average respectively.
$M.D.$	Mean Deviation.
n	The number of observations or items in a sample.
$n_1, n_2, \dots$	The numbers of individuals in sub-groups of a sample.
n_{ij}	Number of individuals in the cell defined by the ith row and the jth column of a two-way cross-classification.

N (i) The number of individuals in a sample which has been sub-divided, or the total number of individuals in a group of samples.

(ii) The number of individuals in a finite Population, from which a sample of n is drawn for examination.

π (i) Ratio of the circumference to the diameter of a circle, $\pi = 3 \cdot 14159\ldots$

(ii) Can be used for the proportion of individuals having a particular attribute in a Population.

$\Pi(\ldots)$ Product of the values of the quantity enclosed in brackets.

p The proportion, in the Population, of individuals having a particular attribute. Parameter of the binomial distribution.

p (i) The proportion, in a sample, of individuals having a particular attribute, e.g. the proportion defective in Quality Control. (Some writers, however, use p for pro-portion effective.)

(ii) Number of variables in multiple regression analysis.

p_a, p_r Acceptable and rejectable quality levels, respectively, for pro-portions in a sample.

$p(x)$ Probability density associated with the value x of a statistical variate.

$p_1(x), p_2(x)$ Prior and posterior probability densities of x, respectively.

P Probability (cumulative).

ϕ Number of degrees of freedom.

$\phi_0, \phi_1, \ldots$ Degrees of freedom for sources of variation in Analysis of Variance. ϕ_0 is usually associated with the error variance.

ϕ_D, ϕ_N Degrees of freedom for denominator and numerator respec-tively in an F-test.

χ Can be used for $1 - \pi$, the Population parameter corresponding to an observed proportion q.

χ^2 "Chi-squared", the sum of squares of a number of independent Normally distributed variates with zero mean and unit standard deviation [Table B].

q The proportion, in the Population, corresponding to an observed proportion q in a sample, e.g. the true proportion effective in Quality Control:

$$q = 1 - \mathrm{p}$$

q (i) The complement of p, and in particular the proportion effective in a sample drawn for quality control:

$$q = 1-p$$

(ii) Studentised range, i.e. the ratio of the range of k random observations from a Normal Population to an independent estimate of the standard deviation.

Q (i) Cash value or cost.

(ii) Ratio of upper confidence limit to the estimate of a parameter.

ρ Population coefficient of correlation.

r (i) Estimate of the coefficient of correlation [Table E].

(ii) Number of rows in a two-way cross-classification.

R (i) Sum of ranks.

(ii) Coefficient of multiple correlation or its estimate.

(iii) Ratio of rejectable to acceptable quality levels in process control of attributes.

R_i Sum of values in the ith row of a two-way cross-classification.

$R.Q.L.$ Rejectable quality level.

σ Population standard deviation.

σ_x or $\sigma(x)$ Standard deviation (or error) of x. The standard error of a mean $\bar{x}$ may, however, be denoted by σ_m to avoid the typographical difficulty of setting $\bar{x}$ as a subscript.

σ^2 Variance. (See also V.)

σ_x^2 or $\sigma^2(x)$ Variance of x. The variance of a mean may be denoted by σ_m^2.

$\Sigma(\ldots)$ Summation of the values of the quantity enclosed in brackets. The brackets may be omitted in expressions like Σx or Σxy.

s Estimate of standard deviation.

s^2 Estimate of variance. (See also V.)

s_x^2 or $s^2(x)$ Estimate of the variance of x. The estimated variance of a mean $\bar{x}$ may be denoted by s_m^2 to avoid the difficulty of setting $\bar{x}$ as a subscript.

S A total of a set of observed values.

$S_1, S_2, \ldots$ (i) Subtotals of a number of groups of observations.

(ii) Cumulative sums of a sequence of individual values about a reference value k:

$$S_r = \sum_{i=1}^{r} (x_i - k) \qquad r = 1, 2, \ldots.$$

S_{ij} (i) Sub-group total in hierarchic Analysis of Variance with three sources of variation.

(ii) In multiple regression, sum of cross-products of the two variables x_i and x_j, each about its mean:

$$S_{ij} = \Sigma(x_i - \bar{x}_i)(x_j - \bar{x}_j)$$

S_{xx} Sum of squares of the variable x about the mean:

$$S_{xx} = \Sigma(x - \bar{x})^2$$

S_{xy} Sum of cross-products of the two variables x and y, each about its mean:

$$S_{xy} = \Sigma(x - \bar{x})(y - \bar{y})$$

$S.D.$ Standard deviation.

$S.E.$ Standard error.

$S.S.$ Sum of squares.

τ Average time at a stated quality level before an out-of-control signal is given in a Quality Control scheme.

t "Student's t", the ratio, without sign, of a variate Normally distributed about zero, to an independent estimate of its standard deviation [Table C].

θ (i) General symbol for a distribution parameter.

(ii) Semi-vertex angle of V-mask used in double-sided Quality Control based on cumulative sums.

u A Normal variate with zero mean and unit standard deviation [Table A]:

$$u = (x - \mu)/\sigma$$

V Variance or estimate of variance. V is a useful alternative to σ^2 or s^2, particularly in typewritten documents.

$V(x)$ Variance of x.

$V.C.(\mathbf{x})$ Variance-covariance matrix of vector $\mathbf{x}$.

w The range of values in a sample.

$w_1, w_2, \ldots$ Weighting factors associated with observations $x_1, x_2, \ldots$

x Any variable quantity, such as an observation or measurement or any quantity calculated from a number of observations or measurements. Used more especially for a continuous variable or variate. For quantities which can only take integral values, the symbols a, n or f are more usual.

x_0 (i) A standard value of the variable x.

(ii) A constant, "working origin", subtracted from values of a variable x to reduce the labour of subsequent arithmetical calculations.

$x_1, x_2, \ldots$ (i) Particular values of the variable x.

(ii) Two or more independent variables, for example in a multiple regression analysis.

x_{ij} Value in the cell defined by the ith row and jth column of a two-way cross-classification.

$\bar{x}$ The arithmetic mean of values of x:

$$\bar{x} = \Sigma x / n$$

y Another variable, more especially one whose expected value is a function of x.

Y The expected or predicted value of y given by a regression equation.

z "Fisher's z", the difference between the natural logarithms of two independent estimates of standard deviation:

$$z = \tfrac{1}{2} \log_e F$$

Tables and Charts of Statistical
Functions

List of Tables and
Their Origins

Page

Table A: Normal Distribution (Single-sided)   434
Condensed and adapted from the *Biometrika Tables for Statisticians*, **1**, Table 1, with permission from the Trustees of *Biometrika*.

Table B: Probability Points of the χ^2 Distribution   436
Condensed and adapted from the *Biometrika Tables for Statisticians*, **1**, Table 8, with permission of the Trustees of *Biometrika*, and from *Statistical Tables for Biological, Agricultural and Medical Research*.

Table C: Probability Points of the *t*-Distribution (Single-sided)   438
Condensed and adapted from the *Biometrika Tables for Statisticians*, **1**, Table 12, with permission of the Trustees of *Biometrika*, and from *Statistical Tables for Biological, Agricultural and Medical Research*.

Table D: Probability Points of the Variance Ratio (*F*-Distribution)   440
Condensed and adapted from the *Biometrika Tables for Statisticians*, **1**, Table 18, with permission of the Trustees of *Biometrika*, and from *Statistical Tables for Biological, Agricultural and Medical Research*.

Table E: Values of the Correlation Coefficient for Different Levels of Significance   444
Condensed and adapted from *Statistical Tables for Biological, Agricultural and Medical Research*.

Table F: Binomial and Poisson Distributions: Confidence Limits   445
Condensed and adapted from *Statistical Tables for Biological, Agricultural and Medical Research* (due to *W. L. Stevens*).

Table G: Control Chart Limits for Average   448

Tables G.1 and G.2: Control Chart Limits for Range   450
Reproduced from B.S.600R, with recensions from *Biometrika*, **32**, 301–310 (1942).

Page

Table H: Multipliers L_1 and L_2 for the Confidence Limits of the Ratio of Two Standard Deviations 452

Table I: Values of Outlier Ratio at Two Levels of Significance 460
Condensed and reproduced with permission from F. E. Grubbs, *Technometrics*, **11**, 1–21, Table 1 (1969) and **14**, 847–854, Table 1 (1972).

Table J: Random Permutations 461
Generated by the method described by E. S. Page in *Applied Statistics*, **16**, 273–274 (1967).

Table K: Table of Random Numbers 462
Extracted from *Tables of Random Sampling Numbers*, by M. G. Kendall and B. Babington Smith, and *Tracts for Computers*, XXIV, with permission of University College, London. Table K comprises page 40 of these tables.

Table L: Critical Values for the Sign Test 463
Reproduced with permission of the publishers from *Introduction to Statistical Analysis* by W. J. Dixon and F. J. Massey. McGraw-Hill (Second edition, 1957).

Table M: Critical Values for the Rank-Sum Test 464
Reproduced with permission of the publishers from *Nonparametric and Short-cut Statistics* by M. W. Tate and R. C. Clelland. Interstate (1957).

Table N: Attribute Sampling Schemes for Process Control 465
Reproduced with permission from K. W. Kemp, *Applied Statistics*, **11**, 16–31, Table IX (1962).

Intersection Chart: Single-sided Cusum Control Schemes 466
Constructed from data listed in Table 5 of *Cumulative Sum Tests: Theory and Practice* by C. S. Van Dobben de Bruyn (Griffin, 1968) according to a layout suggested by A. F. Bissell.

Nomogram: Interpolation of Statistical Tables for Other Probability Levels 469

TABLE A.   NORMAL DISTRIBUTION (SINGLE-SIDED)

Proportion (P) of whole area lying to right of ordinate through $u = (x-\mu)/\sigma$

Deviate u	0·00	0·01	0·02	0·03	0·04	0·05	0·06	0·07	0·08	0·09
0·0	·5000	·4960	·4920	·4880	·4840	·4801	·4761	·4721	·4681	·4641
0·1	·4602	·4562	·4522	·4483	·4443	·4404	·4364	·4325	·4286	·4247
0·2	·4207	·4168	·4129	·4090	·4052	·4013	·3974	·3936	·3897	·3859
0·3	·3821	·3783	·3745	·3707	·3669	·3632	·3594	·3557	·3520	·3483
0·4	·3446	·3409	·3372	·3336	·3300	·3264	·3228	·3192	·3156	·3121
0·5	·3085	·3050	·3015	·2981	·2946	·2912	·2877	·2843	·2810	·2776
0·6	·2743	·2709	·2676	·2643	·2611	·2578	·2546	·2514	·2483	·2451
0·7	·2420	·2389	·2358	·2327	·2296	·2266	·2236	·2206	·2177	·2148
0·8	·2119	·2090	·2061	·2033	·2005	·1977	·1949	·1922	·1894	·1867
0·9	·1841	·1814	·1788	·1762	·1736	·1711	·1685	·1660	·1635	·1611
1·0	·1587	·1562	·1539	·1515	·1492	·1469	·1446	·1423	·1401	·1379
1·1	·1357	·1335	·1314	·1292	·1271	·1251	·1230	·1210	·1190	·1170
1·2	·1151	·1131	·1112	·1093	·1075	·1056	·1038	·1020	·1003	·0985
1·3	·0968	·0951	·0934	·0918	·0901	·0885	·0869	·0853	·0838	·0823
1·4	·0808	·0793	·0778	·0764	·0749	·0735	·0721	·0708	·0694	·0681
1·5	·0668	·0655	·0643	·0630	·0618	·0606	·0594	·0582	·0571	·0559
1·6	·0548	·0537	·0526	·0516	·0505	·0495	·0485	·0475	·0465	·0455
1·7	·0446	·0436	·0427	·0418	·0409	·0401	·0392	·0384	·0375	·0367
1·8	·0359	·0351	·0344	·0336	·0329	·0322	·0314	·0307	·0301	·0294
1·9	·0287	·0281	·0274	·0268	·0262	·0256	·0250	·0244	·0239	·0233
2·0	·0228	·0222	·0217	·0212	·0207	·0202	·0197	·0192	·0188	·0183
2·1	·0179	·0174	·0170	·0166	·0162	·0158	·0154	·0150	·0146	·0143
2·2	·0139	·0136	·0132	·0129	·0125	·0122	·0119	·0116	·0113	·0110
2·3	·0107	·0104	·0102		·00964		·00914		·00866	
2·4	·00820		·00776		·00734		·00695		·00657	
2·5	·00621		·00587		·00554		·00523		·00494	
2·6	·00466		·00440		·00415		·00391		·00368	
2·7	·00347		·00326		·00307		·00289		·00272	
2·8	·00256		·00240		·00226		·00212		·00199	
2·9	·00187		·00175		·00164		·00154		·00144	
	0·00		0·02		0·04		0·06		0·08	

TABLE A (*continued*). EXTENSION FOR HIGHER VALUES OF THE DEVIATE

Deviate u	Proportion of Whole Area P	Deviate u	Proportion of Whole Area P	Deviate u	Proportion of Whole Area P	Deviate u	Proportion of Whole Area P
3·0	·001 35	3·5	·000 233	4·0	·0^4 317	4·5	·0^5 340
3·1	·000 968	3·6	·000 159	4·1	·0^4 207	4·6	·0^5 211
3·2	·000 687	3·7	·000 108	4·2	·0^4 133	4·7	·0^5 130
3·3	·000 483	3·8	·0^4 723	4·3	·0^5 854	4·8	·0^6 793
3·4	·000 337	3·9	·0^4 481	4·4	·0^5 541	4·9	·0^6 479
						5·0	·0^6 287

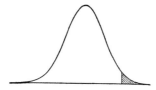

The illustration shows the Normal Curve. The shaded portion is the area P, which is given in the table.

The entries refer to positive values of the argument u. For negative values of u write down the complements $(1-P)$ of the entries.

Examples

Let $u = +1·96$. Area to the right $= 0·0250$. Area to left $= 1-0·0250 = 0·9750$.

Let $u = -3·00$. The tabulated value $= 0·00135$. Since u is negative, this represents the area to the *left*. Area to right $= 1-0·00135 = 0·99865$.

Let $u = +4·50$. Tabulated value $= 0·00000340$. Area to left $= 0·99999660$.

To find the value of u corresponding to a given P we use the table in reverse, thus:

Let area to right (i.e. P) $= 0·10$. The two adjacent tabulated values are $P = 0·1003$ for $u = 1·28$, and $P = 0·0985$ for $u = 1·29$. We interpolate linearly to obtain the required value of u. Thus $u = 1·28+(3)(0·01)/18 = 1·2817$.

TABLE B. PROBABILITY POINTS OF THE χ^2 DISTRIBUTION

ϕ	0·995	0·99	0·975	0·95	0·90	0·75	0·50	0·25	0·10	0·05	0·025	0·01	0·005	0·001	ϕ
1	—	—	—	—	·016	·102	·455	1·32	2·71	3·84	5·02	6·63	7·88	10·8	1
2	·010	·020	·051	·103	·211	·575	1·39	2·77	4·61	5·99	7·38	9·21	10·6	13·8	2
3	·072	·115	·216	·352	·584	1·21	2·37	4·11	6·25	7·81	9·35	11·3	12·8	16·3	3
4	·207	·297	·484	·711	1·06	1·92	3·36	5·39	7·78	9·49	11·1	13·3	14·9	18·5	4
5	·412	·554	·831	1·15	1·61	2·67	4·35	6·63	9·24	11·1	12·8	15·1	16·7	20·5	5
6	·676	·872	1·24	1·64	2·20	3·45	5·35	7·84	10·6	12·6	14·4	16·8	18·5	22·5	6
7	·989	1·24	1·69	2·17	2·83	4·25	6·35	9·04	12·0	14·1	16·0	18·5	20·3	24·3	7
8	1·34	1·65	2·18	2·73	3·49	5·07	7·34	10·2	13·4	15·5	17·5	20·1	22·0	26·1	8
9	1·73	2·09	2·70	3·33	4·17	5·90	8·34	11·4	14·7	16·9	19·0	21·7	23·6	27·9	9
10	2·16	2·56	3·25	3·94	4·87	6·74	9·34	12·5	16·0	18·3	20·5	23·2	25·2	29·6	10
11	2·60	3·05	3·82	4·57	5·58	7·58	10·3	13·7	17·3	19·7	21·9	24·7	26·8	31·3	11
12	3·07	3·57	4·40	5·23	6·30	8·44	11·3	14·8	18·5	21·0	23·3	26·2	28·3	32·9	12
13	3·57	4·11	5·01	5·89	7·04	9·30	12·3	16·0	19·8	22·4	24·7	27·7	29·8	34·5	13
14	4·07	4·66	5·63	6·57	7·79	10·2	13·3	17·1	21·1	23·7	26·1	29·1	31·3	36·1	14
15	4·60	5·23	6·26	7·26	8·55	11·0	14·3	18·2	22·3	25·0	27·5	30·6	32·8	37·7	15
16	5·14	5·81	6·91	7·96	9·31	11·9	15·3	19·4	23·5	26·3	28·8	32·0	34·3	39·3	16
17	5·70	6·41	7·56	8·67	10·1	12·8	16·3	20·5	24·8	27·6	30·2	33·4	35·7	40·8	17
18	6·26	7·01	8·23	9·39	10·9	13·7	17·3	21·6	26·0	28·9	31·5	34·8	37·2	42·3	18
19	6·84	7·63	8·91	10·1	11·7	14·6	18·3	22·7	27·2	30·1	32·9	36·2	38·6	43·8	19
20	7·43	8·26	9·59	10·9	12·4	15·5	19·3	23·8	28·4	31·4	34·2	37·6	40·0	45·3	20
21	8·03	8·90	10·3	11·6	13·2	16·3	20·3	24·9	29·6	32·7	35·5	38·9	41·4	46·8	21
22	8·64	9·54	11·0	12·3	14·0	17·2	21·3	26·0	30·8	33·9	36·8	40·3	42·8	48·3	22
23	9·26	10·2	11·7	13·1	14·8	18·1	22·3	27·1	32·0	35·2	38·1	41·6	44·2	49·7	23
24	9·89	10·9	12·4	13·8	15·7	19·0	23·3	28·2	33·2	36·4	39·4	43·0	45·6	51·2	24
25	10·5	11·5	13·1	14·6	16·5	19·9	24·3	29·3	34·4	37·7	40·6	44·3	46·9	52·6	25
26	11·2	12·2	13·8	15·4	17·3	20·8	25·3	30·4	35·6	38·9	41·9	45·6	48·3	54·1	26
27	11·8	12·9	14·6	16·2	18·1	21·7	26·3	31·5	36·7	40·1	43·2	47·0	49·6	55·5	27
28	12·5	13·6	15·3	16·9	18·9	22·7	27·3	32·6	37·9	41·3	44·5	48·3	51·0	56·9	28
29	13·1	14·3	16·0	17·7	19·8	23·6	28·3	33·7	39·1	42·6	45·7	49·6	52·3	58·3	29
30	13·8	15·0	16·8	18·5	20·6	24·5	29·3	34·8	40·3	43·8	47·0	50·9	53·7	59·7	30

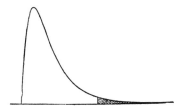

The illustration shows the χ^2 curve for $\phi = 3$. The shaded portion, expressed as a proportion of the total area under the curve, is the columnar heading in the table.

It is of interest to note that χ^2 is related to the bottom row of the table of the F-distribution (Table D), where the number of degrees of freedom associated with the denominator is infinite. These values of F are in fact values of χ^2 divided by the number of degrees of freedom. Thus for $\phi = 8$ and $P = 0.01$ the present table gives $\chi^2 = 20.1$; in Table D with $\phi_N = 8$ and $\phi_D = \infty$ we obtain $F = 2.51 = 20.1/8$.

Examples

Let $\chi^2 = 3.80$, $\phi = 3$. This is between the 0.50 and 0.25 points, and is therefore not significant.

Let $\chi^2 = 20.1$, $\phi = 9$. This is between the 0.025 and 0.01 points, and is therefore significant.

Values of χ^2 for $\phi > 30$

For values of $\phi > 30$ the expression $\sqrt{(2\chi^2)} - \sqrt{(2\phi - 1)}$ may be used as a Normal deviate with unit variance, remembering that the probability for χ^2 corresponds to that of a single tail of the Normal Curve.

Example

Let $\chi^2 = 124.3$, $\phi = 100$. Then $u = \sqrt{248.6} - \sqrt{199} = 1.66$. For $u = 1.66$, the value of $P = 0.0485$. χ^2 is therefore just significant.

TABLE C.   PROBABILITY POINTS OF THE *t*-DISTRIBUTION (SINGLE-SIDED)

ϕ	P				
	0·1	0·05	0·025	0·01	0·005
1	3·08	6·31	12·70	31·80	63·70
2	1·89	2·92	4·30	6·96	9·92
3	1·64	2·35	3·18	4·54	5·84
4	1·53	2·13	2·78	3·75	4·60
5	1·48	2·01	2·57	3·36	4·03
6	1·44	1·94	2·45	3·14	3·71
7	1·42	1·89	2·36	3·00	3·50
8	1·40	1·86	2·31	2·90	3·36
9	1·38	1·83	2·26	2·82	3·25
10	1·37	1·81	2·23	2·76	3·17
11	1·36	1·80	2·20	2·72	3·11
12	1·36	1·78	2·18	2·68	3·05
13	1·35	1·77	2·16	2·65	3·01
14	1·34	1·76	2·14	2·62	2·98
15	1·34	1·75	2·13	2·60	2·95
16	1·34	1·75	2·12	2·58	2·92
17	1·33	1·74	2·11	2·57	2·90
18	1·33	1·73	2·10	2·55	2·88
19	1·33	1·73	2·09	2·54	2·86
20	1·32	1·72	2·09	2·53	2·85
21	1·32	1·72	2·08	2·52	2·83
22	1·32	1·72	2·07	2·51	2·82
23	1·32	1·71	2·07	2·50	2·81
24	1·32	1·71	2·06	2·49	2·80
25	1·32	1·71	2·06	2·48	2·79
26	1·32	1·71	2·06	2·48	2·78
27	1·31	1·70	2·05	2·47	2·77
28	1·31	1·70	2·05	2·47	2·76
29	1·31	1·70	2·05	2·46	2·76
30	1·31	1·70	2·04	2·46	2·75
40	1·30	1·68	2·02	2·42	2·70
60	1·30	1·67	2·00	2·39	2·66
120	1·29	1·66	1·98	2·36	2·62
∞	1·28	1·64	1·96	2·33	2·58

The illustration shows the *t*-curve for $\phi = 3$. The shaded area corresponds to the columnar headings of the table and the unshaded area to their complements.

It is of interest to note that *t* is related to the first column of the *F*-distribution (Table D), where $\phi_N = 1$. Setting $\phi_D = \phi$, *F* is in fact equal to t^2, providing the value of *P* in Table C is doubled to make the comparison double-sided. Thus for $\phi = 8$ and $P = 2 \times 0 \cdot 005$ the present table gives $t = 3 \cdot 36$; in Table D with $\phi_N = 1$, $\phi_D = 8$ and $P = 0 \cdot 01$ we obtain $F = 11 \cdot 3 = 3 \cdot 36^2$.

Example

Single-sided test. For $\phi = 10$ the deviate of the *t*-curve which cuts off a single tail equivalent to $P = 0 \cdot 05$ is given by $t = 1 \cdot 81$. For the Normal Curve the corresponding value of *u* is $1 \cdot 64$.

Table of the variance ratio (F-distribution)

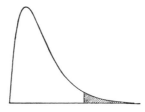

The illustration shows the distribution of the variance ratio for 4 and 16 degrees of freedom. The shaded area, expressed as a proportion of the total area under the curve, is the argument in the first column of Table D.

The variance ratio is always calculated with the *larger* estimate of variance in the *numerator*, and ϕ_N & ϕ_D are the numbers of degrees of freedom in the numerator and denominator respectively.

Example

Let $F = 4{\cdot}60$, $\phi_N = 5$, $\phi_D = 24$. The 5% and 1% points are 2·62 and 3·90, so the result is significant.

In calculating confidence limits for the variance ratio we require the upper and lower tail areas of the F-distribution. The levels actually tabled refer to the single upper tail area $F_\alpha(\phi_N,\phi_D)$. However, the value $F_{1-\alpha}(\phi_N,\phi_D)$ (i.e. the value of F *below which* a proportion α of the whole curve lies) is given by $1/F_\alpha(\phi_D,\ \phi_N)$.

Example

To obtain 90% confidence limits for the variance ratio we require the values of $F_{0\cdot95}(\phi_N,\ \phi_D)$ and $F_{0\cdot05}(\phi_N,\ \phi_D)$. If $\phi_N = 4$ and $\phi_D = 20$, then:

$$F_{0\cdot95}(4,20) = \frac{1}{F_{0.05}(20,4)} = \frac{1}{5{\cdot}80} = 0{\cdot}172 \quad \text{and} \quad F_{0.05}(4,20) = 2{\cdot}87$$

ϕ_N (corresponding to greater mean square)

Probability Point	ϕ_D	1	2	3	4	5	6	7	8	9	10	12	15	20	24	30	40	60	120	∞
0·100	1	39·9	49·5	53·6	55·8	57·2	58·2	58·9	59·4	59·9	60·2	60·7	61·2	61·7	62·0	62·3	62·5	62·8	63·1	63·3
0·050		161	199	216	225	230	234	237	239	241	242	244	246	248	249	250	251	252	253	254
0·025		648	800	864	900	922	937	948	957	963	969	977	985	993	997	1001	1006	1010	1014	1018
0·010		4052	4999	5403	5625	5764	5859	5928	5982	6022	6056	6106	6157	6209	6235	6261	6287	6313	6339	6366
0·100	2	8·53	9·00	9·16	9·24	9·29	9·33	9·35	9·37	9·38	9·39	9·41	9·42	9·44	9·45	9·46	9·47	9·47	9·48	9·49
0·050		18·5	19·0	19·2	19·2	19·3	19·3	19·4	19·4	19·4	19·4	19·4	19·4	19·4	19·5	19·5	19·5	19·5	19·5	19·5
0·025		38·5	39·0	39·2	39·2	39·3	39·3	39·4	39·4	39·4	39·4	39·4	39·4	39·4	39·5	39·5	39·5	39·5	39·5	39·5
0·010		98·5	99·0	99·2	99·2	99·3	99·3	99·4	99·4	99·4	99·4	99·4	99·4	99·4	99·5	99·5	99·5	99·5	99·5	99·5
0·100	3	5·54	5·46	5·39	5·34	5·31	5·28	5·27	5·25	5·24	5·23	5·22	5·20	5·18	5·18	5·17	5·16	5·15	5·14	5·13
0·050		10·1	9·55	9·28	9·12	9·01	8·94	8·89	8·85	8·81	8·79	8·74	8·70	8·66	8·64	8·62	8·59	8·57	8·55	8·53
0·025		17·4	16·0	15·4	15·1	14·9	14·7	14·6	14·5	14·5	14·4	14·3	14·3	14·2	14·1	14·1	14·0	14·0	13·9	13·9
0·010		34·1	30·8	29·5	28·7	28·2	27·9	27·7	27·5	27·3	27·2	27·1	26·9	26·7	26·6	26·5	26·4	26·3	26·2	26·1
0·100	4	4·54	4·32	4·19	4·11	4·05	4·01	3·98	3·95	3·94	3·92	3·90	3·87	3·84	3·83	3·82	3·80	3·79	3·78	3·76
0·050		7·71	6·94	6·59	6·39	6·26	6·16	6·09	6·04	6·00	5·96	5·91	5·86	5·80	5·77	5·75	5·72	5·69	5·66	5·63
0·025		12·2	10·6	10·0	9·60	9·36	9·20	9·07	8·98	8·90	8·84	8·75	8·66	8·56	8·51	8·46	8·41	8·36	8·31	8·26
0·010		21·2	18·0	16·7	16·0	15·5	15·2	15·0	14·8	14·7	14·5	14·4	14·2	14·0	13·9	13·8	13·7	13·7	13·6	13·5
0·100	5	4·06	3·78	3·62	3·52	3·45	3·40	3·37	3·34	3·32	3·30	3·27	3·24	3·21	3·19	3·17	3·16	3·14	3·12	3·10
0·050		6·61	5·79	5·41	5·19	5·05	4·95	4·88	4·82	4·77	4·74	4·68	4·62	4·56	4·53	4·50	4·46	4·43	4·40	4·36
0·025		10·0	8·43	7·76	7·39	7·15	6·98	6·85	6·76	6·68	6·62	6·52	6·43	6·33	6·28	6·23	6·18	6·12	6·07	6·02
0·010		16·3	13·3	12·1	11·4	11·0	10·7	10·5	10·3	10·2	10·1	9·89	9·72	9·55	9·47	9·38	9·29	9·20	9·11	9·02
0·100	6	3·78	3·46	3·29	3·18	3·11	3·05	3·01	2·98	2·96	2·94	2·90	2·87	2·84	2·82	2·80	2·78	2·76	2·74	2·72
0·050		5·99	5·14	4·76	4·53	4·39	4·28	4·21	4·15	4·10	4·06	4·00	3·94	3·87	3·84	3·81	3·77	3·74	3·70	3·67
0·025		8·81	7·26	6·60	6·23	5·99	5·82	5·70	5·60	5·52	5·46	5·37	5·27	5·17	5·12	5·07	5·01	4·96	4·90	4·85
0·010		13·7	10·9	9·78	9·15	8·75	8·47	8·26	8·10	7·98	7·87	7·72	7·56	7·40	7·31	7·23	7·14	7·06	6·97	6·88
0·100	7	3·59	3·26	3·07	2·96	2·88	2·83	2·78	2·75	2·72	2·70	2·67	2·63	2·59	2·58	2·56	2·54	2·51	2·49	2·47
0·050		5·59	4·74	4·35	4·12	3·97	3·87	3·79	3·73	3·68	3·64	3·57	3·51	3·44	3·41	3·38	3·34	3·30	3·27	3·23
0·025		8·07	6·54	5·89	5·52	5·29	5·12	4·99	4·90	4·82	4·76	4·67	4·57	4·47	4·42	4·36	4·31	4·25	4·20	4·14
0·010		12·2	9·55	8·45	7·85	7·46	7·19	6·99	6·84	6·72	6·62	6·47	6·31	6·16	6·07	5·99	5·91	5·82	5·74	5·65

For explanation and example see p. 440.

TABLE D (continued)

ϕ_D	Probability Point	\multicolumn{19}{c}{ϕ_N (corresponding to greater mean square)}																		
		1	2	3	4	5	6	7	8	9	10	12	15	20	24	30	40	60	120	∞
8	0.100	3.46	3.11	2.92	2.81	2.73	2.67	2.62	2.59	2.56	2.54	2.50	2.46	2.42	2.40	2.38	2.36	2.34	2.32	2.29
	0.050	5.32	4.46	4.07	3.84	3.69	3.58	3.50	3.44	3.39	3.35	3.28	3.22	3.15	3.12	3.08	3.04	3.01	2.97	2.93
	0.025	7.57	6.06	5.42	5.05	4.82	4.65	4.53	4.43	4.36	4.30	4.20	4.10	4.00	3.95	3.89	3.84	3.78	3.73	3.67
	0.010	11.3	8.65	7.59	7.01	6.63	6.37	6.18	6.03	5.91	5.81	5.67	5.52	5.36	5.28	5.20	5.12	5.03	4.95	4.86
9	0.100	3.36	3.01	2.81	2.69	2.61	2.55	2.51	2.47	2.44	2.42	2.38	2.34	2.30	2.28	2.25	2.23	2.21	2.18	2.16
	0.050	5.12	4.26	3.86	3.63	3.48	3.37	3.29	3.23	3.18	3.14	3.07	3.01	2.94	2.90	2.86	2.83	2.79	2.75	2.71
	0.025	7.21	5.71	5.08	4.72	4.48	4.32	4.20	4.10	4.03	3.96	3.87	3.77	3.67	3.61	3.56	3.51	3.45	3.39	3.33
	0.010	10.6	8.02	6.99	6.42	6.06	5.80	5.61	5.47	5.35	5.26	5.11	4.96	4.81	4.73	4.65	4.57	4.48	4.40	4.31
10	0.100	3.28	2.92	2.73	2.61	2.52	2.46	2.41	2.38	2.35	2.32	2.28	2.24	2.20	2.18	2.16	2.13	2.11	2.08	2.06
	0.050	4.96	4.10	3.71	3.48	3.33	3.22	3.14	3.07	3.02	2.98	2.91	2.84	2.77	2.74	2.70	2.66	2.62	2.58	2.54
	0.025	6.94	5.46	4.83	4.47	4.24	4.07	3.95	3.85	3.78	3.72	3.62	3.52	3.42	3.37	3.31	3.26	3.20	3.14	3.08
	0.010	10.0	7.56	6.55	5.99	5.64	5.39	5.20	5.06	4.94	4.85	4.71	4.56	4.41	4.33	4.25	4.17	4.08	4.00	3.91
12	0.100	3.18	2.81	2.61	2.48	2.39	2.33	2.28	2.24	2.21	2.19	2.15	2.10	2.06	2.04	2.01	1.99	1.96	1.93	1.90
	0.050	4.75	3.89	3.49	3.26	3.11	3.00	2.91	2.85	2.80	2.75	2.69	2.62	2.54	2.51	2.47	2.43	2.38	2.34	2.30
	0.025	6.55	5.10	4.47	4.12	3.89	3.73	3.61	3.51	3.44	3.37	3.28	3.18	3.07	3.02	2.96	2.91	2.85	2.79	2.72
	0.010	9.33	6.93	5.95	5.41	5.06	4.82	4.64	4.50	4.39	4.30	4.16	4.01	3.86	3.78	3.70	3.62	3.54	3.45	3.36
15	0.100	3.07	2.70	2.49	2.36	2.27	2.21	2.16	2.12	2.09	2.06	2.02	1.97	1.92	1.90	1.87	1.85	1.82	1.79	1.76
	0.050	4.54	3.68	3.29	3.06	2.90	2.79	2.71	2.64	2.59	2.54	2.48	2.40	2.33	2.29	2.25	2.20	2.16	2.11	2.07
	0.025	6.20	4.77	4.15	3.80	3.58	3.41	3.29	3.20	3.12	3.06	2.96	2.86	2.76	2.70	2.64	2.59	2.52	2.46	2.40
	0.010	8.68	6.36	5.42	4.89	4.56	4.32	4.14	4.00	3.89	3.80	3.67	3.52	3.37	3.29	3.21	3.13	3.05	2.96	2.87
20	0.100	2.97	2.59	2.38	2.25	2.16	2.09	2.04	2.00	1.96	1.94	1.89	1.84	1.79	1.77	1.74	1.71	1.68	1.64	1.61
	0.050	4.35	3.49	3.10	2.87	2.71	2.60	2.51	2.45	2.39	2.35	2.28	2.20	2.12	2.08	2.04	1.99	1.95	1.90	1.84
	0.025	5.87	4.46	3.86	3.51	3.29	3.13	3.01	2.91	2.84	2.77	2.68	2.57	2.46	2.41	2.35	2.29	2.22	2.16	2.09
	0.010	8.10	5.85	4.94	4.43	4.10	3.87	3.70	3.56	3.46	3.37	3.23	3.09	2.94	2.86	2.78	2.69	2.61	2.52	2.42

For explanation and example see p. 440.

TABLE D (continued)

ϕ_N (corresponding to greater mean square)

ϕ_D	Probability Point	1	2	3	4	5	6	7	8	9	10	12	15	20	24	30	40	60	120	∞
24	0.100	2.93	2.54	2.33	2.19	2.10	2.04	1.98	1.94	1.91	1.88	1.83	1.78	1.73	1.70	1.67	1.64	1.61	1.57	1.53
	0.050	4.26	3.40	3.01	2.78	2.62	2.51	2.42	2.36	2.30	2.25	2.18	2.11	2.03	1.98	1.94	1.89	1.84	1.79	1.73
	0.025	5.72	4.32	3.72	3.38	3.15	2.99	2.87	2.78	2.70	2.64	2.54	2.44	2.33	2.27	2.21	2.15	2.08	2.01	1.94
	0.010	7.82	5.61	4.72	4.22	3.90	3.67	3.50	3.36	3.26	3.17	3.03	2.89	2.74	2.66	2.58	2.49	2.40	2.31	2.21
30	0.100	2.88	2.49	2.28	2.14	2.05	1.98	1.93	1.88	1.85	1.82	1.77	1.72	1.67	1.64	1.61	1.57	1.54	1.50	1.46
	0.050	4.17	3.32	2.92	2.69	2.53	2.42	2.33	2.27	2.21	2.16	2.09	2.01	1.93	1.89	1.84	1.79	1.74	1.68	1.62
	0.025	5.57	4.18	3.59	3.25	3.03	2.87	2.75	2.65	2.57	2.51	2.41	2.31	2.20	2.14	2.07	2.01	1.94	1.87	1.79
	0.010	7.56	5.39	4.51	4.02	3.70	3.47	3.30	3.17	3.07	2.98	2.84	2.70	2.55	2.47	2.39	2.30	2.21	2.11	2.01
40	0.100	2.84	2.44	2.23	2.09	2.00	1.93	1.87	1.83	1.79	1.76	1.71	1.66	1.61	1.57	1.54	1.51	1.47	1.42	1.38
	0.050	4.08	3.23	2.84	2.61	2.45	2.34	2.25	2.18	2.12	2.08	2.00	1.92	1.84	1.79	1.74	1.69	1.64	1.58	1.51
	0.025	5.42	4.05	3.46	3.13	2.90	2.74	2.62	2.53	2.45	2.39	2.29	2.18	2.07	2.01	1.94	1.88	1.80	1.72	1.64
	0.010	7.31	5.18	4.31	3.83	3.51	3.29	3.12	2.99	2.89	2.80	2.66	2.52	2.37	2.29	2.20	2.11	2.02	1.92	1.80
60	0.100	2.79	2.39	2.18	2.04	1.95	1.87	1.82	1.77	1.74	1.71	1.66	1.60	1.54	1.51	1.48	1.44	1.40	1.35	1.29
	0.050	4.00	3.15	2.76	2.53	2.37	2.25	2.17	2.10	2.04	1.99	1.92	1.84	1.75	1.70	1.65	1.59	1.53	1.47	1.39
	0.025	5.29	3.93	3.34	3.01	2.79	2.63	2.51	2.41	2.33	2.27	2.17	2.06	1.94	1.88	1.82	1.74	1.67	1.58	1.48
	0.010	7.08	4.98	4.13	3.65	3.34	3.12	2.95	2.82	2.72	2.63	2.50	2.35	2.20	2.12	2.03	1.94	1.84	1.73	1.60
120	0.100	2.75	2.35	2.13	1.99	1.90	1.82	1.77	1.72	1.68	1.65	1.60	1.54	1.48	1.45	1.41	1.37	1.32	1.26	1.19
	0.050	3.92	3.07	2.68	2.45	2.29	2.18	2.09	2.02	1.96	1.91	1.83	1.75	1.66	1.61	1.55	1.50	1.43	1.35	1.25
	0.025	5.15	3.80	3.23	2.89	2.67	2.52	2.39	2.30	2.22	2.16	2.05	1.94	1.82	1.76	1.69	1.61	1.53	1.43	1.31
	0.010	6.85	4.79	3.95	3.48	3.17	2.96	2.79	2.66	2.56	2.47	2.34	2.19	2.03	1.95	1.86	1.76	1.66	1.53	1.38
∞	0.100	2.71	2.30	2.08	1.94	1.85	1.77	1.72	1.67	1.63	1.60	1.55	1.49	1.42	1.38	1.34	1.30	1.24	1.17	1.00
	0.050	3.84	3.00	2.60	2.37	2.21	2.10	2.01	1.94	1.88	1.83	1.75	1.67	1.57	1.52	1.46	1.39	1.32	1.22	1.00
	0.025	5.02	3.69	3.12	2.79	2.57	2.41	2.29	2.19	2.11	2.05	1.94	1.83	1.71	1.64	1.57	1.48	1.39	1.27	1.00
	0.010	6.63	4.61	3.78	3.32	3.02	2.80	2.64	2.51	2.41	2.32	2.18	2.04	1.88	1.79	1.70	1.59	1.47	1.32	1.00

For explanation and example see p. 440.

TABLE E. VALUES OF CORRELATION COEFFICIENT FOR DIFFERENT
LEVELS OF SIGNIFICANCE

ϕ	α					ϕ	α				
	0·10	0·05	0·02	0·01	0·001		0·10	0·05	0·02	0·01	0·001
1	·998	·997	1·00	1·00	1·00	16	·400	·468	·543	·590	·708
2	·900	·950	·980	·990	·999	17	·389	·456	·529	·575	·693
3	·805	·878	·934	·959	·991	18	·378	·444	·516	·561	·679
4	·729	·811	·882	·917	·974	19	·369	·433	·503	·549	·665
5	·669	·754	·833	·875	·951	20	·360	·423	·492	·537	·652
6	·621	·707	·789	·834	·925	25	·323	·381	·445	·487	·597
7	·582	·666	·750	·798	·898	30	·296	·349	·409	·449	·554
8	·549	·632	·715	·765	·872	35	·275	·325	·381	·418	·519
9	·521	·602	·685	·735	·847	40	·257	·304	·358	·393	·490
10	·497	·576	·658	·708	·823	45	·243	·288	·338	·372	·465
11	·476	·553	·634	·684	·801	50	·231	·273	·322	·354	·443
12	·457	·532	·612	·661	·780	60	·211	·250	·295	·325	·408
13	·441	·514	·592	·641	·760	70	·195	·232	·274	·302	·380
14	·426	·497	·574	·623	·742	80	·183	·217	·257	·283	·357
15	·412	·482	·558	·606	·725	90	·173	·205	·242	·267	·338
						100	·164	·195	·230	·254	·321

ϕ is the number of degrees of freedom.

The significance points for r are related to the corresponding (double-sided) critical levels of t by the formula:

$$r = t/\sqrt{(\phi + t^2)}$$

Thus when $\phi = 10$, the deviate of the t-curve which cuts off a double tail equivalent to $P = 0·10$ is given by $t = 1·81$, and the equivalent significance point for r is:

$$1·81/\sqrt{(10 + 1·81^2)} = 1·81/3·64 = 0·497$$

Table of confidence limits for binomial and Poisson distributions
(*See next page*)

To obtain the confidence limits, corresponding to a given probability level *P*, of an event observed to occur *a* times out of *n*, enter the table with *a*, *P* and $p = a/n$. Interpolate linearly, if necessary, between values of *p*. Confidence limits for the Poisson distribution are given directly, taking $p = 0$. For $a > n/2$, enter the table with $b = n - a$ instead of *a*.

Examples

A sample of 25 articles is drawn at random from a bulk and 3 of them are found to be defective. What conclusions can we draw about the proportion of defective articles in the bulk?

The value of *a* is 3 and the value of $p = a/n$ is $3/25 = 0.12$. This value of *p* is not tabulated and we have to interpolate between the arguments 15 and 30 for *n*, or 0·2 and 0·1 for *p*. For $P = 0.025$ the confidence limits are $0.634 + (0.2)(0.016) = 0.637$, and $7.96 - (0.2)(0.75) = 7.81$. Multiplying by 4 to bring these figures to a percentage basis, we assess the percentage defective in the bulk as lying between 2·55% and 31·2%. For $P = 0.10$ the confidence limits are 4·48% and 24·8% (see § **9.152**).

A specification lays down that 12 articles shall be drawn at random and the consignment passed if none of the articles is defective. Discuss the protection given to the consumer.

Take $a = 0$, probability *P* of *a* or fewer (i.e. probability that $a = 0$) = 0·1, and interpolate at sight between $n = 10$ and $n = 20$, or interpolate with respect to $1/n = 1/12 = 0.083$ between (0·1) and (0·05):

$$2.06 + (0.33)(0.11) = 2.10$$

This is the confidence limit in a sample of 12. The corresponding limit of the percentage defective in the consignment is $(100)(2.10)/12 = 17.5\%$. Hence there is a one-in-ten chance that the consumer would accept a consignment containing as much as 17·5% defective material.

Note that accurate interpolation with respect to *p* is rarely necessary in practice: it is usually sufficient to make a rough interpolation at sight or even take the nearest value of *n* tabulated. When *a* is less than one-twentieth of *n*, it is usually sufficiently accurate to enter the table with $n = \infty$ ($p = 0$).

a	n	$p = a/n$	\multicolumn{6}{c}{Confidence limits for $\mu = np$ corresponding to probabilities of}
			0·005	0·025	0·1	0·1	0·025	0·005
			\multicolumn{3}{c}{Lower limit}	\multicolumn{3}{c}{Upper limit}
0	5		For $a = 0$		(·2)	1·84	2·61	3·27
			interpolate		(·15)	1·95	2·83	3·66
	10		upper limit		(·1)	2·06	3·09	4·11
	20		with respect		(·05)	2·17	3·37	4·65
	∞	0	to $1/n$		(0)	2·30	3·69	5·30
1	2	·5	·005	·025	·103	1·90	1·97	1·99
	3		·005	·025	·104	2·41	2·72	2·88
	4	·25	·005	·025	·104	2·72	3·22	3·56
	5	·2	·005	·025	·104	2·92	3·58	4·07
		·15	·005	·025	·105	3·14	3·99	4·69
	10	·1	·005	·025	·105	3·37	4·45	5·44
	20	·05	·005	·025	·105	3·62	4·97	6·34
	∞	0	·005	·025	·105	3·89	5·57	7·43
2	4	·5	·118	·270	·570	3·43	3·73	3·88
	5	·4	·114	·264	·561	3·77	4·27	4·59
	6		·112	·260	·556	4·00	4·67	5·14
	7		·111	·257	·552	4·17	4·97	5·54
	8	·25	·110	·255	·549	4·31	5·21	5·94
	10	·2	·108	·252	·545	4·50	5·56	6·48
		·15	·107	·249	·542	4·69	5·94	7·09
	20	·1	·106	·247	·538	4·90	6·34	7·74
	40	·05	·105	·245	·535	5·11	6·77	8·44
	∞	0	·103	·242	·532	5·32	7·22	9·27
3	6	·5	·398	·709	1·21	4·79	5·29	5·60
	7		·387	·693	1·19	5·05	5·71	6·18
	8		·380	·682	1·17	5·24	6·04	6·64
	9		·374	·674	1·17	5·39	6·31	7·03
	10	·3	·370	·667	1·16	5·52	6·52	7·35
	15	·2	·358	·650	1·14	5·89	7·21	8·41
	30	·1	·348	·634	1·12	6·28	7·96	9·61
	∞	0	·338	·619	1·10	6·68	8·77	11·0
4	8	·5	·799	1·26	1·92	6·08	6·74	7·20
	10	·4	·768	1·22	1·88	6·46	7·38	8·09
		·3	·741	1·18	1·84	6·83	8·04	9·06
	20	·2	·716	1·15	1·80	7·21	8·73	10·1
	40	·1	·694	1·12	1·77	7·60	9·47	11·3
	∞	0	·672	1·09	1·74	7·99	10·2	12·6
5	10	·5	1·28	1·87	2·67	7·33	8·13	8·72
		·4	1·23	1·81	2·61	7·72	8·79	9·66
		·3	1·19	1·76	2·56	8·10	9·47	10·7
	25	·2	1·15	1·71	2·52	8·49	10·2	11·7
	50	·1	1·11	1·66	2·47	8·88	10·9	12·9
	∞	0	1·08	1·62	2·43	9·27	11·7	14·1
6	12	·5	1·83	2·53	3·46	8·54	9·47	10·2
	15	·4	1·75	2·45	3·38	8·95	10·2	11·2
	20	·3	1·69	2·38	3·32	9·35	10·9	12·2
	30	·2	1·63	2·31	3·26	9·74	11·6	13·3
	60	·1	1·58	2·26	3·20	10·1	12·3	14·4
	∞	0	1·54	2·20	3·15	10·5	13·1	15·7

For examples see p. 445.

a	$p = a/n$	Confidence limits for $\mu = np$ corresponding to probabilities of					
		0·005	0·025	0·1	0·1	0·025	0·005
			Lower limit			Upper limit	
7	·5	2·41	3·23	4·26	9·74	10·8	11·6
	·4	2·32	3·12	4·17	10·1	11·5	12·6
	·3	2·24	3·03	4·09	10·6	12·2	13·7
	·2	2·16	2·95	4·02	11·0	12·9	14·8
	·1	2·10	2·88	3·96	11·4	13·7	15·9
	0	2·04	2·81	3·89	11·8	14·4	17·1
8	·5	3·04	3·94	5·09	10·9	12·1	13·0
	·4	2·92	3·82	4·98	11·3	12·8	14·0
	·3	2·82	3·72	4·89	11·8	13·5	15·1
	·2	2·73	3·62	4·80	12·2	14·3	16·2
	·1	2·65	3·53	4·73	12·6	15·0	17·4
	0	2·57	3·45	4·66	13·0	15·8	18·6
9	·5	3·68	4·68	5·92	12·1	13·3	14·3
	·4	3·55	4·54	5·80	12·5	14·1	15·4
	·3	3·43	4·42	5·70	13·0	14·8	16·5
	·2	3·32	4·31	5·60	13·4	15·6	17·6
	·1	3·22	4·21	5·51	13·8	16·3	18·8
	0	3·13	4·12	5·43	14·2	17·1	20·0
10	·5	4·35	5·44	6·76	13·2	14·6	15·6
	·4	4·20	5·28	6·63	13·7	15·3	16·8
	·3	4·06	5·14	6·51	14·1	16·1	17·9
	·2	3·93	5·02	6·41	14·6	16·9	19·0
	·1	3·82	4·90	6·31	15·0	17·6	20·2
	0	3·72	4·80	6·22	15·4	18·4	21·4
11	·5	5·04	6·21	7·62	14·4	15·8	17·0
	·4	4·87	6·03	7·47	14·8	16·6	18·1
	·3	4·71	5·88	7·34	15·3	17·4	19·2
	·2	4·57	5·74	7·23	15·7	18·1	20·4
	·1	4·44	5·61	7·12	16·2	18·9	21·6
	0	4·32	5·49	7·02	16·6	19·7	22·8
12	·5	5·75	6·99	8·48	15·5	17·0	18·2
	·4	5·55	6·80	8·32	16·0	17·8	19·4
	·3	5·37	6·63	8·18	16·5	18·6	20·6
	·2	5·22	6·47	8·05	16·9	19·4	21·7
	·1	5·07	6·33	7·94	17·3	20·2	22·9
	0	4·94	6·20	7·83	17·8	21·0	24·1
13	·5	6·47	7·78	9·34	16·7	18·2	19·5
	·4	6·25	7·57	9·17	17·1	19·0	20·7
	·3	6·06	7·38	9·02	17·6	19·8	21·9
	·2	5·88	7·22	8·89	18·1	20·6	23·1
	·1	5·72	7·06	8·76	18·5	21·4	24·3
	0	5·58	6·92	8·65	19·0	22·2	25·5
14	·5	7·20	8·58	10·2	17·8	19·4	20·8
	·4	6·96	8·36	10·0	18·3	20·3	22·0
	·3	6·75	8·15	9·87	18·8	21·1	23·2
	·2	6·56	7·97	9·73	19·2	21·9	24·4
	·1	6·39	7·81	9·59	19·7	22·7	25·6
	0	6·23	7·65	9·47	20·1	23·5	26·8

For examples see p. 445.

TABLE G. CONTROL CHART LIMITS FOR AVERAGE $(\bar{x})$

No. in Sample n	For Inner Limits $A_{0.025}$	For Outer Limits $A_{0.001}$	For Inner Limits $A'_{0.025}$	For Outer Limits $A'_{0.001}$
2	1·386	2·185	1·229	1·937
3	1·132	1·784	0·668	1·054
4	0·980	1·545	0·476	0·750
5	0·877	1·382	0·377	0·594
6	0·800	1·262	0·316	0·498
7	0·741	1·168	0·274	0·432
8	0·693	1·093	0·244	0·384
9	0·653	1·030	0·220	0·347
10	0·620	0·977	0·202	0·317
11	0·591	0·932	0·186	0·294
12	0·566	0·892	0·174	0·274
13	0·544	0·857		
14	0·524	0·826		
15	0·506	0·798		
16	0·490	0·773		
17	0·475	0·750		
18	0·462	0·728		
19	0·450	0·709		
20	0·438	0·691		
21	0·428	0·674		
22	0·418	0·659		
23	0·409	0·644		
24	0·400	0·631		
25	0·392	0·618		
26	0·384	0·606		
27	0·377	0·595		
28	0·370	0·584		
29	0·364	0·574		
30	0·358	0·564		

Samples containing more than 12 individuals should not be used when utilising the range of the results.

These factors should only be used when it is not necessary to calculate s for the samples and when sufficient test data are available to make an accurate estimate of σ from $\bar{w}$.

Note. For larger values of n use the general formulae:

$$A_{0.025} = 1·96/\sqrt{n}$$
$$A_{0.001} = 3·09/\sqrt{n}$$

$$\text{To obtain limits} \begin{cases} \text{multiply } \sigma \text{ by the appropriate value of } A_{0.025} \text{ and} \\ \quad A_{0.001}, \text{ or} \\ \text{multiply } \bar{w} \text{ by the appropriate value of } A'_{0.025} \text{ and} \\ \quad A'_{0.001} \end{cases}$$

Then add to and subtract from the average value (μ).

Example

Let $\mu = 18{\cdot}0$ min., $\bar{w} = 3{\cdot}63$ min., and $n = 4$.

$$\text{Inner Control Limits} = 18{\cdot}0 \pm 3{\cdot}63 \times 0{\cdot}476$$
$$= 19{\cdot}7 \text{ and } 16{\cdot}3 \text{ min.}$$
$$\text{Outer Control Limits} = 18{\cdot}0 \pm 3{\cdot}63 \times 0{\cdot}750$$
$$= 20{\cdot}7 \text{ and } 15{\cdot}3 \text{ min.}$$

Or alternatively, since the estimate of $\sigma = \bar{w}/d_n = 3{\cdot}63/2{\cdot}059 = 1{\cdot}76$:

$$\text{Inner Control Limits} = 18{\cdot}0 \pm 1{\cdot}76 \times 0{\cdot}980$$
$$= 19{\cdot}7 \text{ and } 16{\cdot}3 \text{ min.}$$
$$\text{Outer Control Limits} = 18{\cdot}0 \pm 1{\cdot}76 \times 1{\cdot}545$$
$$= 20{\cdot}7 \text{ and } 15{\cdot}3 \text{ min.}$$

Note. The two sets of figures coincide because the value of σ has been estimated from the value of $\bar{w}$. Had the value of σ been calculated directly the figures would have differed slightly.

TABLE G.1. CONTROL CHART LIMITS FOR RANGE (w)

(For use when the limits are calculated from the standard deviation)

No. in Sample	For Lower Limits		For Upper Limits		For Average Value of w ($\bar{w}$)
n	Outer $D_{0.001}$	Inner $D_{0.025}$	Inner $D_{0.975}$	Outer $D_{0.999}$	d_n
2	0·00	0·04	3·17	4·65	1·128
3	0·06	0·30	3·68	5·06	1·693
4	0·20	0·59	3·98	5·31	2·059
5	0·37	0·85	4·20	5·48	2·326
6	0·54	1·06	4·36	5·62	2·534
7	0·69	1·25	4·49	5·73	2·704
8	0·83	1·41	4·61	5·82	2·847
9	0·96	1·55	4·70	5·90	2·970
10	1·08	1·67	4·79	5·97	3·078
11	1·20	1·78	4·86	6·04	3·173
12	1·30	1·88	4·92	6·09	3·258

To obtain the limits, multiply σ by the appropriate values of D. To obtain σ, divide $\bar{w}$ (the average value of w) by the appropriate value of d_n.

Example

Let $\sigma = 1\cdot76$ min. and $n = 4$.

Inner Control Limits: Outer Control Limits:

   Lower: $1\cdot76 \times 0\cdot59 = 1\cdot0$ min.   Lower: $1\cdot76 \times 0\cdot20 = 0\cdot4$ min.

   Upper: $1\cdot76 \times 3\cdot98 = 7\cdot0$ min.   Upper: $1\cdot76 \times 5\cdot31 = 9\cdot3$ min.

The constants d_n may be used to obtain an estimate of σ from the mean range ($\bar{w}$) (see § **3.35**) from the formula:

$$\text{Estimate of } \sigma = \bar{w}/d_n$$

Let mean range $\bar{w} = 47\cdot9$ and $n = 5$. From table, $d_n = 2\cdot326$.

$$\therefore \quad \text{Estimate of } \sigma = 47\cdot9/2\cdot326 = 20\cdot6$$

TABLE G.2.   CONTROL CHART LIMITS FOR RANGE (w)

(For use when control limits are calculated from average range)

No. in Sample	For Lower Limits		For Upper Limits	
	$D'_{0.001}$	$D'_{0.025}$	$D'_{0.975}$	$D'_{0.999}$
2	0·00	0·04	2·81	4·12
3	0·04	0·18	2·17	2·99
4	0·10	0·29	1·93	2·58
5	0·16	0·37	1·81	2·36
6	0·21	0·42	1·72	2·22
7	0·26	0·46	1·66	2·12
8	0·29	0·50	1·62	2·04
9	0·32	0·52	1·58	1·99
10	0·35	0·54	1·56	1·94
11	0·38	0·56	1·53	1·90
12	0·40	0·58	1·51	1·87

To obtain limits, multiply $\bar{w}$ by the appropriate values of D'.

Example

Let $\bar{w} = 3\cdot63$ min. and $n = 4$.

Inner Control Limits: Outer Control Limits:

Lower: $3\cdot63 \times 0\cdot29 = 1\cdot1$ min.   Lower: $3\cdot63 \times 0\cdot10 = 0\cdot4$ min.

Upper: $3\cdot63 \times 1\cdot93 = 7\cdot0$ min.   Upper: $3\cdot63 \times 2\cdot58 = 9\cdot4$ min.

TABLE H.　MULTIPLIERS L_1 AND L_2 FOR THE CONFIDENCE LIMITS OF THE RATIO OF TWO STANDARD DEVIATIONS

PROBABILITY LEVEL 0·10 (SINGLE SIDE)

ϕ_N \ ϕ_D	1	2	3	4	5	6	7	8	9	10	12	15	20	24	30	40	60	120	∞
1	0·16 / 6·31	0·14 / 2·92	0·14 / 2·35	0·13 / 2·13	0·13 / 2·02	0·13 / 1·94	0·13 / 1·89	0·13 / 1·86	0·13 / 1·83	0·13 / 1·81	0·13 / 1·78	0·13 / 1·75	0·13 / 1·72	0·13 / 1·71	0·13 / 1·70	0·13 / 1·68	0·13 / 1·67	0·13 / 1·66	0·13 / 1·64
2	0·34 / 7·04	0·33 / 3·00	0·33 / 2·34	0·33 / 2·08	0·33 / 1·94	0·33 / 1·86	0·33 / 1·80	0·33 / 1·76	0·33 / 1·73	0·33 / 1·71	0·33 / 1·68	0·33 / 1·64	0·33 / 1·61	0·33 / 1·59	0·33 / 1·58	0·33 / 1·56	0·32 / 1·55	0·32 / 1·53	0·32 / 1·52
3	0·42 / 7·32	0·43 / 3·03	0·43 / 2·32	0·43 / 2·05	0·43 / 1·90	0·44 / 1·81	0·44 / 1·75	0·44 / 1·71	0·44 / 1·68	0·44 / 1·65	0·44 / 1·61	0·44 / 1·58	0·44 / 1·54	0·44 / 1·53	0·44 / 1·51	0·44 / 1·49	0·44 / 1·48	0·44 / 1·46	0·44 / 1·44
4	0·47 / 7·47	0·48 / 3·04	0·49 / 2·31	0·49 / 2·03	0·50 / 1·88	0·50 / 1·78	0·50 / 1·72	0·50 / 1·68	0·50 / 1·64	0·51 / 1·61	0·51 / 1·57	0·51 / 1·54	0·51 / 1·50	0·51 / 1·48	0·51 / 1·46	0·51 / 1·45	0·51 / 1·43	0·51 / 1·41	0·52 / 1·39
5	0·50 / 7·57	0·51 / 3·05	0·53 / 2·30	0·53 / 2·01	0·54 / 1·86	0·54 / 1·76	0·54 / 1·70	0·55 / 1·65	0·55 / 1·62	0·55 / 1·59	0·55 / 1·55	0·56 / 1·51	0·56 / 1·47	0·56 / 1·45	0·56 / 1·43	0·56 / 1·41	0·56 / 1·39	0·57 / 1·38	0·57 / 1·36
6	0·51 / 7·63	0·54 / 3·05	0·55 / 2·30	0·56 / 2·00	0·57 / 1·85	0·57 / 1·75	0·58 / 1·68	0·58 / 1·63	0·58 / 1·60	0·58 / 1·57	0·59 / 1·53	0·59 / 1·49	0·59 / 1·45	0·60 / 1·43	0·60 / 1·41	0·60 / 1·39	0·60 / 1·37	0·60 / 1·35	0·61 / 1·33
7	0·53 / 7·68	0·55 / 3·06	0·57 / 2·29	0·58 / 1·99	0·59 / 1·84	0·59 / 1·74	0·60 / 1·67	0·60 / 1·62	0·61 / 1·58	0·61 / 1·55	0·61 / 1·51	0·62 / 1·47	0·62 / 1·43	0·62 / 1·41	0·63 / 1·39	0·63 / 1·37	0·63 / 1·35	0·63 / 1·33	0·64 / 1·31
8	0·54 / 7·71	0·57 / 3·06	0·58 / 2·29	0·60 / 1·99	0·61 / 1·83	0·61 / 1·73	0·62 / 1·66	0·62 / 1·61	0·62 / 1·57	0·63 / 1·54	0·63 / 1·50	0·64 / 1·46	0·64 / 1·41	0·64 / 1·39	0·65 / 1·37	0·65 / 1·35	0·65 / 1·33	0·66 / 1·31	0·66 / 1·29
9	0·55 / 7·74	0·58 / 3·06	0·60 / 2·29	0·61 / 1·98	0·62 / 1·82	0·63 / 1·72	0·63 / 1·65	0·64 / 1·60	0·64 / 1·56	0·64 / 1·53	0·65 / 1·49	0·65 / 1·44	0·66 / 1·40	0·66 / 1·38	0·67 / 1·36	0·67 / 1·34	0·67 / 1·32	0·68 / 1·30	0·68 / 1·28
10	0·55 / 7·76	0·58 / 3·06	0·61 / 2·29	0·62 / 1·98	0·63 / 1·82	0·64 / 1·71	0·64 / 1·64	0·65 / 1·59	0·65 / 1·55	0·66 / 1·52	0·66 / 1·48	0·67 / 1·44	0·67 / 1·39	0·68 / 1·37	0·68 / 1·35	0·68 / 1·33	0·69 / 1·31	0·69 / 1·29	0·70 / 1·26

TABLE H (continued)

ϕ_D \ ϕ_N	1	2	3	4	5	6	7	8	9	10	12	15	20	24	30	40	60	120	∞
12	0·56 7·79	0·60 3·07	0·62 2·28	0·63 1·97	0·65 1·81	0·65 1·70	0·66 1·63	0·67 1·58	0·67 1·54	0·68 1·51	0·68 1·47	0·69 1·42	0·70 1·38	0·70 1·35	0·71 1·33	0·71 1·31	0·71 1·29	0·72 1·27	0·72 1·24
15	0·57 7·82	0·61 3·07	0·63 2·28	0·65 1·97	0·66 1·80	0·67 1·69	0·68 1·62	0·69 1·57	0·69 1·53	0·70 1·50	0·70 1·45	0·71 1·40	0·72 1·36	0·73 1·34	0·73 1·31	0·74 1·29	0·74 1·27	0·75 1·24	0·75 1·22
20	0·58 7·86	0·62 3·07	0·65 2·28	0·67 1·96	0·68 1·79	0·69 1·68	0·70 1·61	0·71 1·56	0·71 1·52	0·72 1·48	0·73 1·44	0·74 1·39	0·75 1·34	0·75 1·32	0·76 1·29	0·77 1·27	0·77 1·24	0·78 1·22	0·79 1·19
24	0·58 7·87	0·63 3·07	0·66 2·28	0·67 1·96	0·69 1·79	0·70 1·68	0·71 1·60	0·72 1·55	0·72 1·51	0·73 1·48	0·74 1·43	0·75 1·38	0·76 1·33	0·77 1·30	0·77 1·28	0·78 1·25	0·79 1·23	0·80 1·20	0·81 1·18
30	0·59 7·89	0·63 3·08	0·66 2·27	0·68 1·95	0·70 1·78	0·71 1·67	0·72 1·60	0·73 1·54	0·74 1·50	0·74 1·47	0·75 1·42	0·76 1·37	0·77 1·32	0·78 1·29	0·79 1·27	0·80 1·24	0·81 1·21	0·82 1·19	0·83 1·16
40	0·59 7·91	0·64 3·08	0·67 2·27	0·69 1·95	0·71 1·78	0·72 1·67	0·73 1·59	0·74 1·54	0·75 1·49	0·75 1·46	0·76 1·41	0·78 1·36	0·79 1·31	0·80 1·28	0·81 1·25	0·81 1·23	0·83 1·20	0·84 1·17	0·85 1·14
60	0·60 7·92	0·65 3·08	0·68 2·27	0·70 1·95	0·72 1·77	0·73 1·66	0·74 1·59	0·75 1·53	0·76 1·49	0·77 1·45	0·78 1·40	0·79 1·35	0·81 1·29	0·81 1·27	0·82 1·24	0·83 1·21	0·85 1·18	0·86 1·15	0·88 1·11
120	0·60 7·94	0·65 3·08	0·69 2·27	0·71 1·94	0·73 1·77	0·74 1·66	0·75 1·58	0·76 1·52	0·77 1·48	0·78 1·44	0·79 1·39	0·80 1·34	0·82 1·28	0·83 1·25	0·84 1·22	0·86 1·19	0·87 1·16	0·89 1·12	0·92 1·08
∞	0·61 7·96	0·66 3·08	0·69 2·27	0·72 1·94	0·74 1·76	0·75 1·65	0·76 1·57	0·77 1·51	0·78 1·47	0·79 1·43	0·80 1·38	0·82 1·32	0·84 1·27	0·85 1·24	0·86 1·21	0·88 1·17	0·90 1·14	0·93 1·09	1·00 1·00

Values of ϕ_N for the right portion of the table: 12, 15, 20, 24, 30, 40, 60, 120, ∞.

L_1 is given by the upper entry in each cell and L_2 by the lower.

ϕ_N represents the degrees of freedom of the standard deviation in the numerator and ϕ_D the degrees of freedom of the standard deviation in the denominator.

The confidence limits for σ_1/σ_2 are $L_1 s_1/s_2$ and $L_2 s_1/s_2$.

The multipliers for the confidence limits of one standard deviation are given in the row $\phi_D = \infty$.

TABLE H. MULTIPLIERS L_1 AND L_2 FOR THE CONFIDENCE LIMITS OF THE RATIO OF TWO STANDARD DEVIATIONS (continued)

PROBABILITY LEVEL 0·05 (SINGLE SIDE)

ϕ_D \ ϕ_N	1	2	3	4	5	6	7	8	9	10	12	15	20	24	30	40	60	120	∞
1	·079 / 12·7	·071 / 4·30	·068 / 3·18	·067 / 2·78	·066 / 2·57	·065 / 2·45	·065 / 2·36	·065 / 2·31	·064 / 2·26	·064 / 2·23	·064 / 2·18	·064 / 2·13	·063 / 2·09	·063 / 2·06	·063 / 2·04	·063 / 2·02	·063 / 2·00	·063 / 1·98	·063 / 1·96
2	0·23 / 14·1	0·23 / 4·36	0·23 / 3·09	0·23 / 2·64	0·23 / 2·41	0·23 / 2·27	0·23 / 2·18	0·23 / 2·11	0·23 / 2·06	0·23 / 2·03	0·23 / 1·97	0·23 / 1·92	0·23 / 1·87	0·23 / 1·84	0·23 / 1·82	0·23 / 1·80	0·23 / 1·77	0·23 / 1·75	0·23 / 1·73
3	0·31 / 14·7	0·32 / 4·38	0·33 / 3·05	0·33 / 2·57	0·33 / 2·33	0·33 / 2·18	0·34 / 2·08	0·34 / 2·02	0·34 / 1·97	0·34 / 1·93	0·34 / 1·87	0·34 / 1·81	0·34 / 1·76	0·34 / 1·73	0·34 / 1·71	0·34 / 1·68	0·34 / 1·66	0·34 / 1·64	0·34 / 1·61
4	0·36 / 15·0	0·38 / 4·39	0·39 / 3·02	0·40 / 2·53	0·40 / 2·28	0·40 / 2·13	0·41 / 2·03	0·41 / 1·96	0·41 / 1·91	0·41 / 1·86	0·41 / 1·81	0·41 / 1·75	0·42 / 1·69	0·42 / 1·67	0·42 / 1·64	0·42 / 1·61	0·42 / 1·59	0·42 / 1·56	0·42 / 1·54
5	0·39 / 15·2	0·42 / 4·39	0·43 / 3·00	0·44 / 2·50	0·44 / 2·25	0·45 / 2·09	0·45 / 1·99	0·46 / 1·92	0·46 / 1·87	0·46 / 1·82	0·46 / 1·76	0·47 / 1·70	0·47 / 1·65	0·47 / 1·62	0·47 / 1·59	0·47 / 1·57	0·48 / 1·54	0·48 / 1·51	0·48 / 1·49
6	0·41 / 15·3	0·44 / 4·40	0·46 / 2·99	0·47 / 2·48	0·48 / 2·22	0·48 / 2·07	0·49 / 1·97	0·49 / 1·89	0·49 / 1·84	0·50 / 1·79	0·50 / 1·73	0·50 / 1·67	0·51 / 1·61	0·51 / 1·58	0·51 / 1·56	0·51 / 1·53	0·52 / 1·50	0·52 / 1·47	0·52 / 1·45
7	0·42 / 15·4	0·46 / 4·40	0·48 / 2·98	0·49 / 2·47	0·50 / 2·21	0·51 / 2·05	0·51 / 1·95	0·52 / 1·87	0·52 / 1·81	0·52 / 1·77	0·53 / 1·71	0·53 / 1·65	0·54 / 1·59	0·54 / 1·56	0·54 / 1·53	0·55 / 1·50	0·55 / 1·47	0·55 / 1·44	0·56 / 1·42
8	0·43 / 15·5	0·47 / 4·40	0·50 / 2·97	0·51 / 2·46	0·52 / 2·20	0·53 / 2·04	0·53 / 1·93	0·54 / 1·85	0·54 / 1·80	0·55 / 1·75	0·55 / 1·69	0·56 / 1·63	0·56 / 1·56	0·57 / 1·54	0·57 / 1·51	0·57 / 1·48	0·58 / 1·45	0·58 / 1·42	0·58 / 1·39
9	0·44 / 15·5	0·48 / 4·40	0·51 / 2·97	0·52 / 2·45	0·54 / 2·18	0·54 / 2·02	0·55 / 1·92	0·56 / 1·84	0·56 / 1·78	0·56 / 1·74	0·57 / 1·67	0·58 / 1·61	0·58 / 1·55	0·59 / 1·52	0·59 / 1·49	0·59 / 1·46	0·60 / 1·43	0·60 / 1·40	0·61 / 1·37
10	0·45 / 15·6	0·49 / 4·40	0·52 / 2·96	0·54 / 2·44	0·55 / 2·18	0·56 / 2·01	0·56 / 1·91	0·57 / 1·83	0·58 / 1·77	0·58 / 1·73	0·59 / 1·66	0·59 / 1·59	0·60 / 1·53	0·60 / 1·50	0·61 / 1·47	0·61 / 1·44	0·62 / 1·41	0·62 / 1·38	0·63 / 1·35

TABLE H (continued)

φ_N \ φ_D	1	2	3	4	5	6	7	8	9	10	12	15	20	24	30	40	60	120	∞
12	0·46 / 15·6	0·51 / 4·41	0·54 / 2·96	0·55 / 2·43	0·57 / 2·16	0·58 / 2·00	0·59 / 1·89	0·59 / 1·81	0·60 / 1·75	0·60 / 1·71	0·61 / 1·64	0·62 / 1·57	0·63 / 1·51	0·63 / 1·48	0·64 / 1·45	0·64 / 1·42	0·65 / 1·38	0·65 / 1·35	0·66 / 1·32
15	0·47 / 15·7	0·52 / 4·41	0·55 / 2·95	0·57 / 2·42	0·59 / 2·15	0·60 / 1·98	0·61 / 1·87	0·62 / 1·79	0·62 / 1·73	0·63 / 1·69	0·64 / 1·62	0·65 / 1·55	0·65 / 1·48	0·66 / 1·45	0·67 / 1·42	0·67 / 1·39	0·68 / 1·36	0·69 / 1·32	0·70 / 1·29
20	0·48 / 15·7	0·54 / 4·41	0·57 / 2·94	0·59 / 2·41	0·61 / 2·14	0·62 / 1·97	0·63 / 1·86	0·64 / 1·77	0·65 / 1·71	0·65 / 1·67	0·66 / 1·59	0·67 / 1·53	0·69 / 1·46	0·69 / 1·42	0·70 / 1·39	0·71 / 1·36	0·72 / 1·32	0·73 / 1·29	0·74 / 1·25
24	0·48 / 15·8	0·54 / 4·41	0·58 / 2·94	0·60 / 2·40	0·62 / 2·13	0·63 / 1·96	0·64 / 1·85	0·65 / 1·77	0·66 / 1·70	0·67 / 1·65	0·68 / 1·58	0·69 / 1·51	0·70 / 1·44	0·71 / 1·41	0·72 / 1·37	0·73 / 1·34	0·74 / 1·30	0·75 / 1·27	0·76 / 1·23
30	0·49 / 15·8	0·55 / 4·41	0·58 / 2·94	0·61 / 2·40	0·63 / 2·12	0·64 / 1·95	0·65 / 1·84	0·66 / 1·75	0·67 / 1·69	0·68 / 1·64	0·69 / 1·57	0·70 / 1·50	0·72 / 1·43	0·73 / 1·39	0·74 / 1·36	0·75 / 1·32	0·76 / 1·28	0·77 / 1·25	0·79 / 1·21
40	0·49 / 15·8	0·56 / 4·41	0·59 / 2·93	0·62 / 2·39	0·64 / 2·11	0·65 / 1·94	0·67 / 1·83	0·68 / 1·74	0·69 / 1·68	0·69 / 1·63	0·71 / 1·56	0·72 / 1·48	0·74 / 1·41	0·75 / 1·38	0·76 / 1·34	0·77 / 1·30	0·78 / 1·26	0·80 / 1·22	0·81 / 1·18
60	0·50 / 15·9	0·56 / 4·41	0·60 / 2·93	0·63 / 2·38	0·65 / 2·11	0·67 / 1·93	0·68 / 1·82	0·69 / 1·73	0·70 / 1·67	0·71 / 1·62	0·72 / 1·54	0·74 / 1·47	0·76 / 1·40	0·77 / 1·36	0·78 / 1·32	0·79 / 1·28	0·81 / 1·24	0·83 / 1·20	0·85 / 1·15
120	0·51 / 15·9	0·57 / 4·41	0·61 / 2·92	0·64 / 2·38	0·66 / 2·10	0·68 / 1·92	0·69 / 1·81	0·70 / 1·72	0·71 / 1·66	0·72 / 1·61	0·74 / 1·53	0·76 / 1·45	0·78 / 1·38	0·79 / 1·34	0·80 / 1·30	0·82 / 1·26	0·84 / 1·21	0·86 / 1·16	0·89 / 1·11
∞	0·51 / 15·9	0·58 / 4·42	0·62 / 2·92	0·65 / 2·37	0·67 / 2·09	0·69 / 1·92	0·71 / 1·80	0·72 / 1·71	0·73 / 1·65	0·74 / 1·59	0·76 / 1·52	0·77 / 1·44	0·80 / 1·36	0·81 / 1·32	0·83 / 1·27	0·85 / 1·23	0·87 / 1·18	0·90 / 1·12	1·00 / 1·00

L_1 is given by the upper entry in each cell and L_2 by the lower.

ϕ_N represents the degrees of freedom of the standard deviation in the numerator and ϕ_D the degrees of freedom of the standard deviation in the denominator.

The confidence limits for σ_1/σ_2 are $L_1 s_1/s_2$ and $L_2 s_1/s_2$.

The multipliers for the confidence limits of one standard deviation are given in the row $\phi_D = \infty$.

TABLE H. MULTIPLIERS L_1 AND L_2 FOR THE CONFIDENCE LIMITS OF THE RATIO OF TWO STANDARD DEVIATIONS (continued)
PROBABILITY LEVEL 0·025 (SINGLE SIDE)

Each cell gives L_1 (upper) / L_2 (lower).

ϕ_D \ ϕ_N	1	2	3	4	5	6	7	8	9	10	12	15	20	24	30	40	60	120	∞
1	·039 / 25·5	·035 / 6·20	·034 / 4·18	·033 / 3·50	·033 / 3·16	·033 / 2·97	·032 / 2·84	·032 / 2·75	·032 / 2·69	·032 / 2·63	·032 / 2·56	·032 / 2·49	·032 / 2·42	·032 / 2·39	·032 / 2·36	·032 / 2·33	·031 / 2·30	·031 / 2·27	·031 / 2·24
2	0·16 / 28·3	0·16 / 6·24	0·16 / 4·01	0·16 / 3·26	0·16 / 2·90	0·16 / 2·69	0·16 / 2·56	0·16 / 2·46	0·16 / 2·39	0·16 / 2·34	0·16 / 2·26	0·16 / 2·18	0·16 / 2·11	0·16 / 2·08	0·16 / 2·05	0·16 / 2·01	0·16 / 1·98	0·16 / 1·95	0·16 / 1·92
3	0·24 / 29·5	0·25 / 6·26	0·25 / 3·93	0·26 / 3·16	0·26 / 2·79	0·26 / 2·57	0·26 / 2·43	0·26 / 2·33	0·26 / 2·25	0·26 / 2·20	0·26 / 2·12	0·26 / 2·04	0·27 / 1·96	0·27 / 1·93	0·27 / 1·89	0·27 / 1·86	0·27 / 1·83	0·27 / 1·80	0·27 / 1·77
4	0·29 / 30·0	0·31 / 6·26	0·32 / 3·89	0·32 / 3·10	0·33 / 2·72	0·33 / 2·50	0·33 / 2·35	0·33 / 2·25	0·34 / 2·17	0·34 / 2·11	0·34 / 2·03	0·34 / 1·95	0·34 / 1·87	0·34 / 1·84	0·34 / 1·80	0·34 / 1·77	0·35 / 1·73	0·35 / 1·70	0·35 / 1·67
5	0·32 / 30·4	0·34 / 6·27	0·36 / 3·86	0·37 / 3·06	0·37 / 2·67	0·38 / 2·45	0·38 / 2·30	0·38 / 2·19	0·39 / 2·12	0·39 / 2·06	0·39 / 1·97	0·39 / 1·89	0·40 / 1·81	0·40 / 1·78	0·40 / 1·74	0·40 / 1·70	0·40 / 1·67	0·41 / 1·64	0·41 / 1·60
6	0·34 / 30·6	0·37 / 6·27	0·39 / 3·84	0·40 / 3·03	0·41 / 2·64	0·41 / 2·41	0·42 / 2·26	0·42 / 2·16	0·43 / 2·08	0·43 / 2·02	0·43 / 1·93	0·44 / 1·85	0·44 / 1·77	0·44 / 1·73	0·44 / 1·69	0·45 / 1·66	0·45 / 1·62	0·45 / 1·59	0·45 / 1·55
7	0·35 / 30·8	0·39 / 6·27	0·41 / 3·82	0·43 / 3·01	0·43 / 2·62	0·44 / 2·39	0·45 / 2·23	0·45 / 2·13	0·46 / 2·05	0·46 / 1·99	0·46 / 1·90	0·47 / 1·81	0·47 / 1·73	0·48 / 1·70	0·48 / 1·66	0·48 / 1·62	0·48 / 1·58	0·49 / 1·55	0·49 / 1·51
8	0·36 / 30·9	0·41 / 6·27	0·43 / 3·81	0·44 / 3·01	0·46 / 2·60	0·46 / 2·37	0·47 / 2·21	0·47 / 2·11	0·48 / 2·03	0·48 / 1·96	0·49 / 1·87	0·49 / 1·79	0·50 / 1·71	0·50 / 1·67	0·51 / 1·63	0·51 / 1·59	0·51 / 1·55	0·52 / 1·52	0·52 / 1·48
9	0·37 / 31·0	0·42 / 6·28	0·44 / 3·80	0·46 / 2·98	0·47 / 2·58	0·48 / 2·35	0·49 / 2·20	0·49 / 2·09	0·50 / 2·01	0·50 / 1·94	0·51 / 1·85	0·52 / 1·77	0·52 / 1·68	0·53 / 1·64	0·53 / 1·60	0·53 / 1·57	0·54 / 1·53	0·54 / 1·49	0·55 / 1·45
10	0·38 / 31·1	0·43 / 6·28	0·46 / 3·80	0·47 / 2·97	0·49 / 2·57	0·50 / 2·34	0·50 / 2·18	0·51 / 2·07	0·51 / 1·99	0·52 / 1·93	0·53 / 1·84	0·53 / 1·75	0·54 / 1·67	0·55 / 1·62	0·55 / 1·58	0·55 / 1·55	0·56 / 1·51	0·56 / 1·47	0·57 / 1·43

TABLE H (continued)

ϕ_D \ ϕ_N	1	2	3	4	5	6	7	8	9	10	12	15	20	24	30	40	60	120	∞
12	0·39 31·3	0·44 6·28	0·47 3·79	0·49 2·96	0·51 2·55	0·52 2·32	0·53 2·16	0·53 2·05	0·54 1·97	0·54 1·90	0·55 1·81	0·56 1·72	0·57 1·64	0·58 1·59	0·58 1·55	0·59 1·51	0·59 1·47	0·60 1·43	0·61 1·39
15	0·40 31·4	0·46 6·28	0·49 3·78	0·51 2·94	0·53 2·54	0·54 2·30	0·55 2·14	0·56 2·02	0·57 1·94	0·57 1·88	0·58 1·78	0·59 1·69	0·60 1·60	0·61 1·56	0·62 1·52	0·62 1·48	0·63 1·44	0·64 1·39	0·65 1·35
20	0·41 31·5	0·47 6·28	0·51 3·76	0·53 2·93	0·55 2·52	0·57 2·27	0·58 2·11	0·59 2·00	0·59 1·91	0·60 1·85	0·61 1·75	0·62 1·66	0·64 1·57	0·64 1·53	0·65 1·48	0·66 1·44	0·67 1·39	0·68 1·35	0·69 1·31
24	0·42 31·6	0·48 6·28	0·52 3·76	0·54 2·92	0·56 2·51	0·58 2·26	0·59 2·10	0·60 1·99	0·61 1·90	0·62 1·83	0·63 1·74	0·64 1·64	0·66 1·55	0·66 1·51	0·67 1·46	0·68 1·42	0·69 1·37	0·71 1·33	0·72 1·28
30	0·42 31·6	0·49 6·28	0·53 3·75	0·55 2·91	0·57 2·50	0·59 2·25	0·60 2·09	0·61 1·97	0·62 1·89	0·63 1·82	0·64 1·72	0·66 1·63	0·67 1·53	0·68 1·49	0·69 1·44	0·71 1·39	0·72 1·35	0·73 1·30	0·75 1·25
40	0·43 31·7	0·50 6·28	0·54 3·75	0·57 2·90	0·59 2·49	0·60 2·24	0·62 2·08	0·63 1·96	0·64 1·87	0·65 1·80	0·66 1·70	0·68 1·61	0·70 1·51	0·71 1·46	0·72 1·42	0·73 1·37	0·74 1·32	0·76 1·27	0·78 1·22
60	0·43 31·8	0·50 6·28	0·55 3·74	0·58 2·89	0·60 2·47	0·62 2·23	0·63 2·06	0·64 1·95	0·65 1·86	0·66 1·79	0·68 1·69	0·70 1·59	0·72 1·49	0·73 1·44	0·74 1·39	0·76 1·34	0·77 1·29	0·80 1·24	0·82 1·18
120	0·44 31·8	0·51 6·28	0·56 3·73	0·59 2·88	0·61 2·46	0·63 2·21	0·65 2·05	0·66 1·93	0·67 1·84	0·68 1·77	0·70 1·67	0·72 1·57	0·74 1·47	0·75 1·42	0·77 1·37	0·79 1·31	0·81 1·26	0·84 1·20	0·87 1·13
∞	0·45 31·9	0·52 6·28	0·57 3·73	0·60 2·87	0·62 2·45	0·64 2·20	0·66 2·04	0·68 1·92	0·69 1·83	0·70 1·75	0·72 1·65	0·74 1·55	0·77 1·44	0·78 1·39	0·80 1·34	0·82 1·28	0·85 1·22	0·89 1·14	1·00 1·00

L_1 is given by the upper entry in each cell and L_2 by the lower.

ϕ_N represents the degrees of freedom of the standard deviation in the numerator and ϕ_D the degrees of freedom of the standard deviation in the denominator.

The confidence limits for σ_1/σ_2 are $L_1 s_1/s_2$ and $L_2 s_1/s_2$.

The multipliers for the confidence limits of one standard deviation are given in the row $\phi_D = \infty$.

TABLE H. MULTIPLIERS L_1 AND L_2 FOR THE CONFIDENCE LIMITS OF THE RATIO OF TWO STANDARD DEVIATIONS (continued) PROBABILITY LEVEL 0·01 (SINGLE SIDE)

ϕ_N \ ϕ_D	1	2	3	4	5	6	7	8	9	10	12	15	20	24	30	40	60	120	∞
1	·016 / 63·7	·014 / 9·92	·014 / 5·84	·013 / 4·60	·013 / 4·03	·013 / 3·71	·013 / 3·50	·013 / 3·36	·013 / 3·25	·013 / 3·17	·013 / 3·05	·013 / 2·95	·013 / 2·85	·013 / 2·80	·013 / 2·75	·013 / 2·70	·013 / 2·66	·013 / 2·62	·013 / 2·58
2	0·10 / 70·7	0·10 / 9·95	0·10 / 5·55	0·10 / 4·24	0·10 / 3·64	0·10 / 3·31	0·10 / 3·09	0·10 / 2·94	0·10 / 2·83	0·10 / 2·75	0·10 / 2·63	0·10 / 2·52	0·10 / 2·42	0·10 / 2·37	0·10 / 2·32	0·10 / 2·28	0·10 / 2·23	0·10 / 2·19	0·10 / 2·15
3	0·17 / 73·5	0·18 / 9·96	0·18 / 5·43	0·19 / 4·09	0·19 / 3·47	0·19 / 3·13	0·19 / 2·91	0·19 / 2·76	0·19 / 2·64	0·19 / 2·56	0·19 / 2·44	0·19 / 2·33	0·19 / 2·22	0·19 / 2·17	0·19 / 2·12	0·19 / 2·08	0·19 / 2·03	0·20 / 1·99	0·20 / 1·94
4	0·22 / 75·0	0·24 / 9·96	0·24 / 5·36	0·25 / 4·00	0·25 / 3·38	0·26 / 3·02	0·26 / 2·80	0·26 / 2·65	0·26 / 2·53	0·26 / 2·45	0·26 / 2·33	0·27 / 2·21	0·27 / 2·10	0·27 / 2·05	0·27 / 2·00	0·27 / 1·96	0·27 / 1·91	0·27 / 1·87	0·27 / 1·82
5	0·25 / 75·9	0·27 / 9·96	0·29 / 5·31	0·30 / 3·94	0·30 / 3·31	0·31 / 2·96	0·31 / 2·73	0·31 / 2·58	0·31 / 2·46	0·32 / 2·37	0·32 / 2·25	0·32 / 2·13	0·32 / 2·03	0·33 / 1·97	0·33 / 1·92	0·33 / 1·87	0·33 / 1·83	0·33 / 1·78	0·33 / 1·74
6	0·27 / 76·5	0·30 / 9·97	0·32 / 5·28	0·33 / 3·90	0·34 / 3·27	0·34 / 2·91	0·35 / 2·68	0·35 / 2·52	0·35 / 2·41	0·36 / 2·32	0·36 / 2·20	0·36 / 2·08	0·37 / 1·97	0·37 / 1·91	0·37 / 1·86	0·37 / 1·81	0·38 / 1·77	0·38 / 1·72	0·38 / 1·67
7	0·29 / 77·0	0·32 / 9·97	0·34 / 5·26	0·36 / 3·87	0·37 / 3·23	0·37 / 2·87	0·38 / 2·64	0·38 / 2·49	0·39 / 2·37	0·39 / 2·28	0·39 / 2·15	0·40 / 2·04	0·40 / 1·92	0·41 / 1·87	0·41 / 1·82	0·41 / 1·77	0·41 / 1·72	0·42 / 1·67	0·42 / 1·62
8	0·30 / 77·3	0·34 / 9·97	0·36 / 5·24	0·38 / 3·85	0·39 / 3·21	0·40 / 2·85	0·40 / 2·62	0·41 / 2·46	0·41 / 2·34	0·41 / 2·25	0·42 / 2·12	0·43 / 2·00	0·43 / 1·89	0·44 / 1·83	0·44 / 1·78	0·44 / 1·73	0·45 / 1·68	0·45 / 1·63	0·45 / 1·58
9	0·31 / 77·6	0·35 / 9·97	0·38 / 5·23	0·39 / 3·83	0·41 / 3·19	0·42 / 2·82	0·42 / 2·59	0·43 / 2·43	0·43 / 2·31	0·44 / 2·22	0·44 / 2·09	0·45 / 1·97	0·46 / 1·86	0·46 / 1·80	0·46 / 1·75	0·47 / 1·70	0·47 / 1·65	0·48 / 1·60	0·48 / 1·55
10	0·32 / 77·8	0·36 / 9·97	0·39 / 5·22	0·41 / 3·81	0·42 / 3·17	0·43 / 2·81	0·44 / 2·57	0·44 / 2·41	0·45 / 2·29	0·45 / 2·20	0·46 / 2·07	0·47 / 1·95	0·48 / 1·84	0·48 / 1·78	0·49 / 1·73	0·49 / 1·67	0·49 / 1·62	0·50 / 1·57	0·51 / 1·52

TABLE H (*continued*)

ϕ_D \ ϕ_N	1	2	3	4	5	6	7	8	9	10	12	15	20	24	30	40	60	120	∞
12	0·33 / 78·1	0·38 / 9·97	0·41 / 5·20	0·43 / 3·79	0·44 / 3·14	0·46 / 2·78	0·46 / 2·54	0·47 / 2·38	0·48 / 2·26	0·48 / 2·17	0·49 / 2·04	0·50 / 1·91	0·51 / 1·80	0·51 / 1·74	0·52 / 1·69	0·53 / 1·63	0·53 / 1·58	0·54 / 1·53	0·55 / 1·48
15	0·34 / 78·5	0·40 / 9·97	0·43 / 5·18	0·45 / 3·77	0·47 / 3·12	0·48 / 2·75	0·49 / 2·51	0·50 / 2·35	0·51 / 2·23	0·51 / 2·14	0·52 / 2·00	0·53 / 1·88	0·54 / 1·76	0·55 / 1·70	0·56 / 1·64	0·57 / 1·59	0·57 / 1·53	0·58 / 1·48	0·59 / 1·43
20	0·35 / 78·8	0·41 / 9·97	0·45 / 5·17	0·48 / 3·74	0·49 / 3·09	0·51 / 2·72	0·52 / 2·48	0·53 / 2·32	0·54 / 2·19	0·54 / 2·10	0·56 / 1·96	0·57 / 1·84	0·58 / 1·71	0·59 / 1·65	0·60 / 1·60	0·61 / 1·54	0·62 / 1·48	0·63 / 1·43	0·64 / 1·37
24	0·36 / 79·0	0·42 / 9·97	0·46 / 5·16	0·49 / 3·73	0·51 / 3·08	0·52 / 2·70	0·53 / 2·46	0·55 / 2·30	0·55 / 2·17	0·56 / 2·08	0·57 / 1·94	0·59 / 1·81	0·60 / 1·69	0·61 / 1·63	0·62 / 1·57	0·63 / 1·51	0·65 / 1·45	0·66 / 1·40	0·67 / 1·34
30	0·36 / 79·1	0·43 / 9·97	0·47 / 5·15	0·50 / 3·72	0·52 / 3·06	0·54 / 2·69	0·55 / 2·45	0·56 / 2·28	0·57 / 2·16	0·58 / 2·06	0·59 / 1·92	0·61 / 1·79	0·63 / 1·67	0·64 / 1·61	0·65 / 1·54	0·66 / 1·48	0·67 / 1·42	0·69 / 1·36	0·71 / 1·30
40	0·37 / 79·3	0·44 / 9·97	0·48 / 5·14	0·51 / 3·71	0·53 / 3·05	0·55 / 2·67	0·57 / 2·43	0·58 / 2·26	0·59 / 2·14	0·60 / 2·04	0·61 / 1·90	0·63 / 1·77	0·65 / 1·64	0·66 / 1·58	0·67 / 1·52	0·69 / 1·45	0·70 / 1·39	0·72 / 1·33	0·74 / 1·26
60	0·38 / 79·5	0·45 / 9·97	0·49 / 5·13	0·52 / 3·69	0·55 / 3·03	0·57 / 2·66	0·58 / 2·41	0·60 / 2·24	0·61 / 2·12	0·62 / 2·02	0·63 / 1·88	0·65 / 1·75	0·67 / 1·61	0·69 / 1·55	0·70 / 1·49	0·72 / 1·42	0·74 / 1·36	0·76 / 1·29	0·79 / 1·21
120	0·38 / 79·6	0·46 / 9·97	0·50 / 5·12	0·54 / 3·68	0·56 / 3·02	0·58 / 2·64	0·60 / 2·40	0·61 / 2·22	0·63 / 2·10	0·64 / 2·00	0·65 / 1·86	0·68 / 1·72	0·70 / 1·59	0·72 / 1·52	0·73 / 1·45	0·75 / 1·38	0·78 / 1·31	0·81 / 1·24	0·85 / 1·15
∞	0·39 / 79·8	0·47 / 9·98	0·51 / 5·11	0·55 / 3·67	0·58 / 3·00	0·60 / 2·62	0·62 / 2·38	0·63 / 2·20	0·64 / 2·08	0·66 / 1·98	0·68 / 1·83	0·70 / 1·69	0·73 / 1·56	0·75 / 1·49	0·77 / 1·42	0·79 / 1·34	0·82 / 1·27	0·87 / 1·17	1·00 / 1·00

L_1 is given by the upper entry in each cell and L_2 by the lower.

ϕ_N represents the degrees of freedom of the standard deviation in the numerator and ϕ_D the degrees of freedom of the standard deviation in the denominator.

The confidence limits for σ_1/σ_2 are $L_1 s_1/s_2$ and $L_2 s_1/s_2$.

The multipliers for the confidence limits of one standard deviation are given in the row $\phi_D = \infty$.

TABLE I.   VALUES OF OUTLIER RATIO AT TWO LEVELS
OF SIGNIFICANCE

No. in Sample	α		No. in Sample	α	
	0·05	0·02		0·05	0·02
3	1·15	1·15	18	2·65	2·82
4	1·48	1·49	19	2·68	2·85
5	1·71	1·75	20	2·71	2·88
6	1·89	1·94	21	2·73	2·91
7	2·02	2·10	22	2·76	2·94
8	2·13	2·22	23	2·78	2·96
9	2·21	2·32	24	2·80	2·99
10	2·29	2·41	25	2·82	3·01
11	2·36	2·48	30	2·91	3·10
12	2·41	2·55	35	2·98	3·18
13	2·46	2·61	40	3·04	3·24
14	2·51	2·66	45	3·09	3·29
15	2·55	2·71	50	3·13	3·34
16	2·59	2·75	60	3·20	3·41
17	2·62	2·79	70	3·26	3·47
18	2·65	2·82	80	3·31	3·52
19	2·68	2·85	90	3·35	3·56
20	2·71	2·88	100	3·38	3·60

An outlier may be regarded as significant if the ratio:

$$\frac{|\,\text{Extreme} - \text{Overall Mean}\,|}{\text{Overall Standard Deviation}}$$

exceeds the values tabulated above. The significance levels are double-sided, i.e. they assume that the outlier may be either the lowest or the highest observation. If it is known that only one of these types is possible, the values of α may be halved.

The critical values in Table I are based on the assumption that the non-suspect results are Normally distributed. When this is not even approximately true, calculate the probability $P = 1 - (1 - \alpha/2)^{1/n}$ corresponding to a result being judged extreme at a single drawing from the alternative Population. Plot the non-suspect results on suitable probability paper and compare the position of a suspected low outlier with the $100P$th percentile, or of a suspected high outlier with the $100(1 - P)$th percentile, the percentiles being estimated by extrapolation.

							N									
10	15	20	25	30	40	50	60		70		80		90		100	
9	5	10	17	10	22	13	28	17	7	52	54	34	24	30	35	22
5	9	8	22	15	31	29	54	18	48	64	48	46	45	66	50	5
6	6	7	18	18	37	7	59	51	10	43	66	4	36	85	53	24
4	2	5	20	8	26	19	23	25	29	28	36	17	21	25	14	84
10	8	18	2	21	6	49	55	50	9	61	7	27	70	5	63	97
7	4	14	16	26	40	50	24	33	15	55	32	35	79	38	34	80
1	7	1	3	23	11	23	43	22	47	3	74	49	32	81	96	61
3	1	3	13	11	17	25	56	26	40	37	65	50	49	52	83	98
2	10	15	8	25	9	9	52	32	46	58	70	56	56	12	11	9
8	11	16	24	9	3	21	40	30	8	27	53	75	54	47	64	79
	13	2	6	28	35	42	4	15	63	11	72	63	86	61	77	8
	15	11	23	22	10	3	11	31	31	17	69	18	77	2	99	10
	3	17	12	20	24	2	14	41	24	5	33	8	48	76	26	82
	12	9	25	7	28	47	16	13	66	36	68	30	1	73	51	37
	14	13	11	25	8	48	48	57	30	54	14	38	16	29	55	92
		19	9	14	1	36	29	5	65	60	11	47	41	57	21	57
		20	10	6	13	18	42	34	26	68	10	23	3	26	33	30
		6	14	3	36	46	38	39	16	4	64	76	59	68	32	44
		4	4	17	27	37	49	12	45	34	40	55	72	44	29	60
		12	21	2	30	10	21	3	50	56	44	19	9	58	68	94
			5	12	7	28	2	20	23	59	15	58	64	63	69	38
			15	4	14	11	46	60	6	49	57	73	20	27	45	28
			7	30	32	12	1	44	38	20	42	71	88	80	88	7
			19	5	34	43	9	45	69	18	21	2	15	87	95	89
			1	29	2	27	35	10	70	67	25	1	4	74	56	16
				19	20	14	7	58	14	19	12	39	34	55	23	4
				27	29	38	47	8	42	2	78	26	71	84	85	42
				13	15	33	53	37	57	35	28	13	37	69	20	18
				24	9	30	6	19	13	1	59	77	28	8	54	58
				1	38	4	36	27	39	44	79	31	42	23	74	67
					21	39			32	62	9	6	7	78	2	36
					23	6			41	22	37	3	50	39	73	46
					5	40			53	33	5	51	51	89	31	27
					16	20			51	12	67	45	33	17	91	1
					12	17			25	21	41	29	67	60	47	3
					18	44					20	62	35	6	12	76
					33	34					16	80	13	46	41	93
					39	5					22	24	31	62	86	71
					19	24					61	43	65	22	49	48
					4	22					52	60	53	75	62	39
						8							18	43	72	59
						41							83	11	75	40
						15							10	82	87	100
						1							40	14	66	43
						32							90	19	70	81
						16									13	15
						45									65	25
						31									78	17
						35									90	6
						26									52	19

For other values of N use the next higher column and discard the superfluous entries. Successive columns have been generated independently. Hence replicate permutations for a given value of N can also be obtained from higher columns.

83 28	78 05	18 98	49 22	54 11	92 37	45 11	63 60	19 05	91 26
84 73	82 58	01 90	55 37	85 68	98 15	99 52	99 84	51 91	73 81
00 79	20 99	42 57	55 67	93 39	99 25	65 10	94 54	84 65	16 23
94 48	02 99	71 08	50 84	66 10	10 34	92 30	89 28	30 74	24 24
54 37	52 43	87 22	21 34	20 15	07 67	64 98	36 01	33 34	04 42
47 68	59 90	98 90	27 71	89 89	98 20	24 19	85 02	34 38	26 71
76 16	58 55	51 85	44 00	28 28	38 91	70 70	16 81	13 49	46 54
37 64	90 35	64 45	47 72	82 03	01 65	05 97	13 90	90 57	51 97
92 78	39 12	48 01	83 46	39 29	98 71	39 56	97 66	97 70	05 77
24 50	29 02	71 28	53 99	75 07	13 18	76 97	72 54	85 79	71 60
01 72	71 23	86 40	70 05	35 36	15 64	11 01	11 18	90 14	95 05
43 28	52 77	22 80	49 89	79 65	91 17	80 94	34 02	17 61	00 42
29 09	19 54	67 67	88 54	62 09	07 97	35 19	31 25	06 92	25 02
27 95	74 89	62 45	75 39	06 89	58 96	64 65	81 84	85 20	01 47
52 43	54 97	75 80	00 38	20 38	57 46	57 33	87 19	66 06	40 32
78 11	60 42	09 83	28 40	93 57	61 22	27 27	47 80	44 34	47 27
03 74	36 27	13 19	14 76	35 73	66 29	95 65	12 87	61 91	34 30
82 25	35 57	16 29	21 27	51 23	06 52	40 00	28 11	47 23	63 01
09 91	87 20	33 76	61 55	79 21	74 36	21 36	05 47	28 42	92 51
19 82	00 40	15 52	45 35	13 48	74 10	97 36	22 85	44 57	91 72
69 41	17 07	11 54	36 81	57 38	55 39	85 74	48 05	06 43	10 63
48 80	36 26	28 95	03 79	54 31	41 55	48 84	78 63	09 05	69 07
80 02	51 78	94 07	88 62	85 82	80 37	56 15	59 30	46 42	84 02
19 51	95 22	72 72	95 51	57 73	04 68	00 95	04 30	66 52	60 74
50 36	31 76	75 39	04 95	69 47	95 23	01 70	95 04	04 18	68 14
60 03	34 57	41 76	35 06	75 60	21 58	86 36	02 33	00 59	63 13
59 40	60 83	61 73	45 18	08 23	54 86	64 57	76 70	00 89	43 24
29 51	12 43	14 24	35 78	76 22	82 50	68 02	13 19	07 00	19 07
57 07	34 86	57 96	99 57	44 54	90 87	33 76	71 71	23 28	88 37
81 73	29 08	96 62	34 26	52 32	23 74	17 49	45 62	17 88	50 50
40 20	21 54	17 65	99 31	09 72	67 87	16 34	00 76	26 23	42 40
81 26	86 30	79 17	93 45	74 50	50 24	65 52	06 59	04 60	73 63
13 65	31 57	36 88	98 35	04 96	41 37	45 87	57 57	21 15	34 59
23 41	47 66	24 73	31 96	72 07	09 43	88 63	33 80	54 79	84 18
79 62	53 27	85 43	51 69	83 81	90 85	84 72	18 48	41 20	81 59
13 40	75 73	19 92	12 01	91 95	23 99	99 30	30 58	46 22	64 41
54 87	97 55	83 91	42 61	41 02	40 18	39 20	56 19	56 35	04 32
09 29	30 63	75 86	85 29	15 34	68 92	34 06	81 60	32 16	05 37
61 99	27 99	73 18	94 29	25 74	22 20	70 46	30 38	26 91	59 16
31 84	93 27	40 23	25 86	68 30	10 11	91 59	61 07	41 97	10 39
35 86	11 25	98 38	27 14	79 68	77 60	63 34	23 80	75 43	48 79
40 42	68 85	23 40	27 56	54 56	75 65	70 49	24 08	10 44	75 59
25 14	94 00	99 80	81 44	49 08	98 93	71 74	11 14	54 69	71 69
56 18	75 63	56 68	25 36	75 98	00 18	19 15	24 28	56 80	75 97
79 61	54 67	58 38	93 69	45 95	61 19	17 35	89 90	98 70	26 20
92 91	85 49	33 32	46 67	28 20	40 99	88 73	56 33	29 13	41 89
01 79	85 45	45 36	05 67	56 17	59 77	59 34	35 01	15 21	00 35
55 84	71 36	40 39	47 25	25 73	69 14	55 73	35 86	61 17	98 69
38 36	66 66	19 40	90 83	06 31	24 67	91 74	54 14	87 24	61 80
01 69	50 70	31 02	98 86	42 01	94 98	07 85	28 38	37 30	72 76

TABLE L.   CRITICAL VALUES FOR THE SIGN TEST

	α for Double-Sided Test				α for Double-Sided Test		
	·10	·05	·01		·10	·05	·01
	α for Single-Sided Test				α for Single-Sided Test		
n	·05	·025	·005	n	·05	·025	·005
6	0	0	–	41	14	13	11
7	0	0	–	42	15	14	12
8	1	0	0	43	15	14	12
9	1	1	0	44	16	15	13
10	1	1	0	45	16	15	13
11	2	1	0	46	16	15	13
12	2	2	1	47	17	16	14
13	3	2	1	48	17	16	14
14	3	2	1	49	18	17	15
15	3	3	2	50	18	17	15
16	4	3	2	52	19	18	16
17	4	4	2	54	20	19	17
18	5	4	3	56	21	20	17
19	5	4	3	58	22	21	18
20	5	5	3	60	23	21	19
21	6	5	4	62	24	22	20
22	6	5	4	64	24	23	21
23	7	6	4	66	25	24	22
24	7	6	5	68	26	25	22
25	7	7	5	70	27	26	23
26	8	7	6	72	28	27	24
27	8	7	6	74	29	28	25
28	9	8	6	76	30	28	26
29	9	8	7	78	31	29	27
30	10	9	7	80	32	30	28
31	10	9	7	82	33	31	28
32	10	9	8	84	33	32	29
33	11	10	8	86	34	33	30
34	11	10	9	88	35	34	31
35	12	11	9	90	36	35	32
36	12	11	9	92	37	36	33
37	13	12	10	94	38	37	34
38	13	12	10	96	39	37	34
39	13	12	11	98	40	38	35
40	14	13	11	100	41	39	36

TABLE M. CRITICAL VALUES FOR THE RANK-SUM TEST

n_2	α for 2-sided Test	α for 1-sided Test	1	2	3	4	5	6	7	8	9	10	11	12	13	14	15	16	17	18	19	20
															n_1 (Smaller Sample)							
3	0·10	0·05		6																		
	0·05	0·025																				
	0·01	0·005																				
4	0·10	0·05		6	11																	
	0·05	0·025			10																	
	0·01	0·005																				
5	0·10	0·05	3	7	12	19																
	0·05	0·025		6	11	17																
	0·01	0·005				15																
6	0·10	0·05	3	8	13	20	28															
	0·05	0·025		7	12	18	26															
	0·01	0·005			10	16	23															
7	0·10	0·05	3	8	14	21	29	39														
	0·05	0·025		7	13	20	27	36														
	0·01	0·005			10	16	24	32														
8	0·10	0·05	4	9	15	23	31	41	51													
	0·05	0·025	3	8	14	21	29	38	49													
	0·01	0·005			11	17	25	34	43													
9	0·10	0·05	4	10	16	24	33	43	54	66												
	0·05	0·025	3	8	14	22	31	40	51	62												
	0·01	0·005		6	11	18	26	35	45	56												
10	0·10	0·05	4	10	17	26	35	45	56	69	82											
	0·05	0·025	3	9	15	23	32	42	53	65	78											
	0·01	0·005		6	12	19	27	37	47	58	71											
11	0·10	0·05	4	11	18	27	37	47	59	72	86	100										
	0·05	0·025	3	9	16	24	34	44	55	68	81	96										
	0·01	0·005		6	12	20	28	38	49	61	73	87										
12	0·10	0·05	5	11	19	28	38	49	62	75	89	104	120									
	0·05	0·025	4	10	17	26	35	46	58	71	84	99	115									
	0·01	0·005		7	13	21	30	40	51	63	76	90	105									
13	0·10	0·05	5	12	20	30	40	52	64	78	92	108	125	142								
	0·05	0·025	4	10	18	27	37	48	60	73	88	103	119	136								
	0·01	0·005		7	13	22	31	41	53	65	79	93	109	125								
14	0·10	0·05	6	13	21	31	42	54	67	81	96	112	129	147	166							
	0·05	0·025	4	11	19	28	38	50	62	76	91	106	123	141	160							
	0·01	0·005		7	14	22	32	43	54	67	81	96	112	129	147							
15	0·10	0·05	6	13	22	33	44	56	69	84	99	116	133	152	171	192						
	0·05	0·025	4	11	20	29	40	52	65	79	94	110	127	145	164	184						
	0·01	0·005		8	15	23	33	44	56	69	84	99	115	133	151	171						
16	0·10	0·05	6	14	24	34	46	58	72	87	103	120	138	156	176	197	219					
	0·05	0·025	4	12	21	30	42	54	67	82	97	113	131	150	169	190	211					
	0·01	0·005		8	15	24	34	46	58	72	86	102	119	136	155	175	196					
17	0·10	0·05	6	15	25	35	47	61	75	90	106	123	142	161	182	203	225	249				
	0·05	0·025	5	12	21	32	43	56	70	84	100	117	135	154	174	195	217	240				
	0·01	0·005		8	16	25	36	47	60	74	89	105	122	140	159	180	201	223				
18	0·10	0·05	7	15	26	37	49	63	77	93	110	127	146	166	187	208	231	255	280			
	0·05	0·025	5	13	22	33	45	58	72	87	103	121	139	158	179	200	222	246	270			
	0·01	0·005		8	16	26	37	49	62	76	92	108	125	144	163	184	206	228	252			
19	0·10	0·05	1	7	16	27	38	51	65	80	96	113	131	150	171	192	214	237	262	287	313	
	0·05	0·025		5	13	23	34	46	60	74	90	107	124	143	163	183	205	228	252	277	303	
	0·01	0·005		3	9	17	27	38	50	64	78	94	111	129	148	168	189	210	234	258	283	
20	0·10	0·05	1	7	17	28	40	53	67	83	99	117	135	155	175	197	220	243	268	294	320	348
	0·05	0·025		5	14	24	35	48	62	77	93	110	128	147	167	188	210	234	258	283	309	337
	0·01	0·005		3	9	18	28	39	52	66	81	97	114	132	151	172	193	215	239	263	289	315

For larger values of n_1 and n_2, critical values are given to a good approximation by the formula:

$$\frac{n_1}{2}(n_1+n_2+1)-u\left\{\frac{n_1 n_2(n_1+n_2+1)}{12}\right\}^{\frac{1}{2}}$$

where $u = 1·28$ for $\alpha = 0·20$ (double-sided test) $u = 1·96$ for $\alpha = 0·05$ (double-sided test)
    $u = 1·64$ for $\alpha = 0·10$ (double-sided test) $u = 2·58$ for $\alpha = 0·01$ (double-sided test)

TABLE N.   ATTRIBUTE SAMPLING SCHEMES FOR PROCESS CONTROL

R	L_a	L_r								
		5·0			7·5			10·0		
		m_a	k	h	m_a	k	h	m_a	k	h
2·5	500	1·18	2·0	5·00	0·64	1·2	3·75	0·50	0·9	3·75
	250	0·93	1·5	4·50	0·52	0·9	3·50	0·42	0·8	3·00
	125	0·71	1·2	3·75	0·47	0·7	3·25	0·32	0·6	2·25
3·0	500	0·66	1·2	4·00	0·46	0·9	3·50	0·32	0·7	3·00
	250	0·56	0·9	3·00	0·40	0·8	3·00	0·27	0·6	2·50
	125	0·48	0·8	3·00	0·31	0·6	3·00	0·15	0·3	2·00
3·5	500	0·54	1·2	3·00	0·35	0·8	3·00	0·24	0·6	2·75
	250	0·41	0·9	2·50	0·27	0·6	2·50	0·18	0·4	2·50
	125	0·34	0·7	2·25	0·18	0·4	2·00	0·13	0·3	1·75
4·0	500	0·38	0·9	2·75	0·24	0·6	2·75	0·16	0·4	2·50
	250	0·32	0·8	2·25	0·21	0·6	2·00	0·12	0·3	2·00
	125	0·28	0·7	1·75	0·16	0·4	1·75	0·07	0·2	1·50

m_a is the acceptable mean level of faults; $R(= m_r/m_a)$ is the ratio of the rejectable to the acceptable levels. The control scheme is characterised by its reference level k and decision limit h. These are selected according to the desired average run lengths L_a and L_r at the acceptable and rejectable fault levels respectively.

Example

Suppose it is desired to control the fraction of defective articles in a process where $p_a = 0·01$ and $p_r = 0·04$ and we want $L_a = 500$ and $L_r = 5$. Then for $R = 0·04/0·01 = 4$ we find from the table:

$$m_a = 0·38 \qquad k = 0·9 \qquad h = 2·75$$

We should therefore take samples of $0·38/0·01 = 38$ articles, count the number of defectives in each, subtract 0·9 from the figures and accumulate the deviations. Action would be taken if the cusum rose to 2·8 or higher.

Intersection Chart for Single-Sided Cusum Control Scheme

This chart gives average run lengths ($A.R.L.$s) for various cumulative sum schemes for controlling a process average, based on the decision interval procedure [§ **11.6**]. The notation is:

μ_0 Acceptable Quality Level ($A.Q.L.$), often the same as the target level, μ

μ_1 Rejectable Quality Level ($R.Q.L.$)

k Central reference value $= \frac{1}{2}(\mu_0 + \mu_1)$

σ Population standard deviation

n Sample size

h Decision limit

The chart spans a range of $A.R.L.$s from 100 to 1,000 at $A.Q.L.$ and from 3 to 14 at $R.Q.L.$, which should cover most practical requirements. The process averages $\bar{x}$ are assumed to be independent and Normally distributed. For skew distributions see reference [11.6].

The method of using the chart depends on whether or not the sample size is fixed beforehand. If, as often happens, n is fixed, then only one of the $A.R.L.$s can be defined arbitrarily, say at $A.Q.L.$ After calculating $|\mu_1 - \mu_0|\sqrt{n}/\sigma$, read off $h\sqrt{n}/\sigma$ and the $A.R.L.$ at $R.Q.L.$ If the latter is regarded as too high for a satisfactory scheme, some reduction will also have to be made in the $A.R.L.$ at $A.Q.L.$

If n is not fixed however, the $A.R.L.$s at both $A.Q.L.$ and $R.Q.L.$ can be chosen arbitrarily. Entering the chart at the selected combination, read off the value of the ordinate and use it to calculate n. It will then be necessary to round the sample size to a convenient whole number. The chart is then re-entered as described in the previous paragraph for fixed n. Because of the rounding, the final $A.R.L.$s will differ slightly from those selected initially.

A numerical example is given on p. 468.

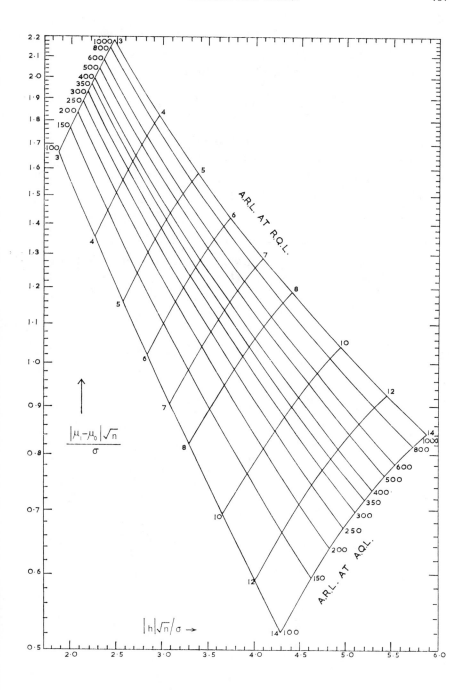

Example

Given: $A.Q.L. = 4\cdot5$ *units*, $R.Q.L. = 6\cdot5$ *units*; $\sigma = 2\cdot5$ *units*. *Require*: $A.R.L. \simeq 500$ *at* $A.Q.L.$; $A.R.L. \simeq 5$ *at* $R.Q.L.$

Reading from the chart at the (500, 5) combination:

$$\frac{(6\cdot5 - 4\cdot5)\sqrt{n}}{2\cdot5} = 1\cdot47, \text{ whence } n = 3\cdot38$$

Assuming that it is practicable to round up (to achieve greater discrimination) a sample size of 4 would be selected and cusums would be formed about a reference value of $5\cdot5$ units. Then $(\mu_1 - \mu_0)\sqrt{n}/\sigma = 1\cdot6$. Reading again from the chart, a choice is now available anywhere between

$$h\sqrt{4}/2\cdot5 = 2\cdot9, \text{ giving } A.R.L.\text{s of 500 \& } 4\cdot4 \text{ approx.}$$

and $\qquad h\sqrt{4}/2\cdot5 = 3\cdot4$, giving $A.R.L.$s of 1,200 & 5 approx.

The rounded value $h = 4\cdot0$ is probably the most suitable, giving final $A.R.L.$s of about 800 at $A.Q.L.$ and $4\cdot7$ at $R.Q.L.$

Nomogram for Interpolating Statistical Tables for other Probability Levels

Modern tables of statistical criteria, t, χ^2, F, etc., usually show the values corresponding to a series of levels of significance. The chosen levels commonly belong to the following sequence, though the sequence is often incomplete and levels not belonging to the sequence are also encountered:

$$P = 25 \quad 10 \quad 5 \quad 2.5 \quad 1 \quad 0.5 \quad 0.25 \quad 0.1 \quad ... \%$$

For most practical purposes it is sufficient to quote the significance of the result within limits, e.g. using the table of F given in this book, one can always state that P is greater than 10%, between 10 and 5% or 5 and 1%, or less than 1%.

Some people, however, prefer to make a rough interpolation between the two tabular entries in order to quote the actual significance of the result. Direct linear interpolation is unreliable, particularly when the two values of P are widely spaced. Linear interpolation is more accurate when performed against a transformed scale of P, such as $\log P$, or $u(P)$, i.e. the Normal deviate corresponding to P.

The nomogram on p. 471 provides a rapid method for interpolating linearly with reference to $u(P)$, which, although not necessarily the best method for any given criterion, is probably the best general method to be applied to all of them. The nomogram gives, in principle, an exact interpolation for any criterion with a Normal distribution, and for other criteria it is at least likely to be considerably more reliable than simple linear interpolation with reference to P.

To use the nomogram proceed as follows. Obtain from the tables values of the function at two standard probabilities, one higher and the other lower than the value to be examined. Find the differences between this value and each of the standard values. These differences can be multiplied or divided by any common factor to bring them to a convenient size. On the nomogram find the horizontal line corresponding to one of the standard probabilities and measure the associated difference along it from the central line. Repeat this for the other difference on the corresponding probability level, measuring the two differences in opposite directions. A straight line through the two points so obtained intersects the central line at the required probability.

Examples

1. *Find the significance level of* $\chi^2 = 14 \cdot 240$ *for* $\phi = 8$.

From the tables for $\phi = 8$, $\chi^2 = 13 \cdot 362$ $P = 10\%$

$\chi^2 = 15 \cdot 507$ $P = 5\%$

Differences are $+0 \cdot 878$ from $P = 10\%$

$-1 \cdot 267$ from $P = 5\%$

To obtain a convenient scale move the decimal point one place to the right and divide by 2, giving differences of $+4 \cdot 4$ and $-6 \cdot 3$ respectively.
Measure $4 \cdot 4$ to the left on the 10% line
     and $6 \cdot 3$ to the right on the 5% line.
Join these two points with a straight-edge.
Then the point of intersection on the central line is $7 \cdot 60\%$, which is the required probability.
This result is correct to two decimal places.

2. *Find the significance level of* $F = 3 \cdot 20$ *for* $\phi_1 = 8$, $\phi_2 = 15$.

From the tables $F = 2 \cdot 64$ $P = 5\%$

$F = 4 \cdot 00$ $P = 1\%$

Differences are $+0 \cdot 56$ from $P = 5\%$

$-0 \cdot 80$ from $P = 1\%$

To obtain a convenient scale move decimal point one place to right.
Measure $5 \cdot 6$ to left on 5% line
     and $8 \cdot 0$ to right on 1% line.
The point of intersection on the central line is $2 \cdot 7\%$, which is the required probability.
The true probability is $2 \cdot 50\%$. The error is not large in view of the wide range used in this case for interpolation.

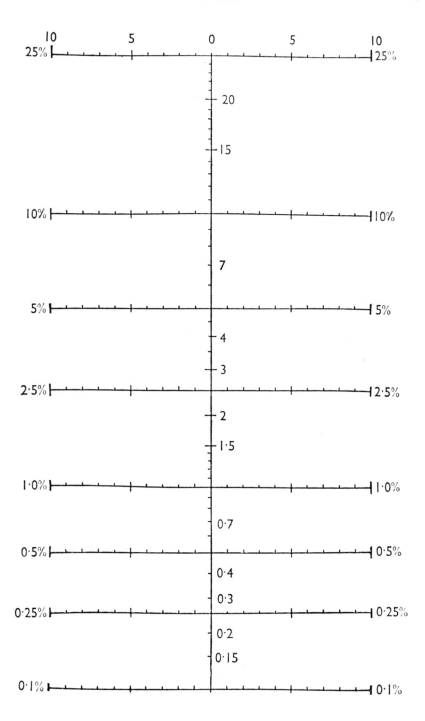

Index

Acceptance testing, 95
Accident rates, 306, 320
Accuracy, 29
Action limits, 342
Additive property of:
  sums of squares, 128, 133, 147
  variances, 122
Analysis of covariance, 225
Analysis of Variance, 121-177
  comparison of means, 114
  cross-classification, 123, 143
  hierarchic classification, 123, 125, 162
  in linear regression, 190
  missing values, 147, 176
  mixed classification, 154, 158
  in multiple regression, 259
  types of classification, 123
  unequal numbers of repeat analyses, 130, 141, 159
Analogue computers, 411
Analytical error, 122, 126
Assays, 206, 213
Attributes, 376
  double and multiple sampling, 382
  plan for sampling, 387, 465
  sequential sampling, 383
  single sampling, 377
Average:
  generally, 30
  outgoing quality, 380
  run length, 349, 389, 402

Balancing a quadratic model, 284
Barnard, G. A., 234
Bartlett, M. S., 173, 210
Bayes, T., 76
Berkson, J., 212, 298
Bias, 28
Binomial distribution, 5, 304
  approximation to, 313
  confidence limits, 309, 445
  mean and variance, 305, 309
  sampling, 376
  significance test for proportions, 307
Binomial index of dispersion, 329
Bivariate Normal distribution, 230

Blending, 409, 419
Blocking of data, 274

Canonical analysis, 288
Cause and effect, 198, 292
Centred data, 246
Chance causes of variation, 1, 337
Chance. *See* Probability
Chemical reaction, first-order, 300
Chi-squared distribution, 317, 436
Coefficient of variation, 40
  combination of, 45
Combination of results, 161
Comparison:
  with a control, 70, 151
  global, 152
  of means, 61, 72, 149, 174
  of means, number of observations, 114
  of several variances, 173
  of two variances, 75, 440
  of two variances, number of observations, 117
Computers, 6, 13, 42, 246, 266, 283, 302, 418
Concomitant variation, 218, 225
Confidence coefficient, 59
Confidence ellipse, 256
Confidence interval, 2, 59
Confidence limits, 2, 58
  binomial distributions, 309, 445
  components of variance, 166
  correlation coefficients, 234
  differences between means, 62
  mean, 59, 61
  multiple regression coefficients, 253
  paired comparisons, 65
  Poisson distribution, 445
  predicted value in regression, 196, 258
  proportions, 308
  ratio of two means, 236
  ratio of two standard deviations, 66, 452
  regression coefficients, 193, 253
  single- and double-sided, 59
  standard deviation, 68, 88
  variance estimates, 129, 140, 166
Confidence regions, 255
Consumer's and producer's risks, 377, 394

Contingency tables, 323
  2×2, 330
  $2 \times k$, 329
  3×4, 323
  significance test for small expected frequencies, 329
  three-dimensional, 333
Continuity correction, 309, 332
Contour diagrams, 242, 286
Control charts, 5, 336-372, 381
  construction, 341
  interpretation, 344
  for mean and standard deviation simultaneously, 359
  purpose, 337
  for sample average, 342, 448
  for sample range, 343, 450
Control limits, 336, 342
Correlation, 229
  coefficient, 230, 444
  matrix, 247, 271
  multiple, 259, 271
  spurious, 234, 292
Covariance, 52, 231
Cross-classification:
  missing values, 147, 176
  two-way without repeats, 143, 174
Cumulative distribution, 12
Cumulative scores, 406
Cumulative sum (cusum), 350
Cusum charts, 5, 350
  control scheme for variables, 401, 466
  double, 360
  for number of defectives, 359
Curvilinear regression, 276
Cyclic pattern in data, 273, 340, 346

Decision:
  function, 94
  interval, 389
  limit, 389
  theory, 76
Defect charts, 358
Defectives, 310, 358
Defects, number of, 359
Degrees of freedom, 38, 44, 128, 190, 253, 317
Design matrix, 247
Design of experiments, 3, 123, 141, 168, 250, 302
Desk calculator, 31, 186
Difference method, 64
Discrete distribution, 17, 304
Dispersion, measures of, 2, 37
Distribution:
  form of, 19
  specific. See under respective headings

Distribution-free tests, 69, 73
Dot diagram, 180
Double cusum chart, 360
Double dichotomy, 331
Double-sided test, 73, 115

Economic considerations, 2, 76, 101, 141, 161, 168, 375
Effects, practical importance of, 71
Efficiency of designs, 168
Element analysis, 266
Error:
  first and second kind, 110, 151, 378
  sources of, 50, 121
Errors of analysis, 122, 126
Estimation, nonlinear, 299
Expectation, 57, 82, 127
Experiments:
  controlled and uncontrolled, 212
  See also Design of experiments
Extrapolation, 277
Extreme values:
  Normal, 24
  outliers, 35, 49, 269, 272, 460

F-distribution, 440
F-test, 75, 134
Fiducial distribution, 60
Fieller, E. C., 236
Fisher, R. A., 90, 234
Floating decimal point notation, 229
Frequency:
  cumulative, 13
  curve, 14
  data, 304-335
  diagrams, 11
  distributions, 2, 9-29
  expected, 329
  polygon, 11
  table, 9
Functional relationship:
  linear. See Linear functional relationship
  multiple, 297

Gain, 82
Gamma distribution, 315
Gauging, 334, 376
Gaussian distribution, 20, 434
Gomes J functions, 301
Goodness of fit:
  of distributions, 322
  in regression, 199, 270
Gosset, W. S., 411
Grouped data, 9
  mean from, 32
  variance from, 41

Hierarchic classification, 123
  two sources of variation, 125, 162
  three sources of variation, 135, 164
Hierarchic designs, comparison of efficiencies, 168
Hill climbing, 302
Histogram, 2, 10, 272
Homogeneity of variances, 173
Homogeneous data, 337
Hypotheses, null and alternative, 70, 110, 134

Increments, 393
Information matrix, 247
Inliers, 50
Inner and outer control limits, 342
Instrument calibration, 205
Interaction, 148, 275, 283
Intercept, standard error of, 194
Intermediate stage responses, 295
Interpolation of statistical tables, 72, 327, 469
Intersection chart, 466
Inverse diagonals in regression, 270
Inverse matrix, 247

Kurtosis, 31, 49

Lack-of-fit, 199, 270
Lead distance, 353
Least squares:
  in curvilinear regression, 276
  in linear regression, 185, 203
  in multiple regression, 244
Level of control of a process, 347
Level of significance, 71
Likelihood, 78, 90, 185
Limits:
  action, 342
  confidence. *See* Confidence limits
  decision, 389
  single- and double-sided, 59
  inner and outer control, 342
Linear functional relationship, 179, 203
  both variables subject to error, 208
  controlled and uncontrolled experiments, 212
  error variance unknown, 210
  one variable with negligible error, 204
  prediction from, 205
  ratio of variances known, 209
Linear regression, 178-236
  analysis of variance, 190, 201
  cause and effect, 198
  coefficient, 186
  comparison of several lines, 213, 225
  comparison of two coefficients, 193

  estimation of line, 185
  relation to correlation, 235
  significance test for coefficient, 193
  standard error of coefficients, 192, 202
  standard error of estimate, 195
  sum of squares about and due to, 189, 202
  through the origin, 201
  validity of, 197
  v. functional relationship, 179
  weighted, 202
Location, measures of, 2, 31
Logarithmic probability paper, 25
Loss, 82
Lot size, 375

Mathematical symbols, 423-430
Maclaurin's theorem, 283
Matrix:
  correlation, 247, 271
  design, 247
  information, 247, 271
  inverse, 247
  notation for least-squares estimation, 246
  variance-covariance, 248
Mean:
  arithmetic, 2, 18, 31
  geometric, 34
  harmonic, 34
  correction for, 128, 186
  deviation, 48
  square, 129
  squares, combination of, 158
  squares, expectation of, 57, 129, 135, 146, 163, 165, 191
  weighted, 161
Means:
  combination of, 161
  comparison of, 72, 149, 174
  difference between, 61
  distribution of, 21
  significance of, 71
  t-tests on, 71
Medial lines, 232
Median, 35
Missing values in a two-way classification, 176
Mitscherlich equation, 301
Mode, 36
Model building, 264, 269
Models I and II in Analysis of Variance, 125
Moments:
  third, 49
  fourth, 49
Monte Carlo simulation, 5, 411
Multiple confidence intervals, 152
Multiple functional relationships, 297

Multiple regression, 237-303
  analysis of variance, 259
  applicability of method, 275
  automatic selection of variables, 266
  backwards elimination of variables, 265
  calculation of coefficients, 245
  correlation coefficient, 259, 271
  forward selection of variables, 265
  initial editing of data, 294
  interpretation, 284
  method of least squares, 244
  model building, 264, 269
  omission of a variable, 250, 257
  partial coefficient, 241
  of plant records, 291
  prediction from, 258
  presentation, 284
  quadratic, 282
  qualitative variables in, 274
  selection of best subset of variables, 264
  separation of effects, 240
  significance of an additional variable, 261
  significance testing, 254, 259
  stabilized elimination of variables, 266
  standard errors of coefficients, 253, 259
  stepwise, 266
  total coefficient, 241
  validation of model, 270
Multiplicative congruential method, 414
Murographic display, 288

Nature of variation, 125
Negative binomial distribution, 5, 315, 377
Nested classification, 123
Next-event simulation technique, 412
Nomogram for interpolation, 469
Non-central t, 360
Non-Normal data, standard deviation of, 88
Non-Normal distribution, tendency to normality, 21
Non-Normality, effect on confidence limits, 68
Non-parallel linear regressions, 223, 225
Normal:
  distribution, 20, 79, 434
  equations, 245
  probability paper, 25
  pseudo-random samples, 415
  surface, 230
Normality:
  assumptions of, 68, 400
  test for, 27
Number average, 37
Number of observations, 94-120
  comparison of means, 64, 114
  comparison of standard deviations, 117

Odds, 24, 76
Ogive, 12
One-sided test, 73, 111, 114
Operating characteristic curve, 97, 378
Orthogonality, 249
Outgoing quality curve, 97, 380
Outliers, 35, 49, 269, 272, 460

Paired comparisons, 64, 74
Parallel line assays, 206, 213
Partial regression coefficients, 241
Path of steepest ascent, 290
Percentage defective, 312
Percentage fit, 190, 260
Periodicity, 273, 340, 346
Plant records, 291
Poisson distribution, 5, 312, 377, 445
Polynomial regression, 276
Population, 17
Posterior distribution, 78
Power curve, 99, 378
Precision, 19, 39
  of sampling and testing schemes, 140
Predictions, 195, 258, 268, 285, 297, 338
Principal axes, 289
Prior distribution, 78, 95, 381
Probability:
  density, 16
  discrete, 304
  distribution, 15
  distribution, subjective, 77
  interpolation for other levels, 72, 327, 469
  paper, 25
  prior, 79, 81
Process control, 339
  of attributes, 388
  by decision limit schemes, 405
  hunting, 357
  level of, 347
  of variables, 401, 466
Products, sum of, 53, 186
Proportion defective, 358
  precision of estimate of, 308
Proportions:
  comparison of, 318
  confidence limits for, 308
  homogeneity of two proportions, 331
  sequential comparison of, 383
  2×2 table estimation of, 332
  testing for difference in, 306
Pseudo-random numbers, 414
Public opinion survey, 309

Qualitative variables, 274
Quality control, 310, 336
  corrective action in, 357
  chart limits for average, 448
  chart limits for range, 450

choice of charting method, 357
decision limit schemes in, 357, 402
rate of detection of changes in average
    level, 355
V-mask schemes in, 353
Quality curve:
  incoming, 96
  outgoing, 97, 380
Quality level:
  acceptable and rejectable, 379, 402
  indifference, 380
Quality specifications, 5, 347, 388

Random:
  allocation, 47
  numbers, 414, 462
  permutations, 48, 461
  sample, 28, 275
  sampling, 28, 374, 391
  samples from non-rectangular distri-
    butions, 415
Range, 2, 44, 45
  chart, 341, 343, 450
  standard deviation from, 47, 415, 450
  Studentized, 152
Rank-sum test, 74, 464
Rate of detection of changes in average
    level, 355
Rectangular distribution, 21, 27
  random samples from, 414, 462
Reference level, 389, 401
Regression:
  curvilinear, 276
  linear. See Linear regression
  multiple. See Multiple regression
  multiple quadratic, 282
  parabolic, 277
Replication, 199, 270
Residuals, 201, 272
Response surface, 276, 285
Root mean square deviation. See Standard
    deviation

Sample, 26
  average size, 384, 396
  random, 28, 275
Sampling, 4, 310, 373-408
  acceptance, 375
  of attributes, 376
  clauses in specifications, 312, 388, 398
  design of testing schemes and of, 141
  double, 382
  economics, 141, 161, 168, 375
  errors of, 122
  inspection, 94
  interval, 356
  multiple, 382

random, 28, 374, 391
sequential, 383, 395
single, 377
size, 356
stratified, 376, 393
systematic, 374
of variables, 390
Scaling, 23, 247, 352, 404
Scatter diagram, 180, 273
Sensitivity analysis, 418
Sequential:
  inspection, 385
  testing, 120
Sheppard, W. F., 41
Shewhart, W. A., 338, 362
Sign-test, 73, 463
Significance, 3, 69
  of correlation coefficient, 232, 444
  levels of, 71
  of means, 71
  of variances, 75
  tests, use and misuse in regression, 268
Simulation, 5, 409-422
  continuous, 411
  experiments, 417
  validation of model, 417
Single-sided test, 73, 111, 114
Skewness, 29, 31, 48
Sources of variation, 121, 337
Specifications, 111, 347, 388
Specific values model, 149
Squares, sum of:
  additive property of, 128, 133, 147
  between samples, 127
  corrected, 41, 45
  crude, 45
  lack-of-fit, 199, 270
  remainder, 175
  residual, 260, 271
  total, 127
  within samples, 126
Standard deviation, 2, 18, 40, 341
  confidence limits for single, 68, 88
  estimate from range, 47, 415, 450
  ratio of, 66, 117, 452
Standard error, 50
Standardized variable, 23, 247, 276
Standard Normal curve, 24
Statistic, 30
Statistical:
  inference, 58-93
  methods, 1, 28
  symbols, 423-430
  tables and charts, 431-471
Symmetry, centre of, 288

t-test, 61, 71, 438

Tables and charts of statistical functions, 431-471
Testing:
  economics of scheme, 101
  frequency of, 348
Time series, 292, 337
  correlation between, 234
  periodic, 273, 340, 346
  step changes in, 273, 352
Time-step simulation technique, 412
Transformations:
  logarithmic, 45, 52, 69, 206, 214, 220
  of models, 300
  to Normality, 29, 49
  reciprocal, 34
  of variables, 215, 269
Triangular distribution, 23
Tukey, J. W., 152, 232
Two-sided test, 73, 115

Uniformity of variances, 173
Utility, 83, 91, 101

Validity checks, 7, 270, 272
Variance, 38
  analysis of. *See* Analysis of Variance
  about regression, 189
  computation of, 41
  confidence limits for components of, 166
  -covariance matrix, 248

estimate, 129, 146, 162
estimate, expected value of, 38, 57
  of general function, 54
  of linear function, 54
  mean, 27, 51, 54
  of a product, 55
  of a quotient, 55
  reducing techniques, 417
  residual, 199, 253, 270
  significance of, 75
  of simple function, 52
  standard error of, 51
Variances:
  additive property of, 122
  combination of, 43
  comparison of two, 75, 440
  homogeneity of, 173
Variate, 16
Variation:
  coefficient of, 40
  nature of, 125
  types of, 121, 337
V-mask, 353

Weighted:
  average, 37
  linear regression, 202

Zones of acceptance, rejection, etc., 387
z-test, 234